URBAN AIR QUALITY:
MEASUREMENT, MODELLING AND MANAGEMENT

URBAN AIR QUALITY: MEASUREMENT, MODELLING AND MANAGEMENT

Proceedings of the Second International Conference on Urban Air Quality: Measurement, Modelling and Management Held at the Computer Science School of the Technical University of Madrid 3–5 March 1999

Guest edited by

Ranjeet S. Sokhi

Atmospheric Science Research Group (ASRG), Department of Environmental Sciences, University of Hertfordshire, College Lane, Hatfield, Hertfordshire, AL10 9AB, UK; E-mail: r.s.sokhi@herts.ac.uk

Roberto San José

Environmental Software and Modelling Group, Computer Science School, Technical University of Madrid, Campus de Montegancedo, Boadilla del Monte–28660; E-mail: roberto@fi.upm.es

Nicolas Moussiopoulos

Laboratory of Heat Transfer and Environmental Engineering, Box 483, Aristotle Unievrsity, GR-54006, Thessalonika, Greece; E-mail: moussio@vergina.eng.auth.gr

Ruwim Berkowicz

National Environmental Research Institute (NERI), Frederiksborgvej 399, P.O. Box 358, DK-4000, Roskilde, Denmark; E-mail: rb@dmu.dk

Organised by

The Environmental Physics Group of the Institute of Physics, U.K., The Technical University of Madrid, Spain, and The University of Hertfordhire, U.K.

in collaboration with

International Union of Air Pollution Prevention and Environmental Protection Associations (IUAPPA) and the Air & Waste Management Association (A&WMA)

Reprinted from Environmental Monitoring and Assessment, Volume 65, Nos. 1–2, 2000

KLUWER ACADEMIC PUBLISHERS

Dordrecht/Boston/London

A C.I.P. Catalogue record for this book is available from the Library of Congress.

ISBN 0-7923-6676-X

Published by Kluwer Academic Publishers
P.O. Box 17, 3300 AA Dordrecht, The Netherlands

Sold and distributed in North, Central and Latin America
by Kluwer Academic Publishers
101 Philip Drive, Norwell, MA 02061, U.S.A.

In all other countries, sold and distributed
by Kluwer Academic Publishers
P.O. Box 322, 3300 AH Dordrecht, The Netherlands

Printed on acid-free paper

Printed in the Netherlands

TABLE OF CONTENTS

Urban Air Quality: Measurement, Modelling and Management

Air Quality Measurements

AEROSOLS

Complex Geometry and Physical Modelling

Modelling

Roadside Air Pollution

Air Quality Management

Conference Organising Committee

Dr Ruwim Berkowicz, *NERI and TRAPOS, Denmark*
Dr Larry Cravey, *European Section, AWMA*
Professor Rosa M. Gonzalez Barras, *Complutense University of Madrid, Spain*
Mr. Richard Mills, *International Union of Air Pollution, Prevention and Environmental Protection Associations (IUAPPA), UK*
Professor Nicolas Moussiopoulos, *Aristotle University Thessaloniki and SATURN, Greece*
Mr John Sadler, *European Section, AWMA*
Professor Maria L. Sanchez, *European Section, AWMA and University of Valladolid, Spain*
Professor Roberto San Jose, *Technical University of Madrid, Spain*
Dr Ranjeet S. Sokhi, *Chairperson, University of Hertfordshire, UK*

International Scientific and Advisory Committee (ISAC)

Dr John G. Bartzis, *NCSR DEMOKRITOS, Greece*
Dr Ruwim Berkowicz, *NERI, Denmark*
Dr Trond Bohler, *NILU, Norway*
Professor Carlos Borrego, *Unievrsity of Aveiro, Portugal*
Dr Norbert Gonzalez-Flesca, *INERIS, France*
Professor Steven R. Hanna, *George Mason University, USA*
Professor Mark Z. Jacobson, *University of Stanford, USA*
Dr Jaakko Kukkonen, *FMI, Finland*
Dr Doug R. Middleton, *Meteorological Office, UK*
Professor Nicolas Moussiopoulos, *Aristotle University Thessaloniki, Greece*
Professor Roberto San Jose, *Technical University of Madrid, Spain*
Professor Michael Schatzmann, *University of Hamburg, Germany*
Professor Jack Slanina, *ECN, Netherlands*
Dr Andreas Skouloudis, *JRC, Ispra, Italy*
Dr Ranjeet S. Sokhi, *Chairperson, University of Hertfordshire, UK*

Conference Secretariat

Lucy Hamilton, Institute of Physics, 76 Portland Place, London, W1N 4AA, UK

Preface

Since the first international conference on urban air quality, held at the University of Hertfordshire in 1996, significant advances have taken place in the field of urban air pollution. In addition to the scientific advances in the measurement, modelling and management of urban air quality, significant progress has been achieved in relation to the establishment of major frameworks to ensure a more effective mechanism for international collaboration. Two such frameworks are SATURN (Studying Atmospheric Pollution in Urban Areas) and TRAPOS (Optimisation of Modelling Methods for Traffic Pollution in Streets). In response to such advances, the second international conference was held at the Technical University of Madrid in March 1999 with active participation of SATURN and TRAPOS investigators. The organisation of the conference was headed by the Institute of Physics in collaboration with the Technical University of Madrid and the University of Hertfordshire. The support of IUAPPA and AWMA ensured a truly worldwide promotion and participation. The meeting attracted 140 scientists from 26 different countries establishing it as a major forum for exchanging and discussing the latest research findings in this field.

Very few scientific international conferences specifically address the area of air pollution in cities and towns. This is despite the growing concern about the health and environmental impacts that result from air pollutants in urban areas. The need to have a forum for discussing and exchanging scientific results in this field is hence overwhelming. It is clear from the response of scientists and other experts working in this area that these UAQ conferences are helping to address this need. Nearly 120 papers were presented in areas ranging from emissions to indoor air quality and exposure. All papers submitted for publication were subjected to a thorough peer review process involving over 30 referees with a minimum of two independent referees per article.

The success of this conference can be attributed to the excellent organisation by the main sponsoring and collaborating bodies and especially the local host institution, the Technical University of Madrid. The dedication of the referees is fully acknowledged for their diligence during the review process. I am grateful to the IOP and in particular Lucy Hamilton for providing the excellent administration for the conference. I would also like to thank Anna Tod and Anja Tremper for their assistance during the preparation of this special issue. Most of all, I would like to extend my sincere thanks to the researchers in this field whose continued efforts have led to major scientific advances which are helping to better understand urban atmospheric pollution and its impact on health and the environment.

Ranjeet S Sokhi
University of Hertfordshire, UK

Environmental Monitoring and Assessment **65**: 1, 2000.

USING MEASUREMENTS OF NITROGEN OXIDES TO ESTIMATE THE EMISSION CONTROLS REQUIRED TO MEET THE UK NITROGEN DIOXIDE STANDARD

J. DIXON[1], D.R. MIDDLETON*[2] AND R.G. DERWENT[2]

[1] *Institute for Environmental Policy, University College London, 29-30 Tavistock Square, London, WC1H 9EZ*

[2] *Meteorological Office, London Rd, Bracknell, Berkshire, RG12 2SZ *Corresponding author*

Abstract. In this paper we introduce a new method of analysing the relationship between nitrogen dioxide (NO_2) and oxides of nitrogen (NO_x) concentrations using data from the UK National Air Quality Archive. The study includes analyses of measurements from two different types of site in London, a kerbside site: Cromwell Rd, and three background sites: Bridge Place, London Bloomsbury and West London, over several years (1991-7). The data in some years showed that hourly NO_2 concentrations exceeded the UK Standard of 150 ppb. Data were binned, averaged, and polynomials fitted at each site. Analysis of the resulting polynomials was used to estimate reductions in NO_x emission required to achieve the National Air Quality Strategy Objective. Examination of the empirical ratio NO_2:NO_x (the 'yield') gives an indication of the sensitivity of the NO_2 to NO_x controls and the amount of NO_2 that would arise from modelled values of total oxides of nitrogen. The response of NO_2 to emission changes is very non-linear, implying 30-45% controls on NO_x may be required.

Key words: emissions control, nitrogen dioxide, hourly concentration, empirical function, London, sensitivity

1. Introduction and Aims

There are currently a large number of initiatives, both at a national and international level, to reduce emissions of total oxides of nitrogen (NO_x) in the UK. Hourly and annual air quality Standards and Objectives are contained within the National Air Quality Strategy and the consultation document of its subsequent review (NAQS, 1997; NAQS, 1999) and are defined in Air Quality Regulations (1997). The NAQS (1997) has a health based Objective for hourly maximum nitrogen dioxide (NO_2) of 150 ppb to be achieved by 2005. The consultation on the review of the NAQS suggested that this be changed to 104.6 ppb, with a maximum of 18 exceedences by 2005 (NAQS, 1999). However, at the time of writing the outcome of the consultation was unknown. The first Air Quality Daughter Directive has been adopted and discussions on the proposed Acidification Strategy and Emission Ceilings Directive are continuing. It seems prudent therefore to study the sensitivity of NO_2 concentrations to controls on NO_x concentrations, where $NO_x = NO + NO_2$.

Environmental Monitoring and Assessment **65**: 3–11, 2000.

The UK National Air Quality Archive contains all of the ambient air quality data from monitoring stations supported by the DETR, providing extensive information on the behaviour of air pollutants in urban areas across the UK. During a very high episode in London, 351 ppb NO_2 was reached at Exhibition Road, on Friday 13 December 1991 (QUARG, 1993). The highest hourly average concentration of 423 ppb NO_2 was measured at Bridge Place, with 382 ppb at Cromwell Rd and 388 ppb at West London. Previous empirical assessment of relationships between $[NO_2]$ and $[NO_x]$ has relied on a polynomial produced by Derwent and Middleton (1996). This relationship (Equation 1) was based on binned data from a kerbside site at Exhibition Road (Ex Rd), from May 1991 to June 1992.

$$[NO_2] = 2.166 - [NOx]\,(1.236 - 3.348A + 1.933A^2 - 0.326A^3) \tag{1}$$

where square brackets indicate the hourly mean concentration in ppb and $A = \log_{10}([NO_x])$ in ppb. This function applies in the range 9.0 ppb to 1141.5 ppb of $[NO_x]$ (Derwent and Middleton, 1996). Spurious negative values are produced by the polynomial below 9.0 ppb $[NO_x]$, whilst above 1145.5 ppb $[NO_x]$ the shape of the curve is open to debate, for there are few data here. A significant practical advantage of this or other polynomial functions in describing measurements of $[NO_2]$ and $[NO_x]$ is that it allows values of urban $[NO_2]$ to be estimated in the absence of information about future ozone levels (without ozone concentrations, simple chemical kinetics modelling is severely handicapped).

Once a good polynomial fit to the measurements is established, the rate of change of $[NO_2]$ as $[NO_x]$ varies is then the gradient of the polynomial, i.e.

$$slope = \frac{d[NO_2]}{d[NOx]} \tag{2}$$

The aims of this study were: (i) to fit revised polynomials to a set of four study sites over 7 years of consecutive data; (ii) to assess the sensitivity of $[NO_2]$ to small changes in $[NO_x]$ concentrations by studying the slope of the polynomial at each $[NO_x]$ concentration; and (iii) to estimate the amount of NO_x reduction required for NO_2 to approach the hourly Standard of 150 ppb (NAQS, 1997).

2. Methods

Automatic Urban Network (AUN) sites[1] included in the study were: Bridge Place (Bri Pl; Background), London Bloomsbury (Bloom; Background), West London

[1] Site details on the site information archive www.environment.detr.gov.uk/airq/aqinfo.htm

(W Lon; Background) and Cromwell Road (Cr Rd; Kerbside). Cromwell Rd is ozone deficient due to high local nitrogen oxide (NO) from vehicles. Fully ratified hourly average data for $[NO_2]$ and $[NO_x]$ was downloaded from the National Air Quality Archive[2]. Years were included from 1991 onwards with a data capture >75%.

The raw data (Figure 1) are difficult to analyse due to naturally occurring scatter and the small number of points at higher concentrations (where <5% of NO_x values are >500 ppb, where NO_2 exceedences may arise); a binning process was therefore used. Hourly air quality data for NO_x were sorted into 10 ppb bins, e.g. 0-10, 11-20. We denote each upper bin limit as $[NO_x]_{upper}$ e.g. 10, 20. NO_2 values in each bin were averaged and denoted as $[NO_2]_{bin}$. Values of $[NO_2]_{bin}$ were plotted against $[NO_x]_{upper}$ (Figure 2). For each bin, their ratio was calculated. The power series Equation 3 was fitted to this ratio, using $A = \log_{10}([NO_x]_{upper})$:

$$\frac{[NO_2]_{bin}}{[NOx]_{upper}} = a + bA + cA^2 + dA^3 + eA^4 \tag{3}$$

where a, b, c, d and e are coefficients. The fitting used a standard multiple regression program and for all the data sets studied returned multiple correlation coefficients R^2 greater than 0.9 and were highly statistically significant. By using Equation 3, annual hourly datasets for NO_2 and NO_x could be efficiently compressed with minimal loss of detail. Once a-e have been established, equation 3 is easily rearranged to estimate $[NO_2]$ for a given value of $[NO_x]$.

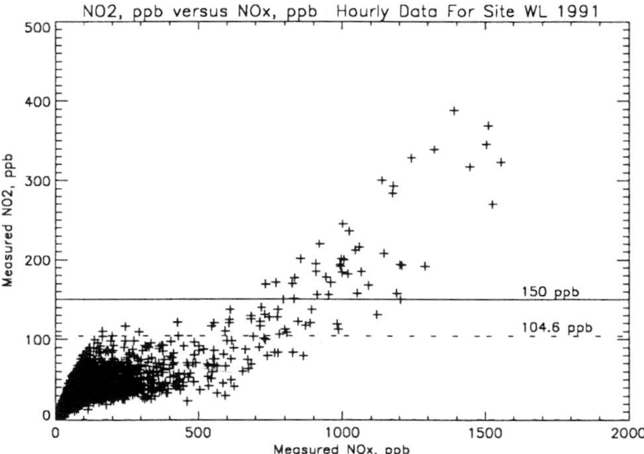

Fig. 1. Hourly concentrations of NO_2, ppb, versus NOx, ppb, for West London, 1991.

[2] National Air Quality Archive www.environment.detr.gov.uk/airq/aqinfo.htm

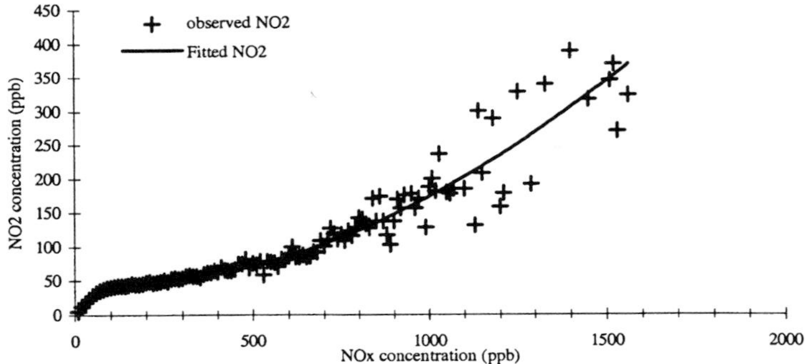

Fig. 2 Bin average NO_2 concentrations versus bin NO_x concentration for West London, 1991 and the fitted function (Equation 3; derived from data in Figure 1).

3. Results

3.1 BIN CURVE FITTING

In our initial studies during this project, we found that polynomials produced for each year (Figure 3) at each site had quite different shapes, especially at the upper range of $[NO_x]$ concentrations. It was therefore decided to combine all the data at each site to produce an average polynomial for each site (Table 1 and Figure 4). Equation 3 is sensitive to the values of its coefficients: it is therefore necessary to use the full number of significant figures shown in Table 1. Figure 4 also shows that, relative to our new curves, the Derwent and Middleton (1996) function for Exhibition Road underpredicts the $[NO_2]$ at $[NO_x]$ <600 ppb and overpredicts at $[NO_x]$ >900 ppb.

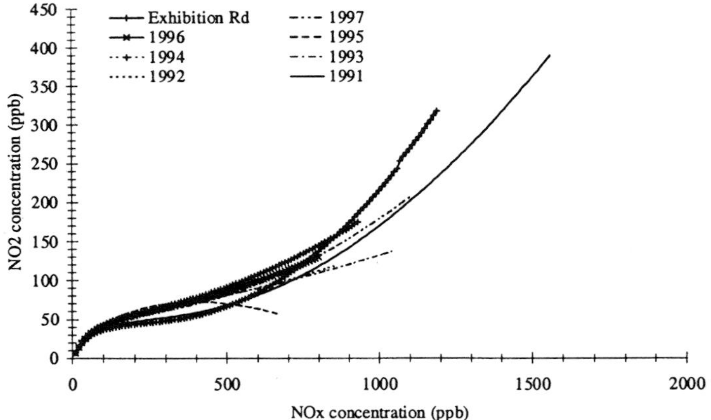

Fig. 3. Functions fitted to West London data showing inter-year variation. The original Exhibition Rd function is also shown for comparison.

TABLE 1.
Constants for Equation 3 at each site.

SITE	a	b	c	d	e
Bridge Place	-2.492160	5.990032	-3.809571	0.914665	-0.070282
Cromwell Rd	2.241479	-3.240099	2.371635	-0.830800	0.107137
Bloomsbury	-2.419469	6.543467	-4.602157	1.251494	-0.116134
West London	-2.975882	7.152116	-4.818636	1.268526	-0.113448

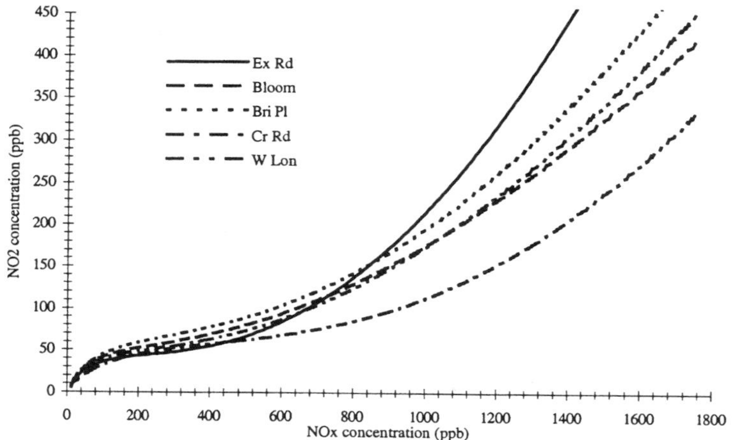

Fig. 4. Comparison of new curves for London (1991-7) with Exhibition Rd (1991-2).

The shape of the rearranged polynomial is shown by an example, the combined 1991-1997 function for West London which is on Figure 5. The initial portion of the curve was relatively stable up to around 200 ppb NO_x (Figure 5) corresponding to approximately 70 ppb NO_2. Between 200 and 600 ppb the curve responds to the episodic concentrations at the sites and is therefore more variable. The results also depend on the positioning of the site (cf. QUARG, 1993, p. 50): kerbside monitors have higher [NO_x] relative to [NO_2] due to less time for reaction chemistry to take place, whereas background sites are, by their nature, further away from the main source of [NO_x] (Figure 4). The importance of site position to the polynomials will be explored in later work (Dixon *et al*, in prep). Scatter increased as the number of values in the bins decreased, so that adjacent bin averages of [NO_2] showed more erratic behaviour. In these cases, functions fitted through the [NO_2] values were not well behaved. These were easily identified in the inter-year plots because the functions veered off to the x axis (e.g. Figure 3, 1995 function) and the plots of their gradients behaved most oddly.

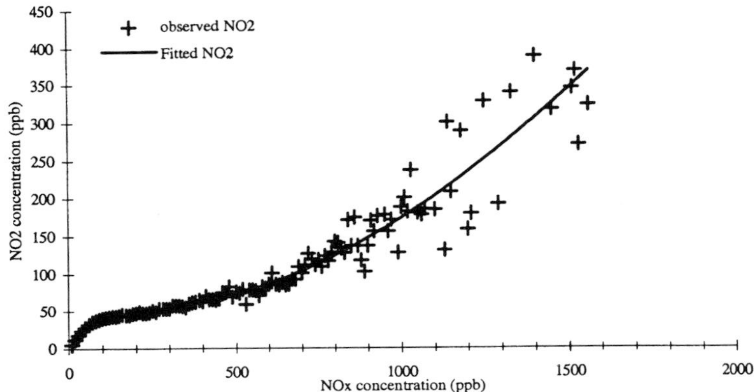

Fig. 5. Binned data with function for West London (1991-7).

3.2 SENSITIVITY TO SMALL MEASURES

When looking at the sensitivity of [NO_2] to control of [NO_x], the slope or rate of change of the polynomial was used. The slope of the curve through the bins was calculated using Equation 2 and is illustrated in Figure 6. For West London, the initially high slope quickly falls to a minimum value of 0.051 as [NO_x] increases. This shows that the lowest sensitivity of [NO_2] to [NO_x] occurs at around 50 ppb NO_2. Here small changes in [NO_x] will have the smallest impact upon [NO_2]. Beyond this minimum, the slope increases towards 0.500 at large values of [NO_x]. Table 2 gives values for the slope where [NO_2] equals the UK 1 hour maximum Objective of 150 ppb NO_2. These show that the relationship between NO_2 and NO_x at 150 ppb NO_2 was changing less rapidly at the other London sites than for Exhibition Rd. Cromwell Rd is a kerbside site with a similar slope to that of the background site at Bloomsbury. To summarise, Figure 6 reveals that the sensitivity of hourly mean nitrogen dioxide to small changes in total oxides of nitrogen is largest at the very lowest concentrations, quickly decreases, and is noticeably smaller in the region of larger concentrations where policy measures have to be targeted.

3.3 MEASURES FOR 1 HOUR MAXIMUM

The 1997 NAQS is aimed at preventing exceedences of a 1 hour mean value. Having examined the response to small changes, we now turn our attention to a much larger goal of policy: by how much might emissions of NO_x need to be reduced in order to bring NO_2 down from its maximum 1 hour value to not exceed the NAQS Objective. With just a few polynomials, as here, a spread sheet can be used for these estimates. Later work by Dixon *et al.* (in prep) presents a more

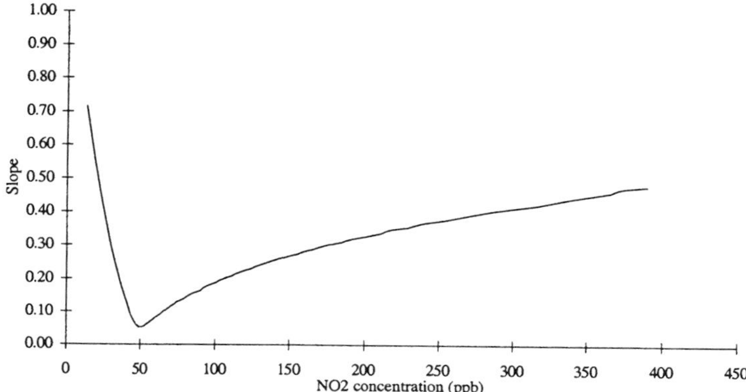

Fig. 6. Gradient of the polynomial versus NO_2 for West London (1991-7).

formal approach that builds on this study. The essential point is that for large changes in $[NO_x]$, the curvature of the polynomial must be taken into account. The slope calculated in section 3.2, and plotted in figure 6, represents the tangent to the curve, whilst a policy of large emissions reduction would represent a chord from one point on the curve (the maximum $[NO_x]$) to another (that where the curve crosses the $[NO_2]$ Standard).

At each site, the reduction from maximum $[NO_x]$ to reach the $[NO_x]$ at $[NO_2]$ = 150 ppb was calculated. The difference between the $[NO_x]$ at 150 ppb NO_2, and the maximum $[NO_x]$ that was recorded in the measurements, represents the amount of reduction indicated in ppb for $[NO_x]$. This was then converted to a percentage reduction (it was divided by the maximum $[NO_x]$ and multiplied by 100). These results (Table 2) show that, on average, a reduction of between 32 and 43% of the maximum NO_x concentrations seen in these data would be required to achieve the UK Objective of 150 ppb by 2005. It is important, however, to remember that the maximum hourly concentration is sensitive to meteorology and so varies from year to year. This work demonstrates a method to estimate the NOx reductions indicated by a sample of monitoring data.

This paper discusses peak $[NO_2]$, not the annual mean which is more likely to produce problems of non compliance during the review and assessment of local air quality (Derwent, 1999). Previous unpublished studies have shown that in the UK peak 1 hour mean nitrogen dioxide is largely a London problem.

Figure 4 indicates that the three London background sites considered in the present study were reasonably similar to each other, although they are in different areas of the city. Our initial unpublished study had shown inter-year and inter-site variability, and so aggregating all available years (up to 61320 points), averaging the data in bins and fitting a polynomial to the binned data, reduced the problems of curve-fitting.

TABLE 2.
Slope values at 150 ppb nitrogen dioxide at 4 monitoring sites in London. Reductions required as percentages in the maximum recorded NO_x concentration to reach the NO_x concentration at 150 ppb NO_2 at each site.

SITE	Year	Slope at 150 ppb nitrogen dioxide	Reduction % in maximum [NO_x] to reach [NO_x] at 150 ppb NO_2
Exhibition Rd	1991-2	0.350	
Bridge Place	1991-7	0.240	42.8
Cromwell Rd	1991-6	0.229	38.7
Bloomsbury	1992-7	0.227	32.0
West London	1991-7	0.250	41.3

4. Conclusions

We found that:

- The magnitude of the slope in the vicinity of 150 ppb indicates the likely response of the NO_2 concentrations to small reductions in NO_x. The sensitivity passes through a minimum of 0.05 ppb NO_2 per ppb NO_x where NO_2 is 50 ppb.
- Peak NO_2 concentrations in any one year determine whether the 1 hour Standard was exceeded, and their corresponding reductions in NO_x required to achieve the UK Objective have been estimated to be 32-43% in this study. They depend on the difference between the magnitudes of [NO_x] in the peak hour, and where the fitted function crosses the NO_2 Standard, 150 ppb.

Later work plans to extend and develop the study of the yield NO_2:NO_x as a parameter when analysing monitoring data at a wider range of sites. It will present a much fuller analysis (Dixon *et al*, in prep).

Acknowledgements

The authors would like to acknowledge the funding of the Department of the Environment, Transport and the Regions (contract no. EPG 1/3/128).

References

Derwent, R. G.: 1999, *Meteorological Office Report to the Department of the Environment, Transport and the Regions*.

Derwent, R.G. and Middleton, D.R.: 1996, *Clean Air* **26**(3/4), 57-60

Dixon, J., Middleton, D.R., and Derwent, R.G.: in preparation, *To be submitted to Atmospheric Environment*

NAQS.: 1997, *The United Kingdom National Air Quality Strategy*, ISBN 0 10135872-5

NAQS.: 1999, *Review of the United Kingdom National Air Quality Strategy*, The Stationery Office, London.

QUARG.: 1993, *Urban air quality in the United Kingdom,* First report of the Quality of Urban Air Group, ISBN 0 9520771 6, Department of the Environment, Transport and the Regions.

EMISSION AND PERFORMANCE CHARACTERISTICS OF A 2 LITRE TOYOTA DIESEL VAN OPERATING ON ESTERIFIED WASTE COOKING OIL AND MINERAL DIESEL FUEL.

M. E. GONZALEZ GOMEZ, R. HOWARD-HILDIGE, J. J. LEAHY, T. O'REILLY, B. SUPPLE, M. MALONE.
University of Limerick, Ireland.

Abstract. Exhaust emission and performance characteristics were evaluated in a Toyota van, powered by a 2l indirect injection (IDI) naturally aspirated diesel engine, operating on vegetable based waste cooking oil methyl ester (WCOME).
Tests were performed on a chassis dynamometer and the data were compared with previous results conducted on the same vehicle using mineral diesel fuel. The data obtained includes smoke opacity, carbon monoxide (CO), carbon dioxide (CO_2), oxygen (O_2), nitrogen dioxide (NO_2), nitric oxide (NO), sulfur dioxide (SO_2) and brake power. Engine lubricating oil samples were also taken. Results from this study indicated a difference of approximately 9% in brake power between the two fuels. WCOME developed a significant lower smoke opacity level and reduced CO, CO_2, SO_2 emissions. However, O_2, NO_2 and NO levels were higher with the vegetable oil based fuel. Power values were comparable for both fuels. Lubricating oil analysis gave little change of viscosity and wear metal concentrations after 2887km were: Silicon 35ppm, Chromium 3.3ppm, Iron 33.8ppm, Copper 14.1ppm and lead 78.6ppm.

Key words: mineral diesel fuel, biodiesel, vegetable oils, exhaust emissions, engine wear

1. Introduction

There has been much interest in vegetable oil based fuels as alternative diesel fuel for years. Rudolf Diesel first experimented with peanut oil in a diesel engine around 1900. Interest has increased due to decreasing petroleum reserves and environmental consequences of exhaust gases. Using a fuel that can be obtained from renewable resources with characteristics similar to those of mineral diesel fuel is attractive especially to countries rich in oleaginous plants. The needs of being energy self-sufficient and the consciousness of the deterioration of the environment make vegetable oil based fuels a good option to mineral diesel fuel. Many research programs have been carried out to study different aspects of the fuels: variety of sources, physical and chemical characterisation, engine performance, the cost of processing the oil, etc. (Rice *et al.*, 1997; Graboski and McCormick, 1998).

An advantage of these fuels is their potential to reduce emissions. Experience has shown that vegetable oil based fuels can significantly reduce exhaust gas emissions, including carbon monoxide (CO), carbon dioxide (CO_2), and particulate matter (PM) (Graboski and McCormick, 1998). Because of their insignificant sulfur content, the sulfur dioxide (SO_2) emissions are low (Murayama, 1994). However, emissions of oxides of nitrogen (NO_x) are in general, higher than those for mineral diesel fuel. From most of the reported studies on vegetable oil based fuels, it can be seen that the values of regulated

Environmental Monitoring and Assessment **65**: 13–20, 2000.

emissions are dependent on the engine technology employed and the type of emission test used. These parameters often have an effect on the emissions data larger than the composition and properties of the fuel under test. This causes difference in results reported by different authors.

The power obtained from mineral diesel fuel and the vegetable based fuels are comparable, although some vegetable oil based fuels give higher values than others, depending on the oil used as source. These differences in performance can be related to the difference in chemical structure of the vegetable oils (Ryan III *et al.*, 1985).

There is a variety of vegetable oil based fuels due to the wide variety of vegetable oils, the most common classification is the distinction between raw and esterified vegetable oils. The latter category is usually called biodiesel. Biodiesels are produced from vegetable oils through different processes. The most common is transesterification and its most remarkable effect is the reduction of viscosity (Tahir *et al.*, 1989). Biodiesels consist mostly of fatty acids triglycerides and the most important compositional difference between mineral diesel fuel and biodiesels is the oxygen content. Biodiesels have a high oxygen content (10 - 11%) and this has an effect on PM and NO_x emissions (Graboski and McCormick, 1998; Mittelbach *et al.*, 1988).

Some of the biodiesel properties such as high viscosities, low volatilities and the presence of highly unsaturated fatty acids in their composition lead to problems such as poor atomisation of the fuel, injector fouling, thickening of the lubricating oil and incomplete combustion, which causes undesirable exhaust emissions (Yahya *et al.*, 1994).

The most economical of the biodiesels is WCOME because in collecting and using waste cooking oil as raw material, the cost of seed production is eliminated.

The present study was based on the comparison of the performance characteristics and exhaust emissions produced by the combustion of esterified waste cooking oil and mineral diesel fuel in an IDI diesel engine, typical of light commercial vehicles and passenger cars.

2. Experimental apparatus and procedure

A 1991 2litre IDI Toyota Liteace was powered by WCOME for a five month trial. Tests were carried out before and during the trial period in order to compare the exhaust emissions and performance of WCOME and mineral diesel fuel.

A Sun Road-A-Matic XI chassis dynamometer was utilised to measure the vehicle speed and power exerted at the road wheels of the vehicle during the tests. The exhaust emissions of CO, CO_2, O_2, SO_2, NO_2 and NO were measured using two different gas analysers, a Sun MCA 3000 and an IMR 2800P. The PM related exhaust gas opacity (smoke) was also measured using a Hartridge Smokemeter MK3.

The power was recorded at various vehicle speeds, which were subsequently converted to engine speeds. At the same time, the emissions from the exhaust pipe of the vehicle were measured and the readings from the gas analysers and the smokemeter were recorded.

3. Results and discussion

3.1. EMISSIONS.

Since a large amount of data was collected, only summary data is reported in this paper. Statistical analysis of the vehicle performance and the exhaust emissions was computed.

The trend of exhaust emissions showed that there was essentially no change in the exhaust emissions with time, during the trial period. The calculated standard deviation values for the different emissions were approximately: 0.03% for CO, 4% for CO_2, 5% for O_2, 21ppm for NO, 1.5ppm for NO_2, 8ppm for SO_2 and 17 HSU for the smoke, which did tend to vary more than the other emissions.

Fig. 1 shows a comparison of the mean power values obtained for each fuel. The maximum power value for both fuels occurred at the maximum speed. Mineral diesel fuel showed a maximum of 26kW, and WCOME showed a maximum of 25.1kW.

The power developed by WCOME was higher (approximately 9%) than that for mineral diesel fuel at low speeds, although it was lower at higher speeds. It seems that WCOME had better performance characteristics than mineral diesel fuel at low speeds. This could explain the lower level of CO (fig. 2) and the slightly higher level of CO_2 (fig. 3) emitted by WCOME at those speeds. Mineral diesel fuel gave higher CO and CO_2 emissions at high speed than WCOME. There were differences of approximately 64% and 7.5%, respectively between both fuels. Mineral diesel CO emissions seemed to be an artifact of this particular engine that perhaps reduced air at 4000rpm provoked a rise in CO and smoke emissions.

Figure 4 shows smoke emissions for both fuels. Smoke is usually due to the presence of unburnt fuel, WCOME developed approximately 48% less smoke than mineral diesel. This agrees with other studies that have reported a reduction of smoke with the use of biodiesel in vehicles (Graboski and McCormick, 1998).

NO emissions are shown in figure 5. NO emissions tended to increase with engine speed for both fuels, which was expected due to the dependency of NO emissions on temperature. The values were higher for WCOME (around 20%), probably due to its high oxygen content (Graboski and McCormick,

1998). NO emissions tended to be lower for those speeds at which O_2 emissions (fig. 6) were higher.

NO_2 emissions were very low for both fuels. Even though WCOME gave slightly higher values than mineral diesel fuel.

Some possible reasons for the increase of NO_x emissions of biodiesels are their high oxygen content and viscosity. The latter characterisitic alters the spray properties at the time of the injection of the fuel and can delay the combustion process in the engine. This delay can improve emissions in terms of smoke opacity and CO levels, but it has the reverse effect on NO_x emissions. On the other hand, the high oxygen content increases the oxygen level available to oxidise more nitrogen and increases the combustion temperature which gives higher NO_x emission levels (Graboski and McCormick, 1998).

Figure 6 shows O_2 emissions. The values recorded for WCOME were higher than those for mineral diesel fuel, which can be expected due to the higher oxygen content of biodiesel fuels.

SO_2 emissions are shown in figure 7. At low engine speeds SO_2 can be absorbed onto PM. The lower SO_2 value for mineral diesel fuel at low engine speed may be due to the higher level of PM produced by this fuel. At higher speeds more SO_2 is produced and possibly exceeds the capture capability of PM increasing the SO_2 emissions. At very high speeds an induction choking effect can reduce the amount of air and so the availability of oxygen resulting in lower SO_2 emissions. It is to be noted that as SO_2 increases CO falls.

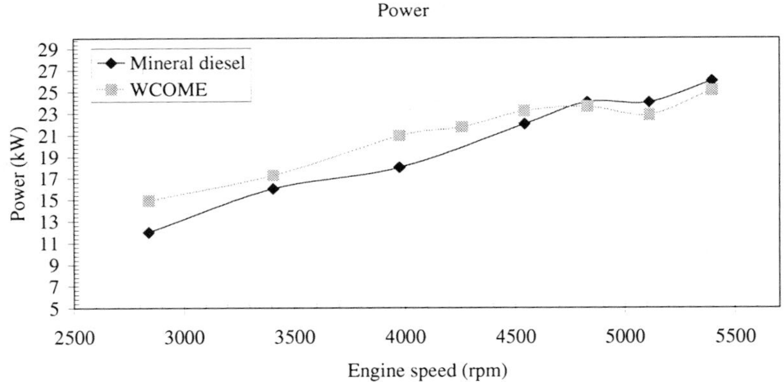

Figure 1. Power relationship with engine speed.

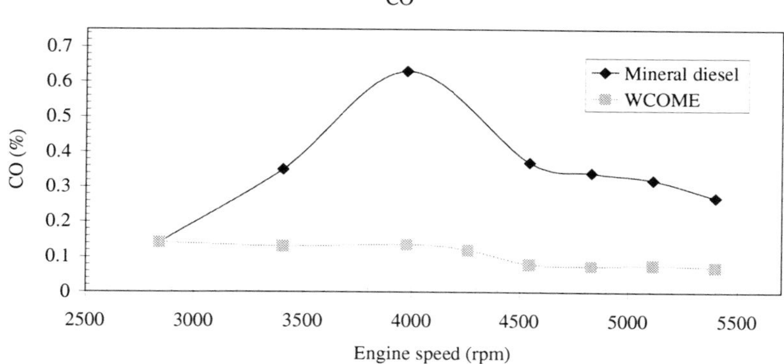

Figure 2. CO relationship with engine speed.

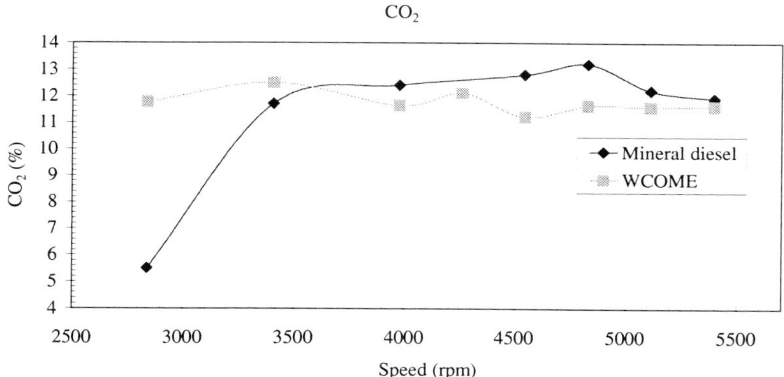

Figure 3. CO_2 relationship with engine speed.

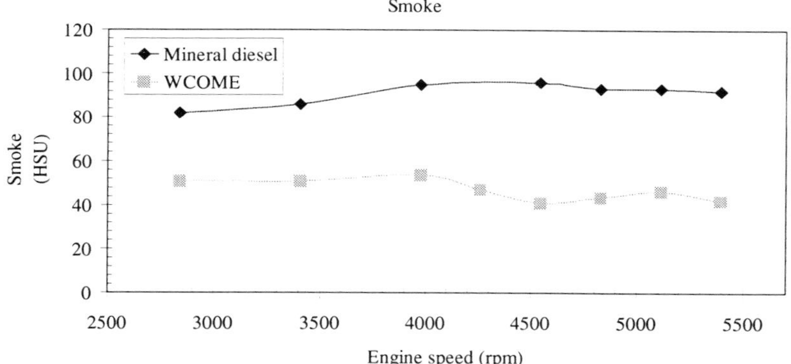

Figure 4. Smoke relationship with engine speed.

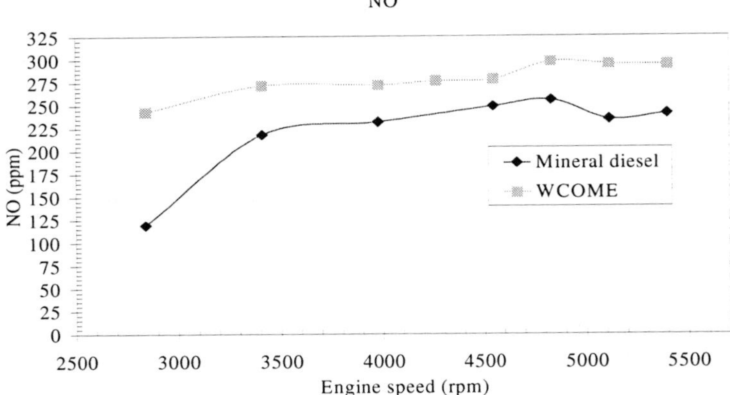

Figure 5. NO relationship with engine speed.

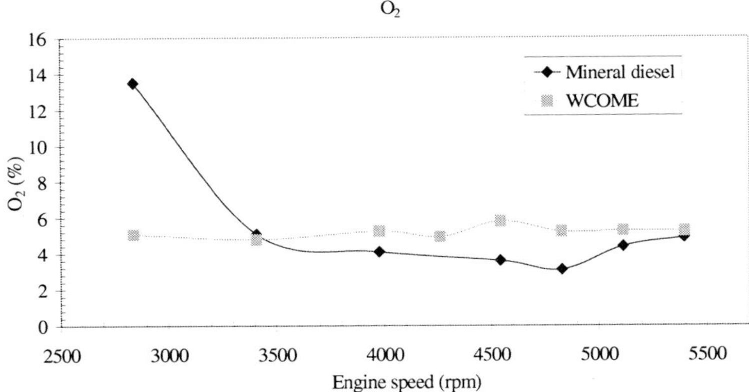

Figure 6. O_2 relationship with engine speed.

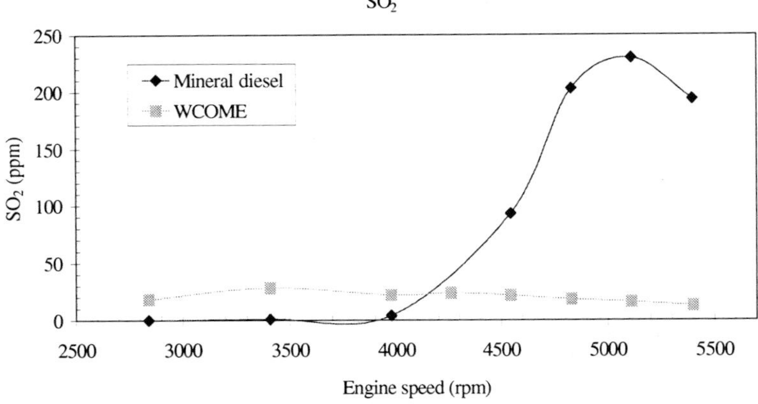

Figure 7. SO_2 relationship with engine speed.

3.2 ENGINE LUBRICATING OIL ANALYSIS.

In order to avoid engine problems due to polymerisation by biodiesel contamination, the engine lubricating oil was monitored during the trial.

The engine wear was evaluated on the basis of concentration of wear metals in the lubricating oil. The metals considered were silicon, chromium, iron, copper and lead, which give indication of wear of different engine parts such as cylinders, pistons, rings, bearings.

A summary of the analysis of the results is presented in Table I. The sample numbers and the kilometers travelled by the vehicle at the time of collection of the sample are indicated in the first and second columns respectively. The sample number 1 was clean oil and was used as reference to compare with the rest of the samples.

Table I
Lubricating oil analysis results

Sample Number	kilometers travelled	Kinematic viscosity (40°C) (cSt)	Silicon (ppm)	Chromium (ppm)	Iron (ppm)	Copper (ppm)	Lead (ppm)
1 (Clean oil)	0	98.41	-	-	-	-	-
2	1112	107.54	24	1.3	26.8	10.8	16.6
3	2035	109.34	6	0.9	24.6	5.1	7
4	2887	111.42	35	3.3	33.8	14.1	78.6

There was a significant change in the contents of sample 3, this was due to the method of collection of the sample. This was taken when the engine was cold, so that the contaminants were not well mixed with the oil and the levels measured were low.

There was not a significant change in viscosity between samples 2 and 4. On the other hand, the levels of chromium, iron and lead, indicators of engine wear, were higher in sample 4 than in sample 2. However, only silicon and lead levels were high compared to typical values (Ali *et al.*, 1996).

4. Conclusions

Exhaust emissions and performance characteristics were compared for WCOME and mineral diesel fuel.

Exhaust emission results showed that lower levels of CO, CO_2, smoke (approximately 64%, 7.5% and 48% respectively) and SO_2 can be attained with WCOME. On the other hand, NO_x emissions were higher (approximately 20%) for WCOME.

Engine performance was satisfactory for WCOME. The power values were similar to those of mineral diesel fuel. There was a difference of approximately 9% between the fuels.

Polymerisation of the lubricating oil did not occur. The viscosity was still in grade at the end of the trial. The wear metals were higher when the trial finished.

WCOME is a good option as alternative fuel due to the similarities with mineral diesel fuel and its improvement in exhaust emission levels. However, it is certain that further research into the reduction of NO_x emissions is needed.

Acknowledgements

The authors thank to Enterprise Ireland (FORBAIRT) for their support.

References

Ali, Y., Hanna M. A.: 1996, Beef tallow as a biodiesel fuel. Liquid fuels and industrial products from renewable sources. Proceedings of the 3rd liquid fuels conference. 15-17 September 1996. Nashville, Tennessee.

Graboski, M. S.and McCormick, L.: 1998, Combustion of fat and vegetable oil derived fuels in diesel engines. Prog. Energy Comb. Sci., vol. 24, pp. 125-164.

Mittelbach, M., Tritthart, P. Diesel fuel derived from vegetable oils, III: 1988, Emission tests using methyl esters of used frying oil. JAOCS, 65(7).

Murayama, T.: 1994, Evaluating vegetable oils as a diesel fuel. INFORM, 5(10).

Rice, B., Frohlich, A., Leonard, R. Bio-diesel production based on waste cooking oil: promotion of the establishment of an industry in Ireland: Sept. 1997, Teagasc, Oak Park Research Centre. ALTENER contract No.XVII/4.1030/AL/77/95/IRL. Final report.

Ryan III, T. W., Bagby, M. O.: 1985, Identification of chemical changes occurring during the transient injection of selected vegetable oils. 76th Annual AOCS Meeting. Philadelphia, May 5-9,1985.

Tahir, A R; Zafar Khan, M and Lapp, H M: 1989, Vegetable oil derivatives: fuel for diesel engines, Agricultural mechanisation in Asia, Africa and Latin America. 20(1), pp.69-72.

Yahya, A and Marley, S J.: 1994, Performance and exhaust emissions of a compression ignition engine operating on ester fuels at increased injection pressure and advanced timing, Biomass and Bioenergy 6(4), pp.297-319.

MODELLING OF MOTOR VEHICLE FUEL CONSUMPTION AND EMISSIONS USING A POWER-BASED MODEL

D.Y.C. LEUNG[1] and D.J. WILLIAMS[2]

[1] *Department of Mechanical Engineering, The University of Hong Kong, Hong Kong, China, email: ycleung@hkucc.hku.hk;* [2] *Division of Coal & Energy Technology, CSIRO, NSW 2113, Australia, e-mail: david.williams@det.csiro.au*

Abstract. The performance of a power based fuel consumption and exhaust emissions model for spark ignition vehicles has been evaluated using a large Australian database derived from testing a wide range of in-use cars on a chassis dynamometer. It was also applied to results of on-road fuel consumption measurement using a "floating" car which was driven back and forth on hilly roadways in Sydney with a length of 8.6 km. The model is found to predict the fuel consumption well over the standard drive cycles and also for the floating car. Average exhaust emissions were also well predicted, but, as would be expected, vehicle-to-vehicle correlation is impossible due to the well-known high variability of emissions between nominally identical vehicles.

Keywords: modelling, spark ignition vehicle, fuel consumption prediction, exhaust emissions, instantaneous power model

1. Introduction

With the increased concern over urban air pollution due to motor vehicles, there is a need for models of vehicle fuel consumption and exhaust emissions. One common approach of modelling vehicle generated pollution is to take emissions data from standard drive cycle tests on a chassis dynamometer. Application of the data to a given situation is accomplished by a number of adjustment factors to allow for the different running modes, speeds, temperatures, fuels used etc. This approach is used, for example, in the US models MOBILE6 and EMFAC, which incorporate average in-use fleet emission factors for different vehicle technology groups based on test cycle data using the Federal Test Procedure (FTP), a standard urban drive cycle used in the USA.

In Australia the standard urban test procedures for light vehicle emissions are specified by the Australian Design Rule (ADR) 27 and 37, which also use an FTP test cycle. The ADR37 test cycle extends the ADR27 test by repeating the first 505 s of the 1172 s ADR27 cycle after a 10 min hot soak. This extended test came into force with the introduction of exhaust catalysts in 1986.

A limitation of many of the emission models is that they are more suitable for estimating emissions within an airshed or over a number of links rather than at a specific location, particularly near significant inclines. Furthermore, many models do not provide reliable estimates of on-road pollutant emissions. Analysis of air quality in well defined situations such as road tunnels shows that

CO and HC concentrations are about twice those anticipated from the traffic type and density, while NOx levels are 50% higher (Robinson et al., 1993).

In this paper, we report on the further development of an instantaneous power model to exhaust emissions. In this approach, the total power required to propel a vehicle at any instant is taken as the basis for estimating the instantaneous fuel consumption and pollutant emission rates. These quantities are compared against data obtained from chassis dynamometer testing of in-use motor vehicles and on-road fuel consumption measurement.

2. Power-based emission model

2.1 POWER ESTIMATION

Power is required for a vehicle to overcome all of resistive forces and to run its accessories. The instantaneous power, Z_t, to propel a vehicle along a road can be expressed as:

$$Z_t = Z_d + Z_r + Z_a + Z_e + Z_m \quad (1)$$

where Z_d , Z_r , Z_a , Z_e and Z_m are, respectively, the power required to overcome vehicle drive-train resistance, tyre rolling resistance, aerodynamic drag, inertial and gravitational resistance, and for the accessories (such as air-conditioning). According to Richardson (1982), the major contributions (in kW) are:

$$Z_d = 2.36 \times 10^{-7}\, v^2 M \quad (2)$$
$$Z_r = (3.72 \times 10^{-5} v + 3.09 \times 10^{-8} v^2)\, M \quad (3)$$
$$Z_a = 1.29 \times 10^{-5}\, C_d\, A v^3 \quad (4)$$
$$Z_e = 2.78 \times 10^{-4}\, (a + g \sin\theta)\, Mv \quad (5)$$

where M (kg) is the inertia mass of vehicle, v (km/hr) the vehicle speed, a (m/s^2) the vehicle acceleration, C_d the aerodynamic drag coefficient, A (m^2) the vehicle frontal area, and θ the road gradient. For testing on a chassis dynamometer the tyre pressure has to be increased resulting in a reduction in the rolling resistance. Z_r is assumed to reduce 0.3 of the value given by Eq. (3). It should be noted that Z_m depends on the type of accessories used and is assumed zero in this study.

2.2 ESTIMATION OF FUEL CONSUMPTION

Because detailed information of vehicle behaviour on the road is not generally available a number of procedures have been developed for estimating fuel consumption under different road and traffic conditions (Watson, 1980; Biggs and Akcelik, 1986). However, when detailed speed time profiles are known, the instantaneous power model can be applied. Good correlation between fuel consumption and power demand was found for vehicles under a standard driving cycle (Post et al. 1981). It is generally accepted that the fuel consumption rate (in ml/min) is linearly related to the total instantaneous power (in kW) by:

$$F_c = \alpha + \beta Z_t = \gamma EC + \beta Z_t \quad (6)$$

where α is the idle fuel consumption rate and β a measure of the thermodynamic efficiency of power generation. For spark ignition (SI) engines, Post et al. (1981) found that the idle fuel flow rate was proportional to engine capacity (EC, litres). The idle fuel consumption data from literature are used to determine γ as shown in Fig. 1. A straight line with a slope of 8.5 can be drawn through the data. This slope (γ) will be used in subsequent regression analysis to determine β.

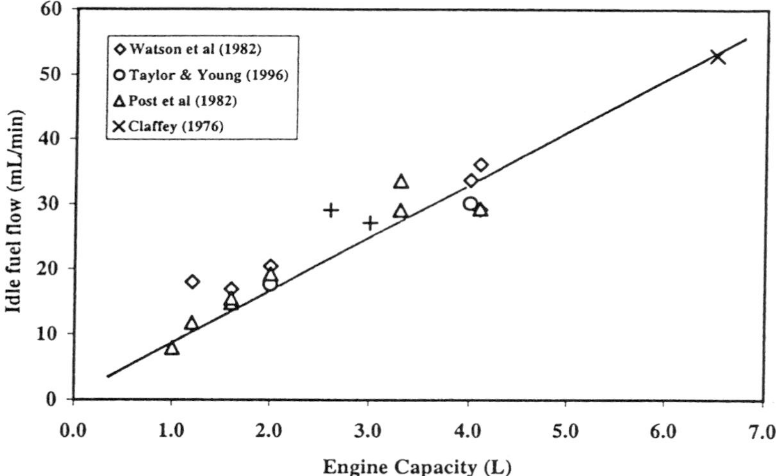

Fig.1. Idle fuel consumption at different engine capacities.

2.3 ESTIMATION OF EXHAUST EMISSIONS

The current study is limited to three pollutants i.e. CO, HC and NOx, which had been measured in the dynamometer tests. Post et al. (1981) showed that CO emissions could be estimated in similar manner to fuel consumption. Williams et al. (1994) extended this concept to HC and NO_x. This is not unreasonable as the HC speciation in the exhaust pipe is highly correlated with the fuel composition and NO_x would be expected to increase with power. On this basis:

$$\text{Pollutant emission rate} = \alpha' + \beta' Z_t = \gamma' EC + \beta' Z_t \quad (7)$$

where α' represents the idle emission rate and β' emission rate per unit power output, respectively. As the emission rate increases with the fuel consumption rate, α' is expected to be a function of EC in similar fashion to the fuel consumption equation and thus can be expressed as a product of γ' and EC.

Equation (7) is used to model the emission under normal operating conditions. However, it is known that more CO and HC are emitted when the engine is cold. To simulate this we allow the engine to warm-up and exponentially achieve normal operating conditions (Eq. 8).

$$F_{cold} = a_1 - a_2\,(1\text{-}e^{-SFn/t}) \tag{8}$$

where SFn is the fuel consumed (ml) since engine started (normalised to 2.5L EC) and t is the characteristic amount of fuel required for warm-up, chosen to be 120 ml (as idle fuel consumption rates are ~ 0.5 ml/s, t can also be viewed as a warm-up time of 4 mins). The values of a_1 and a_2 are respectively set to 12 and 11 for modelling CO, and 8 and 7 for HC, The higher a_1 and a_2 values for CO are due to the observed higher CO emission when the engine is cold. For CO and HC emission from catalyst-equipped vehicles a further factor, F_{cat}, which combines catalyst performance, E_{cat}, and warm up from cold start has been incorporated in Eq. (7):

$$F_{cat} = 1/(1+E_{cat})e^{-SFn/t} + E_{cat} \tag{9}$$
$$E_{cat} = 0.3\,(1- e^{-Fc/2}) \tag{10}$$

where t has a value of 60 ml, as the catalyst warms up more rapidly than the engine and coolant systems. For NOx emissions,

$$F_{cat} = 0.45\,(1+e^{-Fc/2}) \tag{11}$$

The forms of F_{cat} in Eqs. (9) and (11) are designed to reduce the catalyst efficiency as the exhaust flow rate increases and hence the residence time decreases. This appears necessary, based on the data of Carnovale et al. (1991).

3. Dynamometer testing and field measurement

Two sets of data have been gathered to determine the unknown coefficients of Eqs. (6) and (7) and to test their performance. The first database is obtained from the Federal Office of Road Safety (FORS) of Australia. A total of 611 passenger cars were tested on chassis dynamometers, both in the "as delivered condition (pre-tune) and again after minor tuning and repairs to the fuel and ignition system (post-tune).

The second set of data was obtained by driving a "floating" car (3.8L Holden Commodore sedan) back and forth on a major commuter route, 8.6 km long in urban Sydney. The measurements were conducted over a two-week period in July 1997. The car was fitted with a GPS whose output was logged by a computer at 1/3 Hz. A 2L measuring cylinder was fitted for measuring fuel consumption, which could be switched into line in place of the normal fuel delivery system. The fuel level in the measuring cylinder was read manually at the beginning and end of the journey whilst the car was stationary. It was also read, with less accuracy, at intermediate locations whilst the car was traversing.

4. Results and discussions

4.1 FUEL CONSUMPTION

The measured and predicted fuel economy of the vehicles is shown in Fig. 2 for the ADR37 test cycle. The predicted values are obtained from Eq. (6) with β firstly determined to be 8.8 by regression using the FORS measured data and the pre-determined γ value (8.5).

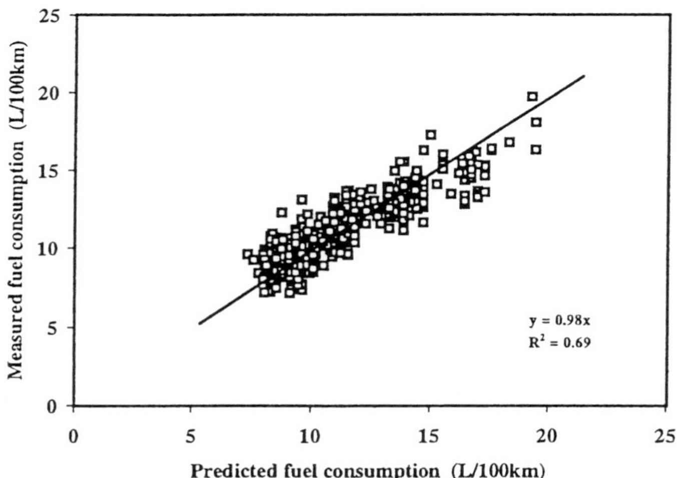

Fig.2. Predicted and measured fuel consumptions.

As shown, the fuel consumption rate of all the vehicles tested fall onto a straight line with slope ≈ 1. The regression coefficient (R^2) is 0.74 and 0.69 respectively for ADR27 and ADR37 test cycle while the corresponding average error is 8.0 and 8.3%. This indicates that the fuel consumption of all the vehicles can be well predicted by Eq. (6). It is noted that larger data scattering occurs for the set of cars before tuning than those after tuning. This is reasonable since the car conditions after tuning could more or less return to normal running conditions. More than 60% of the vehicles had an average of 5% improvement in fuel economy after tuning. This was particularly evident on older vehicles that do not have closed loop engine management systems. We restrict ourselves here to the post-tune. Fuel consumption for any one vehicle is influenced by the particular frictional resistances inherent to the vehicle and by the engine compression status that affects the thermodynamic efficiency, β. Thus, significant variations would be expected between nominally identical vehicles.

The model is further validated by using the on road measurement data of the floating car. The forward trip of the measurement was towards city direction which is subject to down hill slope with an overall descent of about 100m. As expected, more fuel was consumed in the return journey as shown in Fig. 3. Despite the difference in the running conditions the fuel consumption of the forward and return trip fall onto a straight line with slope about 1. The overall model accuracy for this measurement is ± 10% with an average of 5%.

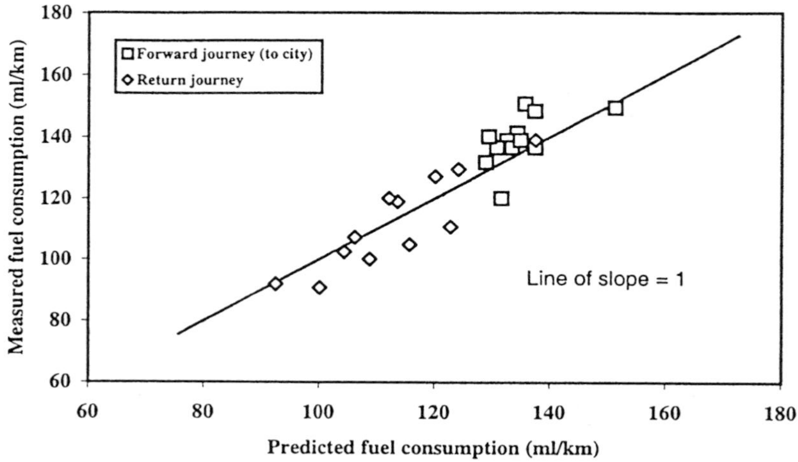

Fig. 3. Fuel consumption results of on-road measurement.

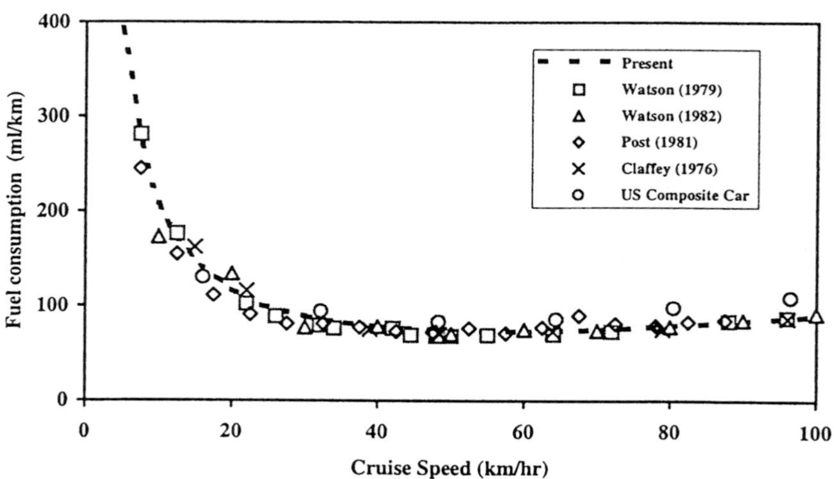

Fig.4. Fuel consumption at different cruise speeds.

As indicated from Eqs. (1)-(5) the vehicle mass and cruise speed are two main parameters affecting the fuel economy. Therefore, in comparing the fuel economy of different vehicles the vehicle mass should be considered. The fuel consumption per unit distance, corrected to a mass of 1000 kg, is calculated at various speeds for a hypothetical car with 2.5L EC. As shown in Fig. 4 the fuel consumption is very high at low cruise speed, due to the comparatively larger friction, and reduces dramatically as increases in cruise speed. This trend agrees with those obtained by other researchers after normalising their data to the same mass basis (Claffey, 1976; Watson, 1980; Watson, 1982; Post et al., 1981).

4.2 POLLUTANT EMISSIONS

A similar procedure as the fuel consumption is carried out to determine the coefficients of the emission equation. The respective value of γ' and β' are 1.75 and 0.75 for CO, 0.14 and 0.10 for HC, and 0.01 and 0.30 for NOx. With these values the predicted CO, HC and NOx levels are compared with the measured values for ADR37 to reveal the overall performance of the model (Fig. 5).

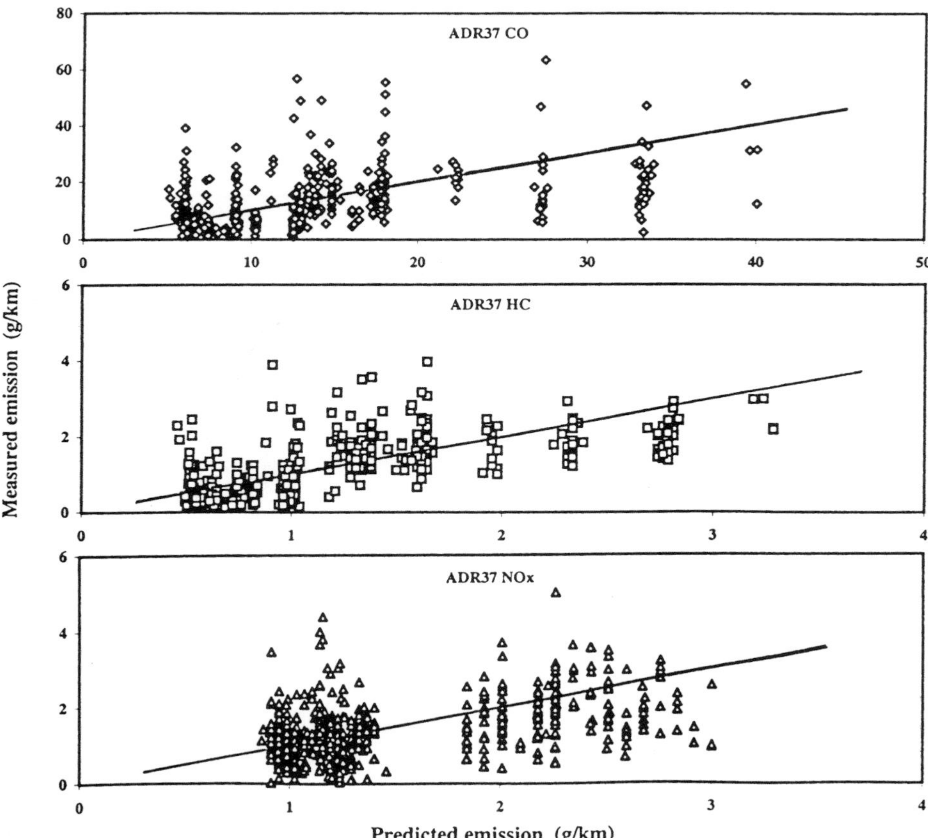

Fig. 5. Predicted and measured pollutant concentrations under the ADR37 drive cycle.

There is a much larger data scattering of the emission data than the fuel consumption data, which is in line with those discrepancies observed between the roller bank test results and the outcome of tunnel experiments (Robinson et al., 1993). This is common to all such databases as the minor exhaust components depend to a much greater extent on the conditions of the vehicle than does fuel consumption. Another reason may be due to the ageing of vehicles. As found by Carnovale et al. (1991) the deterioration factors for CO and HC due to vehicle mileage are quite substantial for both catalytic and non-catalytic cars. This factor has not been considered in the present analysis due to the lack of acceptable correction formulae. Overall speaking, the predicted values were in the "ballpark". As shown in Table 1 the average predicted pollutant emissions are of the same order of magnitude as the measured values and lie between the pre-tune and post-tune values. Comparing the result of the two drive cycles, those vehicles undergone ADR37 drive cycle have less emissions, due to the inclusion of the less polluted warm start phase. This is also revealed by the much greater pollutant emissions in the cold start C505 (first 505 seconds) than the transitional T867 (middle 867 seconds) and hot start H505 (last 505 seconds) phases.

It should be noted that there is a range of values for the coefficients that can give a reasonable simulation of the test data. This work is on-going and further refinement can be expected as more on-road work is performed.

Table 1. Comparison of predicted and measured mean pollutant emissions.

		CO		HC		NOx	
		Non-cat	Catalyst	Non-cat	Catalyst	Non-cat	Catalyst
ADR27	Pre-tune	29.8	-	2.8	-	2.1	-
(g/km)	Post-tune	23.0	-	2.0	-	1.9	-
	Predicted	24.9		2.1		2.1	
ADR37	Pre-tune	26.3	10.3	2.1	0.7	2.1	1.3
(g/km)	Post-tune	19.4	7.8	1.8	0.5	1.9	1.2
	Predicted	20.7	8.9	1.8	0.7	2.2	1.1
C505	Pre-tune	183.6	116.7	15.5	7.5	15.1	9.9
(g)	Post-tune	161.4	100.5	13.9	6.2	13.7	9.4
	Predicted	210.7	119.9	16.6	8.8	14.6	8.1
T867	Pre-tune	173.3	48.4	12.8	3.3	9.5	5.7
(g)	Post-tune	115.5	32.1	10.7	2.4	8.9	5.1
	Predicted	87.7	19.6	8.4	1.9	10.9	5.2
H505	Pre-tune	111.2	42.8	10.5	3.0	15.1	8.9
(g)	Post-tune	84.0	32.0	9.4	2.4	13.6	8.3
	Predicted	71.4	20.0	7.6	2.2	14.6	7.0

5. Conclusions

Fuel consumption and exhaust emissions are found to relate to the instantaneous power required to propel vehicles. A power-based fuel consumption and exhaust emission model has been developed and evaluated. The predicted fuel consumption matches well with the measured values over the ADR27 and ADR37 drive cycles, and with the on-road measurement. Whilst modelled fuel consumption correlates quite well with measured data from vehicle to vehicle, this is not the case for pollutant emissions, which is heavily dependent on the variability of vehicle engines, tuning and maintenance conditions. Thus the comparison consists only of 'scatter' grams which show that the predictions fall within measured range and have similar average values. The coefficients used in the models may well be refined in the light of further work.

Acknowledgements

The authors wish to thank Ms J. Cheng, Mr A. Quintanar and Mr B. Halliburton (CSIRO Energy Technology) for their assistance. DYC Leung is grateful to the Croucher Foundation for supporting part of this study.

References

Biggs, D.C. and Akcelik, R.:1986, Estimation of car fuel consumption in urban traffic. Proc 13th Australian Road Research Board ARRB Conf. 13(7), 124-132.

Carnovale, F., Alviano, P., Carvalho, C., Deitch, G., Jiang, S., Macaulay, D. and Summers, M. :1991, Air emissions inventory Port Phillip Control Region. EPA (Vic) report SRS 91/001.

Claffey, P.J.:1976, Passenger car fuel conservation. US Fed. Highw. Admin. Rep. No. FHWA-PL-77009. Washington, DC.

Guensler, R., Washington, S., Koenig, B. and Sperling, D.:1993, The impact of speed correction factor uncertainty on mobile source emission inventories in the South Coast. Proc. Int. Speciality Conf. "The Emission Inventory-Perception and Reality", Air & Waste Mgt. Assoc.

Post, K., Tomlin, J., Pitt, D., Carruthers, N., Maunder, A., Kent, J.H. and Bilger, R.W.:1981, Fuel Economy and Emissions Res. Report. Charles Kolling Res. Lab., Univ. of Sydney. TN ER-36.

Richardson A.J.:1982, Stop-start fuel consumption rates, Second Conference on Traffic, Energy and Emission, 19-20 May 1982 at the National Science Centre, Melbourne.

Robinson, N.F., Pierson, W.R. and Gerler, A.W.:1993, Comparison of real world CO, VOC and NOx emission rates with motor vehicle emission models. Proc. Int. Speciality Conf. "The En/mission Inventory-Perception and Reality", Air & Waste Mgt. Assoc.

Taylor, M.A.P. and Young, T.M.:1996, Fuel consumption and emission models for traffic engineering and transport planning applications. Proc. Roads 96 Conf., Part 6, Christchurch, New Zealand.

Watson, H.C.:1980, Sensitivity of fuel consumption and emissions to driving patterns and vehicle design, SAE Aust. ARRB Conf: Can traffic management reduce fuel consumption and emissions and affect vehicle design requirements, Australia.

Watson, H.C.:1982, Calibration and application of two fuel consumption models. In: R. Akcelik (Ed). Fuel consumption modelling for urban traffic management, Australian Road Research Board, Research Report No. 124.

Williams D.J., Shenouda D.A. and Carras J.N.:1994, Modelling air toxic emissions from motor vehicles. Proc. 1994 Air Toxics Conf., Aug. 1994 at National Measurement Lab., Sydney.

VALIDATION OF URBAN EMISSION INVENTORIES

CLEMENS MENSINK
Vito, Boeretang 200, B-2400 Mol, Belgium
E-mail: mensinkc@vito.be

Abstract. Two emission validation methods are presented. The first method focuses on the precision of the emission factors and the accuracy of modelled traffic flows. Emission factors derived from the COPERT II methodology are compared with on-board emission measurements and modelled traffic flow rates are compared with observations. The second validation method focuses on the completeness of the inventory, i.e. coverage of all sources. The method compares measured pollutant fluxes in the urban plume with the downwind transported and dispersed emissions integrated over plume width and mixing height. Both methods seem to indicate that traffic emission factors used in the urban emission inventories show large uncertainties. Besides the lack of measurement precision this is mainly induced by external influence factors like driving behaviour and vehicle maintenance.

Key words: emission validation, emission inventories, road transport, urban traffic, emission factors, emission measurements

1. Introduction

In order to improve the quality of urban emission inventories and estimate more accurately the uncertainties associated with the different pollutants considered in urban emission inventories, it has been recognised (Sturm *et al.*, 1998) that validation of urban emission inventories has become a necessity in terms of quality assurance and quality control (QA/QC). A question which arises immediately is *how* the quality of urban emission inventories and models can be assessed. The Atmospheric Emission Inventory Guidebook (Mc Innes, 1996) provides a general overview of verification procedures representing activities that can be applied to demonstrate the applicability and reliability of emission inventory data. Important elements in the verification process are completeness analyses, uncertainty analyses and validation processes.

The completeness analysis addresses the question if all activities which generate relevant emissions are included. The uncertainty analysis looks at the degree of accuracy and precision of the data. Accuracy is defined as the measure of truth of a measurement or estimate. The term precision is used to express the repeatability of multiple measurements of the same event. In this context validation is defined as a verification method which allows a comparison of the results of an emission inventory or model to an independent approach.

In this paper two validation methods are presented contributing to this topic. The first method can be characterised as a *punctual validation method* and focuses on precision (applied to the emission factors) and accuracy (applied to

Environmental Monitoring and Assessment **65**: 31–39, 2000.

modelled traffic flow numbers). The validation is carried out for individual locations in the city of Antwerp, where measured traffic flow numbers and emission factors are compared with the quantities modelled in function of place, time and temperature. The second validation method is called *integrated validation* and focuses on completeness of the inventory, i.e. the coverage of all activities generating emissions. The method is applied for the Brussels area by comparing measured pollutant fluxes obtained from aircraft monitoring in the urban plume of Brussels with the downwind transported and dispersed emissions. The emission flux is obtained from the emission inventory by integrating its contribution over the plume width and the mixing height.

2. Punctual validation method for the Antwerp Emission Inventory

A detailed urban emission inventory (SO_2, NO_x, CO, VOC, PM and heavy metals) for the Antwerp area (20 x 20 km^2) has been developed which includes industrial sources, road transport and space heating (Mensink *et al.*, 1999). The emission inventory for the Antwerp area can be characterised by the presence of a relatively large contribution of the industrial point sources. In 1996, 72 % of the $NO_x(NO_2)$ emissions and 94% of the SO_2 emissions within the urban area of Antwerp could be allocated to industrial point sources.

Since 1993, industrial companies are obliged to report their emission yearly (VMM, 1998). Based on these registrations, 208 point sources were included in the inventory emitting $NO_x(NO_2)$ and 174 point sources were included emitting SO_2 in 1996. Figure 1 shows the cumulative distribution of the industrial point source emissions, both for NO_x and for SO_2. It can be seen that within the Antwerp area, the largest 15% of the industrial point sources are responsible for 95% of the industrial SO_2 emissions. For the $NO_x(NO_2)$ emissions, 95% of the industrial emissions can be allocated to only 30% of the point sources.

Emission due to space heating are provided per km^2 and are based on the number buildings, type of heating system, fuel consumption and temperature variations expressed in terms of degree days.

Much attention is paid to road transport emissions. They are obtained from a transport emission model. For arbitrary periods the emission model computes the accumulated hourly emissions coming from passenger cars, light duty vehicles, heavy duty vehicles and busses in the Antwerp area. The emission calculation itself is based on an urban traffic flow model (Nys, 1995), which is actually used by the Antwerp City authorities. The urban traffic flow model for the City of Antwerp contains a network with 1963 road segments and was implemented in a GIS environment. Additional statistical data (vehicle classes, vehicle age distributions, fuel data, traffic counts, etc.) were partly obtained from the Antwerp Traffic Police (1998) and the Ministry of the Flemish Community (1997). These data are assimilated in the transport emission model.

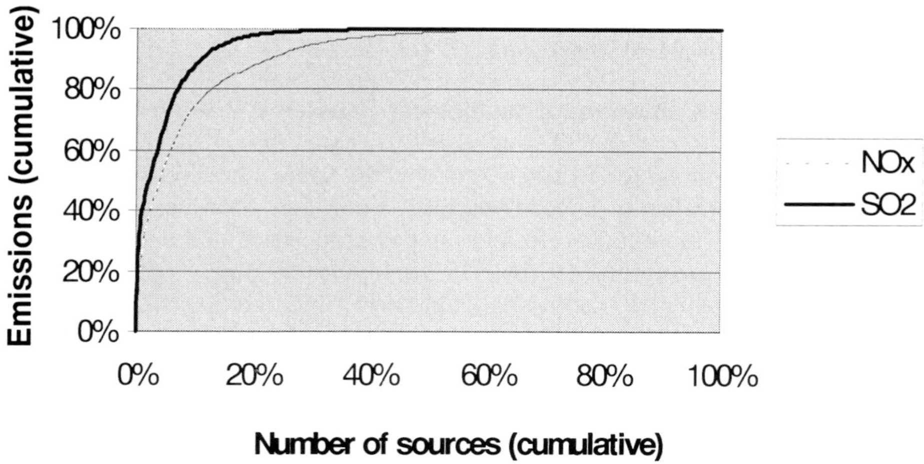

Figure 1: Cumulative distribution of industrial point sources in the Antwerp area

The emission factors used in the model are derived from the COPERT-II methodology (Ahlvik *et al.*, 1997). The time dependent emission variations are obtained by applying one set of three emission time factors describing monthly, daily and hourly variations (Mensink *et al.*, 1999).

The actual emission in function of time E(t) calculated for pollutant i, vehicle class j and road type k on road segment n, is given by:

$$E_{i,j,k,n}(t) = EF_{i,j,k} \cdot F_{m,d,h}(n, t) \cdot L(n) \qquad (1)$$

EF is the emission factor ($g{\cdot}km^{-1}$), F is the time dependent traffic flow rate (h^{-1}) for month m (m = 1…12), day d (d = 1…7) and hour h (h = 1…24). L is the road length (km) per road segment n (n = 1… 1963). E(t) is expressed in $g{\cdot}h^{-1}$.

For an assessment of the accuracy of the emission model, the modelled traffic flow rate F was compared with the observed F in function of n and t. This was carried out for various hourly values obtained in 1996 at various locations in Antwerp (Van Langenhove *et al.*, 1996).

For a limited part of the vehicle fleet, the precision of the emission factors EF was assessed by comparing the modelled emission factors EF with measured EF in function of pollutant i, vehicle type j and road type k. In this case emission factors were measured for 6 gasoline passenger cars with a closed-loop controlled three-way-catalyst (TWC) and an urban driving cycle (De Vlieger, 1997). The six different models that were selected represent types that are common in Belgium. Each car was submitted to 4 tests under normal driving conditions, with average accelerations on city journeys ranging from 0.65 to 0.80 m s^{-2}. In 1996, 55% of the passenger cars in Belgium were equipped with a closed-loop controlled TWC.

3. Integrated validation method for the Brussels Emission Inventory

The emission inventory for Brussels includes the same source categories as the Antwerp emission inventory. Road transport emissions, however, are described in less detail. They are obtained from a top down approach and derived from the national emission inventory (VMM, 1998, Institut Wallon, 1995). Temporal variations are obtained by applying emission time factors (Mensink *et al.*, 1999).

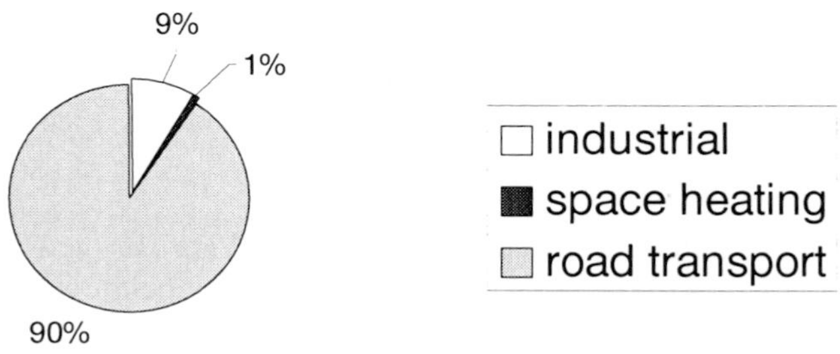

Figure 2 : Emission source distribution for Brussels on 28 August 1998, 12h00 (local time)

Figure 2 shows the distribution of the emissions over the source categories on 28 August 1998 at 12h00, when the emission flux in the urban plume was measured. In contrast with the emission inventory for Antwerp, where industrial point sources form the dominating contribution (see section 2), the Brussels emission inventory is dominated by traffic emissions. Note that at the moment of the flux measurements, emissions due to space heating were negligible.

Measured pollutant fluxes for Ozone, CO, NO and NO_2 have been obtained for Brussels during the STAAARTE T-FLUX flight measurement campaign over Flanders on 28 Augustus 1998. The aim of the campaign was to measure and calculate transboundary mass fluxes (Ozone, CO, NO_x) in order to evaluate photochemical transport phenomena and their possible impact on smog reduction measures (Mensink and Debruyn, 1998). The flight was performed on one single day and consisted of two circuits around the borders of Flanders at "high" altitude (750 m) and one circuit at "low" altitude (300 m), supplemented with 7 vertical profiles (15 m – 750 m) across these layers. Concentrations of Ozone, CO, NO NO_2 and NO_x together with the standard meteorological parameters were monitored continuously on board of an C130 Hercules (MRF, Met Office, UK).

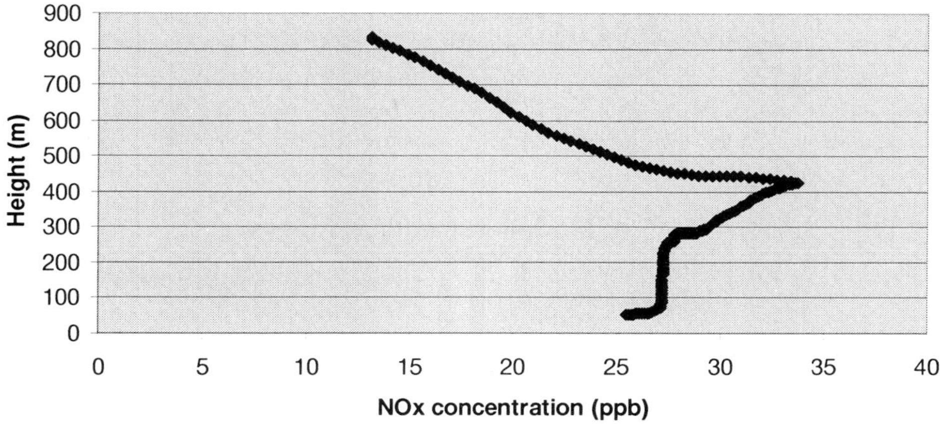

Figure 3 : Vertical NO_x profile measured southeast of Brussels

Due to the northwesterly winds, the flight campaign allowed the measurement of the emission flux south of Brussels. The measurements clearly showed an urban plume of Brussels with a distinct mixing height, as shown by the NO_x profile taken southeast of Brussels (Figure 3). In combination with backward trajectories, the concentration measurements were used to calculate the emission contributions from the Brussels area. The measurements of NO, NO_2 and NO_x were obtained by a Trace Level Chemiluminescence analyser with a detection limit of 50 ppt

4. Results and discussion

4.1 ANTWERP CASE (PUNCTUAL VALIDATION)

Figure 4 shows the comparison between modelled and observed traffic flow numbers for four different road types (highways, national roads, main urban and secondary urban roads). Except for one point, it can be observed that the

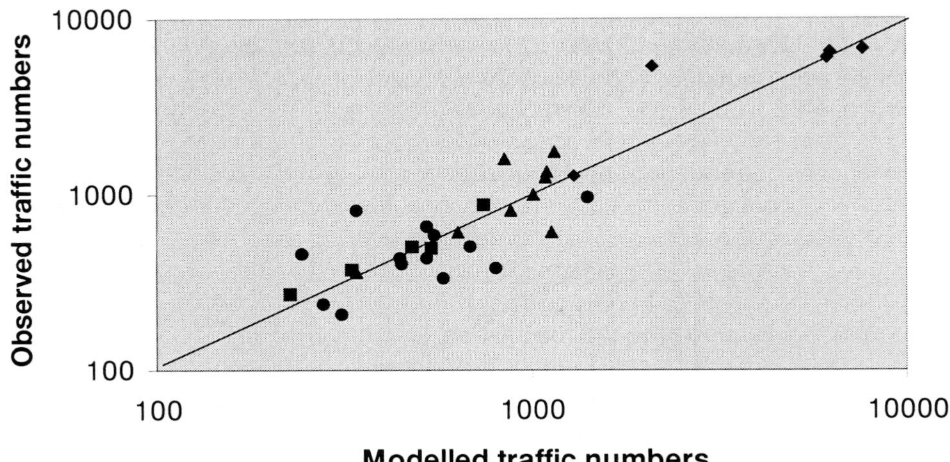

Figure 4: Modelled versus observed traffic flow numbers on highways (◆), national roads (▲), main urban roads (■) and secondary urban roads (●).

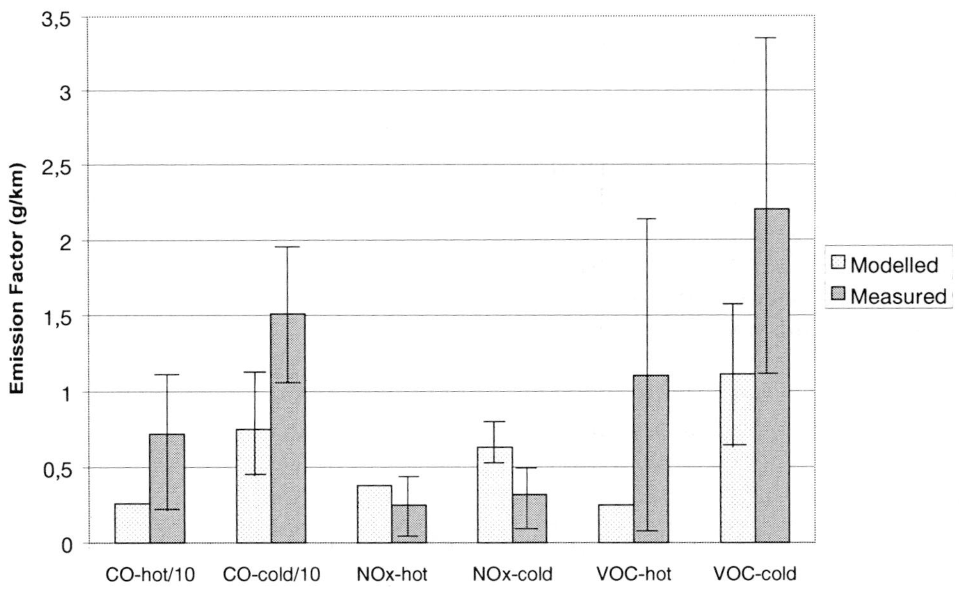

Figure 5: Modelled versus measured emission factors for closed-loop controlled TWC passenger cars

modelled values are relatively accurate for the highways, but less accurate for the urban road segments.

The results in Figure 4 seem to indicate that applying only one set of the emission time factors does not provide accurate results for the urban roads. There is clearly a different temporal distribution on urban roads when compared to highways. Transport to and from schools, local commercial and business transport are probably not correctly represented in the emission time factors.

Figure 5 shows a comparison between the modelled emission factors obtained for closed-loop TWC gasoline cars as averaged over the whole year 1996 and those measured on the road for a limited number of closed-loop TWC gasoline cars. Both hot and cold start emission factors are presented. The error bar on the modelled cold start emissions shows the range of values obtained over the year due to temperature variations. The error bar on the measured emission factors shows the standard deviation obtained for the measurements carried out for 6 gasoline passenger cars with a closed-loop controlled TWC. The uncertainty for the hot emission factors could not be indicated in Figure 5, since no information is given in the COPERT II methodology. The atmospheric emission inventory guidebook (Mc Innes, 1996) gives only a qualitative indication of the precision of the emissions estimates. For NO_x, CO and VOC the emission factors used are categorised as "statistically significant based on sufficiently large set of measured and estimated data".

Despite of the fact that the standard deviation in the measurements is rather large, it can be observed that the model results in Figure 5 show an underestimation of the CO and VOC emission factors, whereas the NO_x emission factors seem to be overestimated. A very similar trend was reported by De Vlieger (1997) when the on-the-road measurements were compared with other emission factors derived from chassis dynamometer tests under the same normal driving conditions on city journeys.

The large standard deviation in the measurements points out the difficulty in measuring emission factors in real, i.e. on-the-road circumstances. This is mainly caused by external influences like driving behaviour and vehicle maintenance, rather than by the measurement equipment itself. There exists only limited amount of maintenance and inspection programs to adjust and control individual vehicle emissions. During the measurement campaign CO and VOC emission factors measured for aggressive driving in urban traffic were found to be up to 4 times higher than those obtained for normal driving (De Vlieger, 1997). Also the influence of traffic jams is not considered in the computation of the emissions factors.

4.2 BRUSSELS CASE (INTEGRATED VALIDATION)

For the integrated validation the emission fluxes were calculated from:

$$\Phi = (C_{pl} - C_{bg}) \cdot h \cdot w \cdot v \qquad (2)$$

where Φ is the pollutant flux (g/s), C_{pl} is the concentration in the plume (g/m^3), C_{bg} the back ground concentration (g/m^3), h the mixing height (m), w the plume width (m) and v the wind speed (m/s). Integrated over a mixing height of 420 ± 30m (Figure 3) and an estimated plume width of 20 ± 2.5 km with an average velocity of 7.2 ± 0.1 m/s, the measured NO_x flux derived from (2) was $\Phi_{measured}$ = 2.6 ± 0.5 kg/h. The flux calculated from the emission inventory was found to be Φ_{emis} = 1.2 kg/h.

Uncertainties in the emission inventory are not known. Besides these uncertainties, it is very difficult to estimate the plume width and the mixing height in an accurate way. It was also assumed that in the calculation the mixing height is constant over the plume width.

5. Conclusions

Punctual emission validation seems to be a valuable and reliable QA/QC technique, but does not provide an overall picture. Integrated emission validation, on the other hand, seems to be a less reliable QA/QC technique, because the technique itself introduces many uncertainties. From the results obtained by the punctual emission validation it could be concluded that the emission time factors have to be differentiated and specified in function of the road type (highway, national way, urban roads) in order to improve the accuracy. Traffic emission factors show large uncertainties. This is expressed by the lack of measurement precision obtained for on-the-road emission measurements. The uncertainties are caused by external influence factors like driving behaviour and maintenance. Also the impact of traffic jams is not adequately reflected by the emission factors. The results of the integrated validation indicate that the actual measured transport emissions are higher than modelled by means of the proposed methodology.

Acknowledgements

The author wants to acknowledge The City of Antwerp and the Antwerp Traffic Police for providing the required data and the Flemish Environmental Agency for their financial support. MRF at Farnborough is acknowledged for supporting the flight campaign. The measurement campaign was funded by the EC Large Scale Facilities Program (STAAARTE). The work was carried out in the framework of the EUROTRAC-2 subprojects SATURN and GENEMIS

References

Ahlvik P., Eggleston E., Gorissen N., Hassel D., Hickman A.-J., Joumard R., Ntziachristos L., Rijkeboer R., Samaras Z. and Zierock K.-H.: 1997, COPERT II Computer Programme to Calculate Emissions from Road Transport, Methodology and emission factors, 2nd edition, European Environmental Agency, European Topic Center on Air Emissions.

Antwerp Traffic Police: 1998, Yearly report 1996, Antwerp (in Dutch).

De Vlieger I.: 1997, On-board emission and fuel consumption measurement campaign on petrol-driven passenger cars, *Atmospheric Environment* **31**(22), 3753-3761.

Institut Wallon: 1995, Bilan energetique de la région de Bruxelles-capitale 1994 (Energy balance of the Brussels capital region), Namur (in French).

Mc Innes G. (ed.): 1996, Atmospheric emission inventory guide book, A joint EMEP/CORINAIR Production, EEA, B710/9-11, Copenhagen.

Mensink C., Van Rensbergen J., Viaene P., De Vlieger I., Beirens, F.: 1999, Temporal and spatial emission modelling for urban environments using emission measurement data, **in:** Borell, P.M. and Borell, P. (eds.) *Proceedings of EUROTRAC Symposium '98*, Vol 2, p. 711-714, WIT Press, Southampton, 1999

Mensink, C. and Debruyn, W.: 1998, Transboundary flux measurements for photochemical model validation in Flanders, *Annales Geophysicae IV*, **16**, C 1158 (abstract).

Ministry of the Flemish Community: 1997, Traffic counts 1996, Report no. 196. Department of Roads and Traffic, Brussels (in Dutch).

Nys J.: 1995, Computer program Urban Traffic Planning, City of Antwerp, Antwerp (in Dutch).

Sturm, P., Blank, P., Böhler, T., Lopes, M., Mensink, C., Volta, M. and Winiwater, W. : 1998, Harmonised method for the compilation of urban emission inventories for urban air modelling, SATURN/GENEMIS Workshop on Urban Emission Inventories, Graz, February 10-11.

Van Langenhove, H. , Moortgat M., Geukens, M. and Kerremans, W. : 1996, Volatile hydrocarbons in the urban air of Antwerp, Measurement campaign September 1994 – June 1996, Report G96/CLW/072, University of Ghent and Centre for Air and Water pollution, Antwerp (in Dutch).

Vlaamse Milieumaatschappij (VMM): 1998, Lozingen in de lucht (Air emission report) 1996-1997, Flemish Environmental Agency, Aalst (in Dutch).

A THREE-DIMENSIONAL MODEL STUDY OF THE IMPACT OF AVOC AND BVOC EMISSIONS ON OZONE IN AN URBAN AREA OF THE EASTERN SPAIN

DIAMANDO VLACHOGIANNIS, SPYROS ANDRONOPOULOS, ARTEMIS PASSAMICHALI, NIKOS GOUNARIS and JOHN G. BARTZIS

Environmental Research Laboratory, NCSR "Demokritos", Ag-Paraskevi, Attikis, 15310, Greece

Abstract. This study concentrated on the effects of Biogenic Volatile Organic Compounds (BVOC) emissions on ozone (O_3) in an area of the Eastern Spain on June 12, 1997, a day characterised by sea breeze. Simulation of meteorology was performed with the three-dimensional model ADREA-I. Comparisons of the model results with observations have revealed overall a good agreement in temperature and wind velocity. Two runs were performed with UAM-IV for the photochemical calculations. The first simulated the effects of the anthropogenic emissions only (run A) and the second the combined effects of anthropogenic and biogenic emissions, (run B). Comparisons of the model O_3 concentrations with measurements showed a general agreement with the experimental data. Discrepancies between the calculated results and the observations during the early morning hours could be attributed to inaccuracies in nitrogen oxides (NO_x) from the anthropogenic emissions inventory. Comparisons between runs A and B yielded differences up to 30% in the morning, over inland areas. It was deduced that the inclusion of BVOC in total emissions could result in an increase or decrease of tropospheric O_3, depending on the available amounts of anthropogenic emissions.

Key words: Three-dimensional model, anthropogenic emissions, BVOC, photochemistry, ozone.

1. Introduction

Ozone (O_3) formation in the troposphere is controlled photochemically by reactions of NO_x and by precursor emissions of volatile hydrocarbons. The concentrations of NO_x have risen substantially in rural areas due to human activities. The increased production of Anthropogenic and Biogenic Volatile Organic Compounds, (AVOC and BVOC, respectively), is considered to contribute to the elevated levels of tropospheric O_3 in the Mediterranean area.

This work was part of the Biogenic Emissions in the Mediterranean Area (BEMA) campaign and it involved the computational simulation of the meteorological and photochemical conditions prevailing during June 12, 1997, in a coastal area over eastern Spain. The particular area is characterised by a relatively highly populated zone parallel to the coast, where industry and agricultural land activities are important. The computational study was carried out using three-dimensional (3D), mesoscale meteorological and photochemical atmospheric models. Topography, land-use and anthropogenic emissions data, required as model input, were manipulated with the aid of a Geographical Information System (GIS). Three computational domains were constructed in

Environmental Monitoring and Assessment **65**: 41–48, 2000.

total. The meteorology simulation was performed by means of nesting, firstly, in the Macro domain and subsequently into the Local domain, chosen specifically to include all stations that provided meteorological data. The photochemistry simulation was carried out on a domain that covered totally the available anthropogenic emissions inventory with a grid resolution of 1×1 km^2 (Figure 1).

This study aims at assessing the impact of BVOC and anthropogenic emissions on the spatial and temporal distribution of O_3 in relation to the meteorological conditions of the region, which are generally characterised by thermal wind flows (Millan *et al.*, 1996).

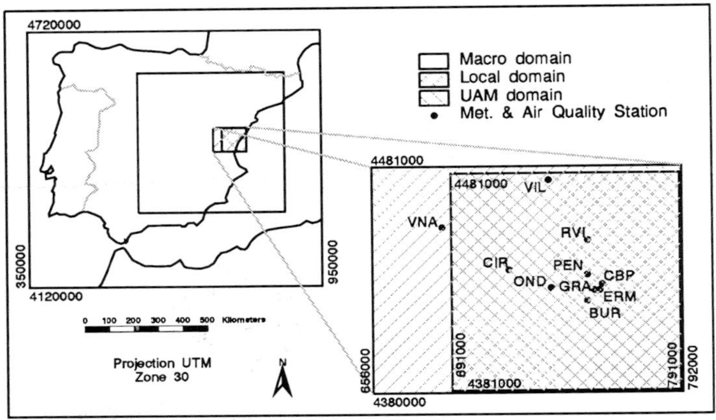

Figure 1. The computational domains.

2. Materials and methods

The computational simulation of the topography and land use of the selected domains (Macro and Local) was carried out using DELTA code (Discretisation with ELements of Triangle Approach) (Catsaros *et al.*, 1993). The vertical extent of both domains was ~10 km, with a non-equidistant resolution in the range of 30m to 500m. The topography was described by triangular surfaces, each characterised by a land cover type, in terms of roughness and albedo, according to the percentage of land cover derived for each cell by the GIS technique. The ADREA-I (Atmospheric Dispersion of Pollutants over Irregular Terrain) classification system for the 21 land cover classes was adopted.

ADREA-I was employed for the simulation of the meteorology. The model is especially designed for wind flow predictions over complex topography (Bartzis *et al.*, 1991). It is a non-hydrostatic, fully compressible model, utilising the conservation laws of mass, energy, momentum, humidity and in-ground heat conduction. Initial model data of temperature, wind speed, wind velocity and specific humidity were provided as vertical profiles of aircraft measurements (Campagne Aeroportee BEMA, June 1997). The total simulated period was 24 hours (June 12, 1997). The ADREA-I model provided the meteorological variables (wind, temperature, mixing layer height, humidity and pressure) required for the calculation of the biogenic emissions and the photochemistry.

A biogenic emissions inventory for the photochemistry domain was derived, using the Biogenic Emissions Inventory System (BEIS) (Pierce *et al.*, 1990). BEIS is a stand alone pre-processor that calculates hourly biogenic emissions from forests (canopy areas) and agricultural crops (non-canopy vegetation) as a function of temperature, sunlight and vegetation type. Emission factors, combined with land cover data following the appropriate 26 BEIS vegetation categories, along with environmental correction factors were used to derive isoprene, a-pinene, monoterpene and unknown BVOC emission rates for the area under study. Since citrus grove is the dominant plantation in the area of interest, experimental data from citrus trees have been combined with BEIS data with the aid of a GIS system to appropriately represent the land cover type. Solar radiation is calculated by BEIS based on given geographical co-ordinates of the area and time.

The model used in this study for the photochemical calculations was the Urban Airshed Model, version IV, (UAM-IV). UAM is a 3D Eulerian model, which calculates the concentrations of both inert and chemically reactive pollutants. It is based on the atmospheric diffusion equation. The UAM-IV employs the Carbon Bond Mechanism (CB-IV) for solving chemical kinetics. This mechanism contains over 80 reactions and over 30 species (UAM Manual, 1990). The total vertical extent of the UAM domain was 3100m, sufficient to cover the circulation patterns in the coastal area of Eastern Spain (Millan et al. 1996). In total, 16 vertical layers were chosen with 8 layers above and 8- below the mixing height.

Two experiments were designed to investigate differences in the resulted concentrations of O_3 due to anthropogenic and biogenic emissions. The first simulated the photochemistry including the anthropogenic emissions only (run A) while the second experiment involved the total emissions (run B). The runs were started at model hour 00:00 and covered a full 24 hr day interval. Initial air quality data to the UAM model were taken from available observations at the stations of Ermita (ERM), Burriana (BUR), Penyeta (PEN), Vila-Franca (VIL) and Grau (GRA) (Figure 5). These data consisted of O_3, nitrogen dioxide (NO_2) and nitrogen oxide (NO) concentrations at the time interval 00-0100 LST. Boundary values of 40 ppb were assumed for O_3, 0.75 ppb for NO_2 and 0.25 ppb for NO, as suggested by measurements (EUR, 1999).

3. Results

The model calculated near-ground horizontal wind-fields in the Local domain at 1200 LST are depicted in Figure 2 as an example of the simulated meteorological conditions. A well-formed sea breeze could be seen to penetrate considerably inland through valleys. The wind flow changed direction following the topography features.

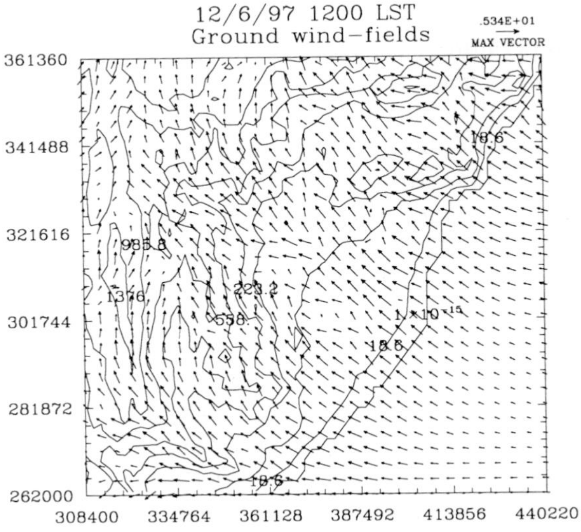

Figure 2. Near-ground wind-field at 12:00 LST and topography contours in the Local domain.

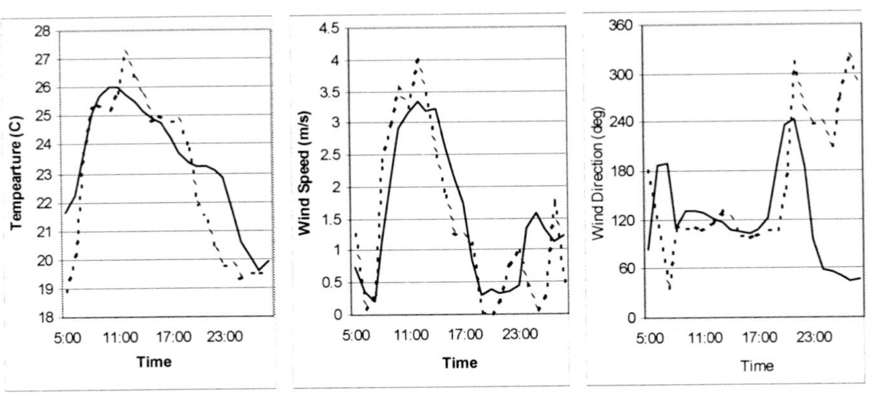

Figure 3. Comparisons of model results (solid line) with observations (dotted line) of temperature (°C), wind speed (m/s) and wind direction (°), at Burriana (12/6/97).

The model results of hourly averaged time history profiles in temperature, wind speed and direction were compared with observations. The stations were grouped according to their relative distance from the coastal line. Hence, BUR, ERM and GRA stations were characterised as coastal while Onda (OND), Cirat (CIR), Rambla de la Viuda (RVI), PEN and VIL as inland (Figure 5). The calculated temperature profiles compared reasonably well with the measurements at all stations. An example for the BUR station can be seen in Figure 3. In late evening, higher temperatures were predicted by approximately 3-4 °C. This behaviour of the model in the evening needs to be investigated further. Noon temperatures reached values in the range between 25 to 30°C. Early in the morning, all stations experienced a very weak flow from a Southeast direction. The wind speed became nearly zero at the time interval 0700-0900 LST, when the pressure gradient was balanced. Then, the sea breeze started to blow inland. The model captured well the magnitude of the wind speed from morning until early evening compared to the measurements. The calculated wind direction was in agreement with the observations at all stations in the morning and afternoon, when the sea breeze was formed. Any discrepancies, found at night, in the comparisons between measurements of the wind direction and the calculated results, could be attributed to the limitations imposed by uncertainties in the model boundary conditions due to lack of observations. In addition, the measured wind speed values were low at night and consequently, the error introduced in the measurements was not insignificant.

The time history profiles of measured O_3 concentrations available from the various stations were compared to model results from runs A and B. Figure 4 shows an example of predicted (run B) and measured O_3 concentrations at the coastal and inland stations BUR and PEN, respectively. The concentrations at the exact sensor location as well as the minimum and maximum predicted concentrations within one grid cell of the sensor are examined. This analysis was carried out to eliminate any effects due to inaccuracies imposed by uncertainties in the emission inventories and the topography, following the recommendations of the UAM Manual (EPA, 1990).

The model predicted O_3 values agreed generally well with the measurements at most stations. At the grid points of the sensors, the O_3 profiles obtained with run B did not differ significantly with those calculated using anthropogenic emissions only (run A). Differences in the comparisons between O_3 model results and measurements, in the early morning hours, could be attributed to discrepancies in NO_x risen from the anthropogenic emissions inventories. This conclusion is in agreement with that by Andronopoulos *et al.*, 2000. All the time history profiles exhibited generally the same shape: low values in early morning, a gradual increase towards maximum around noon and a fall towards minimum values in the evening. The abrupt increase in O_3 values seen at the onset of the sea breeze in the observational profiles was not predicted at the coastal stations by the model. The rather higher O_3 values simulated by the model during early morning did not allow for a steep gradient but a more gradual one. Differences of about 5-15 ppb in O_3 found during the afternoon hours between the model results and the measurements could be attributed to uncertainties in the computational simulation of the mixing of the air masses by UAM.

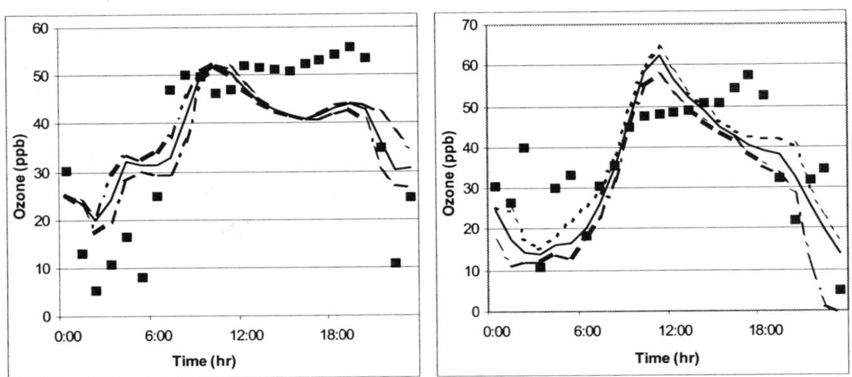

Figure 4. Comparison of model calculated results (Run B) in O_3 (ppb) with observations at Burriana (left) and Penyeta (right). Observed (■)), predicted at exact sensor location (—) and minimum and maximum predicted neighbouring cells (--).

Contour plots of O_3 concentrations (in ppb), resulted from run B were plotted at the time intervals 0600-0700 LST and 1300-1400 LST over the topography of the region (Figure 5). Ozone contours in early morning hours were of low values, in the range of 20 to 45 ppb. Maximum O_3 concentrations occurred at noon, near and along the coast. In the afternoon, the region experienced increased levels of O_3 of approximately 50 to 70 ppb apart from the sea area, where values were between 40 and 45 ppb. The power plant located at GRA and the nitric plant at Sagunto in the south constituted two major pollutant sources of the region. In the afternoon time interval 1300-1400 LST, NO_x concentrations remained still significant in those areas (~ 20 ppb) and in parts of the national road along the coast. Subsequently, at the locations of the power and nitric plant, the O_3 values were very low. In the afternoon, the sea breeze caused the downwind transport of O_3 rich air masses inland, increasing the ozone levels there and relieving the coastal areas.

The impact of BVOC emissions on O_3 was studied by means of plotting contours of the percentage changes in O_3 between runs B and A using as reference concentration values those obtained from run A. Those changes were studied for selected time periods of the day to investigate how the diurnal changes of BVOC emissions may affect corresponding changes in ozone (Figure 6).

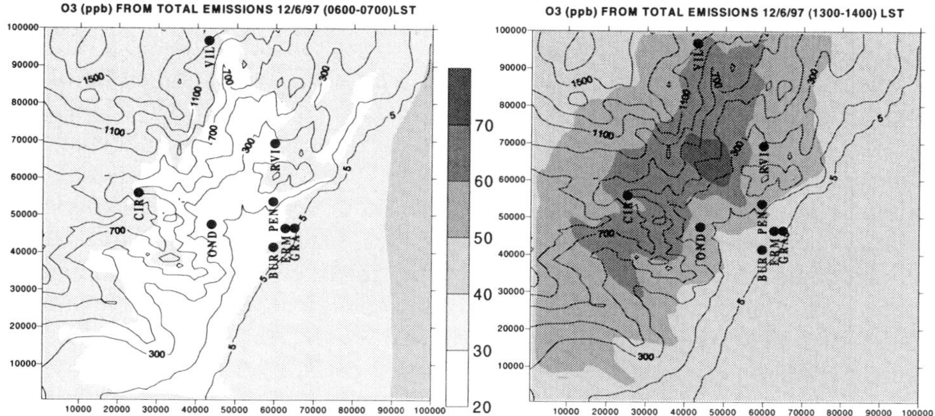

Figure 5. Model O_3 concentrations (ppb) near ground at 0600-0700 LST (left) and 1300-1400 (right). Topography contours are depicted with solid lines. The locations of the ground observational stations are also indicated.

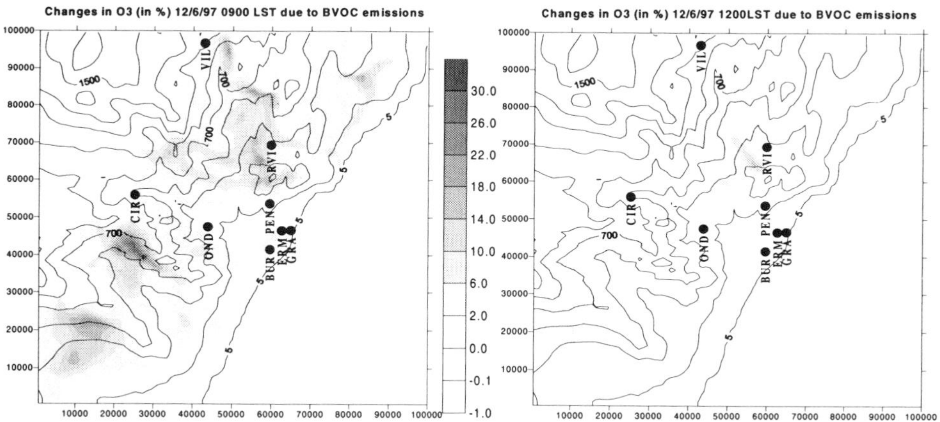

Figure 6. Percentage O_3 changes (in %) between runs B and A ((run B-run A)·100/run A) at 0900 LST (left) and 1200 LST (right). Topography contours depicted with solid lines.

First of all, increases in O_3 concentrations were found at 0900 LST. Those increases were as high as 30%. Generally, increases in O_3 concentrations occurred in a zone parallel to the coast, covering a large part of the domain. The same result could be seen at noon but the maximum change was smaller than the morning value (~13%) and it occurred between OND and RVI. Moreover, the spatial extent of the O_3 increases was reduced in a direction away from the coast. A small decrease in O_3 was calculated at noon (~2%), in the northwestern corner of the domain, where BVOC were higher than the anthropogenic ones. In late afternoon, although less, the O_3 increase was still seen at the area, located between OND and RVI, while the spatial extent of depleted ozone had increased. The changes in O_3 found at night were very small of the order of 2%.

Certain conclusions could be drawn from the qualitative examination of the above mentioned graphs. The inclusion of model calculated BVOC in total emissions can result in an increase or decrease of tropospheric ozone, depending on the available amounts of AVOC. Places where BVOC emissions are significant but anthropogenic emissions are absent or very small may experience an ozone decrease due to the biogenic emissions. Ozone formation may result from biogenic emissions over areas with relatively high anthropogenic emissions. In such a case, the magnitude of the change depends on the available amounts of BVOC since the higher the emitted BVOC, the larger the increase. Biogenic emissions can cause ozone increases at certain periods of the day and in places, where both BVOC and anthropogenic emissions are important.

Acknowledgements

This project was funded by the Institute of Environment, JRC Ispra, Italy. The authors thank CEAM (Spain) for providing meteorological and air-quality data and Dr. G. Seufert and Dr. P. Ciccioli (JRC, Ispra) for supplying experimental citrus data.

References

Andronopoulos S., Passamichali A., Gounaris N., and Bartzis J.G.:2000, Evolution and transport over a Mediterranean coastal area and Biogenic VOC emissions influence on ozone levels, *.J Appl. Meteor.*, **39**, No 4, 526-545.

Bartzis J. G., Venetsanos A., Varvayanni M., Catsaros N., and Megaritou A.: 1991, ADREA-I, A transient three-dimensional transport code for complex terrain and other applications, *Nuclear Technology*, **94**, 135-148.

Campagne Aeroportee BEMA, Juin 1997, Rapport D'Avancement des travaux, Decembre 1997.

Catsaros N., Robeau D., Bartzis J.G., Varvayianni M., Megaritou A., and. Konte P: 1993, A computer system for simulating the transfer of pollutants over complex terrain-some recent applications, *Radiation Prot. Dosimetry*, **50**, 257-263.

EPA: 1990, User's guide for the Urban Airshed Model, Vols I-III (1990). Reports EPA-450/4-90-007A, EPA-450-90-007B and EPA-450/4-90-007C.

EUR: 1999, Report on the BEMA main measuring campaign and modelling at Burriana (Valencia, Spain), EUR Report, Luxembourg, Office for the Official Publications of the European Commission, in press.

Millan M., Salvador R., Mantilla E., and Artinano B.: 1996, Meteorology and photochemical air pollution in Southern Europe: Experimental results from EC research projects, *Atmos. Environ.*, **30**, 1909-1924.

Pierce T. E., Lamb B. K., Van Meter A. R.: 1990, Development of a Biogenic Emissions Inventory System for Regional Scale Air Pollution Models, Paper No 90-94.3, The 83rd Air and Waste Management Annual Meeting, June 24-29, Pittsburgh Pennsylvania.

EFFECTS OF TRAFFIC MANAGEMENT AND TRANSPORT MODE ON THE EXPOSURE OF SCHOOLCHILDREN TO CARBON MONOXIDE

M R Ashmore[1], K Batty[1], F Machin[2], J Gulliver[2], A Grossinho[3], P Elliott[3], J Tate[4], M Bell[4], E Livesley[2] & D Briggs[2]

1 Department of Environment Science, University of Bradford, West Yorkshire BD7 1DP, UK; 2 Nene Centre for Research, University College Northampton NN2 7AH, UK; 3 Department of Epidemiology and Public Health, Imperial College of Science Technology & Medicine, London W2 1PG, UK; 4 Institute of Transport Studies, University of Leeds LS2 9JT, UK

Abstract The exposure to CO of schoolchildren was assessed in the town of Northampton, UK, both by direct measurement and by GIS-based activity modelling. Personal measurement of CO showed that exposures when travelling by car were significantly greater than those when walking, although journey times by car were shorter. However, journey exposures had little effect on maximum 8h mean CO exposures. CO concentration fields in the study area were modelled from current traffic flows, and those expected under different traffic management scenarios. These fields were then used, in combination with children's home and school location, and their activity profiles, to simulate frequency distributions of exposure for different transport modes and traffic management scenarios. The results show a large variability in the effect of traffic management interventions, depending on the child's home and school location.

Key words: carbon monoxide, traffic management, transport mode, GIS

1. Introduction

A range of traffic management schemes is being implemented across Europe to reduce pollutant emissions from motor vehicles. In the UK, a National Air Quality Strategy has been adopted, within which local air quality management will aim to reduce concentrations measured by fixed-site monitors to below guideline concentrations (DETR, 1997). However, pollutant impacts on people's health depend on their exposure in different locations as they move around the city, rather than measured or modelled ambient concentrations (Loth and Ashmore, 1994). The effects of traffic management measures on these personal exposures are not well understood, as changes in the use of different traffic modes and time-activity patterns, as well as pollutant emissions, may affect exposure. New methodological approaches are therefore required which link conventional atmospheric dispersion models to time-activity-location models of exposure.

The research described in this paper forms part of a multi-centre interdisciplinary study to develop new methods to assess the impact of traffic

Environmental Monitoring and Assessment **65**: 49–57, 2000.

management on personal exposures to air pollution. The study used a GIS framework to model the effects of traffic management on vehicle flows, speeds and emissions, to link these to an atmospheric dispersion model, and to assess personal exposures by linking information on the spatial distribution of the population and of pollutant concentrations, based on time-activity patterns. The study was based in one section of the town of Northampton, which was selected because it represented a self-contained traffic management area, as part of the Transport Package plan, in which the commuting routes into the town are bounded on both sides by open spaces. The annual mean CO concentration at a background monitoring site in the area was 0.36ppm, with a maximum 8 hr mean concentration of 2.82ppm, which is well below the EU guideline of 10ppm.

Carbon monoxide (CO) was chosen for this study because over 90% of emissions in U.K. urban areas are from road traffic, and because previous studies of CO exposure have demonstrated the importance of commuting journeys and transport mode for daily exposures (e.g. Akland *et al.*, 1985; Fernandez-Bremauntz and Ashmore, 1995; Vellopoulou and Ashmore, 1998). The same pattern of commuting journeys contributing significantly to daily exposure, and of significant differences in exposure between modes, has been demonstrated for VOCs such as benzene (e.g. Chan *et al.*, 1991; Dor *et al.*, 1995; Barrefors and Petersson, 1996). However, for NO_2 and particle exposure, these features are less clearly established; for example, van Wijnen *et al.* (1995) found significantly higher exposures to CO and benzene in car drivers compared with cyclists in Amsterdam, but no significant difference between the two modes for NO_2.

A major limitation of the models previously used to model exposure to CO (e.g. the SHAPE model; Ott, 1984) and other pollutants is that they have treated variability within a city statistically, and have had no explicit spatial component. The modelling approach we have adopted in this study aims to provide the spatial element which is essential if the impact of local traffic management schemes on exposure is to be properly assessed. This paper focuses on the exposure of schoolchildren to CO, describing the results both of direct measurement of journey exposures and of network analysis and time-activity modelling to simulate exposures.

2 Materials and Methods

CO concentrations were determined using two Draeger Pac III personal monitors, which were calibrated at the start of each week. Monitoring was carried out over a two-week period during March 1998. Two routes to each of

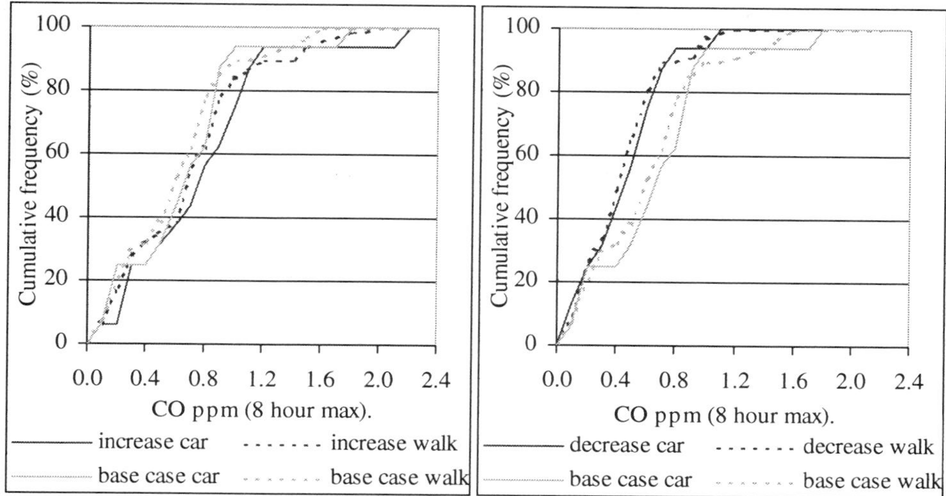

Fig. 4: Calculated frequency distributions of maximum 8h mean exposures of children travelling to school by car or walking without journey exposures.

Figure 5 shows the results of the network analysis, expressed as mean exposure for travel by car and walking, for current emissions and the two traffic management scenarios, during the morning and afternoon travel-to-school periods. For the current emissions, exposures of children are estimated to be higher when walking than when travelling by car in the morning, but the difference between modes was not statistically significant, because of the large variability between children, both in the morning and afternoon.

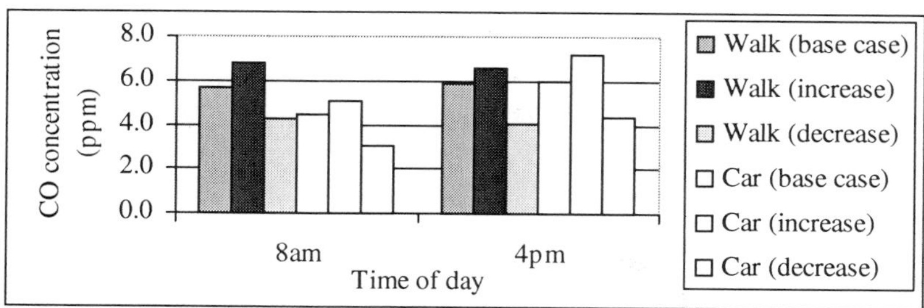

Fig 5: Mean exposures during journeys to school by car and walking, with different traffic flows, estimated using network analysis.

The effects of changes in traffic density, when averaged over all modes and periods, were to increase mean journey CO exposure by 16.8% under the 10% increase in traffic flows and to decrease mean journey CO exposure by 28.4% under the 20% decrease in traffic flows. The greater increase in journey exposures than in mean 8h exposure in the increased traffic flow scenario may reflect the greater influence of altered traffic flows on CO exposures during journeys themselves than on CO exposures at home or school.

4. Discussion

This study has primarily focussed on establishing new methodological approaches to assess the effects of traffic management on personal exposures. The results demonstrate the potential of these methods to provide new insights into the links between traffic emissions and human exposure. However, our model simulations are based on limited field data and contain many important assumptions; more field testing and validation in a wider range of traffic management schemes is needed to improve the model design and parameterisation.

The results obtained from direct measurement in this study indicated that travel by car did result in higher journey CO concentrations, as has been shown by other studies (van Wijnen *et al.*, 1995). This is consistent with the expected gradient of exposure from the roadway to the pavement. The reason that this difference was not apparent in the network analysis is unclear. However, key uncertainties identified are differences between the simulated network routes and the actual routes followed by children, and the journey times, for which significant differences existed between times recorded in the questionnaires and those estimated in the network analysis. The results from the different aspects of the study demonstrate that the effect of modal choice on exposure to CO depends on the duration of the journey and the routes selected. It is clear that the estimation of journey time needs more attention, both in direct observation and GIS analysis, in future studies. Network analysis provides a powerful tool for assessment of traffic management interventions on journey exposures, but it is clear that direct measurement and observation is essential to ensure that its baseline predictions are realistic.

The effect of journey exposure on 8h mean CO exposure was insignificant in our study, primarily because of the relatively short journey times. However, short-term peaks during journeys to school may be more significant for other pollutants (e.g. NO_2), for which short-term exposures are of greater health concern than is the case for CO. Furthermore, the impact of journey exposures may be greater in areas with longer journey times, and greater CO emissions. The mean changes in 8h mean exposures, and journey exposures, under the different traffic scenarios were generally greater proportionally than the changes in traffic density. There were also large differences in the effect on individual children, suggesting that traffic management schemes may have differential benefits across a population, depending on the home and school location of individual children, and their use of transport. There is a need for more detailed spatial analysis of the effects of different transport interventions on exposure patterns, in order to develop methods of optimising their benefits in term of exposure.

Acknowledgements

This work was funded by a grant from the Sustainable Cities programme of the UK Engineering and Physical Sciences Research Council (EPSRC) to D Briggs, M Bell and P Elliott. We gratefully acknowledge the support and advice of Northampton Borough Council and Northamptonshire County Council in carrying out the study.

References

Akland, G.G., Hartwell, T.D., Johnson, T.R. and Whitmore, R.W.: 1985, 'Measuring human exposure to carbon monoxide in Washington DC and Denver, Colorado, during the winter of 1982-1983', *Enviromental. Science and Technology* **19**, 911-918.

Barrefors, G. and Petersson, G.: 1996, 'Exposure to volatile hydrocarbons in commuter trains and diesel buses', *Environmental Technology* **17**, 643-647.

Carruthers, D.J., Edmunds, H.A., Lester, A.E., McHugh, C.A. and Singles, R.J.: 1998, 'Use and validation of ADMS-Urban in contrasting urban and industrial locations', in: *Proceedings of the fifth workshop on harmonisation with atmospheric dispersion modelling for regulatory purposes,* Rhodes.

Chan C-C, Spengler, J.D., Ozkaynak, H. and Lefkopoulou, M.: 1991 'Commuter exposures to VOCs in Boston, Massachussets', *J. Air and Waste Management Association* **41**, 1594-1600.

Crump, D.R.: 1997, 'Indoor air pollution', in: Davison, G. and Hewitt, C.N. (eds), *.Air Pollution in the United Kingdom*, Royal Society of Chemistry, London. pp. 1-21.

Department of Environment, Transport and the Regions (DETR): 1997, 'The United Kingdom Air Quality Strategy', HMSO, London.

Dor, F., le Moullec, Y. and Festy, B.: 1995, 'Exposure of city residents to carbon monoxide and monocyclic aromatic hydrocarbons during commuting trips in the Paris metropolitan area.' *J. Air and Waste Management Association* **45**, 103-110.

Fernandez-Bremauntz, A.A. and Ashmore, M.R.: 1995, 'Exposure of commuters to carbon monoxide in Mexico City – I. Measurements of in-vehicle concentrations', *Atmospheric Environment* **29**, 525-532.

Loth, K.W. and Ashmore, M.R.: 1994, 'Personal exposure to air pollution', *Clean Air* **24**, 114-122.

Ott, W.R.: 1984, 'Exposure estimates based on computer generated activity patterns', *J. Toxicology – Clinical Toxicology* **21**, 97-128.

Van Vliet, D.: 1982, 'SATURN – a modern assignment model', *Engineering and Control* **23**, 578-581.

Van Wijnen, J.H., Verhoeff, A.P., Jans, H.W.A. and van Bruggen, M.: 1995, 'The exposure of cyclists, car drivers and pedestrians to traffic-related air pollutants', *International Archives of Occupational and. Environmental Health* **67**, 187-193.

Vellopoulou, A.V. and Ashmore, M.R.: 1998, 'Personal exposures to carbon monoxide in the City of Athens: I. Commuters' exposures', *Environ. Int.* **24**, 713-720.

BENZENE EXPOSURE ASSESSMENT AT INDOOR, OUTDOOR AND PERSONAL LEVELS. THE FRENCH CONTRIBUTION TO THE LIFE MACBETH PROGRAMME

NORBERT GONZALEZ-FLESCA [1*], MATTHEW S. BATES [1,2], VERONIQUE DELMAS [3], VINCENZO COCHEO [4]

[1] *INERIS, Parc Technologique Alata, B.P 2, 60550 Verneuil en Halatte, France*

[2] *University of Hertfordshire, Hatfield, Herts, AL10 9AB, U.K.*

[3] *AIR NORMAND, 21 Av. de la Porte des Champs 76000 Rouen, France*

[4] *Fondazione Salvatore Maugeri IRCCS, Padova, I-35127, Italy*

E-mail: Norbert.Gonzalez-Flesca@ineris.fr

Abstract. Many VOC represent hazards to human health through chronic exposure. Recent European and world-wide legislation proposes limit values for ambient concentrations of these compounds. However, very little experimental data exists for true population exposure. In 1996, the European MACBETH initiative set out to measure population exposure to benzene in six European cities. This study details the French contribution to this program. Six campaigns were carried out, each comprising measurements at 100 outdoor sites and the participation of 50 non-smoking volunteers who wore personal samplers and had passive monitors installed in their homes. Iso-concentration maps were drawn for each campaign and the results showed that outdoor concentrations were significantly lower than indoors. Almost 75% of the volunteers were exposed to mean concentrations higher than the limit value of $5 \mu gm^{-3}$. It is demonstrated that personal exposure levels cannot be deduced simply by combining indoor and outdoor background concentrations. It is also shown that there is need for better knowledge of the contributions to overall exposure of outdoor microenvironments and the authors hope that future European directives will take this into account.

Key words. Benzene, population exposure, indoor air, outdoor air, passive sampling, MACBETH, European legislation, limit value exceedance.

1. Introduction

Several organic compounds in the environment have been recognised as having adverse effects on human health even at low concentrations (Cicolella, 1997). Amongst these, benzene is a known carcinogen for man principally inducing leukaemia and is classed in Group 1 by the International Agency for Research on Cancer (IARC, 1997).

In 1996 the European section of the World Health Organisation produced air quality guidelines (WHO, 1996) stating that there was no absolute safety threshold for this compound. Based on a model of linear extrapolation without threshold, they proposed a Unit Risk Excess (URE) of $6x10^{-6}$ $(\mu gm^{-3})^{-1}$. This means that an excess of six cases of leukaemia could be induced for a population of one million that has been exposed continuously, throughout its lifetime, to a benzene concentration of $1 \mu gm^{-3}$.

Benzene concentration in ambient air is under regulation in several countries. In France, the Conseil Supérieur d'Hygiène Publique (CSHP) has adopted a limit value for an annual mean of $10 \mu gm^{-3}$ (CSHP, 1997) and the Loi sur l'air et l'utilisation rationnelle de l'énergie (Loi sur l'air, 1998) fixed a target

Environmental Monitoring and Assessment **65**: 59–67, 2000.

value of 2μgm^{-3}. At a European level the proposal for the council directive (EC, 1998) fixes a limit value of 5μgm^{-3} for mean annual exposure to benzene. Most European countries have set up monitoring networks to assess ambient air quality with respect to several pollutants. However, is the data collected by fixed background monitoring stations appropriate for the calculation of population exposure to benzene? The MACBETH programme (Monitoring Ambient Concentrations of Benzene in European Towns and Homes) was undertaken to give an answer to this question.

This article describes the French contribution to this program based on six monitoring campaigns carried out in 1997-1998 by INERIS (The French National Institute of the Industrial Environment and Risks) in collaboration with AirNormand (The Air Quality Monitoring Network for the Normandy Region of France). The campaigns were carried out in the medium sized French town of Rouen (c. 380000 inhabitants) and entailed the measurement of indoor and outdoor concentrations. The estimated personal exposure calculated from these values was then compared to the actual experimentally measured personal exposure ascertained through direct personal monitoring.

2. Experimental

The study was conducted on fifty, non smokers including thirty outdoor workers. Indoor measurements were carried out in the participants' bedrooms as previous studies (Mann et al., 1997) had shown that it was here that the greatest indoor exposure occurs. This is mainly due to two factors: concentrations are relatively high and a large proportion of time at home is spent in this room. Outdoor monitoring was conducted at one hundred urban background sites evenly distributed throughout the area in which volunteers lived and spent their working day.

The campaigns were conducted over the working week (Monday morning to Friday evening). This guaranteed reasonably regular traffic emission levels and some routine in volunteer behaviour avoiding to some extent those activities that could lead to en extra exposure during the week end (use of solvents ,glues paints, etc.). In this way it is expected to reduce the number of variables on which the results would depend. These criteria were deemed necessary for research involving relatively small sample numbers. Each volunteer completed a daily questionnaire detailing their movements, activities and possible exposure to high concentrations of benzene (such as passive smoking, open hearth fire, gas cooker operation, chemical plant, petrol station, fresh paint etc.).

All sampling was carried out by passive means which allows the determination of an average concentration for the whole exposure period. Passive samplers are highly sensitive, non-encumberant devices and fulfil all the basic requirements for this type of study (CEN, 1997).

Outdoor measurements were made by placing samplers inside purpose built aluminium shelters to protect from direct sunlight and rainfall and attached at a height of 3m above ground level in order to minimise sampler degradation by predators. Most of the sites were of background type for mapping purposes. Nonetheless, some extra measurements were carried out at hot spots close to traffic lights in order to characterise micro-environments that could be responsible for some exceedance in personal exposure. Passive samplers used for indoor monitoring were attached to ad-hoc supports 1m above floor level in the volunteers' bedrooms. Personal samplers were attached within the breathing zone of each of the volunteers and left open in the bedroom during the night. The passive sampler used was the Radiello Perkin Elmer (RPE) assembled and validated at INERIS and loaded with 150 mg of Carbotrap-B adsorbent (Supelco Inc.). This axial diffusive sampler is compatible with thermal desorption, has a typical detection limit of 0.1μgm^{-3} for the sampling duration considered and due to the adsorbent choice its measurements are unaffected by ambient humidity. A full description and details of the RPE are given elsewhere (Bates et al., 1997). Analyses where performed on an automatic thermal desorption apparatus (ATD400, Perkin Elmer) linked to a Chrompack CP-9002 gas chromatograph equipped with an FID and a Chrompack Column (CP-Sil 5CB 50m, 0.32mm, 1.2μm film thickness). Quality Assurance procedures were controlled and supervised by an external laboratory (ERLAP, Ispra, Italy).

3. Results

In figure 1 typical chromatograms of samples taken outdoors (A), indoors (B) and on volunteers (C) are represented. It is interesting to note that the profile of indoor and volunteer samples are similar and exhibit numerous peaks corresponding to compounds with higher molecular weights than benzene (Ph) and toluene (PhMe).

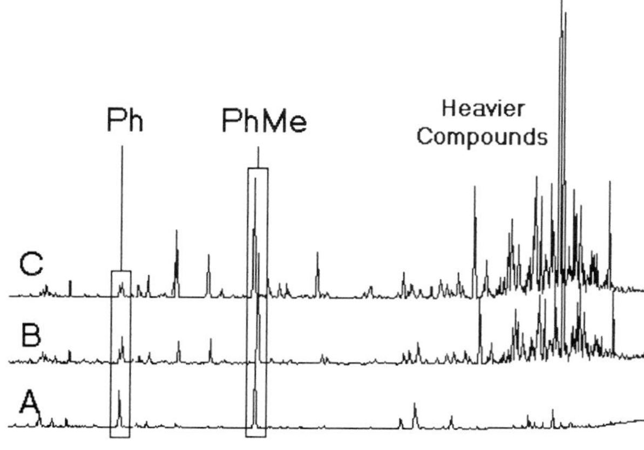

Figure1. Typical chromatograms of outdoor (A), indoor (B) and personal (C) air samples

These substances seem to come from indoor sources as they are not present in the outdoor samples. This characteristic is not unique to French dwellings as studies conducted elsewhere have shown similar results (Brown and Crump, 1998). The box-plot in figure 2 shows the distribution (10th, 25th, median, 75th and 90th percentiles) of all the results from the six campaigns. One can see that there is little dispersion in the outdoor values which is to be expected for background concentrations. It will be shown later how microenvironments, such as those found in the vicinity of traffic lights, where concentrations are much higher, can strongly contribute to overall personal exposure. Indoor and personal concentration levels are higher than those measured outdoors and they are more dispersed. Some candidates and their homes showed mean exposure concentrations of up to ten times the limit value.

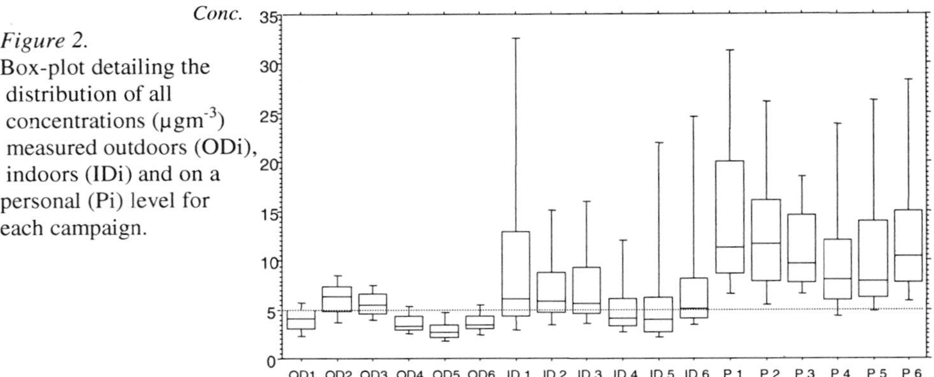

Figure 2. Box-plot detailing the distribution of all concentrations (μgm^{-3}) measured outdoors (ODi), indoors (IDi) and on a personal (Pi) level for each campaign.

OUTDOOR CONCENTRATIONS

Using the six hundred outdoor measurements that were made, it was possible to build up data-sets to establish six iso-concentration maps. Figure 3 shows one of these (the interpolation was carried out using the geostatistical method of Kriging). It may be observed that there is an accumulation of benzene on the northern side of the river. This has been put down to the local topography in this encased and poorly ventilated area of the town.

Table 1. Mean benzene concentrations (μgm^{-3}) excluding all values above the 90th percentile

Campaign	Outdoor	Indoor	Personal
1 (20-24/10/97)	3.8	8.3	13.0
2 (24-28/11/97)	5.7	6.0	11
3 (19-23/1/98)	5.3	6.3	10.2
4 (23-27/3/98)	3.4	4.8	8.3
5 (25-29/5/98)	2.7	4.5	8.9
6 (28/9-2/10/98)	3.4	6.3	10.6
Averages	4.0	6.0	10.3

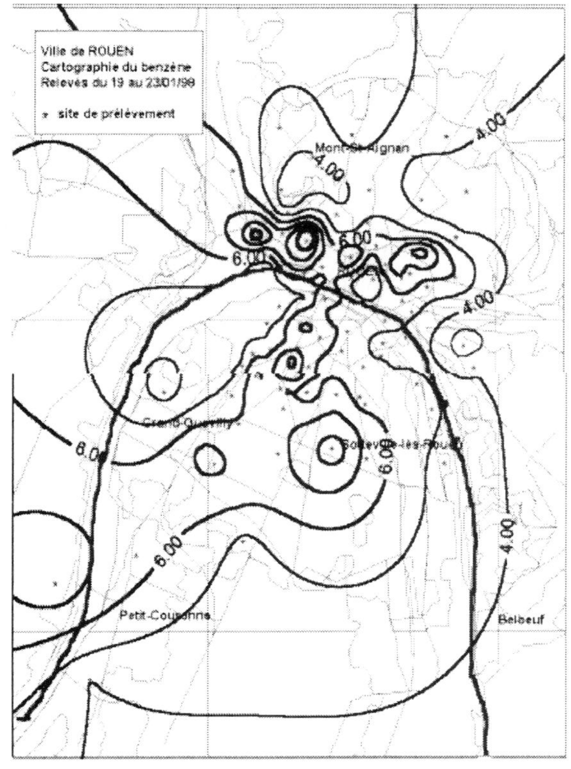

INDOOR CONCENTRATIONS

Table 1 shows the mean concentrations for outdoor, indoor and personal exposure (excluding all values over the 90th percentiles). It can be seen that the average benzene concentration measured indoors is higher than that measured outdoors. This suggests that there are sources other than car emissions for indoor benzene. In addition, the indoor concentrations observed for all campaigns are around, or higher, than the limit value of 5μgm^{-3}.

Figure 3. *Benzene iso-concentration map for Rouen North on top. Concentrations in μgm^{-3}*

PERSONAL EXPOSURE MEASUREMENTS

From table 1 it may also be seen that the values for personal exposure are generally high and in some cases more than twice the limit value.

The MACBETH organisers set out with the preconceived idea that outdoor workers would be exposed to benzene concentrations higher than those for people working indoors. However, the results in table 2 show that there is very little difference between the exposure of the two categories of participants which is no surprise when the high concentrations measured indoors are taken into consideration.

Table 2. Mean personal exposure concentrations to benzene (μgm^{-3}) excluding all values above the 90th percentiles

Campaign	1	2	3	4	5	6
Volun. work indoors	15.3	13.2	9.1	7.9	8.9	13.5
Volun. work outdoors	11.3	10.0	10.5	8.1	8.4	9.4

The overall mean concentration for the six campaigns is 10.3μgm^{-3} (*table 1*) and by simple observation of figure 2 it may be seen that the majority of the volunteers were exposed to levels higher than the limit value.

Total calculated personal exposure (TCPE)

Up till now, benzene concentrations have been examined at three separate levels: ambient outdoor air, indoor air and through personal monitoring. However, if the concentration of pollutants in the different encountered environments is known along with the time spent in those environments, then it should be possible to calculate the TCPE by performing the sum of the products of the time spent in each atmosphere and the concentration of that atmosphere. This approach would be particularly pertinent when considering chronic exposure as it would allow the comparison of the separate contributions from each environment.

Table 3 shows the average TCPE and experimentally measured personal exposure (MPE) to benzene for the thirty volunteers working outdoors (no measurements were made indoors at the work-place).

Table 3. Total calculated personal exposure (TCPE) and measured personal exposure (MPE) for volunteers working outdoors (μghm^{-3})

Campaign	Mean indoor exposure	Mean outdoor exposure	TCPE	MPE	TCPE/MPE
1	642	129	772	1759	2.3
2	507	199	705	1285	1.8
3	533	151	685	1173	1.7
4	595	112	707	1230	1.7
5	598	80	678	1152	1.7
6	515	107	622	1127	1.8

Calculations were carried out as follows:

Indoors

Calculated personal exposure = Mean measured indoor concentration x time spent at home

Recall that the indoor concentration is measured in the bedroom and that this is justified by the large proportion of time spent in this room.

Outdoors

Calculated personal exposure = Mean ambient concentration x time spent outdoors

A more refined calculation could certainly have been carried out by the summation of the products of the encountered ambient concentration (c_i) and the time (t_i) spent in each concentration zone using the daily trajectory of each participant, in this case, the total outdoor exposure for the volunteer would be:($\Sigma c_i t_i$). However due to the relatively small differences in outdoor concentrations encountered, the mean concentration was judged acceptable.

Measured Personal Exposure:

MPE = concentration measured by personal monitoring x total exposure duration

N.B. the exposure calculations were performed for each volunteer and then averaged for the whole population (and not the opposite).
From table 3 it may be observed that indoor exposure in all cases is higher than outdoor exposure; it must be said, however, that outdoor exposure is calculated from the mean concentration of the background sites, and in general, these sites exhibit concentrations lower than those commonly encountered indoors (Begerow, 1995). Finally the TCPE in no way accounts for the MPE which is on average two times higher. One plausible reason for this lack of recovery between calculated and measured exposure is that volunteers are not exposed to average ambient background levels of benzene but to day-time levels that vary both temporally and spatially.This point will be discussed in the following sections.

TEMPORAL VARIATIONS OF BENZENE CONCENTRATIONS

In order to assess the differences between working-week background and daily rush-hour concentrations, the following methodology was applied. At a background site, parallel measurements were made of carbon monoxide, benzene and toluene all of which are emitted from common sources. Carbon monoxide (CO) concentrations were measured continuously using an infra-red absorption monitor throughout the experimental period. Benzene and toluene concentrations were measured in two ways: firstly the time weighted average concentration was monitored by passive means for the whole period (24h/day for 5 days) and secondly the average rush-hour concentrations were measured using passivated canisters (between the hours of 08:30-09:45 and 17:00-18:15). From the results in table 4 it may be seen that during the rush-hour, at the chosen background site, concentrations are roughly double those measured for the 5-day average.

Table 4. Comparison of rush-hour mean concentrations to working week mean concentrations

Compound	**Benzene (μgm-3)**	**Toluene (μgm-3)**	**CO (mgm-3)**
Rush hour mean	15.3	47.2	4.6
Working week mean	7.5	22.7	2.7
Ratio rush:week	2	2.1	1.7

SPATIAL VARIATION OF BENZENE CONCENTRATIONS

Another aspect that must be considered in order to explain the difference between TCPE and MPE is the spatial distribution of pollutants in an urban area. During one campaign, along with the measurements made at the background sites a series of passive samplers were placed at hot-spots

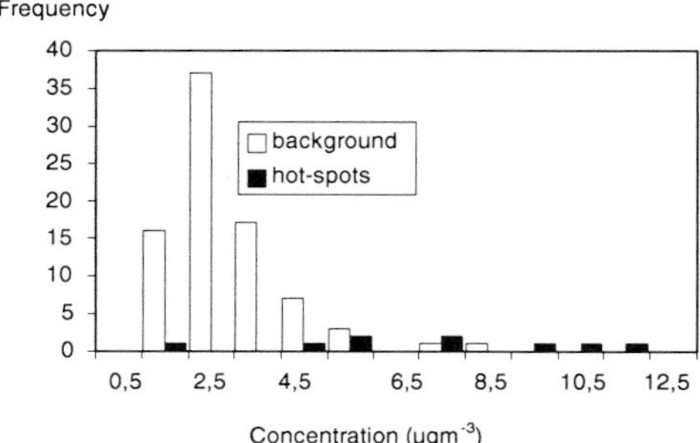

Figure 4. Histogram showing distributio of background and hot pot benzene concentrations

The concentration distributions from these two sets of measurements are shown in figure 4. It can be seen from the picture that one could expect averages values of background sites and hot spots to be very different. In fact, they are also different in nature: An urban background site is an area where pollutants are diluted and their concentrations do not change very much with distance (low gradients). A hot spot however,is a site where concentrations are not only higher but pollutants are in the process of being diluted consequently gradients are high (these sites are generally close to sources)

4. Discussion

The average outdoor background concentrations varied between 3 and 6 μgm^{-3} throughout the six measurement campaigns. The data gathered permitted the generation of iso-concentration maps which may be of great use for environmental management purposes. However, with respect to outdoor personal exposure, microenvironments such as kerbside atmospheres appear to play a dominant role due to the relatively high concentrations encountered there. Therefore the outdoor contribution to total overall exposure cannot be determined from fixed station background site measurements. This observation has been made elsewhere (Gonzalez et al., 1998).

With regards to indoor exposure, table 3 shows that this environment makes a strong contribution to overall personal exposure and thus should be taken into consideration as people typically spend the majority of their day indoors. The high indoor exposure values also suggest that indoor sources exist for benzene. An investigation to identify possible indoor sources of benzene is currently underway.

The comparison of MPE with the annual limit value is supported by the assumption that six, week-long campaigns spread throughout the year are representative of the annual situation. If the limit value of $5\mu gm^{-3}$ indicates that beyond this concentration the carcinogenic risk due to benzene exposure is unacceptable then a public health problem exists for more than 75% of the studied population.

5. Conclusions

This study, carried out in a medium sized French town, was part of a larger European research program designed to quantify population exposure to ambient atmospheric benzene. The data obtained is consistent with other international data dealing with this subjec.The methodology used proved to be practical for widespread multi-site monitoring of extended areas in the urban environment. The urban background concentrations commonly measured do not reflect the true levels to which the population is exposed due to the spatial and temporal distribution of compounds such as benzene and thus they cannot be used alone for population exposure assessment. Further evidence has been provided for the strong contribution of indoor pollution to overall personal exposure, which is of special concern for population sub-groups such as the elderly or young children who spend nearly all their time indoors. European directives should take into account the strong contributions from indoor environments with special attention being paid to potential source materials and ways of life.

Acknowledgements

The authors would like to thank Ms A. Frezier, Mr S. Le Meur, Mr. J-C Pinard and Ms N. François for their analytical and technical assistance.

References

Bates, M., Gonzalez-Flesca, N., Cocheo, V. and Sokhi, R. : Ambient Volatile Organic Compound Monitoring by Diffusive Sampling. Compatibility of High Uptake Rate Samplers with Thermal Desorption. *Analyst*, December 1997, vol. 122, 1481-1484.

Begerow, J.; Jermann, E.; Keles, T.; Ranft, U.; Dunemann, L. Passive sampling for VOC in air at environmentally relevant concentration levels. Fres. J. Anal. Chem. (1995) 351: 549-554

Brown, V.M and Crump, D.R. Diffusive sampling of VOC in ambient air. Environmental Monitoring and Assessment 52: 43-55 1998

CEN 1997 European Standards Publication: Ambient Air Quality. Diffusive Samplers for the determination of concentrations of gases and vapours. CEN-TC 264/ WG 11

Cicolella, A. 1997 Evaluation des risques pour la santé liés au benzène. *INERIS Report*

Conseil Supérieur d'Hygiène Publique de France (CSHP) 1997. Avis relatif au projet de directive concernant la pollution de l'air ambiant par le benzène. 17 septembre 1997.

EC 1998 European council directive relating to the limit value for benzene and CO in ambient air CEN TC 264/ WG13 N 45

Gonzalez-Flesca, N.; Cicolella, A.; Bates, M.; Bastin, E., 1999 Pilot Study of Personal, Indoor and Outdoor Exposure to Benzene, Formaldehyde and Acetaldehyde. Env. Sci. Poll. Res. 6 (2) 95-102

International Agency for Research on Cancer (IARC) 1987. *IARC Monographs* 1987; Suppl. 7, 120-121.

Loi sur l'air et l'utilisation rationnelle de l'energie 1998. Premier decret n° 98-360 du 6 mai 1998.

Mann, H.S., Crump D.R. and Brown V. M.. : Proceedings of Healthy Buildings/IAQ'97, Washington D.C., USA (sept. 27-Oct. 2,1997).

World Health Organisation (WHO) 1993. Updating and revision of the air quality guidelines for Europe. *WHO Meeting Report 11-13 January 1993.*

Characterisation of Particulate Matter Sampled During a Study of Children's Personal Exposure to Airborne Particulate Matter in a UK Urban Environment

A.J. WHEELER, I. WILLIAMS, R.A. BEAUMONT & R.S. HAMILTON
Urban Pollution Research Centre, School of Health, Biological & Environmental Sciences, Middlesex University, Bounds Green Road, London, N11 2NQ, UK
E-mail: A.Wheeler@mdx.ac.uk

Abstract. The personal exposure of children aged 9 - 11 years to particulate matter (PM_{10} and $PM_{2.5}$) was carried out between January and September 1997 in the London Borough of Barnet. Personal sampling along with home, garden and classroom microenvironmental monitoring was completed for all ten children. Each child was monitored for five days during winter, spring and summer. All children completed daily time activity diaries to provide information on any potential activities that could influence their exposure to particulate matter. Each evening a household activity questionnaire was also completed by the parents. Personal Environmental Monitors were used to sample personal exposure to PM_{10} and $PM_{2.5}$. Harvard Impactors were used for the microenvironmental sampling of both size fractions. The children's mean personal exposure concentrations for PM_{10} during winter, spring and summer were 72, 54 and 35 $\mu g/m^3$ respectively and for $PM_{2.5}$ 22, 17 and 18 $\mu g/m^3$ respectively. In order to determine the potential sources of particulate matter, analysis of the Teflon filters has been undertaken. The physical characteristics of the particles have been identified using Scanning Electron Microscopy. The relationships between personal exposure concentrations and the different microenvironments will be discussed.

Key Words: air pollution, personal exposure monitoring, children, particulate matter, seasons, Scanning Electron Microscopy

1. Introduction

The study was designed to identify the expected exposure of ten children, age 9 - 11 years of age, to particulate matter (PM_{10} and $PM_{2.5}$) in an urban environment in the UK. This was conducted between January and September 1997 to identify any seasonal differences in children's exposure to particulate matter. This study provides useful data on the personal exposure concentrations of children as well as any seasonal variations in their exposure patterns. Due to the small sample population selected for this study it will not be possible to infer probable exposure patterns for the wider population or any related health impacts.

Children's susceptibility to health problems as a result of exposure to air pollution is of concern (Scarlett *et al.,* 1996). Children may receive an increased dose of particulate matter to their lungs compared to adults. This may be due to (1) greater fractional deposition with each breath and/or (2) larger minute ventilation relative to lung size (Bennett *et al.* 1998). The children lived and attended school in the London Borough of Barnet, half of the children walked to

Environmental Monitoring and Assessment **65**: 69–77, 2000.

school and only 2 of the children who commuted by vehicle had to travel more than 30 minutes. The children's exposure to particulate matter is therefore more likely to be from local sources. Exposure to environmental Tobacco Smoke (ETS) from smoking, and exposure in other social environments is reduced by sampling non smoking children.

In the review of personal exposure monitoring of particulate matter by Mage *et al.,* (1995) it was shown that personal exposure concentrations were substantially higher than concentrations at corresponding fixed site monitors. The personal exposure studies reviewed represent research primarily conducted in the U.S. It is important to understand the UK relationship before assuming that these epidemiological and exposure assessment research studies can be applied to the UK population.

This paper presents a brief description of the methodology along with an overview of the particulate data. Some Scanning Electron Microscopy (SEM) has been carried out to determine the potential sources of the particulate matter collected.

2. Materials and Methods

The sampling programme monitored each child's exposure over a five day period, three school days and both weekend days. This was repeated on three separate occasions, usually on a 10 week cycle, between January and September 1997. One child was sampled per monitoring period. By monitoring within three different seasons it is anticipated that a representative sample of the children's exposure throughout the year will be obtained. The differences in children's seasonal exposure have also been identified through a study in the Netherlands (Roemer *et al.,* 1993).

Research on indoor air quality has shown that cooking and ETS are major sources of indoor particulate matter (Özkaynak *et al.* 1996). These criteria were included in the selection of suitable families. Other selection criteria included: the children being between 9 - 11 years old; and the use of gas for cooking and heating. All the children attended primary schools where the majority of the lessons were conducted in one classroom.

In all the houses the indoor monitors were located in the communal room used most frequently by the child under investigation. Generally, this was the living-room although in two of the houses a second downstairs room was used. The Particle Total Exposure Assessment Methodology (PTEAM) study (Spengler *et al.* 1981) showed that for both size fractions there was a high correlation between the different rooms sampled in the house and therefore a representative sample could be obtained from sampling in the main living room.

The outdoor monitors were all located in the back garden of the houses, this was primarily a result of the availability of power sources and also to ensure the security of the equipment.

During the day 12 hour personal PM_{10} and $PM_{2.5}$ concentrations were obtained for each child using the Personal Environmental Monitors (MSP Corporation, US). These are small inertial impactors specifically designed for personal monitoring (Thomas *et al.,* 1993). To reduce the weight of the sampler a single pump operating at 5.2 l $min.^{-1}$ was used (Buck HF, Negretti Automation Ltd, Aylesbury, UK). The flow was divided between the PM2.5 and PM10 samplers at 3.2 litres per minute (l $min.^{-1}$) and 2 l $min.^{-1}$ consecutively. The particles were collected onto 37 mm 2 μm pore size Teflon filters (Gelman R2PJ037, Gelman Sciences). The samplers were mounted in two 10 cm long tubes to minimise the particle collection from clothing and any effects of personal cloud (Rodes *et al.,* 1991).The pump and samplers were carried by the children in small backpacks which weighed approximately 1.5 kg.

During the night-time period no personal sampling was carried out, only the home and garden measurements were made. It has been shown in other personal exposure studies that during the night personal exposure to particulate matter reflects the home indoor concentrations (Özkaynak *et al.* 1996). The indoor and outdoor sampling was conducted using Harvard Impactors (HI) which operated at a flow rate of 10 l $min.^{-1}$ each. The garden HI's ran continuously throughout the daytime and night-time monitoring periods. The indoor samplers were set using a timer to operate when the child was present. At the beginning and end of each 12 hour sampling period all flow rates were measured using a calibrated rotameter, the total sampling volume and elapsed time were noted.

Each child completed a Time Activity Diary (TAD) every day that the monitoring took place. The time intervals were 15 minutes each and had space for noting if any smoking had occurred close to the child. Analysis of the TAD and exposure concentration will be assessed in later publications. A questionnaire was completed by the interviewer and the parents to identify any household activities that took place during the previous 24 hour period.

A single air exchange rate was measured at each home using the pressurisation method, also referred to as the 'blower door' technique (Infiltrator Series 900, Retrotec, UK).

The filters were all conditioned in clean room facilities with environmentally controlled temperature and humidity. All weighing was carried out using the Cahn-34 Microbalance (Avery Berkel, Birmingham, UK). The use of calibration weights, laboratory blanks and field blank filters ensured quality control and assurance. All filters were conditioned for at least 24 hours at constant temperature and humidity in the clean room.

A selection of 54 of the total filters from all the children were analysed using the Scanning Electron Microscope (SEM, 240 Stereoscan, Cambridge Instruments, UK linked with a AN1085 EDX, Oxford Instruments, UK). This method is suitable for identifying the physical structure and sizing of the particles. A small section of the filters was cut and removed then gold coated (SEM Coating Unit E5100, Polaron Equipment, UK) for 2 minutes onto a 12 mm stub (Agar Scientific, UK). The filters were magnified to identify the physical structure of the specific particulates.

3. Results and Discussion

A summary of each child's characteristics are shown in Table 1. The differences exhibited for each child's home and personal characteristics may cause their exposure patterns to vary. Further investigation of this will be assessed in other publications.

The Level of Detection (LOD) for the PEM $PM_{2.5}$ blank field filters was estimated to be 10.44 $\mu g/m^3$, 24 samples fall below this value. The PEM PM_{10} LOD was estimated to be 13.08 $\mu g/m^3$ with 3 samples not exceeding this LOD. The HI LOD for $PM_{2.5}$ and PM_{10} was estimated as 5.1 $\mu g/m^3$, 32 samples from the different microenvironments do not exceed this LOD. Those samples that do not exceed the LOD are assigned the value of half the LOD and all analysis is conducted with these values (Phillips *et al.* 1999).

The geometric means of all the children's daytime concentrations for all environments are in Table 2. All of the PM_{10} concentrations exceed the $PM_{2.5}$ concentrations. This is as expected as the $PM_{2.5}$ fraction is also collected by the PM_{10} samplers. The average values of all the children's daytime personal PM_{10} concentrations are greater than the corresponding outdoor Automatic Urban Network (AUN) and garden sites. The UK AUN legislation is based upon a 24 hour running mean of 50 $\mu g/m^3$. This indicates that the AUN sites, if used to directly predict the personal exposure of children to PM_{10}, would underestimate their exposure. This is also the case for the personal $PM_{2.5}$ when compared to the garden $PM_{2.5}$ concentrations. The children's personal exposure concentrations for PM_{10} during winter, spring and summer were 72, 54 and 35 $\mu g/m^3$ and for $PM_{2.5}$ 22, 15 and 17 $\mu g/m^3$ respectively.

The indoor concentrations of PM_{10} from the homes and classrooms both exceed the outdoor ambient measurements. As previously, this suggests that data from AUN sites, which are used to represent the general population's exposure, are not representative of indoor PM_{10} concentrations. The classroom concentrations are consistently the highest of all the sites sampled throughout the three seasons. The ratios in Table 2 show that for the home and garden environments the $PM_{10}/PM_{2.5}$ ratios are approximately 2 with the classroom ratio being 3. The high PM_{10} concentrations in the classroom may be explained by the activity of the children. Similar research conducted by Janssen (1998) found that the concentration of PM_{10} in classrooms were significantly higher than the corresponding outdoor concentrations. The indoor concentrations of $PM_{2.5}$ from the homes and classrooms exceed the personal and garden $PM_{2.5}$ concentrations in all seasons.

A summary of the particulate matter found on the filters of the different microenvironments is given in Table 3. The particulate matter descriptions are derived from comparing the images with references from McCrone (1972). A selection of photographs illustrating the particle types are shown in Plates 1 - 4.

The SEM analysis has shown that the samplers are collecting the intended size fractions and the filters are evenly loaded. Plate 1 shows the blank filter with nothing collected upon it. It can be seen that there are potential difficulties in distinguishing between all of the filters' Teflon threads and the particles. Very careful observation is required to identify the differences between the filter structure and particulates.

In the home the majority of the filters collecting PM_{10} had particles which were skin flakes, furnishing fibres and possibly soil derived coarse particles. The garden PM_{10} particles which were collected exhibited more insect and pollen debris. Both of the indoor and outdoor filters collected many of the smooth globular particles approximately (c.) $<2\mu m$ in size, these may be derived from vehicle combustion processed. Due to their small size these can transfer easily from the outdoor to indoor environments.

The $PM_{2.5}$ filters for both indoor and outdoor environments do not appear to have any specific differences in particulate matter characteristics, suggesting this size fraction is not influenced by particle generating activities in the homes.

There are particles collected that are obviously much larger than the cut size of the samplers however, the sampler specifications are designed with a 50% collection efficiency hence, some larger particulates have been collected.

The differences in the seasonal variation for the different sampled microenvironments are consistent. This suggests that the children's exposure in the winter is higher than in the spring and summer. Factors that could potentially influence this are indoor sources, ventilation of the buildings and time spent within these environments.

Table 1 Characteristics of each participant and their home

Child Identification	*1*	*2*	*3*	*4*	*5*	*6*	*7*	*8*	*9*	*10*
Male / Female (M/F)	M	F	M	M	M	M	M	F	F	M
Smoker in family	No	Yes	No	Yes	No	No	No	No	No	Yes
Parent smokes at home	No	Yes	No	Yes	No	No	No	No	No	Yes
Gas Fire	Yes	Yes	No	No	No	Yes	Yes	Yes	Yes	Yes
Pet at home	Dog	No	No	No	Cats	No	No	No	Bird	No
School travel mode	Walk	Walk	Car	Walk	Bus	Walk	Car	Car	Walk	Car
Window type[1]	D	S	Sec	S	S	S	D	D	S	S
Air Exchange Rates (ACH^{-1})	0.4	0.6	0.8	1.2	0.8	0.6	0.7	0.5	0.5	1.1

Table 2 Winter, Spring and Summer Mean Day Concentrations of all Microenvironments for all Children, PM_{10} and $PM_{2.5}$ ($\mu g/m^3$)

Season	Home $PM_{2.5}$	Home PM_{10}	Garden $PM_{2.5}$	Garden PM_{10}	Class $PM_{2.5}$	Class PM_{10}	Personal $PM_{2.5}$	Personal PM_{10}	Brent AUN[2]	Haringey AUN
Winter	29	54	13	25	30	80	22	72	22	25
Spring	24	54	12	19	28	78	17	54	17	21
Summer	19	42	11	22	19	78	15	35	19	25
Seasonal Average	**23**	**50**	**12**	**22**	**27**	**79**	**18**	**53**	**19**	**24**
Ratios of $PM_{10}/PM_{2.5}$	**2.2**		**1.8**		**2.9**		**2.9**			

[1] Glazing Type, S = Single, D = Double, Sec = Secondary
[2] AUN = Automatic Urban Network

Table 3 Summary of Particulate Matter Characteristics

Environment	Particle Characteristics
Personal $PM_{2.5}$	Small angular particles c. 3.5μm diameter, smooth globular particles <2μm, some agglomerates, no skin flakes, fibres or insect debris. See Plate 2
Personal PM_{10}	Several large particles c. 30-40μm diameter which could be skin flakes, smooth globular particles <2μm, some granular, spherical particles c. 5μm. A small number of furnishing fibres.
Home $PM_{2.5}$	Several agglomerates <4μm diameter, a small number of furnishing fibres, many smooth globular particles <2μm, no skin flakes.
Home PM_{10}	Many smooth globular particles <2μm, several skin flakes and furnishing fibres, granular angular particles <5μm, some agglomerates. See Plate 3
Garden $PM_{2.5}$	Many smooth globular particles <2μm, some spherical sponge textured particles <5μm, no insect debris or fibres present.
Garden PM_{10}	Many smooth globular particles <2μm, some coarse irregular shaped particles c.5-6μm, some insect exoskeleton parts, some sponge like textured particles possibly pollen spores. See Plate 4
Class $PM_{2.5}$	Granular angular particles c.3μm possibly soil derived, some skin flakes, some pollen granules, smooth globular particles <2μm.
Class PM_{10}	Same as Class $PM_{2.5}$ but more heavily loaded, some insect debris and fibres possibly from furniture or clothing.

Plate 1 Blank Teflon Filter

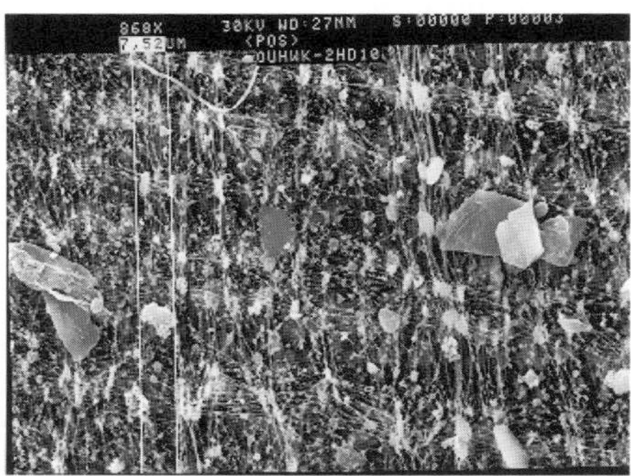

Plate 3 Home PM_{10} Filter

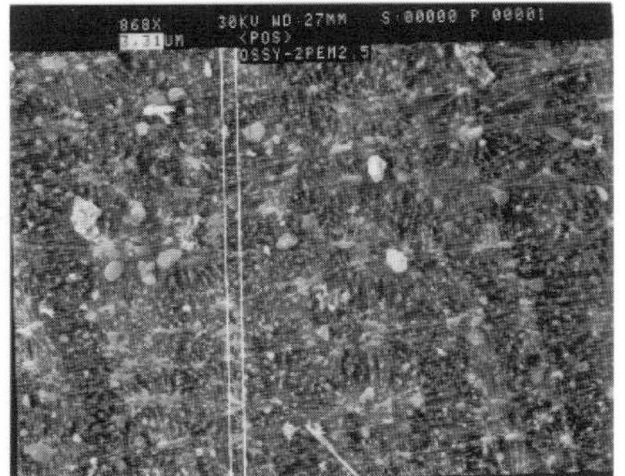

Plate 2 Personal $PM_{2.5}$ Filter

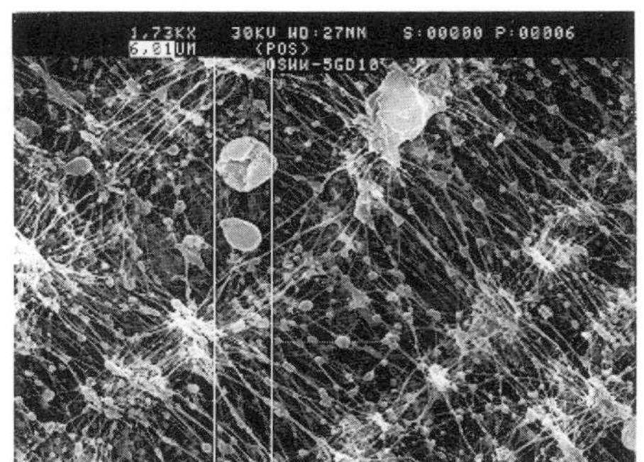

Plate 4 Garden PM_{10} Filter

4. Summary and Conclusion

Children's daytime personal exposures to particulate matter (PM_{10} and $PM_{2.5}$) have been shown to exceed outdoor concentrations in winter, spring and summer. This agrees with the review of personal exposure made by Mage *et al.*, (1995).

The home and classroom concentrations exceed the outdoor concentrations for both size fractions. The classroom environment has the highest concentrations of both PM_{10} and $PM_{2.5}$ in all seasons sampled. The difference in the classroom ratio of $PM_{10}/PM_{2.5}$ when compared to the home and garden environments suggest that the PM_{10} concentrations are related to the children's activity.

The home PM_{10} particles are predominantly furnishing fibres, skin flakes and granular angular particles which could potentially be tracked in soil particles. The garden PM_{10} has components of insect debris, pollen and possibly resuspended soil particles. This suggests that the majority of the indoor and outdoor sources of PM_{10} are of different origin. The $PM_{2.5}$ particulates are smooth and globular, <2 μm in diameter and are found in all of the environments sampled and also on the PM_{10} filters. It is possible that these are from vehicle combustion processes.

The concentrations of $PM_{2.5}$ in each microenvironment are Class> Home> Personal> Outdoor. This suggests that there are important sources of indoor $PM_{2.5}$ which are contributing to personal exposure. Again, if outdoor sites are used to directly predict personal exposure for children living within London they would underestimate exposure.

The seasonal variation of all microenvironments is as expected with the concentrations in all environments for PM_{10} and $PM_{2.5}$ being winter> spring> summer. The QUARG report (1996) states that the highest PM_{10} concentrations in the UK occur during the winter.

Acknowledgments

Funding by an EPSRC & Barnet Health Authority CASE Award. With thanks to P. Koutrakis at Harvard University for the loan of the sampling equipment.

References

Bennett, W.D., and Zeman, K.L. (1998). Deposition of Fine Particles in Children Spontaneously Breathing at Rest *Inhalation Toxicology* **10**(9),831-42.

Harrison R.M. (1996) Airborne Particulate Matter in the United Kingdom *Quality of Urban Air Review Group* 3rd Report.

Janssen, N.A.H., (1998). Personal Exposure to Airborne Particles; Validity of Outdoor Concentrations as a Measure of Exposure in Time Series Studies. *Thesis*, University of Wageningen.

Mage, D. T., and Buckley, T. J. (1995). The Relationship between Personal Exposures and Ambient Concentrations of Particulate Matter. *Air & Waste Management Association, 80th Meeting & Exhibition, Texas*, 2-16.

McCrone, W. C., and Delly, J. G. (1973). The Particle Atlas, An Encyclopedia of Techniques for Small Particle Identification. (Ann Arbor Science Publishers).

Ozkaynak, H., Xue, J., Spengler, J., Wallace, L. A., Pellizzari, E., and Jenkins, P. (1996). Personal Exposure to Airborne Particles and Metals: Results from the Particle TEAM Study in Riverside, California. *J. of Exposure Analysis and Environmental Epidemiology* **6**(1), 57-78.

Phillips, K., Howard, D.A., Bentley, M.C., and Alvan, G. (1999). Assessment of Environmental Tobacco Smoke and Respirable Suspended Particle Exposures for Nonsmokers in Basel by Personal Monitoring. *Atmos. Env.* **33**, 1889-1904.

Rodes, C.E., Kamens, R.M., and Wiener, R.W. (1991). The Significance and Characteristics of the Personal Activity Cloud on Exposure Assessment Measurements for Indoor Contaminants. *Indoor Air* **2**, 123-145.

Roemer, W., Hoek, G., and Brunekreef, B. (1993). Effect of Ambient Winter Air Pollution on Respiratory Health of Children with Chronic Respiratory Symptoms. *American Review of Respiratory Disease* **147**, 118-124.

Scarlett, J.F., Abbott, K.J., Peacock, J.L., Strachan, D.P., and Anderson, H.R. (1996). Acute Effects of Summer Air Pollution on Respiratory Function in Primary School Children in Southern England. *Thorax* **51**, 1109-1114.

Spengler, J.D., Dockery, D.W., Turner, W.A., Wolfson, J.M., and Ferris, B.G. (1981). Long-Term Measurements of Respirable Sulfates and Particles Inside and Outside Homes. *Atmos. Env.* **15**, 23-30.

Thomas, K.W., Pellizzari, E.D., Clayton, C.A., Whitaker, D.A., Shores, R.C., Spengler, J., Ozkaynak, H., Froehlich, S.E., and Wallace, L.A. (1993). Particle Total Exposure Assessment Methodology 1990 Study: Method Performance and Data Quality for Personal, Indoor, and Outdoor Monitoring. *J. of Exposure Analysis & Environmental Epidemiology* **3**(2), 203-226.

ANALYSIS OF THE DAILY VARIATIONS OF WINTERTIME AIR POLLUTION CONCENTRATIONS IN THE CITY OF GRAZ, AUSTRIA

R. A. ALMBAUER*, M. PIRINGER**, K. BAUMANN**, D. OETTL* and P. J. STURM*

* Institute for Internal Combustion Engines and Thermodynamics, Technical University Graz, Austria, Inffeldgasse 25, A-8010 Graz, almbauer@vkmb.tu-graz.ac.at

** Central Institute for Meteorology and Geodynamics, Vienna, Austria

Abstract. Measured air pollution concentrations in a city reflect the influence of different kinds of sources as well as varying meteorological conditions. In the city of Graz in southern Austria, frequent stagnant meteorological conditions can cause elevated levels of air pollution although emission levels are not exceptionally high. With the aid of a detailed emission inventory and an array of sodars and tethersondes as well as lidar systems supplementing the routine meteorological and air chemistry network during a field experiment in January 1998, the daily variations of air pollution concentrations of selected components within the complex topography of the city of Graz are explained. Main results show the almost linear dependence of the morning maximum concentrations on the predicted emission rates. Throughout the day the rising of the well mixed layer reduces concentrations considerably. Concerning NO_X the fast reaction from NO to NO_2 is important due to the down-mixing of O_3 from the residual layer. The maximum in the afternoon is influenced by emission rates and pollution transport due to the mountain wind.

Key words: field measurement campaign, complex topography, air quality, emission inventory

1. Introduction

An urban area can exert modifications of the local meteorological conditions, in particular the temperature (urban heat island effect, e. g. Godowitch et al., 1987; Kuttler et al., 1998) and the wind field (Balling and Cerveny, 1987; Draxler, 1986). For the city of Graz in southern Austria, Piringer and Baumann (1999a) investigated modifications of an existing valley wind system by the city. Such modifications can influence the pollutant concentration pattern of a city.

Air pollution studies for cities in different mountainous areas have been carried out (e.g. Helmis et al., 1997). Results of these studies show, that local wind systems can have different strength according to the unique topography of an area. Local flow regimes as e.g. katabatic and anabatic currents, valley wind systems, temperature inversions and effects induced by a city to its surroundings are best investigated with field campaigns before they are modelled. The city of Graz is located in a prealpine region in the southeast of Austria. Although the city (250.000 inhabitants) houses only small industrial activity, traffic and domestic heating cause poor air quality during stagnant wind conditions. Situations combining low wind speed and strong temperature inversions occur esp. during anticyclonic weather situations in winter. The DATE Graz project

Environmental Monitoring and Assessment **65**: 79–87, 2000.

(Dispersion of Atmospheric Trace Elements taking the city of Graz as an example) is aiming at the investigation of the relations of emissions, meteorology and air quality for the city. The campaign supplemented the routine meteorological measurements of at least 10 ground-based stations at different altitudes by 2 SODAR stations, 4 tethered balloons, 1 eddy correlation measurement device and at the same time 1 DOAS system, 1 DIAL LIDAR, 1 FTIR and 7 air chemistry monitoring stations at ground level were in operation. This experimental effort was used to explain the daily variations of pollution concentrations during the anticyclonic period 10 to 13 January 1998 when local flow conditions, nighttime temperature inversions and low mixing heights at daytime (Piringer a. Baumann, 1999a) dominated the dispersion of air pollutants.

2. Topographical situation and experimental set-up

The topographical situation and the location of the measurement sites in the Graz area are shown in figure 1. The position of the area within Europe is indicated in the small picture in the lower part of figure 1. Ventilation along the Mur river is maintained by the valley wind system of this approx. 160 km long valley of the East-Alps entering the Graz basin from north-west through a narrow gap. The basin is bounded to the west by a steep ridge extending 200 to 400 m above the level terrain; at the eastern side, some tributary valleys run into the basin, separated by east-west extending ridges about 100m high. These valleys develop their own local circulations, leading to better ventilation of the eastern part of the city compared to the western part, where stagnant situations

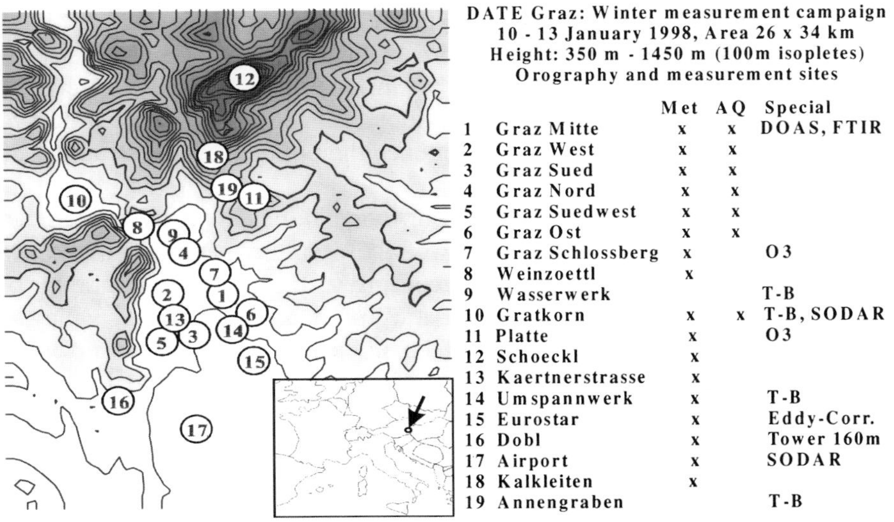

DATE Graz: Winter measurement campaign
10 - 13 January 1998, Area 26 x 34 km
Height: 350 m - 1450 m (100m isopletes)
Orography and measurement sites
Met AQ Special
1 Graz Mitte x x DOAS, FTIR
2 Graz West x x
3 Graz Sued x x
4 Graz Nord x x
5 Graz Suedwest x x
6 Graz Ost x x
7 Graz Schlossberg x O3
8 Weinzoettl x
9 Wasserwerk T-B
10 Gratkorn x x T-B, SODAR
11 Platte x O3
12 Schoeckl x
13 Kaertnerstrasse x
14 Umspannwerk x T-B
15 Eurostar x Eddy-Corr.
16 Dobl x Tower 160m
17 Airport x SODAR
18 Kalkleiten x
19 Annengraben T-B

Figure 1. Measurement stations of DATE Graz winter campaign

are observed frequently. In the centre of Graz, an isolated hill, the "Schloßberg", rises about 150 m above the city plainThe field experiment was carried out between 10 and 13 January 1998. The relatively dense local meteorological network which includes stations at different altitudes was supplemented by vertical soundings to investigate the most important flow regime, the valley wind system along the Mur river. Sodars and tethersondes were positioned on both sides of the gap of the Mur river in the north-west of the city as well as north and south of the city centre. Vertical profiles of horizontal wind speed, wind direction, temperature and ozone (at site 14 only) were measured. Piringer and Baumann (1999b) give a detailed description of the vertical sounding devices used.
Air quality values were measured at six monitoring stations in a height of 4 m a.g.l. in the city (fig. 1: sites 1-6) and in the next basin to the northwest of the city in the Gratkorn basin (fig. 1: site 10). In addition the DOAS measurement took place in the city centre (site 1) over a path-length of 164 m in a height of approx. 10 m a.g.l.. All air quality values are on the basis of half hour mean values. Only the vertical profile of O_3 (site 14) is based on the average of the values gained from the accent and the decent at the same height. All diagrams show the results for UTC (universal time coordinated).

3. Emission inventory for the city of Graz

The emission inventory for the city of Graz (Sturm, 1997) is based on the three source groups traffic, domestic emissions and industry. Pollutants investigated are carbon-monoxide (CO), volatile organic compounds (VOC), nitrogen oxide (NO), nitrogen dioxide (NO_2), sulphur dioxide (SO_2) and dust/soot. In this paper, sulphur dioxide and nitrogen oxide (NO_X) emissions are considered. Emission rates are evaluated with a temporal resolution of 1 hour and a spatial resolution of 250 x 250 m for an area of approx. 13 x 15 km using a GIS system.

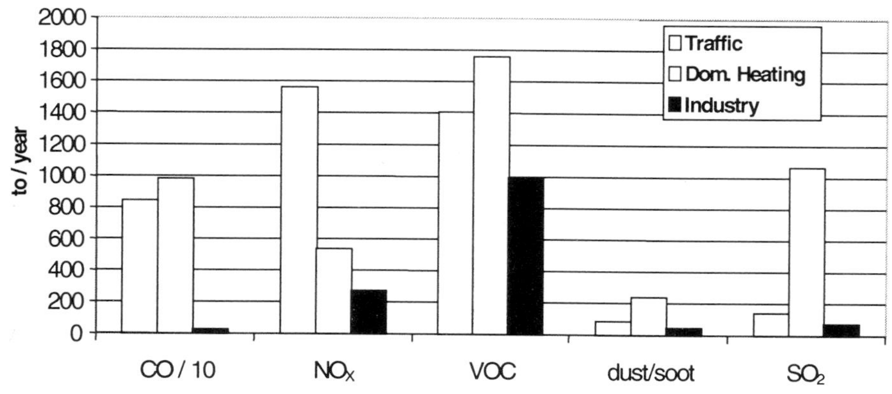

Figure 2. Results of the Graz emission inventory

Figure 2 shows the share of the three source groups for the different pollutants for the year 1995. The diurnal variations of traffic emissions are well known from traffic counts. Emissions from domestic heating depend on the average temperature of the day. The variations in the course of a day described as a fixed percentage of the daily emission per hour are given in Sturm et al., 1999. Industrial emissions vary according to the type of the enterprise. Data is gained from statistics and for large industry from a survey.
Compared to other European cities SO_2 emission are considerably low due to a strict legislation. SO_2 is mainly emitted by domestic heating. Concerning NO_X-emissions Graz can be designated as an average polluted city due to traffic. On an average day more than 3.5 Mill. km are driven (Almbauer et al., 1995).

4. Meteorological results of the winter campaign

The January 1998 field experiment was planned for a typical "winter smog" episode, associated with cold continental air. In these episodes, enhanced levels of primary pollutants are expected from domestic heating and traffic. The measurements started on 10 January, when a pronounced anticyclonic ridge at 500 hPa became established together with a flat anticyclone at ground level. The 500 hPa ridge was unusually intense for the season and prevented the occurrence of ground fog. Due to warm air advection, the period was rather mild, but because of the clear skies, all the particular features associated with local flows in complex terrain could be observed. The measurement campaign ended in the

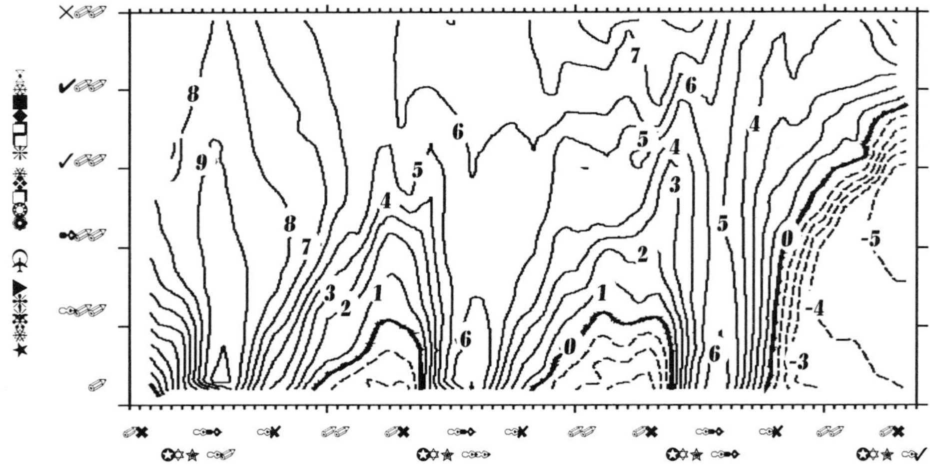

Figure 3. Time-height diagram of temperatures (°C) at site 14 (Umspannwerk)

late evening of January 12 with the advection of ground fog with a height of 300m. The time-height diagram of temperatures at site B (fig. 3) reveal the general course of air temperature. The period starts with a well-mixed boundary

layer on the afternoon of 10 January which extends the tethersonde range (500 m). After sunset on both days, an inversion forms in the lowest 300m, gradually increasing in strength during the night. During daytime on 11 and 12 January, a well mixed boundary layer forms at the ground, extending to about 200 m in the early afternoon. Both tethersondes at site 9 and 14 detect common features of the mountain-valley wind system, i.e. the regular change between daytime (south to south easterly) valley-wind and night-time (north to north-westerly) mountain-wind.
The vertical extend of the up-valley flow covers the whole vertical range of the tethersonde on 10 January, but only about 200 to 300 m on 11 January. This is in accordance with the height of the well-mixed boundary layer (fig. 3). The onset of the up-valley flow occurs about 9 UTC on 11 and 12 January, two hours after sunrise, at the ground. Slowly propagating upwards, the up-valley flow reaches its full height at about 12 UTC. The onset of the down-valley flow occurs at about 16 UTC in the north of the city (site 9), which is approximately one hour after sunset, and about on or two hours later at site 14. The jet effect at the narrow gap in the north-west of the city causes the down-valley flow to touch often to the. At site 14 the down-valley flow touches the ground only for short times at the beginning of the night. Later a weak mostly southerly flow develops near the ground Additional vertical soundings at site 19, in one of the tributary valleys coming from the north-east and east to the Graz basin, showed the expected valley wind system for the tributary valleys in the east of the city.

5. Results and discussion of the field experiment

Considering time and space scales involved, SO_2 is a slow reacting pollutant. Daily variations are determined by emission rates and dispersion parameters which, near ground, vary between well-mixed around noon and very stable between the early evening and the late morning hours (fig. 3). Sunset occurred around 16 UTC, sunrise around 8 UTC. Local winter time is UTC + 1 hour. Fig. 4 shows the temporal variation of the sum of SO_2 emission over the whole city area during the measurement campaign. SO_2-emission during the weekend stem mainly from domestic heating and depends on average ambient temperature. On Monday industry emission increases considerably. In addition the city average of SO_2-concentrations (mean value of sites 1, 2, 3, 5 and 1 (DOAS)) is presented. Comparison of the mean value for the city was chosen, due to the fact that the above mentioned sites showed similar temporal variations during the days. In general time variation show low SO_2-concentrations during the night with values of approx. 15 μg/m³. Apparently, emission reduction dominates the effect of reduced vertical mixing, which could enhance ground-level concentrations. During the early morning hours the SO_2-concentrations increase rapidly up to the daily maximum of 30 – 45 μg/m³. Due to the co-incidence of the emission maximum and the onset of vertical mixing below the slowly rising inversion, the SO_2-concentrations are amplified. Maximum SO_2-emission rate is predicted 7 times higher during the morning than in the night. The maximum on all three days is reached at 10 UTC on Monday according to higher emission maximum, which occurs earlier. Later, the further increase of the mixed layer (approx. 100 m h^{-1}) reduces the concentrations very fast down to the values of the night. In the afternoon, with the transition from well-mixed to stable conditions, the SO_2-concentrations increase again to approx. 25 µg/m³. After 20 UTC, SO_2 concentrations are small due to low emissions and the down-valley wind, which touches down to the city ground just after its beginning at 17 UTC (fig. 4).

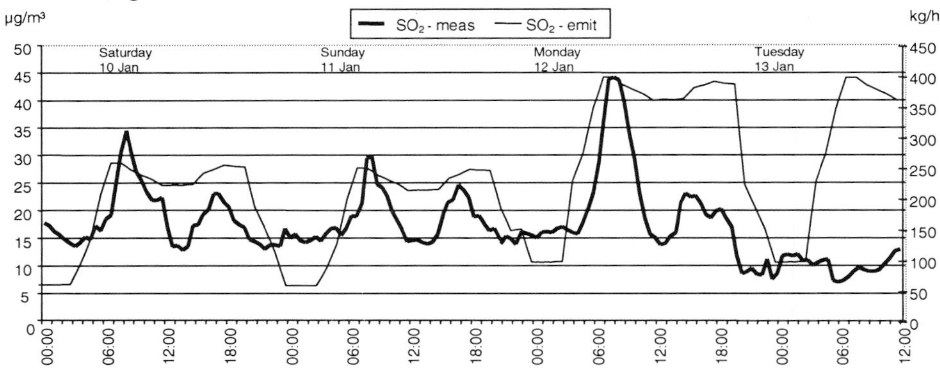

Figure 4. Temporal variation of mean emissions and concentrations of SO_2

According to the emission inventory of Graz approx. 2/3 of NO_X is emitted by traffic and 1/4 by industry (fig.2). Figure 5 shows the temporal variation of NO_X-emissions of the whole city together with the mean NO- and NO_2-concentrations during the measurement campaign. The big differences between individual days can be explained by the fact that the field experiment started on

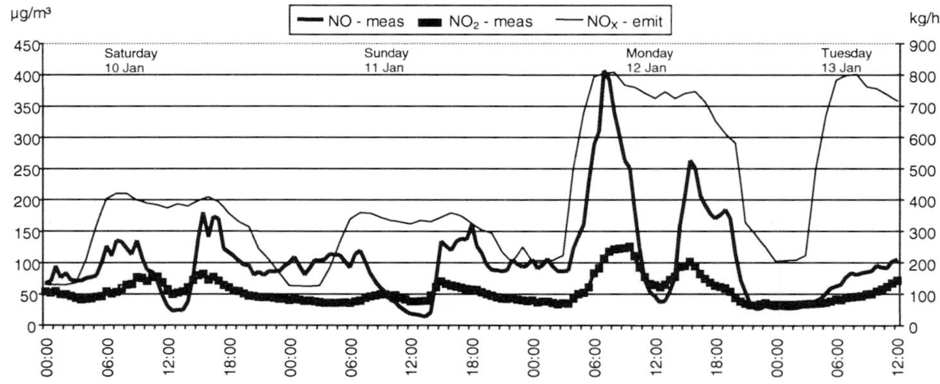

Figure 5. Temporal variation of NO_X- emissions and concentrations of NO and NO_2

10 January, a Saturday, and ended on 13 January, a Tuesday. Traffic of private passenger cars and public buses on a Sunday is assumed to be at approx. 40% of the level of an average labour day. Traffic of heavy duty vehicles is reduced to approx. 14%. Daily variations of traffic during Sunday shows a slow increase during morning and a peak in the afternoon. Industrial emissions of NO_X are also reduced to a value of approx. 20% of a mean labour day. Values for Sunday are statistically less secure than on labour days. On Monday 12 January, NO_X emissions reached 108% of a mean working day with all features of a morning rush hour and a second peak emission during the late afternoon. The mean NO and NO_2-concentrations are calculated from sites 1, 2, 3, 5 and 1 (DOAS)). As the conversion of NO with O_3, HO_2 and RO_2 to NO_2 occurs within minutes (Finlayson-Pitts and Pitts, 1986) both concentration can only be considered together. NO concentrations during both nights from 10 to 11 and from 11 to 12 are on a level of approx. 100 μg/m³. NO_2 concentrations are decreasing from approx. 75 in the evening to 40 μg/m³. in the early morning. Obviously the emission rate of NO is in the order of the dispersion, conversion and deposition. NO_2 concentration is decreasing due to a small conversion rate from NO to NO_2 and due to deposition and further conversion of NO_2.

On Saturday and Sunday NO values increase only little in the morning hours. On Monday a strong increase with time to daily maximum of more than 400 μg/m³ at about 10 UTC was recorded. NO_2 concentrations rise slowly after sunrise with a Monday-maximum of more than 120 μg/m³ short after the NO maximum at about 11 UTC. During that time the conversion rate from NO to NO_2 dominates the increased dispersion according to the crowing of the mixed layer. The fast conversion of NO to NO_2 is caused by the down-mixing of O_3 from the higher

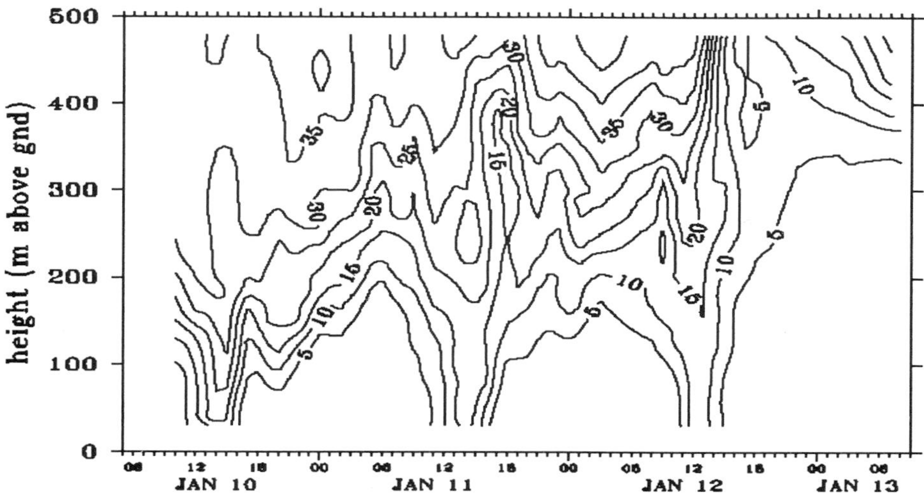

Figure 6. Time-height diagram of O_3-concentrations (ppb) at site 14 (Umspannwerk)

altitudes. The O_3 concentration during all three days in the residual layer above the inversion is about 30 –40 ppb. All three days show a distinct minimum NO concentration during noon with values less than 40 µg/m³. NO_2 concentrations decrease slower to values between 40 to 60 µg/m³. Obviously the losses due to dispersion are higher than the gain due to conversion. During all three days NO-concentrations increase short after the breakdown of the well mixed layer at 16 UTC. On Saturday and Sunday the maxima in the afternoon are higher according to an increased traffic. On Monday the second NO maximum is lower. NO_2-maxima show the same behaviour like NO for their maxima on all three days. The sum of NO and NO_2 concentration decreases after 18 UTC according to the expected emission reduction. NO-concentration show a good co-incidence with predicted emission rates. Vertical sounding of O_3 for the three days (fig. 6) confirm the assumption of O_3 down-mixing. O_3 level during the existence of the well mixed layer is about 10 to 15 ppb.

6. Conclusions

A field experiment conducted in the city of Graz between 10 and 13 January 1998 (section 2) allowed to estimate the influence of emissions (section 3) and meteorological parameters (section 4) on daily variations of selected air pollutants. The detailed emission inventory for the city of Graz (Sturm, 1999) is used to study the diurnal cycles of SO_2 and NO_X emissions during the field experiment. The vertical meteorological soundings conducted at several sites throughout the city area (fig. 1) gave a comprehensive picture of the meteorological situation during the anticyclonic episode (Piringer and Baumann, 1999a). Main results show the almost linear dependence of the morning maximum concentrations on the predicted emission rates. Throughout the day the rising of the well mixed layer reduces concentrations considerably. Concerning NO_X the fast reaction from NO to NO_2 is important due to the down-mixing of O_3 from the residual layer. The maximum in the afternoon is influenced by emission rates and pollution transport due to the mountain wind.

Acknowledgements

DATE Graz is funded by the Austrian Science Fund (FWF) under project numbers P12168-Tec, P12169-TEC and P12170-TEC. We thank the Styrian government, department IKT and FA 1a, for making topography, land-use data and monitoring data available.

References

Almbauer, R. A., K. Pucher, P.J. Sturm; Air quality modeling for the city of Graz, Meteorol. Atmos. Phys. 57, 31 - 42, 1995.

Balling, R. C. Jr., R. S. Cerveny; Long-term associations between wind speeds and the urban heat island of Phoenix, Arizona. J. Clim. Appl. Meteor. 26, 712 – 716, 1987.

Draxler, R. R.; Simulated and observed influence of the nocturnal urban heat island on the local wind field. J. Clim. Appl. Meteor. 25, 1125 – 1133, 1986.

Finlayson-Pitts, B. and J.N. Pitts, Atmospheric Chemistry: Fundamentals and Experimental Techniques. John Wiley and Sons, New York, pp 1098.

Godowitch, J. M., K. S. Ching, J. F. Clarke; Spatial variation of the evolution and structure of the urban boundary layer. Boundary layer meteorol. 38, 241 – 272, 1987.

Helmis, C.G et al., Thessaloniki '91 Field Measurement campaign-I. Wind field and atmospheric boundary layer structure over Graeter Thessaloniki Area, under light background flow, in Atmos. Env. Vol 31, No 8, 1101 - 1114, 1997.

Kuttler, W., D. Dütemeyer, A.-B. Barlag; Influence of regional and local winds on urban ventilation in Cologne, Germany. Meteorol. Z., N.F. 7, 77 – 87, 1998.

Piringer, M., K. Baumann; Modifications of a valley wind system by an urban area – experimental results. Meteorol. Atmos. Phys. 71, 117 - 125, 1999a.

Piringer, M., K. Baumann; Exploring the urban boundary layer by sodar and tethersonde. Submitted to Physics and Chemistry of the Earth, 1999b.

Sturm, P. J., Ch. Sudy, R. A. Almbauer, J. Meinhart: Updated urban emission inventory with a high resolution in time and space for the city of Graz, The Science of the Total Environment 235, 111 - 118, 1999.

ATMOSPHERIC VOLATILE ORGANIC COMPOUND MONITORING. OZONE INDUCED ARTEFACT FORMATION.

MATTHEW S. BATES [1,2] , NORBERT GONZALEZ-FLESCA [1*], RANJEET SOKHI [2] and VINCENZO COCHEO [3]

[1] *INERIS, Parc Technologique Alata, B.P 2, 60550 Verneuil en Halatte, France*

[2] *University of Hertfordshire, Hatfield, Herts, AL10 9AB, U.K.*

[3] *Fondazione Salvatore Maugeri IRCCS, via Svizzera, 16, I-35127, Italy*

Abstract. Assessment of population exposure to VOC in ambient atmospheres is receiving heightened interest as the adverse health effects of chronic exposure to certain of these compounds are identified. Active (pumped) and passive samplers are the most commonly used devices for this type of monitoring. It has been shown, however, that these devices, along with all other pre-concentration techniques, are susceptible to ozone interference. It is demonstrated that this interference occurs even at low ozone concentrations and that it may result in the under-estimation of population exposure. A convenient and effective ozone scrubbing method is identified and successfully applied and validated for both active and passive samplers for a range of VOC.

Key words: Volatile Organic Compounds (VOC), ozone, artefact, passive sampling, active sampling, scrubber, formaldehyde, acetaldehyde, benzene, toluene, styrene.

1. Introduction

Volatile Organic Compounds (VOC) are ubiquitous atmospheric species that originate from both natural and anthropogenic sources (Ciccioli). As direct pollutants many cause adverse health effects through acute and/or chronic exposure and they account for a large proportion of the 189 Hazardous Air Pollutants (HAP) listed in the 1990 US EPA Clean Air Act amendments (USEPA, 1990). As secondary pollutants VOC play a polyvalent role in photochemical smog episodes. They not only represent one of the three key ingredients but they can also, as is the case for carbonyl compounds, be produced as intermediates in the complex reactions that produce high concentrations of species such as ozone, hydroxyl radicals, hydrogen peroxide and peroxy acetyl nitrate (PAN) in the troposphere (Megie *et al.*, 1994).

Short-term exposure to VOC in the work-place has long been monitored, however it is only more recently that the health effects of chronic exposure to ambient levels of these compounds have been considered. This heightened interest in population exposure assessment was reflected, in 1996, when the World Health Organisation (WHO) published air quality guidelines for Europe (WHO, 1996). Alongside the more classical air pollutants, such as carbon monoxide, nitrogen dioxide and ozone, certain VOC also found their place, including formaldehyde and three monocyclic aromatic compounds (MAC):

Environmental Monitoring and Assessment **65**: 89–97, 2000.

benzene, toluene and styrene. Unfortunately, few data exist that quantify population exposure to VOC and although urban background concentrations for compounds such as benzene are often continuously monitored, recent work has shown that these levels do not reflect those to which the population is exposed (Gonzalez *et al.*, 1999). This makes comparison to the air quality guidelines a difficult task.

To determine the magnitude and source of population exposure, widespread multi-site campaigns are necessary, with parallel monitoring indoors, outdoors and on a personal basis. This type of monitoring may be performed either by active (pumped) or passive sampling. However, from a financial and practical point of view, passive samplers are preferable as they are relatively inexpensive, light-weight, non-invasive and may be left for extended periods during which they require no operator interaction (Brown and Wright 1994).

As with all sampling techniques thorough laboratory and field validation to develop a quality assurance protocol are of some guarantee as to the integrity of measurements made in the field (CEN, 1997). However sometimes, unexpected interferences can result in erroneous measurements, one such example is the presence of oxidants like ozone, during VOC sampling.

Much evidence exists to show that ozone interferes when actively sampling aldehydes by chemical derivitization with 2,4-dinitrophenylhydrazine (DNPH) (Arnts and Tejada, 1989; Sirju and Shepson, 1995) and in solid adsorbent sampling of monoterpenes (Hoffmann, 1995). Several ozone removal techniques for use in conjunction with active devices have been proposed (Helmig, 1997). The intention of this work is to extend existing knowledge to cover other VOC and to identify the most suitable scrubbing technique for both active and passive sampling.

2. Experimental

The experiments carried out for this study were all conducted in a dynamic atmosphere generator that permits reconstitution of VOC-containing atmospheres from the ppm through to the ppb range and the introduction of other co-pollutants such as ozone. The device and its validation are described in detail elsewhere (Jaouen *et al.* 1995). Expected VOC concentrations were verified by active sampling prior to ozone introduction and good agreement was found (<5% difference) for all experiments conducted. The investigations always involved simultaneous co-exposure to ozone and to the compounds of interest for both active and passive sampling. The experiments were performed at closely monitored but uncontrolled ambient temperatures (range 4-18°C) and relative humidities (20-60%). Continuous monitoring of ozone concentrations was carried out using an ozone analyser (Model O3-41M, Environnement SA,

France).

OZONE SCRUBBERS

Many different ozone scrubbing techniques for use in conjunction with active sampling of carbonyl compounds and monoterpenes are described in the literature (Helmig, 1997). Four scrubbers were tested by passing a stream of air, containing various concentrations of ozone (10 - 600 ppb), through the devices for 24-hours whilst continuously monitoring the effluent:

- A diffusion denuder consisting of a coiled copper tube coated internally with potassium iodide (KI);
- A glass tube filled with KI crystals;
- A glass tube filled with hydrated iron sulfate crystals ($FeSO_4$, $7H_2O$);
- Manganese dioxide (MnO_2) coated copper gauze.

MnO_2 coated copper gauze (Environnement SA, France) was identified as the most appropriate. This was for several reasons: it destroys ozone by catalytic means, thus requiring no regeneration; it did not interact with the compounds of interest (see results) and it could be incorporated into existing passive devices without otherwise altering sampler performance. In addition, it has been shown (Juttner, 1996) that it is still effective after 3 months continuous use.

For passive aldehyde samplers efficient scrubbing was achieved by inserting a single layer of gauze against the internal face of the HDPE diffusive body. The passive samplers used for MAC were modified Perkin Elmer tube-type devices (PE, Beaconsfield, UK) (Bates *et al.* 1998). Effective ozone removal was accomplished by replacing the original diffusion cap grills by a layer of MnO_2 coated copper gauze.

A PTFE support was manufactured to hold the MnO_2 grills for use in series with both the aldehyde and MAC active samplers. It was found that eight grills were required for 100% ozone removal at 600 ppb however for ambient ozone concentrations (<200 ppb) three or four grills were sufficient.

ALDEHYDE ATMOSPHERE GENERATION, SAMPLING AND ANALYSIS

The atmospheres generated contained various concentrations of formaldehyde (10-100 μgm-3), acetaldehyde (10-100 μgm-3) and ozone (0-600 ppb) and the exposure durations ranged from 4 to 24 hours. This permitted the coverage, either directly or by linear extrapolation, of a wide range of possible ambient atmosphere scenarios up to a five-day exposure duration.

All active sampling was carried out using DNPH impregnated silica cartridges (Sep-Pak, Waters Corp, MA, USA) and pumps set at precisely measured flow rates of around 1 $Lmin^{-1}$. Until recently, due to detection limits, aldehyde sampling at the low concentrations present in the ambient atmosphere was not possible by passive means. However, a novel, radially configured device, is now available (Radiello®, Fondazione Salvatore Maugeri, Italy (1997)) and due to the combination of a high uptake rate without the usually associated local starvation effects, low concentration aldehyde monitoring by passive means is now feasible. All passive sampling was performed using these devices.

It is widely reported that the presence of ozone when sampling formaldehyde by the DNPH derivitization technique can lead to the formation of compounds that interfere with the subsequent HPLC analysis (Jaouen, 1996). Figure 1 shows two HPLC chromatograms for formaldehyde samples taken from the same atmosphere with and without the scrubbers in the presence of 200 ppb of ozone and analysed under HPLC conditions originally used for this type of sample. For the unscrubbed sample a shoulder may be observed on the left-hand side of the HCHO peak. The inset in this same figure shows resolution, under improved HPLC conditions, of the previously co-eluting parasite peaks responsible for the aforementioned interference.

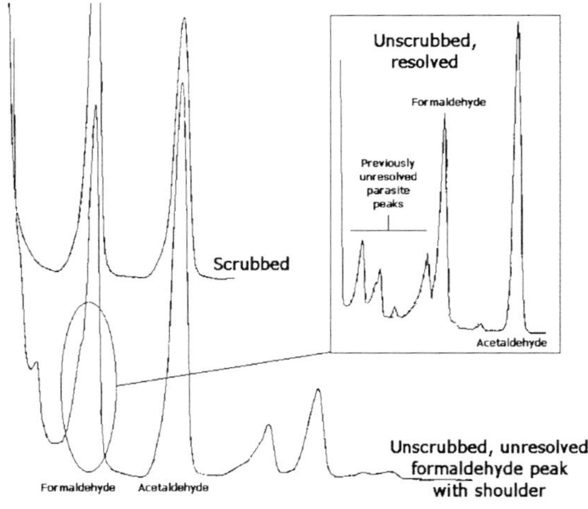

Figure 1. HPLC Chromatograms showing ozone interference in formaldehyde analysis

Derivatised aldehyde samples were extracted from the silica supports using 2 ml of acetonitrile. A 20 μL aliquot was then injected for analysis on a HPLC equipped with a Kromasil column (C-18, 3.5μm pore size, 150 x 3mm) with UV detection (λ = 365 nm). The mobile phase used for the improved separation was tetrahydrofuran, water and acetonitrile in the proportions shown in table 1 and with linear transitions between each concentration change.

Table 1. Mobile phase concentration gradients

Time / min	0	3	22	30	37
THF / %	5	5	0	0	0
Water / %	60	60	40	0	0
Acetonitrile / %	35	35	60	100	100

MAC ATMOSPHERE GENERATION, SAMPLING AND ANALYSIS

The MAC chosen for the study were benzene, toluene, o-xylene and styrene. The reconstituted atmospheres contained various concentrations of each of these four compounds (20-90 µgm-3) and of ozone (0-260 ppb) for exposure durations ranging from 4 to 16 hours to cover, as before, a wide range of possible ambient atmosphere scenarios.

All active sampling was carried out using tubes loaded with approximately 140 mg of Carbotrap 20-40 adsorbent (Supelco, Sigma Aldrich, France) and low-flow rate pumps (Gilian Inc., FL, USA) set to precisely measured flow rates of around 100 mlmin-1.

Passive sampling was carried out using modified Perkin Elmer tube-type samplers loaded with 140 mg of Carbotrap. Analysis of both passive and active samples was performed on an automatic thermal desorption apparatus (ATD 400 Perkin Elmer) connected to a GC equipped with an FID (column: CP-Sil 5CB 50m long / 1.2µm phase thickness (Chrompack, Middelburg, The Netherlands). Temp. gradient: 35-105°C at 2.5°C/min; 105-235°C at 10°C/min; maintain at 235°C for 20 min).

3. Results and Discussion

The graphs in figures 3 and 4 show the percentage recovery of formaldehyde and acetaldehyde in the presence of ozone, on active and passive samplers, with and without MnO_2 scrubbers. It is evident from all figures that the scrubber works efficiently in both sampling modes and that it has no adverse effects on the recovery of the compounds of interest. Care should be taken however as this may not be the case for other MnO_2 coated copper gauze: previous researchers (Kleindienst et al. 1995) found that it quantitatively removed formaldehyde as well as ozone. The authors would therefore strongly recommend testing any proposed MnO_2 scrubber in the laboratory prior to use in the field.

The results demonstrate that interference occurs in all cases in which the scrubbers are not employed even at relatively low concentrations. In the case of passive sampling it would appear that the interference of ozone on acetaldehyde is less pronounced than that on formaldehyde. For active sampling however the effect is of equal intensity. These results reinforce the evidence for ozone interference with respect to formaldehyde sampling and demonstrate a similar interference for acetaldehyde. They also show the effectiveness and adaptability of MnO_2 coated copper gauze as a passive or active ozone scrubber.

In figures 5 and 6 the results for the percentage recovery of the four MAC are given. It is evident that the scrubber is efficient in both active and passive mode with all points lying within ± 5% of full sample recovery. As was the case for aldehyde sampling, the scrubber has no detrimental effect on the recovery of

Figure 3. Recovery (%) of formaldehyde as a function of ozone concentration for a 5-day equivalent exposure passive sample (left) and a 100 litre active sample (right)

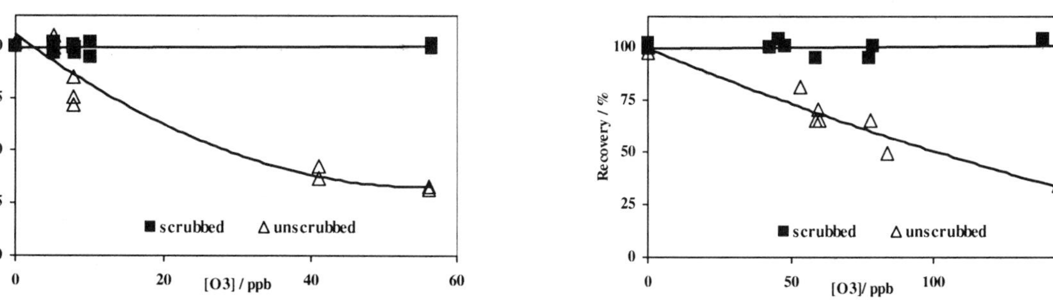

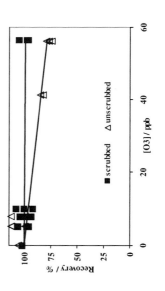

Figure 4. Recovery (%) of acetaldehyde as a function of ozone concentration for a 5-day equivalent exposure passive sample (left) and a 100 litre active sample (right)

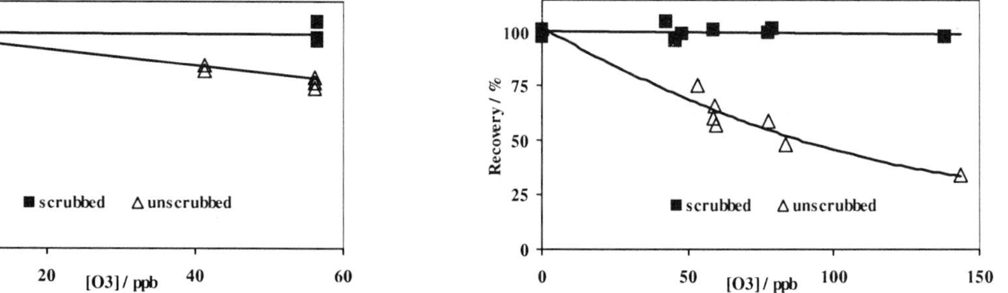

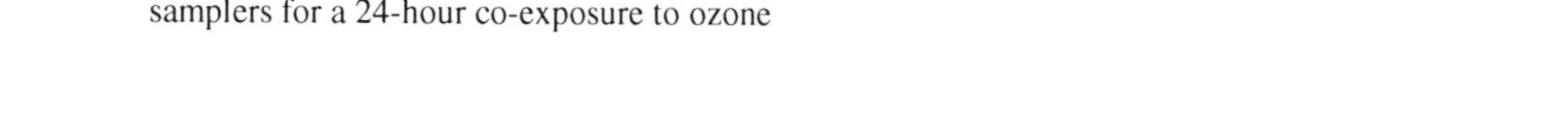

Figure 5. Recovery (%) of benzene (Ph), toluene (PhMe), o-xylene (oxy) and styrene (sty) on scrubbed (left) and unscrubbed (right) passive samplers for a 24-hour co-exposure to ozone

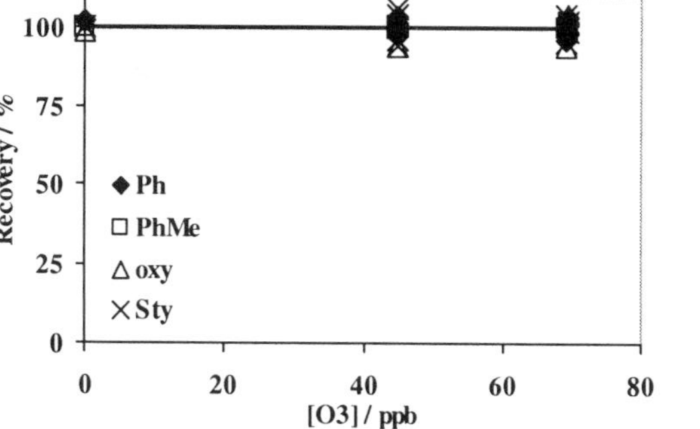

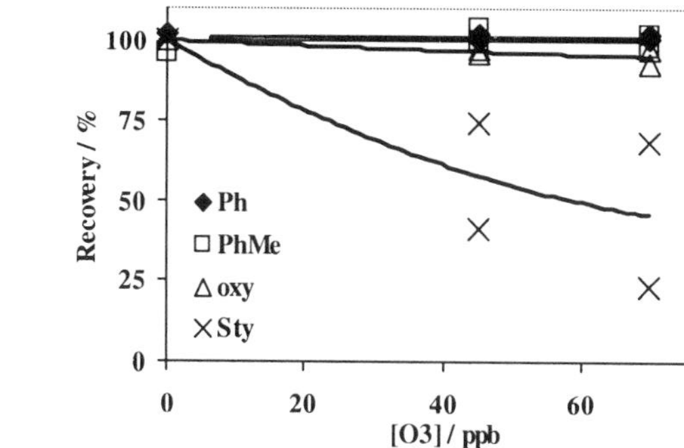

Figure 6. Recovery (%) of benzene, toluene, o-xylene and styrene on scrubbed (left) and unscrubbed (right) active samplers for a 3-litre sample taken in the presence of ozone

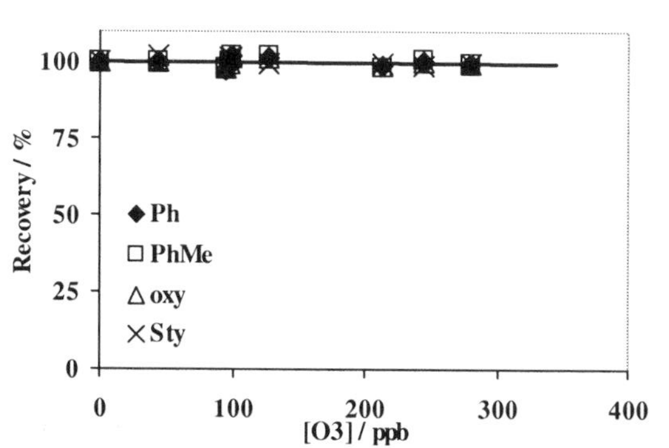

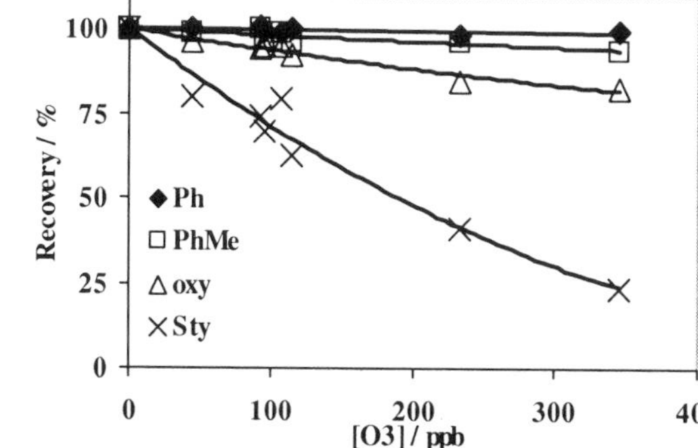

the compounds studied. When the scrubber was not used ozone caused severe and somewhat unpredictable losses of styrene and to a lesser extent loss of o-xylene and toluene. This is in agreement with the expected order of reactivity (susceptibility to electrophilic attack) of the MAC chosen. No new peaks were observed on the GC-FID chromatograms for MAC samples taken without scrubbers and no peaks were detected through adsorbent degradation on samplers exposed to atmospheres containing just ozone.

CORRECTING OLD DATA

A theoretical kinetic study of ozone interference in the case of formaldehyde was carried out (Jaouen, 1996). An expression was derived which, when fitted to experimental data, could be used to correct old results where inadequate separation had occurred at the analysis stage (provided that the corresponding ozone concentration was known). Figure 7 shows the good agreement between results from this work and from earlier research (Arnts and Tejada 1989; Jaouen 1996) and the curve derived from the kinetic study that may be used to correct old data (Recovery = $1/(k \times [O_3] \times t))(1-\exp(-k \times [O_3] \times t))$, where k is an empirical fitting parameter related to the gaseous phase reaction rate constant and equal to 7.5×10^{-5} $ppb^{-1}min^{-1}$; $[O_3]$ is ozone concentration / ppb and t is the exposure time / min).

Figure 7. Formaldehyde recovery as a function of ozone dose

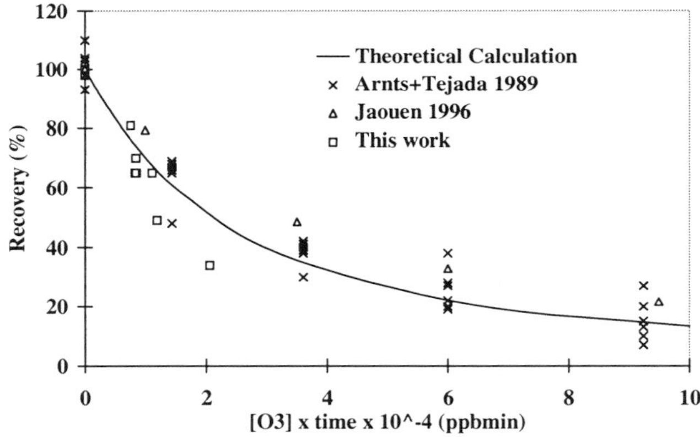

Theoretically, this expression could also work for correction of badly separated old analyses from samples taken in the field. However, many factors such as the presence and/or influence of other oxidant species (such as $HO^{\cdot}$, NO_3, O_2* and $O^{\cdot}$), remain unknown and uncontrolled and thus widespread application may be prone to error.

4. Conclusions

The presence of ozone, even at relatively low concentrations, causes loss of certain VOC when sampling. These losses will lead to underestimation of atmospheric concentrations and thus incorrect assessment of population exposure. Manganese dioxide coated copper gauze is an effective and convenient ozone scrubber that may be used in conjunction with both passive and active samplers. Losses through interference are greatly reduced and even if photochemical activity is low the scrubbers may be left on the devices as they have no apparent detrimental effect on the compounds studied. Some further work is necessary to determine the long-term behaviour of such devices, because, as with many other catalysts, the MnO_2 coated copper gauze may become poisoned by other compounds in ambient air.

Acknowledgements

The authors would like to thank Mr D. Granier, Mr J-C. Pinard and Ms A. Frezier for their technical assistance.

References

Arnts, R.R.; Tejada, S.B.: 1989, 2,4-DNPH coated silica gel cartridge method for determination of formaldehyde in air: identification of an ozone interference. *Env. Sci. Tech.* 23 (11) 1428-1430

Brown, R.H.; Wright, M.D.: 1994, Diffusive sampling using tube type samplers. *Analyst* (119) 75-77

CEN: 1997, Ambient Air Quality: Diffusive samplers for the determination of gases and vapours Requirements and Test Methods. Draft European Standard. CEN/TC 264/WG 11

Ciccioli, P.: Volatile Organic Compounds in the Environment Editors: Bloeman,H.J.Th.; Burn,J.

Gonzalez, N.; Cicolella, A.; Bates, M.; Bastin, E.: 1999, Pilot study of personal, indoor and outdoor exposure to benzene, formaldehyde and acetaldehyde. *Env. Sci. Poll. Res.*, in press

Helmig, D.: 1997, Ozone removal techniques in the sampling of atmospheric organic trace gases. *At. Env.* 31(21) 3635-3651

Hoffmann, T.: 1995, Adsorptive preconcentration technique including oxidant scavenging for the measurement of reactive natural hydrocarbons in ambient air. *Fres. J. of Anal. Chem.* 351, 41-47

Jaouen, P.: 1996, Thesis, Université de Paris VII

Jaouen, P.; Gonzalez-Flesca, N. and Carlier, P., 1995 Dynamic polluted atmosphere generator at low ppbv levels for validating VOC sampling methods. *Enviro. Sci. and Tech.* 29 no. 11 2718-2724

Kleindienst, T.E.; Corse, E.W.; Blanchard, F.T.; Lonneman W.A.: 1998, Evaluation of the performance of DNPH-coated silica gel and C18 cartridges in the presence and absence of ozone *Env. Sci. Tech.* 32 124-130

Megie,G. and others: 1994, Ozone et propriétés oxidantes de la troposphère. *Revue de l'Institut Français du Pétrole* 49 1 83-104

Radiello: 1997, Nota Applicativa Aldeidi, Fondazione Salvatore Maugeri, Italy.

Sirju and Shepson: 1995, Laboratory and field investigation of the DNPH cartridge technique for the measurement of atmospheric carbonyl compounds. *Env. Sci. Tech.* 29, 384-392

USEPA: 1990, Clean Air Act Amendments, List of 189 Hazardous Air Pollutants

WHO: 1996, Updating and revision of the air quality guidelines for Europe. World Health Organisation Publication

LONG PERIOD BENZENE MEASUREMENTS IN URBAN ATMOSPHERE

P. A. BRAGATTO*, S. CORDINER[∇], M. FEOLA[∇], L. LEPORE*, D. SACCO* and I.VENTRONE*
** Laboratorio Inquinamento dell'Aria Centro Ricerche ISPESL*
Via Fontana Candida 00040 Monteporzio (Roma) Italy
[∇] *Dipartimento di Ingegneria Meccanica Università degli Studi di Roma "Tor Vergata"*
Via di Tor Vergata 00133 Roma Italy E_mail:cordiner@uniroma2.it

Abstract: Periodogram analysis has been used as an interpretative tool of long period measurements of benzene concentration in urban atmosphere. By removing the request of evenly sampled data, the periodogram allows to extend spectral analysis to this kind of data, which usually suffer the "missing data" problem due to instrument vacancies. In this way it is possible to identify the periodic behaviour of data concentration and then reproduce it by means of harmonic functions even for very incomplete experimental information allowing a good evaluation of some important statistical properties of data.

Keywords: environmental data analysis, spectral methods, air quality standards, benzene

1. Introduction

Long period measurements of pollutant concentration in urban atmosphere have long been important in Air Quality Standard definition. One of the principal objectives of the politics of Air Quality control is, in fact, to keep pollutants concentration lower than any threshold for effect on human or vegetation. This is particularly true for those chemical species with a well known reactivity either as precursors of photochemical pollution or as carcinogenic components and between these last the benzene is playing a more and more important role. Recognising that, the European policy for air quality attributes to this pollutant more attention then in the past and in December 1998 Commission has adopted a proposal for a Directive which will for the first time set EU limit values for benzene. Italy was one of the first European countries to define air quality objectives for benzene. A Decree of the Environment Ministry in November, 25 1994 (D.M. 21.11.94, 1994) set air quality limits for "annual value" of benzene concentration (15 $\mu g/m^3$ until 1998 and 10 $\mu g/m^3$ from 1999) thus introducing the need of a systematic air monitoring in order to ensure compliance to the quality standards.

However, very long experimental observations sometimes suffer of periods in which instrumentation device failures and/or maintenance vacancies that cause the loss of data. As far as air quality limits are concerned, missing data and their recovery (interpolation, data substitution, etc.) could be extremely relevant and induce wrong results during the averaging procedures used to compare measurement results and Quality Standards.

Pollutants concentration evolution in atmosphere are in nature characterised by

Environmental Monitoring and Assessment **65**: 99–107, 2000.

a strong periodic behaviour due to their interaction with natural phenomena (i.e. the alternation of days and night and seasons) and with anthropic activities which are often periodic themselves. Strong non periodic deviation due to unpredictable local phenomena natural and/or anthropic (local meteorology, thermal stratification of atmosphere, traffic jam), could alter the periodic behaviour determining the actual measured concentration. The relative importance of the latter contribution could, nevertheless, be absolutely comparable or even greater than the periodic behaviour so causing the well known difficulties in predicting the concentration evolution by means of modelling tools.

Nevertheless a strong periodic behaviour is always present and suggests the analysis of time series of pollutants concentration measurements by means of Fourier analysis in the frequency domain to identify the periodic components and their relative importance. It is also possible to investigate any correlation between such components and different natural or anthropic phenomena.

Spectral methods have been used both as an interpretative tool to analyse time series of atmospheric constituents, like carbon dioxide $CO_{2,}$ sulphur dioxide SO_2 aerosols etc. (Enting, 1987), and to compare modelled and observed data (Sirois et al. 1995). However this kind of analysis can be correctly performed only on complete evenly sampled data set. Common ways to get out from unevenly spaced measurements problem are interpolation or the use of fictitious data to recover missing data in the time series. However, both these techniques produce uncorrected predictions in spectral analysis because the 'artificial' data introduce wrong harmonic components in the original time series, so influencing the final result.

In the present work a different approach to spectral analysis, the periodogram analysis, developed by Lomb and improved in (Scargle, 1982) and (Horne and Baliunas, 1986) for unevenly sampled data is used to identify the periodic behaviour of the time series. The method uses the normalised periodogram, which represents the spectral power as a function of the angular frequency, to individuate the most significant harmonic components of the time series.

After periodogram capability have been assessed, one of the main tasks of this work has been to evaluate how this technique performs in the identification of periodic and mean properties of the pollutant atmospheric concentration evolution when significant period of consequent missing data are present in the measure records.

2. Experimental set up

ISPESL has been monitoring since 1992 (Brocco et al., 1997), the concentration of benzene and other pollutants, by means of a long path differential absorption spectrometer (DOAS), in a downtown experimental station in Rome which is characterised by high traffic density. The DOAS equipment consists of an emitter with a 150 W high-pressure xenon lamp, a receiver and an analyser. The distance between emitter and receiver is about 280 m and the light path runs about 10 m above the ground level. The adsorbance of emitted light is continuously measured in the interval 240-340 nm but the aromatic hydrocarbons are detected in the wavelength range between 250 and 290 nm. The cross sensitivity of benzene is then evaluated in accordance to UK National Physics Laboratory method (Partridge et al., 1995) and is less than 2% of the concentration of each interfering gas (in particular ozone).

The evolution of the concentration of this pollutant during the last six years is thus available in a highly populated urban site. The period is particularly significant because catalytic converters became mandatory for new car with spark engine since January '93. The rapid diffusion of unleaded gasoline and the gradual introduction of less polluting cars determined a decrease of benzene atmospheric concentration in urban areas starting from 1993, as shown in fig.1.

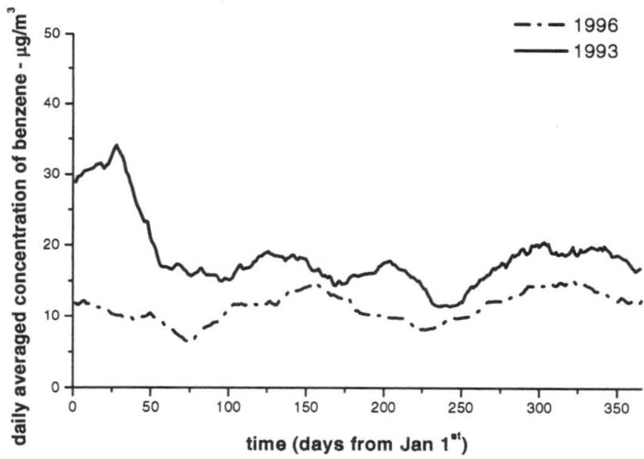

Figure 1 Daily Average Benzene concentration trend in 1993 and 1996 measurements

3. The Periodogram

Spectral methods in environmental pollutants concentration analysis can be efficiently used starting from Parceval relation:

$$\sigma_X{}^2 = \pi \int_0^{fc} f(\omega) d\omega$$

where σ_X is the data variance, $f(\omega)=|FT_X(\omega)|^2$ is the power spectrum, $FT_x(\omega)$ is the Fourier Transform of X(t) and f_c is the Nyquist frequency. Parceval's relation indicates that the power spectrum is the decomposition of the total variance into frequency-dependent components. However, power spectrum could be defined just for continuos function whereas, for a discrete data set $(X(t_i),i=1,2,\ldots.N_0)$, classical periodogram P(ω) as a function of the frequency ω is defined as:

(1)
$$P_x(\omega)=\frac{1}{N_0}\left|FT_x(\omega)\right|^2=\frac{1}{N_0}\left|\sum_{j=1}^{N_0}X(t_j)\exp(-i\omega t_j)\right|^2=$$

$$=\frac{1}{N_0}\left[\left(\sum_{j=1}^{N_0}X(t_j)\cos(\omega t_j)\right)^2+\left(\sum_{j=1}^{N_0}X(t_j)sin(\omega t_j)\right)^2\right]^2$$

as an estimate of the power spectrum of the unknown continuos function X(t). One of the remarkable properties of periodogram is to put into evidence any possible periodic component in the behaviour of the function X(t). In order to make periodogram analysis possible for pollutant concentration in atmosphere, however, it is necessary to introduce a different formulation of equation (1) which can be used for discrete unevenly sampled data sets:

(2)
$$P_x(\omega)=\frac{1}{2}\left\{\frac{\left[\sum_j X_j\cos\omega(t_j-\tau)\right]^2}{\sum_j\cos^2\omega(t_j-\tau)}+\frac{\left[\sum_j X_j sin\omega(t_j-\tau)\right]^2}{\sum_j sin^2\omega(t_j-\tau)}\right\}$$

where τ is defined by:

(3)
$$\tan(2\omega\tau)=\frac{\sum_j sin\ 2\omega t_j}{\sum_j\cos\ 2\omega t_j}.$$

When defined in this manner P(ω) has several useful properties which usual Discrete Fourier Transform does not have. Among these it makes equivalent the periodogram analysis to a least-square fitting of sine waves to the data and also give a convenient way to evaluate if a given peak into the P(ω) curve is a true signal or whether it is the result of some randomly distributed noise. In fact, in (Scargle, 1982) it is demonstrated that each maximum in the P(ω) curve correspond to a frequency ω_f which minimise the difference between the experimental data $\{X(t_j)\}$ and the harmonic function $X_f(t)=A_f\cos\omega_f t+B_f\,sin\,\omega_f t$. In (Scargle 1982) and (Horne and Baliunas, 1986) is demonstrated that if classical periodogram is normalised with reference to the total variance of the data distribution it possible to give a measure of the probability that a peak in the P(ω) curve is signal or noise so giving a tool to discriminate among periodic and non periodic components in the data set. Furthermore, due to the uneven sampling (the missing data) the periodogram defined in (2) shows less aliasing problems and substantially unchanged spectral leakage properties.

Periodogram characteristics have been exploited to recognise the more important frequency contained in measurements and, then, to find, by means of a linear least square fitting, the coefficients A_f and B_f of the best fitting.

As a result of this analysis is then possible to find a model function, constituted by m different harmonic components;

$$X_m(t)=\sum_{j=1}^{m}\left(A_j \cos\omega_j t + B_j sin\omega_j t\right) \tag{4}$$

This model function is defined on the same time set of the original experimental data but is a continuos function allowing to reconstruct an hypothetical evolution of concentration even for the time interval with missing data.

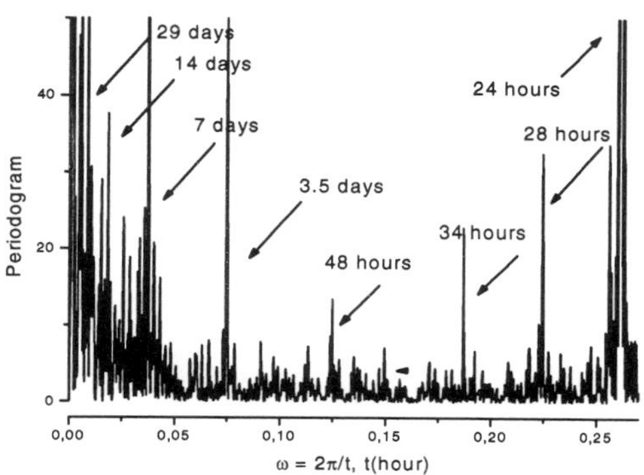

Figure 2 Typical periodogram

Figure 2 shows a typical periodogram extracted by experimental data. Different peaks could be individuated corresponding to different periodic behaviour of pollutant emissions and atmospheric cycles.

Although a noise component is distributed over all, several different and well defined frequencies are well visible. Among them some known periodicity like the expected value of 24 hours are highlighted.

4. Results and Discussion

The almost complete set of benzene hourly concentration records (containing 8611 data out of a total of 8784) measured during 1996 in Rome has been used as a first test bench for the periodogram analysis.

As all spectral analysis also periodogram, can be operated only on stationary data, then the first step is to identify any possible temporal trend in data evolution for every data set. In particular, a linear regression has been used to determine the coefficients a and b of the linear best fit, which, for the 1996 indicates a slightly growing variation of benzene during the whole year.

This trend, together with the mean value is then subtracted from the data to obtain a substantially stationary, zero mean value data set which has been used for the spectral analysis.

In Figure 3 the hourly measured concentration of benzene is compared with the model reconstruction (128 components) as a function of time, for a period of 4 days during January. A good agreement between model predictions and experimental data can be observed.

The approximate functions are able to reproduce the periodicity of data with correct amplitude and phase in the most cases. In particular the 24 hour time periodicity of benzene, is reconstructed very well.

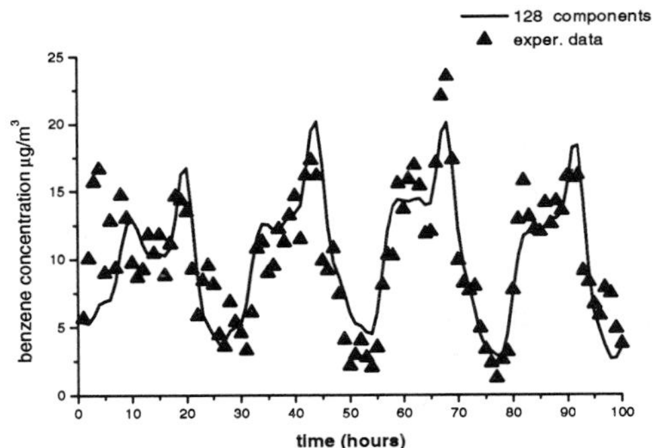

Figure 3 Comparison between experimental data (crossed line) and model reconstruction

The normalised periodogram has been calculated with different number of components to check how the goodness of the approximation varies with an increasing number of components themselves as shown in figure 4.

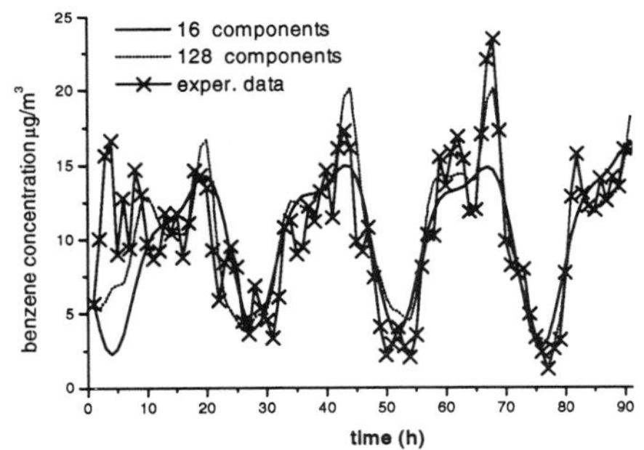

Figure 4 Comparison between experimental data and model functions with an increasing number of components

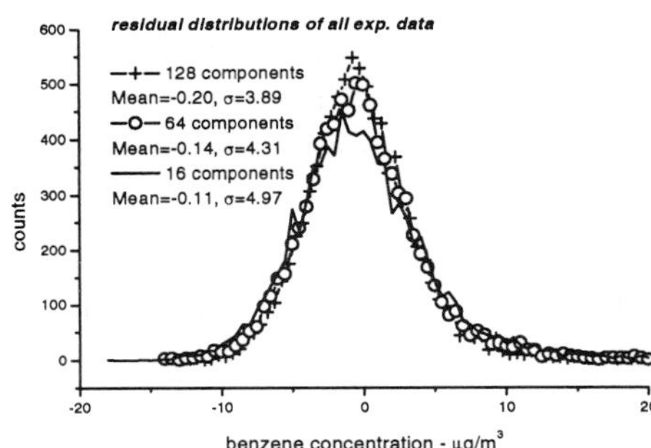

Figure 5 Residual frequency distribution

A comparison of residual distribution is shown in figure 5 and indicates that the difference between data and reconstructed values is narrowly distributed around the zero. An improvement of model reconstruction capability is observed increasing the number of components from 16 to 128.

4.1 ANALYSIS OF UNCOMPLETE DATA SETS

Part of the data were artificially removed in order to evaluate how periodogram analysis performs the identification of periodic and mean properties of the pollutant atmospheric concentration, on non complete and unevenly sampled data set. The following criteria have been adopted in reducing the data set:

- days with missing hourly data have been excluded according to Italian regulations which define a daily average concentration as a mean value of all 24 hourly measurements;
- artificially "missing data" have been created around the above excluded missing days until the 50% limit of valid daily average, prescribed by Italian regulation to define a valid "annual value" for benzene, was reached.

Figure 6, 7 and 8 show, as a function of time, the hourly measured concentration of benzene compared with reconstructed function with 128 components, for a period of one week during 1996, in three different seasons. The complete set of experimental data and the uncompleted "missing data" set used for fit the data with model function are indicated with two different kind of crosses. The lines with circles indicate the model reconstruction, with 128 components. The period of "missed data" is further highlighted by the straight line.

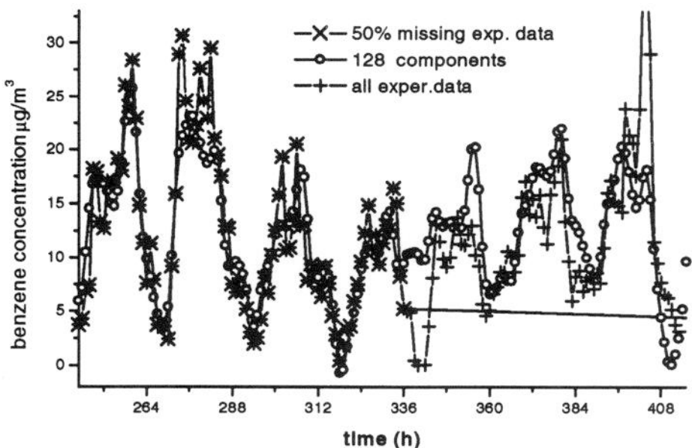

Figure 6 Comparison between experimental data and model reconstruction for an uncomplete data set (winter)

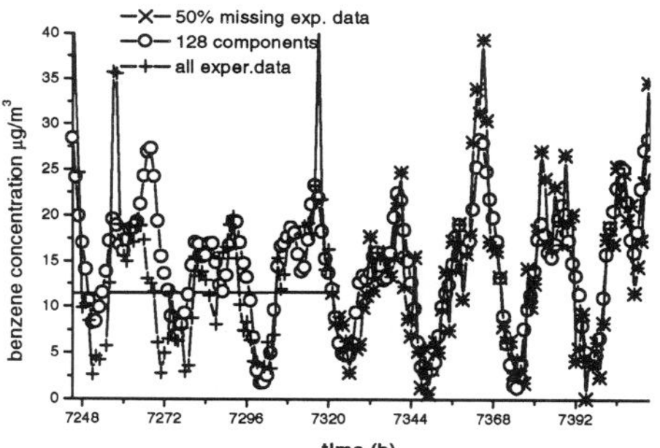

Figure 7 Comparison between experimental data and model reconstruction for an uncomplete data set (autumn)

A good agreement is observed between model predictions and experimental data for that time interval which contain experimental information. These figures show also a quite good agreement for that time interval corresponding to missing data.

Once again the approximate functions are able to reproduce the periodicity of data with correct amplitude and phase in most cases.

The comparison between Quality Standard for benzene, defined by Italian law as "annual value", obtained either by the full set of data or from reduced subsets by spectral analysis is shown in Table 1. The result indicate that, as far as Quality Standards are concerned, the harmonic reconstruction allows to predict the correct annual value with a very good agreement with a maximum difference less than 2% totally comparable with data uncertainty.

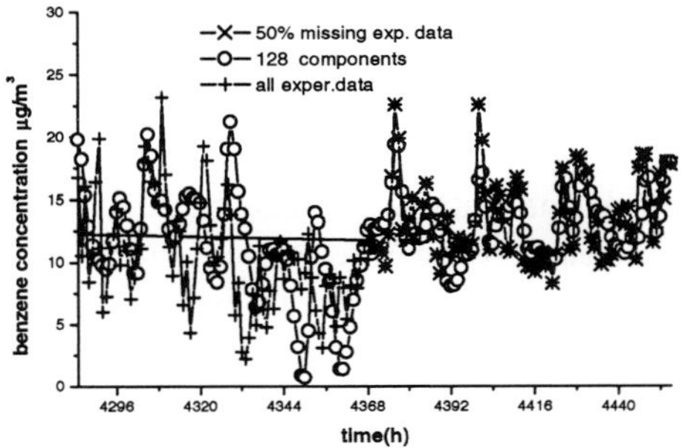

Figure 8 Comparison between experimental data and model reconstruction for an uncomplete data set (summer)

annual value of benzene in 1996 $\mu g/m^3$	complete experimental data set	reduced experimental data set
experimental value	11.36	
model function with 16 components	11.32	11.36
model function with 64 components	11.41	11.36
model function with 128 components	11.42	11.36

Table I

5. Conclusions

In conclusion, by removing the constraint on the completeness of the experimental data set the spectral analysis seems to be a feasible way to interpret time series of pollutant concentration, particularly for the evaluation of Air Quality Standards

Most of the harmonic components given by the spectral analysis fit well with well known periodicity of aromatic hydrocarbons emissions and of their atmospheric behaviour. Some cases have been observed were data have an opposite phase and/or higher frequency oscillations with respect to reconstructed function and could be correlated with occasional meteorological and/or short lasting emissions local events. A study of possible correlation with other pollutants will be made in a future work, most of all to verify if periodogram technique could be used to easily extract special events in experimental data or to estimate extreme values (e.g.98-percentiles). However, as far as integral values like Italian Quality Standards are concerned, the method has proved to be not particular sensitive to the nature of "missing data" and quite good approximations has been obtained for all the examined conditions.

References

Brocco D., Fratarcangeli R., Lepore L., Petricca M., Ventrone I., (1997) "Determination of aromatic hydrocarbon in urban air of Rome" *Atmospheric Environment* vol. 31 n°4, pp 557-566, 1997.

D.M. 24.11.94 (1994) *Gazzetta Ufficiale* della Repubblica Italiana n.290 December 13th 1994

Enting, I.G. (1987) "The interannual variation in the seasonal cycle of Carbon Dioxide concentration in the Mauna Loa", *Journal of Geophysical Research* 92, 5497-5504

Horne J.H. and Baliunas S.L., (1986) " A Prescription for Period Analysis of unevenly Sampled Time Series" *Astrophysical Journal* **302**, 757-763, 1986

Partridge R.H., Curtis I.H., Goody B.A. and Woods P.T. (1995) "An evaluation of the performances of an open path atmospheric air-quality monitor manufactured by Opsis, Sweden" NPL Report Qu109, National Physical Laboratory, UK

Scargle, J.D. "Studies in Astronomical Time Series Analysis II: Statistical Aspects of Spectral Analysis of Unevenly Spaced Data" *Astrophysical Journal* **263**, 835-853, 1982

Sirois, A, Olson, M. and Pabla B. (1995) "The use of spectral analysis to examine model and observed of O3 data" *Atmospheric Environment* 29 n.3, 411-422

Vanicek, P. (1971) "Further development and properties of the spectral analysis by least-squares", *Astrophysics and Space Science* 12, 10-33.

Project of an Air Quality Monitoring Network for Industrial Site in Italy

Corti A.* and Senatore A.**

* Dipartimento di Energetica "Sergio Stecco" di Firenze, Università degli Studi di Firenze, via S. Marta 3, 50139 Firenze – e-mail: corti@pinet.ing.unifi.it

** Dipartimento di Ingegneria Meccanica per l'Energetica, Università degli studi di Napoli Federico II, via Claudio 21, 80125 Napoli

Abstract In this paper results concerning project of an atmospheric monitoring network applied to an area of south of Italy carried out using chemical measurement campaigns (for traffic emissions characterisation couple with diffusional simulation of industrial pollutant emissions will be presented. The area selected is characterised by an high concentration of high environmental impacting industrial activities, with also high concentration of urban settlement and vehicular traffic. A comprehensive definition of all the monitoring system to be installed in order to have a correct monitoring system for the all province will be described and discussed.

Keywords: Atmospheric Monitoring network, atmospheric pollutant measurements, industrial emissions, environmental impact, diffusional models

1 Introduction

The present work refers to the project of an air quality monitoring network applied to an area of the south of Italy characterized by a high concentration of human activities, developed by the University of Naples with the cooperation of the University of Florence.

The air quality network will be developed in order to guarantee an adequate level of monitoring of the urban state of atmospheric pollution as defined by Italian law and as planned by the Campania Region in the official regulations concerning environmental monitoring tools. [1]

Particular attention was devoted to the localization of monitoring stations for the evaluation of the industrial human sources, highly concentrated in the area studied. In this way not only the overall environmental pressure but also the contributions due to the different sources that are present in the area could be foregrounded. For the definition of the optimum point of localization for the air monitoring network, two different approaches were chosen, mixing the different results obtained at the end of the two different methodologies: an analytical approach, measuring the concentration of air pollutant with a mobile analytical laboratory and a modeling approach, studying the diffusional effects of the industrial sources for which sufficient data were found.

Ten monitoring stations with different characteristics of chemical sensors installed inside were defined for the first phase of implementation versus an overall value at a completed stage of the network of 28 stations [2]

Present work was supported from the Salerno Province with a specific contract.

Environmental Monitoring and Assessment **65**: 109–117, 2000.

2 Methodology

In general terms the project of a monitoring system is not regulated by any technical directive for Italian or European law.
In the present case, due to the specific aims, an air quality network was first defined that had to be adequate in order to guarantee [3]:

- a good statistic representation of the real state of air pollution conditions in the urban areas where the study was carried out, to be obtained when the network is installed and running

- an accurate evaluation of the industrial pollutant at ground level loads due to the diffusion of the emitted pollutants to foreground and control the relative weight of industrial emissions with respect to the other sources.

Different approaches for the localization of the air monitoring stations that optimized the results in terms of representation of the real state of urban air pollution have been developed by different authors [4-7]. In all cases large amounts of data measured or representative of real sources are needed. This is possible especially in the case of a new project of an already installed or incomplete network, when amounts of data are obtained from the fixed monitoring stations with the addition of data possibly obtained from measurement campaigns. In the case of a project of a network starting from the condition of absolute absence of stations in the area these methods represent a high cost for sufficient data from measurement campaigns.
For these reasons, in the present case, in order to obtain the above mentioned two aims of the study, a mixed methodology was chosen: chemical measurement campaigns in certain specific areas where high levels of pollutant emissions are present and diffusional modeling study on the main industrial sites using a Gaussian model. The first approach is particularly suitable in the case of primary pollutants, while the second one was used for the characterization of the secondary air pollutant maximum time integrated deposition areas.
The selected sites for measurement where obtained from a preliminary study on individual town traffic planning documents and on the population density of the various urban areas.

3 Area of study

The study was carried out on an overall number of 12 urban settlements in the province of Salerno, corresponding to the highest concentrated urban settlement density, highest concentrated road network and highest concentrated industrial activities. The whole area selected was divided into 6 main territorial sections, satisfying the planning indications given by the Region Authority, so that a smaller number of sensors will be necessary in order to comply with national regulations.

Figure 1 – Mobile analytical laboratory used

The overall level of territorial coverage is very low. The areas studied correspond to 8,8 % of the overall provincial territory and to 7.7% of the overall municipalities. On the other hand a higher level of coverage was obtained, in this way, for population settlement (33.7% with respect to the total) and for industrial activities (72.4 %).

4 Evaluation of the primary air pollutant monitoring sites

For the characterization of the optimum localization for the chemical monitoring of the primary pollutant air quality network, an overall number of 57 sites of investigation were chosen. The air quality characterization of the territory was made using a mobile chemical analysis laboratory for the automatic measurement of sulfur dioxide (SO_2), nitrogen oxides (NO, NO_2 and NO_X), carbon monoxide

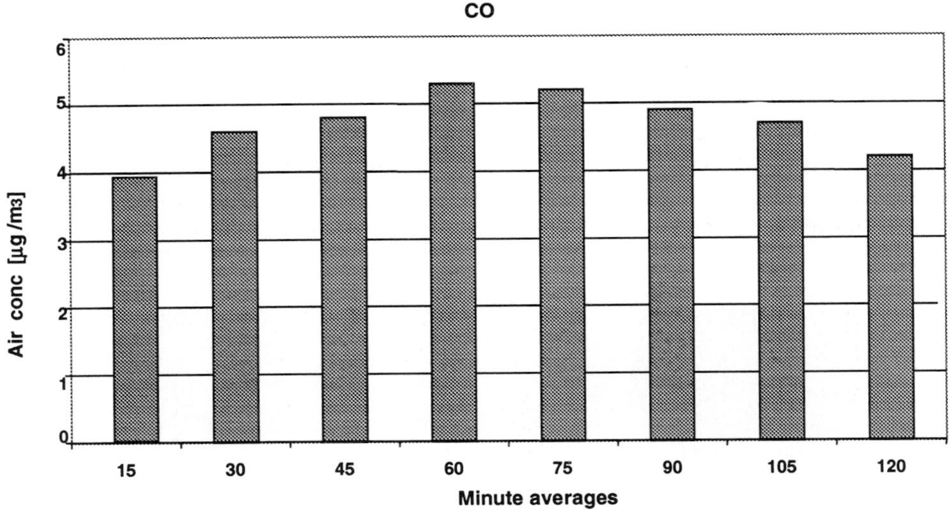

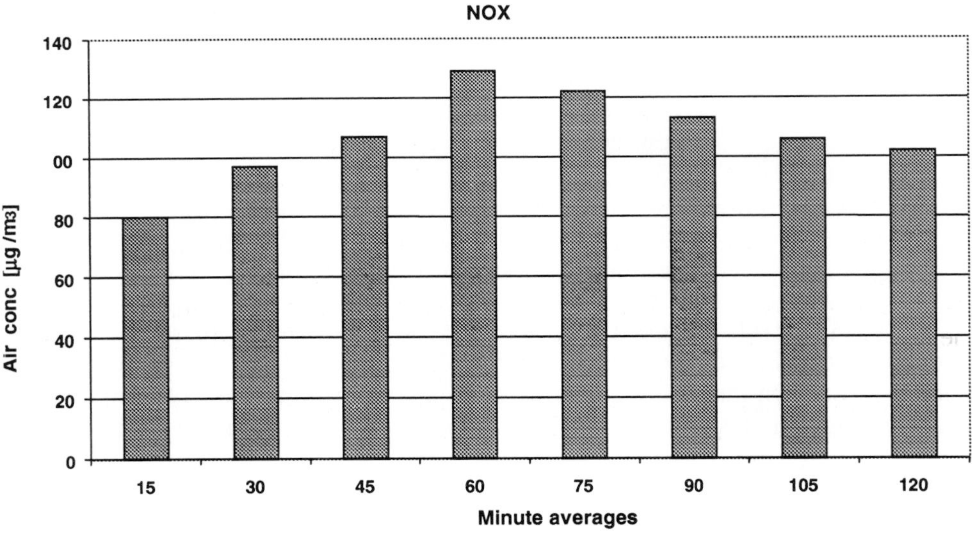

Figure 2 Results of measurements on one site of investigation (Salerno Pontecagnano Area)

(CO) Ozone (O_3), Total Solid Particulate (TSP), Methane and Non Methane Hydrocarbons (H_mC_n) and Benzene (C_6H_6) (figure 1).

Using data obtained from measurement campaigns it was possible to identify the areas where higher concentrations of primary pollutants (mainly due to traffic) were found. With the criteria of "higher pollutants exposition equals better localization of monitoring sensors", it was possible to identify the optimum sites

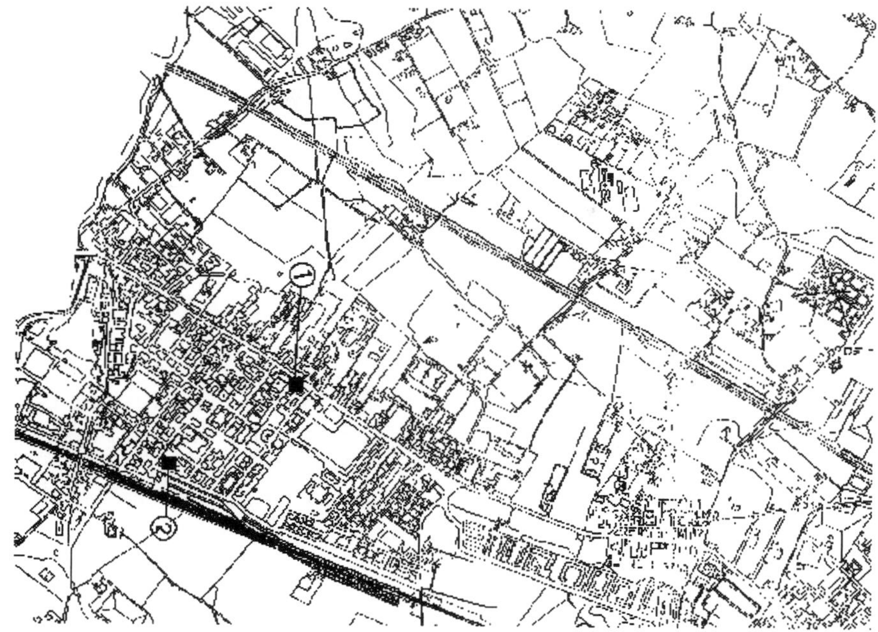

Figure 3 - localization chosen for monitoring network

for the installation of the air monitoring network stations concerned with primary air pollutant high concentration phenomena.

In the present case data reported refer to the case of the urban area of Salerno-Pontecagnano, one of the 6 macro areas on which the study of the 12 urban settlements was based. In figure 2 data referred to measurements made in one of the 6 places where the mobile chemical laboratory was installed, in Salerno-Pontecagnano urban area, are reported. The measurement corresponds to the macro air pollutants CO and NO_X. Data reported in figure 2 refer to a particular peak load condition foregrounded in this urban area. On figure 3 are shown the resulting sites for the two monitoring automatic stations of the network chosen in the specific area.

From data reported in figure 2 it is possible to foreground the high level of concentration obtained for all the pollutants monitored during the period of measurement. This is mainly due to the high traffic load of the roads closer to the mobile laboratory point of measurement. From the trends of the measurements made in all the locations selected for the campaigns it was possible conclude that maximum values of environmental pressure were included in the intervals of time of measurement, and this is very important if a correct evaluation of the sites exposed to higher environmental pressure is the aim of the study.

The same approach was used for all the 57 sites of the measurement campaign analysis, obtaining for all an optimum localization of the sensors of the network devoted to the monitoring of primary air pollutants.

5 Evaluation of the secondary industrial air pollutant monitoring sites

For the evaluation of the optimum localization of chemical air monitoring sensors, a diffusional approach was also used for the secondary air pollutant investigation. For the evaluation of the secondary atmospheric pollutants, a specific gaussian modeling study was made in order to evaluate the effect due to industrially emitted atmospheric air pollutants.

A US-EPA preferred gaussian model was used: ISC3 for this aim, because of the high number of emission points and because of the nature of the non flat terrain typical of the area on which the study was carried out [8]. The gaussian model was used in the long term running mode in order to evaluate the environmental pressure at ground level average on an overall period of one year of meteorological representative data.

With an accurate phase of classification of all the industrial sites present in the area, a data base of industrial gaseous and particulate air pollutant emissions was obtained. The data base built up includes chemical and processing data necessary for the evaluation of pollutant mass flow rates but also of data necessary for the evaluation of the physical dimension of stacks and for the evaluation of downwash effects on plumes. Emission data for industrial stacks considered in the study were obtained form chemical certified measurements or from Emission Factor database. In both cases average emission data are represented without any consideration on different running conditions of emission systems. Meteorological data necessary for the diffusional study were obtained for a historic set of data of more than one year of measurement, obtained from Italian Air Force.

Data relative to the orography of the area studied were obtained after a proper elaboration and interpolation of data, direct from a digitized Italian ground map.

For the evaluation of the optimum localization of air monitoring secondary industrial pollutant automatic sensors, nitrogen oxides were used as a tracing element, due to the high presence of this pollutant in all the emission industrial plumes.

Using a long term diffusional approach it was possible to foreground the areas where the emitted pollutants are most likely to cause high ground level concentrations in terms of secondary pollutants, that means the maximum average environmental pressure. Using relative weight of the average concentration measured by the model related to the maximum value obtained in the specific area studied, for every receptor, it was possible to build up maps that foreground the areas exposed to the maximum average effect of ground level falls of atmospheric pollutants emitted from industrial plants (figure 4) during the overall period of analysis.

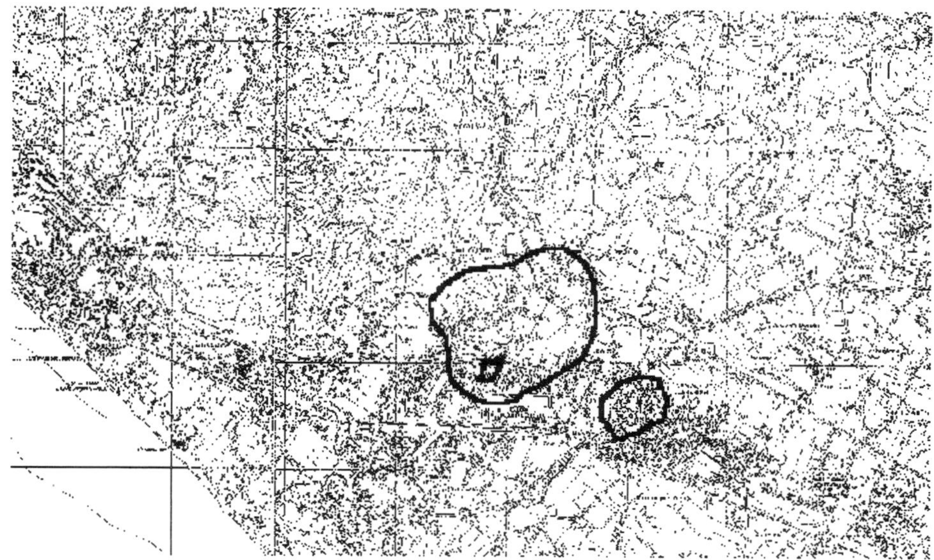

Figure 4 - areas with the maximum average effect of ground level deposition of atmospheric pollutants emitted from industrial sources

For this purpose a relative environmental pressure indicator was created (REP_{LT}) relative to average data obtained over the long period of one year.

$$REP_{LT} = \frac{C_i^{LT}}{C_{\max}^{LT}}$$

'where:

C_i^{LT} = long term average concentration for i-receptor
C_{max}^{LT} = long term maximum concentration measured on the all area of investigation

In figure 5 results are shown from this type of analysis where the area where the higher environment exposure to pollutants are obtained on the long term period considered is clearly foreground. For the specific example data shown are reported concerning the Salerno - Pontecagnano macro project area where more than 0.55 and more than 0.85 for the REP_{LT} indicator are calculated from the model, considering the maximum exposure value registered on the whole representative zone.

On the basis of this data obtained for the whole project network area, it was possible to foreground the areas expose to major at environmental pressure. Inside these areas, also considering some data obtained from some measurement campaigns carried out closer to or inside industrial sites.

6 Conclusions

On the basis of data obtained from mobile laboratory analysis and from diffusional modeling a data base of the optimum sites for the localization of the monitoring network was obtained.

The installation of the monitoring network was divided into two separate phases, an initial one for the installation of a sufficient number of sensors that satisfies the regulation concerning the 6 macro areas (Table 1). Classification of monitoring stations into A, B, C and D classes refer to the specific Italian regulation.

In the second phase of implementation all the urban areas will be completed with the right number of sensors as prescribed by Italian technical regulation.

Table 1 - first phase development monitoring network project (U.A. = Urban Aggregation)

			Number of stations			
	Monitoring stations	***Municipalities***	**A**	**B**	**C**	**D**
U.A. 45	**6**	Angri	**1**	**1**	**1**	
		Scafati		**1**	**1**	**1**
U.A. 46	**6**	Cava dei Tirreni	**1**		**2**	**1**
		Nocera Inferiore		**1**	**1**	
U.A. 48	**3**	Salerno	**1**			**1**
		Vietri sul Mare			**1**	
U.A. 49	**2**	Eboli		**1**	**1**	
U.A. 50	**2**	Sarno		**1**	**1**	
U.A.51	**9**	Battipaglia		**2**	**2**	
		Pontecagnano		**1**	**1**	
		Mercato San Severino			**1**	
		Castel S. Giorgio		**1**	**1**	

References

[1] Delibera n.10413 della Giunta Regionale della Campania (In Italian)

[2] Report DIME University of Neaples "Federico II": “Progetto di Monitoraggio Ambientale della Provincia di Salerno” Convenzione tra il Dipartimento di Ingegneria Meccanica per l’Energetica di Napoli e la Provincia di Salerno, Aprile 1998. (In Italian)

[3] Bøhler, T., “Environmental Surveillance and Information System” in AIR POLLUTION III, ed. da N. Moussiopoulos, H. Power e C.A. Brebbia, Comput. Mech. Publications, Vol.1, Settembre 1995.

[4] Kainuma Y., Shiozawa K., Okamoto S., (1990) “Study of an optimal allocation of ambient air monitoring stations” Atmospheric Environment Vol. 24B, No. 3, pp. 395-406.

[5] Langstaff, Seigneur, Liu, Behar, McElroy, (1987) “Design of an optimum air monitoring network for exposure assessment” Atmospheric Environment Vol. 21 No 6 pp. 1393-1410.

[6] Liu M.K., Avrin J., Pollack R.I., Behar J.V., McElroy J.L., (1986) “Methodology for designing air quality monitoring networks: I.Theoretical Aspects” Environmental Monitoring and Assess. Vol. 6, pp. 1-11.

[7] Shindo J., Oi K., Matsumoto, (1990) “Consideration on air pollution monitoring network design in the light of spatio-temporal variations of data” Atmospheric Environment Vol. 24B (2), pp. 335-342.

[8] Zannetti P., "Air Pollution modeling - Theories, computational methods and available software", Computational Mechanics Publications, 1990

URBAN AIR POLLUTION MONITORING WITH DOAS CONSIDERING THE LOCAL METEOROLOGICAL SITUATION

A.GOBIET, D. BAUMGARTNER, T. KROBATH, R. MADERBACHER and E. PUTZ

Institute for Geophysics, Astrophysics and Meteorology, University of Graz. Universitätsplatz 5, 8010 Graz, Austria

E-mail: ang@igam.kfunigraz.ac.at

Abstract. The concentrations of ozone, NO_2 and SO_2, measured with a DOAS system 70 m above ground level in the city of Graz were compared with data from conventional ground stations. The dependence of vertical trace-gas distributions on stability categories and time of the day or year was investigated. Concerning the maximum ozone concentrations in summer, the DOAS data are representative for the ground-level situation. In average, the concentrations 70 m above ground are more than twice the ground-level concentrations. It has been shown that beside the reaction with NO, dry deposition is an important sink for ozone near the surface. The DOAS NO_2-concentrations are representative for ground-level conditions in summer, except for the morning maximum of NO_2. In winter the DOAS NO_2-concentrations amount for 73% of the ground level values in average. Concerning the slow reacting trace gas SO_2, the DOAS data are always representative for the ground-level conditions.

Key words: DOAS, Differential Optical Absorption Spectroscopy, urban air quality, monitoring, vertical distribution, ozone.

1. Introduction

In urban areas the sources of atmospheric pollutants are distributed very inhomogeneously. Trace gas concentrations measured by conventional air quality monitoring stations are always influenced by small sources in their direct vicinity and surface effects like dry deposition and small-scale wind systems. For these reasons conventional monitoring stations rarely provide representative data for a big area.

Remote sensing techniques like Differential Optical Absorption Spectroscopy (DOAS) (Platt, 1994) can avoid the problems of local influences and surface effects: DOAS can use light paths which range over several kilometres, thus the measured concentrations are averaged over the light path and barely influenced by small-scale variances. Furthermore the light path can be situated high above ground, where the horizontal variances of trace gas concentrations are smaller than at ground level. Long path DOAS measurements produce data that are representative for a big area and could possibly replace several conventional monitoring sites in urban areas. But since the concentrations of atmospheric pollutants depend on the altitude above ground as well as on the local meteorological situation (vertical mixing), DOAS data is not always representative for ground level concentrations.

The objective of this project is to identify the local meteorological conditions at which DOAS data are in agreement with trace-gas concentrations at ground level

Environmental Monitoring and Assessment **65**: 119–127, 2000.

and to obtain more information about the vertical distribution of atmospheric pollutants in urban areas in general.

2. Experimental

The light source of the DOAS system, a xenon high pressure arc lamp (Osram XBO 450 W), was placed at the focus of a parabolic mirror (30 cm diameter, 60 cm focal length). At the other end of the open path the light was received by a telescope, which had the same dimensions as the light telescope to transfer the light by means of a silica fiber to the entrance slit of the spectrograph. The spectrograph is equipped with a holographic flat field grating (550 grooves/mm) and a photomultiplier tube with a rotating slotted disk for scanning the spectra. They were taken in the spectral range from 250 nm to 380 nm with a spectral resolution of about 1 nm. After an integration time from 2–5 minutes the spectra were stored at an external hard disk.

In the evaluation procedure the dark current and background light were subtracted from the spectra. The logarithm of the resulting spectra were then numerically high-pass filtered to remove broadband structures due to the light source and Mie and Rayleigh scattering in the atmosphere. In the last step, high-pass filtered reference spectra of the trace gases were fitted to the measured spectrum using a least squares routine described by Stutz and Platt (1996).

3. Data

3.1. Chemistry

The project is built up on data of four DOAS measurement campaigns which were performed in Graz, Austria (240 000 inhabitants) in the period between July 1996 and July 1998 (Baumgartner *et al.*, 1999). The DOAS instrument provided concentrations of nitrogen dioxide (NO_2), sulphur dioxide (SO_2), ozone (O_3) and in selected periods formaldehyde (HCHO) and monocyclic aromatic hydrocarbons over optical paths of 1, 2 and 3.7 km respectively. The DOAS light source and the receiver/detector unit were placed on high buildings and on a small hill (Schloßberg) in the centre of the town. The average altitude of the optical paths was 70–90 m above ground. The results of the DOAS measurement campaigns were compared with data from three stations of the permanent air quality monitoring network of Graz (Figure 1):

- *Graz West*: Ground level, 370 m asl.
- *Schloßberg*: Level of the DOAS light source or receiver/detector unit respectively, town centre, 450 m asl.
- *Platte*: Top of a hill in the periphery of the town, 661 m asl.

Figure 1: The air quality monitoring network of Graz and the DOAS light-paths.

3.2. METEOROLOGY

Table I: Stability categories based on the vertical temperature gradient and the horizontal wind velocity 10 m above ground. Categories: 2: unstable, 3: slightly unstable, 4: neutral, 5: slightly stable, 6: moderately stable, 7: very stable.

	vertical gradient of temperature dT/dz [°C/100 m]						
wind velocity u_{10} [m/s]	< -1.5	-1.4 to -1.2	-1.1 to -0.9	-0.8 to -0.7	-0.6 to 0.0	0.1 to 2.0	> 2.1
< 0.7	2	2	2	3	4	6	7
0.8 to 1.9	2	2	2	3	4	6	7
2.0 to 2.9	2	2	3	4	4	5	6
3.0 to 3.9	2	2	3	4	4	4	5
4.0 to 4.9	2	3	3	4	4	4	5
5.0 to 6.9	3	3	4	4	4	4	5
> 7.0	4	4	4	4	4	4	4

To classify the stability of the boundary layer we made use of temperature and wind-velocity data of the only station in Graz which is not heavily influenced by surface effects: A chimney of an industrial building with an altitude of 55 m (*Eurostar*). We combined these data with data from a nearby ground station to calculate stability categories (Table I) according to the Austrian Standards Institute (Fachnormenausschuß Luftreinhaltung, 1996). These categories are generally used to determine the dispersion parameters in Gaussian plume models. We used them as a parameters for the stability of the boundary layer in general. They range from 2 (unstable) over 4 (neutral) to 7 (very stable).

4. Results and Discussion

4.1. OZONE

A rough analysis of all four DOAS campaigns (Table II) showed that the median of the DOAS ozone concentrations (42 μg/m³) is more than twice the ground level concentrations at *Graz West* (20 μg/m³). This is not very surprising because ozone decomposition processes like dry deposition and the reaction with NO from ground sources are more pronounced at the surface. Accordingly the median concentration at the station *Platte* (661 m) is highest (64 μg/m³). Similar results were obtained form ozonesonde measurements in other cities (Pisano *et al.*,1997, Günsten *et al.* 1997). The ozone maxima at all stations are very similar (around 150 μg/m³). The differences between the stations are bigger in winter than in summer.

Table II: Medians (upper line) and maxima (lower line) of the ozone concentrations at different altitude levels (given in metres above sea level).

Ozone [μg/m³]	*Graz West* 370 m	*DOAS* 440 m	*Schloßberg* 459 m	*Platte* 661 m
total	20 147	42 147	39 149	64 153
summer June - Sept.	48 147	50 147	58 149	84 153
winter Nov. - Feb.	4 94	35 130	22 99	49 107

4.1.1. *Summer*

According to the vertical distribution of ozone in summer, we determined three different situations: Sunny conditions, cloudy conditions with well mixed boundary layer and cloudy conditions with stable boundary layer.

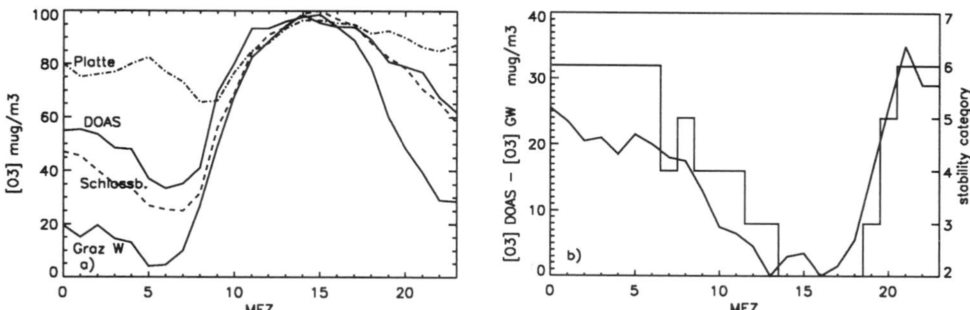

Figure 2 a: Median of diurnal ozone variations for the summer period, sunny conditions (July 96 - Sept. 96). *b:* Median of the stability categories and differences between DOAS (70 m above ground) and ground level ozone concentrations for the same period.

Figure 2 a shows the hourly medians of the ozone concentrations for the period July 96 – September 96 for sunny conditions (days with a sunshine duration less than one third of the maximum duration possible were removed from the data set). The concentrations differ strongly at night and in the morning and the ozone concentration increases with altitude. The time lag of the ozone minima in the morning at the different stations is due to the time the NO enriched convective layer needs to rise. Only the DOAS values in comparison with *Schloßberg* do not exactly behave as one would possibly expect (that is increasing concentrations and later minima at higher altitudes): Even though the average altitude of the DOAS optical path is slightly below the altitude of *Schloßberg* (this corresponds to the earlier ozone minimum along the optical path), the concentrations at *Schloßberg* are lower. This shows that surface effects as dry deposition are important ozone sinks at ground stations.

In the early afternoon, when the ozone-maxima occur, the differences between the individual stations vanish. The cause for this behaviour can be seen in Figure 2 b: It shows the hourly medians of the stability categories for the same period as in Figure 2 a together with the medians of the differences between the ground-level (*Graz West*) and the DOAS ozone concentrations. In the stable boundary layer (in the night and in the morning) the differences are high. In the well mixed boundary layer (at noon and in the afternoon) the differences vanish. As cross correlation analysis shows, the two curves in Fig. 2 b are strongly correlated, best without time lag ($\rho(\tau=-1) = 0.65$, $\rho(\tau=0) = 0.88$, $\rho(\tau=1) = 0.84$). That means that the vertical ozone distribution responds to changes in the stability of the boundary layer within the timescale resolved (1 hour).

In cloudy conditions with a well mixed boundary layer (e.g. when a cold front passes), there are no significant differences in ozone concentration, independent of the time of the day.

In cloudy conditions with a stable boundary layer (e.g. warm front) the ozone concentrations are low all day but remarkable higher at 70 m above ground. The differences are again independent of the time of the day. Figure 3 shows examples of diurnal ozone variations at ground level (*Graz West*) and at 70 m above ground level (DOAS) for all three situations.

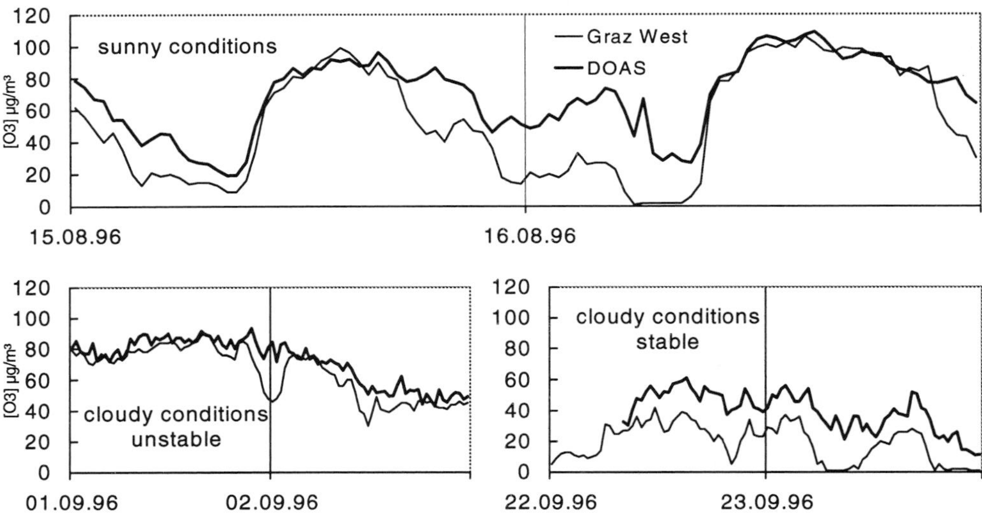

Figure 3: Ozone concentrations (µg/m³) at DOAS (70 m above ground level) and *Graz West* (ground level).

4.1.2. *Winter*

The principal differences between DOAS and conventional ozone-measurements can be seen more clearly in the winter period when the boundary layer tends to be stable (Figure 4 a). The DOAS concentrations are always higher than the concentrations at the stations *Graz West* and *Schloßberg*. The DOAS maximum is even higher than the concentration at *Platte* at the same time. This cannot only be due to ozone destruction by the reaction with NO from ground sources.

Figure 4 b shows the O_x (= $O_3 + NO_2$) concentrations at ground level (*Graz West*) and DOAS for the winter period. Since ozone reacts with NO to O_2 and NO_2 rapidly and more than 90 % of the primarily emitted NO_x is NO, O_x can be regarded as "ozone plus ozone destroyed by NO".

The differences of the O_x concentrations (around 5 ppb) are much smaller than the differences of the ozone concentrations (around 12 ppb). That means that the major part of the difference of the ozone concentration is due to destruction of ozone by NO near the ground. Yet the excess O_x at the DOAS light path must be due to other reasons. Since other chemical reactions can be neglected and the excess O_x at the DOAS-lightpath cannot be attributed to down mixing of ozone from higher altitudes (the ozone concentrations at *Schloßberg,* which is at the same altitude-level as the DOAS-lightpath, are much lower than the DOAS concentrations), the O_x difference (5 ppb) can only be explained by dry deposition of ozone at the surface.

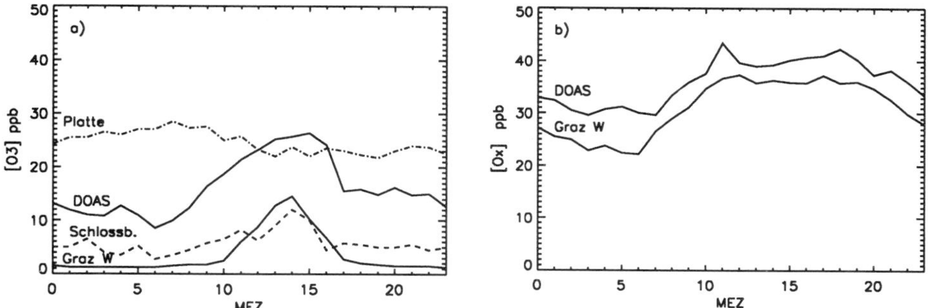

Figure 4 a: Median of diurnal ozone variations for the winter period (Nov. 97 – Feb. 98). *b:* Median of diurnal O_x variations for the same period.

4.1.3 *Correlation with stability classes*

Since one aim of this project was to find a simple relation between atmospheric stability and the ozone concentrations at different altitudes, the differences between DOAS and *Graz West* were individually classified according to stability categories regardless of weather, time of the day, or time of the year. The medians of this classes correlate well with the stability classes. They range from 14 µg/m³ at class 2 (unstable) and are increasing to up 28 µg/m³ at class 7 (very stable). However, the variance inside the classes is large and the Pearson-correlation coefficient only amounts for 0.38 for the whole period and 0.62 for the summer period.

4.2. NO_2

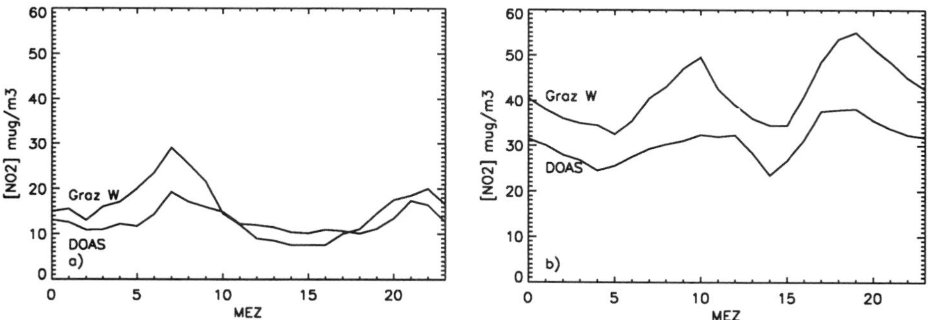

Figure 5 a: Median of diurnal NO_2 variations for the summer period (July 98). *b:* Median of diurnal O_2 variations for the winter period (Nov. 97 – Feb. 98).

In summer the medians of the NO_2 concentrations are almost equal at ground level (*Graz West*: 15 μg/m³) and at 70 m above ground (DOAS: 13 μg/m³). Only in the morning when the NO_2 concentrations at the ground rise rapidly (which is due to the morning rush hour) and the boundary layer is stable the NO_2 concentrations at ground level are significantly higher. In winter the median NO_2 concentration at 70 m above ground (30 μg/m³) amount to 73% of the ground-level concentration (41 μg/m³) in average (Figures 5 a and 5 b).

4.3. SO_2

Figure 6 shows that the medians of the SO_2 concentration in winter are essentially equal at ground level and at 70 m above ground (14 μg/m³). This is due to the long lifetime of SO_2 in the atmosphere. Unfortunately there are not enough data to investigate the situation in the summer period, but since in summer vertical mixing is stronger, significant different SO_2 concentrations 70 m above ground compared to the ground-level are not expected.

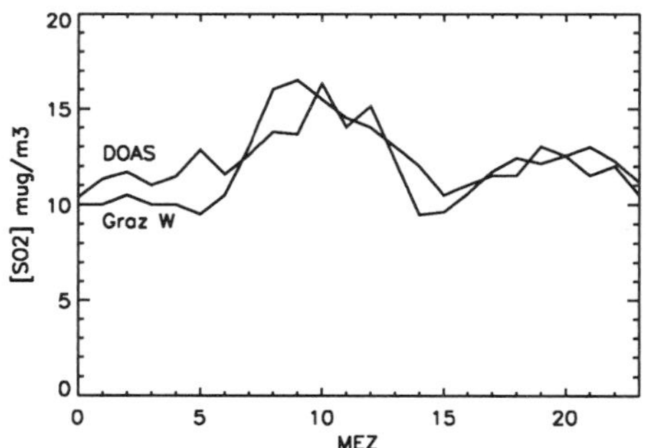

Figure 6: Median of diurnal SO_2 variations for the winter period (Nov. 97 – Feb. 98).

5. Summary and Conclusions

DOAS measurements have been performed in several campaigns in Graz between July 1996 and July 1998 and compared to ground level measurements at different altitudes. The comparisons for ozone show that the differences of the concentrations measured by the DOAS system 70 m above ground and ground-level concentrations clearly correlate with atmospheric stability categories, but no simple method has been found to determine the relation between the ozone concentrations under all conditions. If information like the time of the day, the time of the year and the synoptic meteorological situation is added, the ground-level situation can be estimated: DOAS data describes the summertime ozone maxima at the surface very well because they coincide with well mixed conditions. At stable conditions the ozone concentrations at 70 m above ground

are much higher than at ground level. It has been shown that beside the ozone destruction by NO from ground-level sources, dry deposition at the surface is an important reason for this vertical distribution of ozone.
In summer the NO_2 concentrations 70 m above ground are representative for the ground level, except for the morning maximum. In winter the NO_2 concentration 70 m above ground amount for 73% of the ground level values in average. SO_2 can be regarded as well mixed in the lowest part of the urban boundary layer even in winter. The concentrations 70 m above ground are representative for the ground level concentrations all day.
These results show that the often made assumption that concentrations of pollutants are well mixed below the top of the mixed layer during daytime (e.g. Hayden *et al.*,1997, Pisano *et al.*,1997) is an oversimplification. Not only the amount of vertical mixing in the boundary layer, but also the lifetime of the pollutant near the ground has to be taken into account.

Acknowledgements

The project was supported by the Styrian government and the city council of Graz, Austria.

References

Baumgartner, D., Gobiet, A., Maderbacher, R., Pietsch, H. and Putz, E.: 1999, Improving the air pollution monitoring system in Graz, Austria by additional DOAS measurements. Proceedings of EUROTRAC symposium '98. Vol. 2: 894–899.

Fachnormenausschuß Luftreinhaltung: 1996, Ausbreitung von luftverunreinigenden Stoffen in der Atmosphäre. ÖNORM M9440. Österreichisches Normungsinstitut, Wien.

Günsten, H., Heinrich, G., Mönnich, E., Weppner, J., Cvitaš, T., Klasinc, L., Varotsos, C. A. and Asimakopoulos, D. N.: 1997, Thessaloniki 91 field measurement campaign-II. Ozone formation in the Greater Thessaloniki Area. *Atmospheric Environment*, **37**(8): 1115–1126.

Hayden, K. L., Anlauf, K. G., Hoff, J. W., Strapp, J. W., Bottenheim, J. W., Wiebe, H. A., Froude, F. A., Martin, J. B., Steyn, D. G. and McKendry, I. G.: 1997, The vertical chemical and meteorological structure of the boundary layer in the Lower Fraser Valley during Pacific 93. *Atmospheric Environment*, **31**(14): 2089–2105.

Pisano, J. T., McKendry, I., Steyn, D. G. and Hastie, D. R.: 1997, Vertical nitrogen dioxide and ozone concentrations measured from a tethered balloon in the Lower Fraser Valley. *Atmospheric Environment*, **31**(14): 2071–2078.

Platt, U.: 1994, Differential Optical Absorption Spectroscopy (DOAS). In: Air Monitoring by Spectroscopic Techniques. Sigrist, M. W. (ed.). John Wiley & Sons Inc., New York.

Stutz J. and Platt, U.: 1996, Numerical analysis and estimation of the statistical error of differential optical absorption spectroscopy measurements with least-squares methods. *Applied Optics*, **35**(30): 6041–6053.

A NEW ALL SEASON PASSIVE SAMPLING SYSTEM FOR MONITORING OZONE IN AIR

HONGMAO TANG* and THOMAS LAU
Maxxam Analytics Inc. 9331 48 Street, Edmonton, AB T6B 2R4 Canada
E-mail: htang@edm.maxxam.ca
*To whom correspondence should be addressed

Abstract. A new all season passive sampling system for monitoring O_3 in the atmosphere has been developed in the laboratory and validated in the field. The unique features for this system include a newly designed passive sampler and a rain shelter, which allow the passive sampler to be installed in the field facing downwards. An equation associated with meteorological parameters is used to calculate the passive sampling rates. This system has been extensively tested in the lab (temperature from –18 to 20°C, relative humidity from 13 to 81%, and wind speed from 0.5 to 150 cm/s) and validated in the field in climates of all seasons. The accuracy of the ozone concentrations in the atmosphere obtained with the use of the new passive sampling system was higher than 85% compared to those obtained with continuous ozone analyzers. The new ozone passive sampling system can be used to measure ambient O_3 concentrations ranging from 3 ppb to 1000 ppb based on one-day exposure and 0.1 ppb to 140 ppb for a monthly exposure period. It is also reasonable to conclude that the new passive sampling system can be used for eight-hour exposure study because of the low field blanks and high sampling rates.

Key words: ozone, passive sampler, air pollutant, atmosphere

1. Introduction

Ozone is the most insidious and ubiquitous air pollutant affecting ecological systems and causing health problems for humans and animals in the world. Therefore it is required by regulation to be monitored in many countries. In the assessment of research needs, USEPA has recently identified that rural ozone monitoring network and test value in ecosystem studies of passive monitors should be conducted (Heck et al. 1998).

Over the past few decades, many passive-sampling methods for monitoring O_3 in air have been developed (Monn and Hangartner 1990, Grosjean and Hisham 1992, Kanno and Yanagisawa 1992, Koutrakis et al. 1993). Compared to active samplers, passive samplers do not request electricity; therefore, they are cost-effective. The existing passive samplers developed by different groups are using a fixed sampling rate, and there are no systematic studies for being used outdoors at all seasons.

A key parameter related to correct measurement of O_3 in air using a passive sampler is it's sampling rate. The sampling rate is affected by many factors such as temperature, relative humidity (RH), wind direction, wind speed (WSP), sampler's structure, collection media, etc. It would be highly unreasonable to expect that the sampling rate of a passive sampler would not

Environmental Monitoring and Assessment **65**: 129–137, 2000.

vary from –30 to 30°C and from 90% to 15% RH. Further studies are needed to find the relation between a sampling rate and atmospheric conditions.

It is more difficult to use passive samplers outdoors than indoors Because of highly variable temperatures, relative humidity, wind direction, wind speed, rain, snow, dust, etc. Therefore, special designs of passive sampling systems for outdoor use are needed. Some passive samplers are using an open-face cap to allow ozone to diffuse into the sampler and the ozone is then collected by a collection medium (Ogawa & Company 1994). There are two potential problems for using this type of ozone sampler: dust interference and effect of wind speed.

Although many chemicals can be used to collect ozone, studies have shown that sodium nitrite is a better one (Zhou and Smith 1997). Nitrate, a product from the reaction of ozone and nitrite, is used to assess the concentration of ozone.

With the support of National Research Council of Canada (NRC) and Alberta Environmental Protection (AEP), a new Maxxam all-season passive sampling system (MAPSS) for monitoring SO_2 and NO_2 have been developed (Tang et al. 1997, 1998). The MAPSS for SO_2 has passed an independent validation conducted by Alberta Research Council (ARC 1998). Excellent results have been obtained (Table 1). The unique features for the MAPSS include:

- A newly designed passive sampler (Figure 1) and a rain shelter (Figure 2), which allow the passive sampler to be installed in the field face downwards. Thus, dust and wind problems can be minimized.
- An equation associated with meteorological parameters which is used to calculate the passive sampling rate.

Table 1
Independent MAPSS SO_2 Passive Sampler Validation Results

Site	Exposure Date*	Passive ppb	Analyzer ppb	Error %
TMK	Jul 31 – Aug 28	1.3	1.4	7
TMK	Nov 1 – Dec 3	1.6	1.6	0
EIMU	Apr 30– May 30	1.8	2.0	10
EIMU	Nov 1 – Dec 3	3.0	3.2	6

*All the studies were conducted in 1997

In this paper, a new Maxxam O_3 passive sampling system (MOPSS) is reported. The MOPSS employs the same approach as the Maxxam SO_2 passive sampling system (Tang et al. 1997). Sodium nitrate was used as a collection medium. The MOPSS was first studied in a chamber at different climates. An equation associated with temperature, RH, and WSP was derived from chamber studies.

The MOPSS has been validated for several months in many Canadian locations since 1998. The study period covered summer, fall, and winter. The equation from laboratory studies was used to calculate sampling rates in each location. The passive sampler results were compared to a co- located continuous O_3 analyzer. Very encouraging results have been obtained. Based on one-month exposure, the MOPSS can be used to measure ambient O_3 concentrations ranging from 0.1 ppb to 140 ppb. Compared to continuous O_3 analyzers, the accuracy is 85% higher.

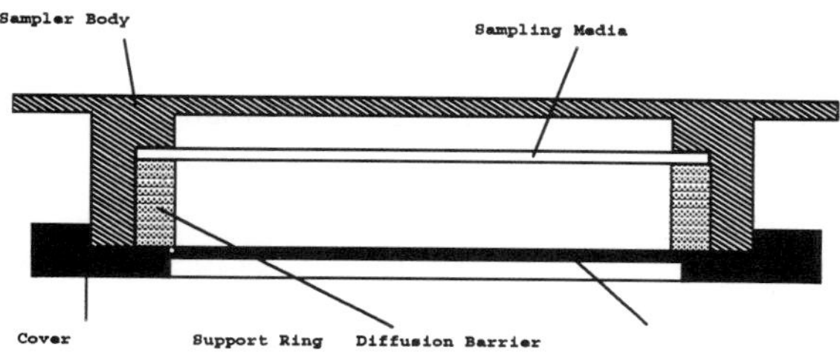

Figure 1. Schematic of the MOPSS Passive Sampler

2. Experimental

2.1 MOPSS PASSIVE SAMPLER

A schematic of the MOPSS passive sampler is shown in Figure 1. It contains a Teflon film as a diffusion barrier and a filter coated with sodium nitrate as a collection medium. An air gap between the diffusion barrier and the collection medium serves as a diffusion zone. The sampler body, support ring, and cover were made from polycarbonate. An edge at the bottom of the passive sampler allows it to be easily installed facing downwards and removed from the MOPSS rain shelter.

2.2 MOPSS RAIN SHELTER

A schematic of the MOPSS rain shelter is shown in Figure 2. It was made from a PVC end-cap for a 6-inch diameter pipe. A slotted plate with three holes was fixed inside the PVC cap. Triplicate passive samplers could be installed in the plate facing downwards. An outside bracket was used to fasten the shelter to fittings.

2.3 CHEMICAL

Chemicals involved in the studies included de-ionized (DI) water, Na_2CO_3, $NaHCO_3$, $NaNO_3$, $NaNO_2$, glycerol, and LiOH (Fisher Scientific, Nepean, CA purified grade).

2.4 TEST PROCEDURE

The MOPSS was first tested in the laboratory using a system previously reported (Tang et al. 1997). An O_3 analyzer (Model 8002, The Bendix

Corporation, Leursburg) was used to continuously monitor O_3 concentrations during the course of the experiment.

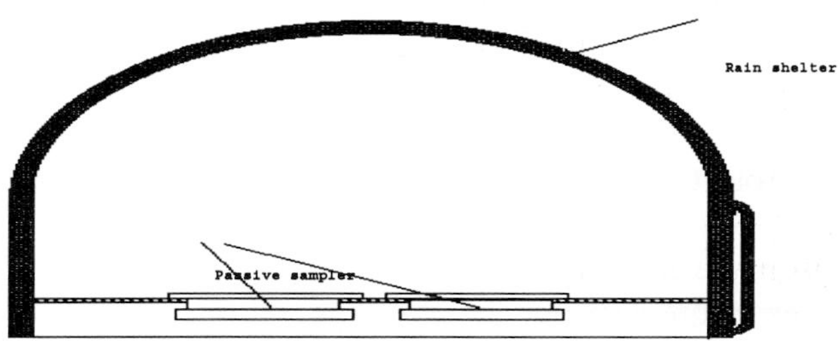

Figure 2. Schematic of MOPSS rain shelter

The MOPSS O_3 passive samplers were exposed in the Maxxam Chamber (Tang et al. 1997) at different conditions. The O_3 concentrations ranged from 20 ppb to 200 ppb, temperatures from –18 to 20°C, RH from 13 to 81%, and face velocities from 0.5 to 170 cm / sec. Triplicate passive samplers and duplicate blanks were studied. The exposure period was approximately 24 hours. After exposure, the medium in the exposed and blank samplers were extracted with DI water and analyzed by ion chromatography.

2.5 FIELD STUDY

The MOPSS passive samplers were installed in six locations in Alberta, Canada. Triplicate passive samplers and duplicate field blanks were used. These stations are equipped with O_3 continuous analyzers (TECO 49, Thermo Environmental Instruments Inc. Franklin, MA), wind speed-monitoring devices (Wind Flo 540, Athabasca Research Ltd. AB), temperature measurement devices (Fluke Model 80TK and 80T-150U, John Fluke MFG CO. INC., Everett, WA), and humidity devices (Vaisula Probe CS500, Vaisula Inc., Woburn, MA).

3. Results and Discussion

3.1 LABORATORY STUDY

A number of studies were conducted in the Maxxam chamber. The test conditions, measured sampling rates, etc., for several studies, are listed in Table 2. The lab triplicate results are very close; average relative standard deviation was 6%. The measured sampling rates range from 43 to 115 ml/min, which reveals the effects of varying temperature, RH, and face velocity on the sampling rate.

3.1.1 *Practical Quantitative Detection Limit*

From laboratory filter blank results, it is found that the pooled standard deviation was 0.6 µg of nitrate per filter based on a 24-hour exposure. The practical quantitative detection limit, thus, can be taken as 6 µg per filter (10 times the standard deviation). This is equivalent to exposure of the passive sampler to 3 ppb O_3 in the atmosphere for 24 hours. If the exposure period were increased to one month (30 days), the method practical quantitative detection limit for O_3 in the atmosphere would be about 0.1 ppb. During the summer, the O_3 concentrations are normally above 30 ppb throughout the day. It is reasonable to conclude that the MOPSS could be used for eight-hour daytime exposure studies.

Table 2
Calculated and Measured R_s Values from Experiments using the Chemex Chamber

Test No.	Temp (°C)	RH (%)	FV (cm/s)	R_s Value (ml/min)		Error %
				Calculated	Measured	
1	-1	13	130	86	85	1
2	-18	19	130	80	81	1
3	19	45	130	103	105	2
4	19	81	130	113	115	2
5	21	19	130	97	100	3
6	19	21	100	89	101	12
7	18	21	40	72	73	1
8	18	20	150	97	103	6
9	20	20	0.5	62	43	44

RH = relative humidity, FV = face velocity, R_s = sampling rate

The MOPSS passive samplers were also exposed for 21 days in the Maxxam chamber at 200 ppb, 130 cm/sec of face velocity, 20°C, and 40% of RH. It was observed that the passive samplers were still not saturated. Therefore, based on a one-month exposure period, it is estimated that the passive sampler can be used to monitor 140 ppb O_3 in the atmosphere.

3.1.2 *Effect of Temperature, Face Velocity, and Relative Humidity*

As shown in Table 2, sampling rates increase with the increase of temperature, wind speed, and relative humidity. For example, at a temperature of –18°C, 19% RH, and 130 cm/sec of face velocity, the measured R_s was 81 ml/min, but at 31°C and 19% RH, the measured R_s was 100 ml/min. The overall R_s increase from –18°C to 21°C was 19 ml/min, which is about a 23% increase. This change is very significant.

Many authors have reported the effect of face velocity on sampling rates (Harper and Purnell 1987). Our studies confirmed findings by these authors. However, we also noticed that even if the face velocity was higher than 100 cm/sec for the MOPSS passive sampler, the sampling rate continuously increased. The threshold face velocity was determined from laboratory studies to be about 130 cm/sec or 4.7 kilometres per hour (km/h).

3.2 EQUATION FOR CALCULATION OF SAMPLING RATE

From the laboratory chamber studies, an equation for calculating the MOPSS passive sampler rates associated with temperature, RH, and face velocity (wind speed) was determined as follows:

$$R_s = 14.8T^{1/2} + 0.259\ RH + 0.275\ WSP - 197 \quad (1)$$

Where R_s = sampling rate, ml/min; T = temperature, K; RH = humidity (%); WSP = windspeed, cm/sec; if WSP>130, then WSP = 130.

The calculated R_ss are shown in Table 2. Compared to measured R_ss, it is seen that the largest percent error is in test 9, in which the face velocity was only 0.5 cm/sec. This test further proved findings by other scientists that a minimum 10 cm/sec face velocity is required for properly operating a passive sampler. Fortunately there are very few cases of such low wind speeds occurring outdoors, and it is therefore unnecessary to worry about such low wind speeds using the MOPSS passive samplers outdoors. The other test errors ranged from 0 to 12%.

3.3 FIELD STUDY

Field study results are listed in Table 3. Table 3 shows the locations, study periods, meteorological conditions, calculated sampling rates, and O_3 concentrations measured by the passive samplers and monitored by the O_3 continuous analyzers and the relative errors in each study. Table 3 clearly shows that the study accuracy is excellent. Field study results indicate that the relative error using the MOPSS passive sampler is within 15% compared to the continuous O_3 analyzer. Figure 3 is a map of Alberta, Canada. All study locations are shown in this map.

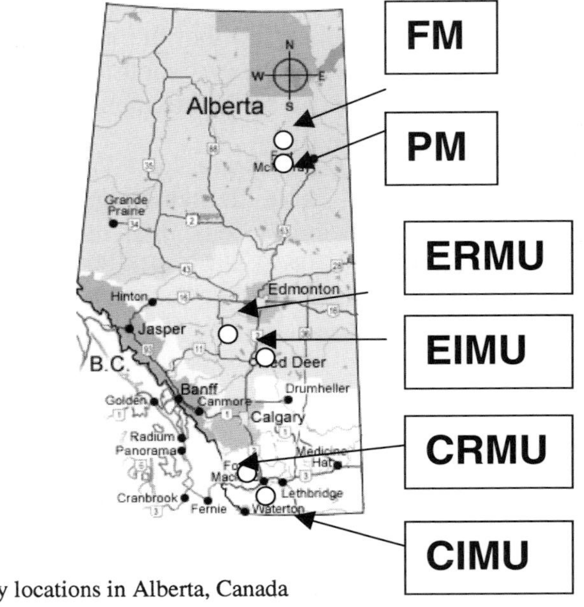

Figure 3. Study locations in Alberta, Canada

Table 3
Field Study Results

#	Location	Day	Date	RH %	T °C	WSP Km/h	R_s cm/min	Ozone ppm Passive	Ozone ppm Analyzer	Error %
1	EIMU	14	10 – 24 Jun	63	17	11	108	22	24	8
2	EIMU	57	10Jun – 5 Aug	69	17	11	109	25	26	4
3	ERMU	2	23 – 25 Jun	60	16	10	107	33	33	0
4	ERMU	23	14 – 30 Dec	70	-20	10	92	19	20	6
5	CIMU	27	4 – 31 Aug	70	20	11	112	24	21	14
6	CIMU	27	4 – 31 Aug	70	20	11	112	28	29	3
7	PM	28	31 Jul – 28 Aug	59	18	10	108	23	24	4
8	FM	31	30Oct– 30 Nov	86	-8	9	103	15	15	0

- EIMU, Edmonton industrial monitoring unit; ERMU, Edmonton residential monitoring unit; CIMU, Calgary industrial monitoring unit; CRMU, Calgary residential monitoring unit; PM, Patricia McLnniss; FM, Ft McKay.

The field study results further prove the importance of Equation 1. Using a fixed sampling rate to conduct calculations in different seasons will cause errors. Study #4, ERMU, is a typical example. Studies #3 and #4 were conducted at the same location but in different seasons. If using Study #3 sampling rates to calculate Study #4 ozone concentrations, the result is 16 ppb. Thus, the relative error compared to the continuous analyzer is 20%, not 6%. When considering the operation of MOPSS at different seasons and different geographical areas in the world, the difference of the sampling rate might be larger.

3.4 SENSITIVITY OF SAMPLING RATE TO METEOROLOGICAL CONDITIONS

The average monthly meteorological parameters used in Equation 1 can be obtained from local weather forecast stations or ECN (1993). It is found that parameters listed in ECN (1993) were close to parameter values monitored by the local weather station based on monthly averages. For example, the temperature for EIMU in Field Study #2 was 17°C obtained from the local weather station; and the temperature listed in ECN was 17.5°C. Even a 5°C decrease in temperature would only reduce the sampling rate from 109 to 107 ml/min (Field Study #2). At different seasons (such as summer and winter in Alberta), or at different locations in the world (such as Florida and Alberta in winter), the temperature difference might be as high as 50°C. This change is very significant. Assuming RH = 60%, WSP = 12 km/h, high temperature is 30°C, and love temperature is –10°C, the sampling rates will change from 111 to 94 ml/min. The difference is 17 ml/min. The change of RH at 5% will not seriously affect the sampling rate either. Let us still using Field Study #1 as an example. If RH changes from 63% to 68%, the R_s changes from 108 to 109 ml/min. The change is only 1 ml/min. If the RH change is large, such as from 63% to 10%, the sampling rate will change from 107 to 95 ml/min. This change is very significant (>12%).

3.5 ACCURACY

Tables 2 and 3 compare the accuracy of the MOPSS with the O_3 continuous analyzer. In both field and laboratory studies, the relative error between the calculated and measured R_s ranged from 1% to 14%.

4. Conclusion

The new Maxxam all season O_3 passive sampling system described in this report is suitable for monitoring O_3 concentrations in the atmosphere for field studies and networks. The unique design of MOPSS minimizes interference of suspended particles on sampling rates. The equation determined from lab and field studies can provide reasonable sampling rates for field measurement. Based on laboratory studies, the MOPSS can be used to measure ambient O_3 concentrations ranging from 3 ppb to 1000 ppb based on one-day exposure and 0.1 ppb to 140 ppb for a monthly exposure period. It is also reasonable to conclude that the MOPSS can be used for eight-hour exposure studies because of the low field blanks and high sampling rate.

Acknowledgements

This study was supported by National Research Council of Canada. Many thanks to Mr. Ray Brassard of Alberta Environmental Protection for technical help. The authors also want to express their appreciation to Dr. Warren B. Kindzierski of the University of Alberta for constructive review.

References

ARC, Alberta Research Council: 1998, Independent Validation of Chemex (Maxxam Analytics Inc.) All Season Passive Sampling System (CSPSS), Agreement #JPD 003.0097, March.

ECN, Environmental Canada: 1993, Canadian Climate Normals 1961 – 1990.

Grosjean, D. and Hisham, Mohamed W.M.: 1992, A Passive Sampler for Atmospheric Ozone, *JA & WMA*, **42,** 169 – 173.

Harper, M. And Purnell, C.J.: 1987, Diffusive Sampling – A Review, *Am. Ind. Hyg. Assoc.,* **48,** 214 – 218.

Heck, W.W., Furiness, C.S., Cowling, E.B. and Sims, C.K.: 1998, Understanding the Effects of Ozone on Ecological Systems: Research Needs Identified, *EM* **October,** 11 – 22.

Kanno, S. and Yanagisawa, Y.: 1992, Passive Ozone Oxidant Sampler with Coulometric Detection Using 12/Nylin – 6 Change – Transfer Complex, *Env. Sci. Technol.,* **26,** 744 – 769.

Koutrakis, P., Wolfson, J.M., Bunyaviroch, A., Froehlich, S.E., Hirono, K. and Mulik, J.D.: 1993, Measurement of Ambient Ozone Using a Nitrite Coated Filter, *Anal. Chem.,* **65,** 209 – 214.

Monn, C. and Hangartner, M.: 1990, Passive Sampling for Ozone, *JA&WMA,* **40,** 357 – 358.

Ogawa & Company, USA, Inc.: 1994, Air Sampling Device.

Tang, H., Brassard, B., Brassard, R., and Peake, E.: 1997, A New Passive Sampling System for Monitoring SO_2 in the Atmosphere, *FACT,* **1(5),** 307 – 314.

Tang, H., Lau, T., Brassard, B., and Cool, W.: 1998, A New All-Season Passive Sampling System for Monitoring NO_2 in Air, *Proceedings, AWMA 91st Annual Meeting,* San Diego, CA, June 14 – 18, Paper #98-TP44.03.

Zhou, J. and Smith, S.: 1997, Measurement of Ozone Concentrations in Ambient Air Using a Badge-Type Passive Monitor, *JA&WMA,* **47,** 697 – 703.

SPATIAL AND TEMPORAL VARIATION OF OZONE CONCENTRATIONS AT HIGH ALTITUDE MONITORING SITES IN GERMANY

TREFFEISEN R. and HALDER M.
Institut für Technischen Umweltschutz, Fachgebiet Luftreinhaltung, Technische Universität Berlin, Straße des 17. Juni 135, D - 10623 Berlin

Abstract. The present study deals with the characterisation and interpretation of hourly taken maximum ozone concentrations of each day at various high altitude monitoring sites in Germany and contributes to the understanding of ozone transport processes. The relation between long range transport of ozone as well as its precursors and high ozone concentrations is of special interest. Aim of this paper is to investigate the influence and importance of large scale meteorological circulation and source regions of anthropogenic ozone precursors to local ozone concentrations using 2-dimensional backward trajectories. Further, reasons for the spatial and temporal variation of ozone concentration levels will be shown. Investigating numerous cases a similar origin of air masses causing high ozone concentrations in Germany and uniform patterns at various sites could be identified.

Key words: higher altitude monitoring sites, high ozone concentration, trajectory, spatial and temporal patterns

1. Introduction

The supporting meteorological conditions to ozone formation covering a great part of Europe and accumulation of ozone in the boundary layer took place during high pressure situations (Guicherit, 1977; Eder, 1994; Beilke, 1991). During those periods of specific synoptic conditions photochemical air pollution occurs simultaneously over extensive areas of Western Europe and describes a great part of the ozone variations. A second reason of the ozone variation is related to transport processes from high polluted areas (Cox, 1977; TOR, 1997). Apart from certain local productions it must be concluded that, due to the transport, precursors are distributed so widely over Europe that increasing ozone concentrations occur over large areas when meteorological conditions are favourable to smog formation. The investigation of these influences with regard to spatial and temporal patterns for Germany are the main concern of the present study.

2. Data base

High altitude monitoring locations are suitable for analysis of long range transport phenomena. Therefore 13 stations of different monitoring networks distributed over whole Germany were analysed in the time period from 1994 - 1997 (April - September) in order to investigate if they fulfil the criteria of high altitude monitoring locations. Some of them were also used within other

Environmental Monitoring and Assessment **65**: 139–146, 2000.

investigation programs (Speth,1994; TOR, 1997). The data set considered consists of hourly average concentrations of ozone, nitrogen oxides and meteorological parameters such as temperature and wind data. Ozone concentrations and nitrogen oxides were measured using the UV-absorption spectroscopy and the chemiluminescence technique, respectively. Figure 1 shows the location of the selected sites. A great geographic coverage was realised. The grey shaded areas indicate industrial and high populated regions in Germany. In order to investigate marine influences we also included to the data base the UBA monitoring site *Westerland*. The site of *Frohnau* is located in the north of Berlin on the top of a 324 m high tower (lattice type tower). Because of the formation of nocturnal inversions it is frequently isolated from the stable nocturnal boundary layer. Since measurements of temperature are not available at this site the nearby site of Grunewald was used instead.

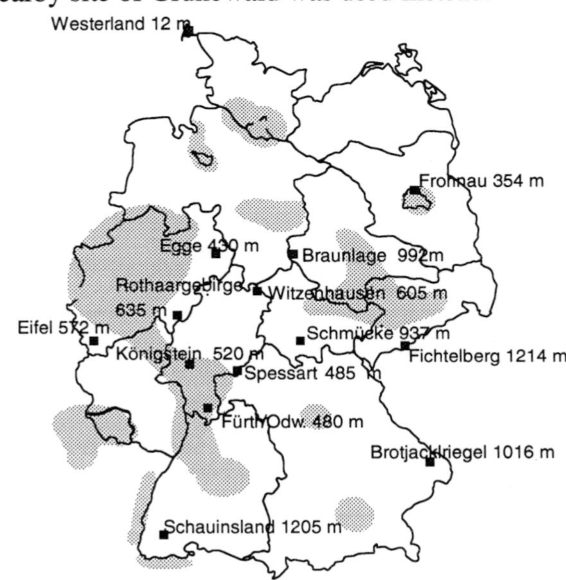

Figure 1. Geographical location of all investigated monitoring sites

With the aid of monitoring site characterisation, in some cases it could be shown that regional influences due to emission areas and local topography exist and therefore some monitoring sites showed a slight mean diurnal variation of the ozone profiles. It turned out that different characterisation techniques yielded similar patterns for each site. The characterisation of the used monitoring sites showed that all sites fulfilled the criteria of high altitude monitoring sites. Therefore they can be applied for long range transport investigations. The results were in accordance with other studies (Angle, 1986; Dietze, 1993; Wunderli, 1990; Zaveri, 1994).

The investigation of the air masses was done by calculating 2 dimensional backward trajectories over the time scale of 96 h by evaluating three hourly taken wind measurement data from 1.600 weather stations in Europe for the time period 1994 - 1997 (Israël, 1992). The time scale of 96 h was chosen according to the work of (Stedman, 1991) and (Lindskog, 1997). Each trajectory was computed backward in time until either 96 h had elapsed since the trajectory began or wind data were not any longer available. Starting point of the backward trajectories was the respective monitoring site. The calculated hourly backward trajectories at each station were saved in a data bank for better handling. In this way an extensive data base of trajectories was available for the present study.

3. Results and Discussion

Independent of the distance between the individual stations they all showed a similar course in the diurnal maximum ozone concentration and the diurnal maximum temperature values (figure 2). The vertical lines always indicate a change in the large scale meteorological weather situation. In order to identify the main synoptic meteorological situation a daily classification according to the Deutscher Wetterdienst was used (Amtsblatt DWD). The abbreviations are given above the x-axes. The ozone levels at various sites within Germany varied almost simultaneously to the meteorological conditions, even though the absolute values may be considerable different. High values of ozone tend to occur concurrently, or within a range of 1-2 days. This behaviour was found during the whole time of investigation indicating that the variation was associated with large scale meteorological fluctuations.

This suggests that on a time scale of several days synoptic scale motions covering Germany affect the O_3 concentrations as a Germanwide phenomenon and can probably be explained by variables that are common to all stations. As expected the ozone profiles at all investigated sites followed clearly the temperature profiles, indicating a close relationship.

The correlation coefficient R for the linear relationship between the diurnal maximum ozone concentration and the diurnal maximum temperature values ranged between 0.67 and 0.77 (N = 732; significant with 0.000 based on a t-test). A further analysis was performed to evaluate if diurnal maximum ozone values at various sites were correlated with each other, as suggested by the ozone profiles (figure 2). The site to site correlation of the ozone values ranged from R = 0.51 to 0.90 (N = 732; significant with 0.000 based on a t-test).

The analysis of case A and B in figure 2 are specified in figure 3a and figure 3b by depicting the respective trajectories. For a given day the time of the highest hourly ozone concentration was used as the starting time of the trajectory calculation. For case A a fast increase of the ozone values within a few days and high maximum ozone concentration occurred when a high pressure system was located over Germany (figure 3a). Hourly ozone concentrations were recorded in the range 200 μg/m³ to 260 μg/m³. The movement of the trajectories was almost circular over Germany due to the high pressure system and was accompanied with a decreasing transport velocity of the air masses. Due to the mentioned high pressure system trajectories were fixed over the north sea and the northern and central part of Germany. The backward trajectories for 9th of May (case B) are shown in figure 3b. In the latter case the weather was not any longer conducive to the ozone formation. The trajectories showed an uniform transport direction from north west with high transport velocities.

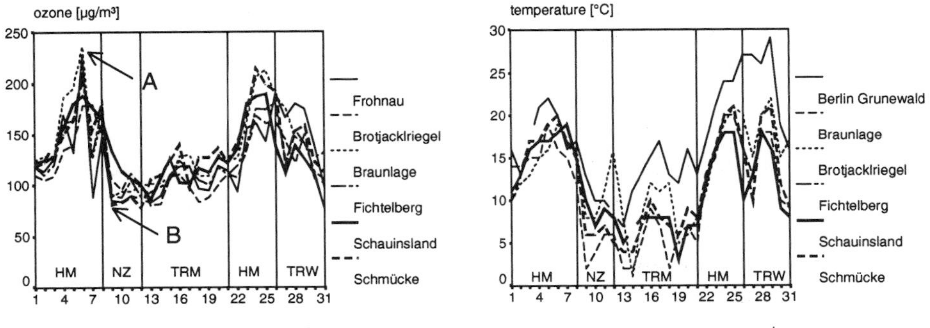

Figure 2. Profiles for the diurnal maximum ozone concentration and the diurnal maximum temperature values at six selected sites (May 1995); with: HM = closed high over Middle Europe; NZ = low pressure over Middle Europe with predominantly northerly flows; TRM = trough of low pressure over Middle Europe; TRW = = trough of low pressure over West Europe;

Analysing comparable meteorological situations it could be demonstrated, that high ozone concentrations (≥ 180 µg/m³) occurred when the trajectories pass over more or less polluted and industrial areas and concomitantly the transport velocity decreased. According to the trajectory analysis air pollutants emitted in this areas can be a key factor leading to higher ozone concentrations. In addition it was demonstrated that meteorological conditions led to a similar ozone concentration level all over Germany.

In certain cases the variability of the ozone concentration between different stations was large. For two sites, one located in the south-west of Germany (dotted line) and one in central Germany (solid line) the maximum ozone concentrations of each day of June 1994 are drawn in figure 4. In both cases (indicated in figure 4) a great difference of the maximum ozone concentrations was found (100 µg/m³ and 70 µg/m³ in case C and in case D, respectively). Both situations were characterised by almost the same temperature conditions. To get a better understanding of the two situations the respective trajectories are depicted in figure 5a and figure 5b.

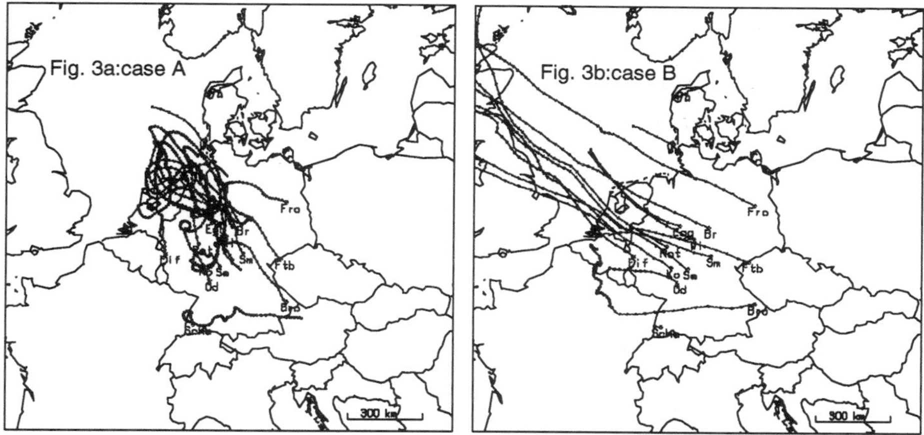

Figure 3a - 3b: Trajectories for 6th May 1995 and the 9th of May

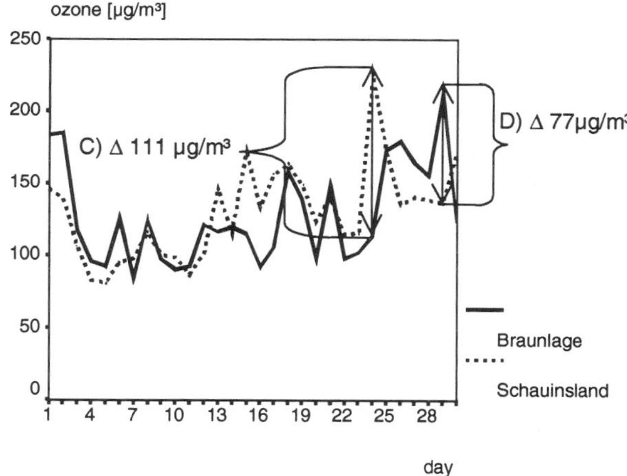

Figure 4: Ozone profile for June 1994 at two selected sites

In figure 5a the trajectories of the 24th of June 1994 are illustrated. In contrast to the central monitoring site *Braunlage* (BRA) the trajectory at the site *Schauinsland* (SCH) in the south-west of Germany traversed source regions with high estimated emissions of nitrogen oxides and organic precursors (e.g. like the south and south west of Germany, Frankfurt area or the Ruhrgebiet). Therefore the hourly maximum ozone concentration measured on this particular day could be explained by highly polluted air masses which reached the station. In figure 5b the situations is reversed. The trajectory of *Schauinsland* did not cross any main industrial areas but the trajectory of the site *Braunlage* showed a circular motion over the Ruhrgebiet. This led to a very high hourly maximum ozone concentration on the 29th of June. In agreement with figure 3 it is obvious that beside the meteorological conditions high ozone concentrations (≥ 180 µg/m³) were associated with the previous path of the air parcels. With respect to the understanding of high ozone concentration particular situations occurred when air masses had passed over industrialised and urban areas. Beside the spatial differences of the ozone concentration the day to day variability was also analysed.

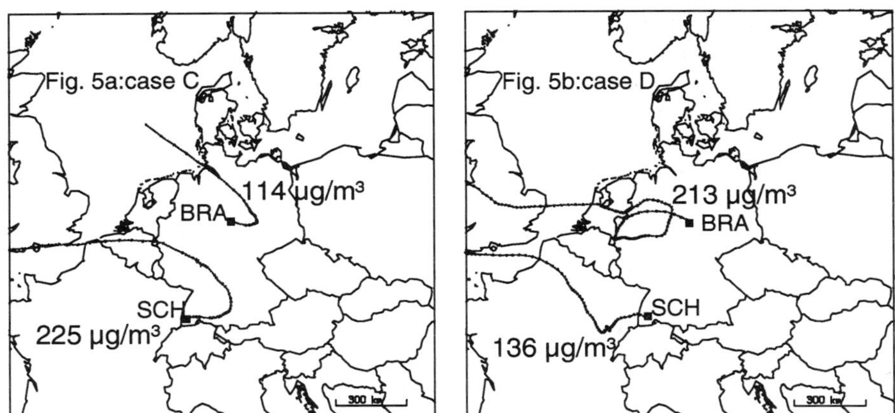

Figure 5a - 5b Trajectories for case C the 24th June 1994 on the left and case D the 29th June 1994 on the right

Temporal variation is a combination of meteorology and the trajectory path over high emission areas and can change within a few days. In order to illustrate this a further example is demonstrated. In figure 6a the trajectories of the time period of 11th to 14th August 1995 are shown for the monitoring site *Braunlage*. The analysed period started on August 11th at the site *Braunlage* with a recorded hourly maximum ozone concentration of 160 µg/m³. Two days later on August 13th the hourly maximum ozone concentration reached 268 µg/m³. As demonstrated by figure 6a the increased ozone concentration is caused by air parcels crossing the industrial areas of south east and central Germany with a low transport velocity and very high temperatures (above 30°C). Cold weather conditions from the north west combined with rain brought clear air from north west and caused a low hourly maximum ozone concentration on August 14th. Almost the same situation occurred two years later. The temperature was also about 30°C during the first two days (figure 6b).

In a similar manner to that described before, all situations with high ozone concentrations were investigated in detail as well as special cases were correlation between the temperature and ozone values showed discrepancies.

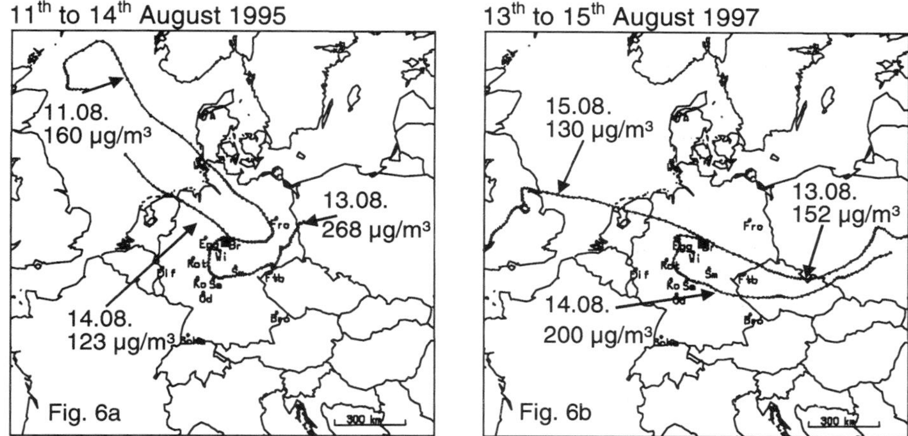

Figure 6a – 6b. Two examples for a fast temporal variation of ozone concentration at one selected monitoring sites (*Braunlage*) for different time periods

In the majority of the cases we found that beside the overall air flow and the meteorological conditions the source regions crossed by the trajectories played an important role for the occurrence of locally increased ozone concentrations. The investigation of numerous cases proved a similar origin of air masses causing high ozone concentrations.

4. Summary

Ozone measurements at 13 higher elevations sites in Germany have been analysed. In order to characterise and compare the ozone concentrations 2-dimensional trajectories were calculated. Meteorological conditions have a decisive impact on surface ozone concentrations but meteorology cannot entirely explain very high maximum ozone concentrations. The origin of air masses play an important role for very high ozone concentrations. The influence of the trajectory path under favourable meteorological situations could be demonstrated. As expected the occurrence of the highest ozone concentrations are explained directly by conditions when air masses originating from Eastern Europe have reached the monitoring sites and high temperatures are measured. Beside the discernible easterly flow the trajectories are frequently anticyclonically bented. In addition they are often associated with air masses arriving from areas which agree reasonably well with those having high anthropogenic emissions of precursors.

Further studies are planned to investigate the trajectory density in order to analyse the trajectories on a statistical base. Moreover the contribution could be quantified of the source regions to local ozone concentrations. In addition the meteorological influence is going to be quantified with the method of spectralanalysis. Our results are expected to be useful for understanding ozone concentrations patterns in Germany and are a potential tool for source apportionment of ozone.

Acknowledgement

We would like to thank all local authorities for providing us with the necessary data. As well we like to thank all students for their help to finish this study. The authors are grateful to Dr. Frenzel for the helpful suggestions. The authors would also like to thank for the helpful and successful remarks of the reviewers.

References

Angle, R., Sandhu H.: 1986, Rural ozone concentrations in Alberta, Canada, Atmospheric Environment, Vol. 20, No. 6, pp. 1221-1228.

Amtsblatt DWD (Deutscher Wetterdienst): 1994-1997, Die Großwetterlagen Europas.

Cox, R., Derwent, R: 1977: Long-range transport of photochemical ozone in north-western Europe. Nature, Vol. 255, May 8. pp. 118 – 121.

Beck, J; Kirchner,E: 1997, Continental ozone issues: monitoring of trace gases, data analysis and modelling of ozone over Europe, in: Transport and chemical transformation of pollutants in the troposphere, Volume 6, Tropospheric ozone research. pp. 448-454.

Beilke, S.: 1991., Meteorologische Voraussetzungen für die Bildung von Ozon und Sommersmog. In: BMU, StMLU 1991: Ozon-Symposium in München 2. – 4. Juli 1991. Handbuch zum gleichnamigen Symposium.

Dietze, G.: 1993, Zu Ursachen für Unterschiede zwischen mittleren Tagesgängen der Ozonkonzentration in Bodennähe, Meteorologische Zeitschrift, N.1, pp. 232-238

Eder,B., Davis, J.: 1994, An automated classification schema designed to better elucidate the dependence of ozone on meteorology. Journal of Applied Meteorology. Vol. 33. pp. 1182-1199.

Guicherit R., Van Dop H.: 1977, Photochemical production of ozone in Western Europe (1971-1975) Atmospheric Environment Vol. 11. No 2. pp 145 – 155.

Israel et al.: 1992, Analyse der Herkunft und Zusammensetzung der Schwebstaubimmission, Fortschrittsberichte VDI Nr. 92, Reihe 15 Umwelttechnik, VDI Verlag Düsseldorf

Lindskog A., Mowrer J., Moldanova J.: 1997, Long-range transport in relation to oxidant occurrence and formation: results from the TOR site at Rörvik, Sweden, in: Transport and chemical transformation of pollutants in the troposphere, Volume 6, Tropospheric ozone research. pp. 333-339.

Stedman J, Williams l.: 1991, A trajectory model of the relationship between ozone and precursor emissions, Atmospheric Environment Vol. 26A, No. 7, pp 1271-1281.

TOR: 1997, Transport and transformation of ozone. Chapter 9. In: Tropospheric Ozone research; Volume 6. Springer Verlag.

Wunderli, S., Gehrig, R., 1990, Surface ozone in rural, urban and alpine regions of Switzerland, Atmospheric Environment Vol. 24A, No. 10, pp 2641-2646.

Zaveri, R., Saylor, R., Peters l., 1994, A model investigation of summertime diurnal ozone behaviour in rural mountainous locations, Atmospheric Environment Vol. 29, No. 9, pp 1043-1065.

SEASONAL DIFFERENCES IN THE LEVELS OF GASEOUS AIR POLLUTANTS IN THE VICINITY OF A WASTE DUMP

V. VAĐIĆ, J. HRŠAK, N. KALINIĆ, M. ČAČKOVIĆ and K. ŠEGA
Institute for Medical Research and Occupational Health
10001 Zagreb, Ksaverska cesta 2, Croatia

Abstract. Hydrogen sulphide, ammonia, nitrogen dioxide, mercaptans and sulphur dioxide (H_2S, NH_3, NO_2, R-SH, SO_2) concentrations were measured at the location in the vicinity of the waste dump to determine the air pollution level of these pollutants prior to the operation of the Mobile Thermal Treatment Plant. Samples were collected over one year period. Seasonal differences, and the influence of meteorological parameters (temperature, relative humidity, pressure and wind direction) on the air pollution levels were studied. Results show relatively low concentrations of H_2S, NO_2, R-SH and SO_2, while NH_3 levels were higher compared to the guideline values. Good weather conditions (high air pressure and low relative humidity) are connected to long range transport of NO_2, while higher temperatures result in elevated NH_3 and R-SH concentrations. Because of the predominant northeast wind direction (the same as the waste dump direction), the contribution of air pollution from the direction of the waste dump at the measuring site is significant, but that does not necessarily mean that the pollutants originated from that source.

Key words: hydrogen sulphide, ammonia, nitrogen dioxide, mercaptans, sulphur dioxide, waste dump, meteorological parameters

1. Introduction

A waste dump located on the right bank of a river and close to a large city possess immediate danger of the groundwater pollution in a water-protected region. A decision was made to stop further aggravation of the present situation and to gradually remedy the entire waste dump by installing a Mobile Thermal Treatment Plant (PUTO). The objective of the air quality monitoring of general and specific air pollutants in the nearby village was to evaluate the impact of the city waste dump on the air quality before the introduction of PUTO.

2. Materials and Methods

This paper presents the results of the monitoring campaign which lasted from October 1997 until September 1998 Samples of hydrogen sulphide, ammonia, nitrogen dioxide, mercaptans (R-SH) and sulphur dioxide were collected over 24-hour period. Measuring site was located several hundred metres south-west from the waste dump, in the opposite direction from the prevailing wind direction. It could be assumed that the winds from 1st quadrant (N-E) bring the air mass directly from the waste dump, influencing the air quality at the sampling site.

Environmental Monitoring and Assessment **65**: 147–153, 2000.

Simultaneously meteorological parameters like average air temperature, air humidity and pressure were recorded on daily basis. Ammonia and sulphur dioxide (ISO-4221, 1980) samples were collected in glass bubbles from 1.9-2.1 m^3, while hydrogen sulphide (ISO-4219, 1979; Vađić *et al.*, 1980) and mercaptans (Moore *et al.*, 1960) were collected on impregnated filters. Nitrogen dioxide (ISO-6768, 1985) was collected by means of passive samplers. Samples were analysed by means of spectrophotometric method.

3. Results

Table 1. shows summary results of pollutant concentrations for whole year and different seasons. Two seasons were chosen, depending on climatological characteristics, the heating or cold season (November-March) and the warm season (May-September). Because of their climatological characteristics (unstable and unpredictable weather) April and October were not included in the seasonal analysis of the results.

Table I
Pollutant concentrations - summary results ($\mu g\ m^{-3}$)

	N	C_{avg}	C_{50}	C_{98}	C_{max}
		October 97 - September 98			
H2S	339	0.90	0.65	3.20	7.00
NH3	327	38.3	28.6	166.3	310.6
NO2	348	28.8	26.7	63.5	88.9
R-SH	340	0.402	0.258	1.486	2.681
SO2	324	4.05	2.72	21.13	48.04
		November - March			
H2S	134	1.03	0.72	5.31	5.44
NH3	136	21.2	19.8	52.2	83.0
NO2	135	32.0	30.3	65.1	75.9
R-SH	136	0.270	0.175	1.171	1.328
SO2	133	5.57	4.62	28.12	48.04
		May - September			
H2S	148	0.89	0.65	3.08	7.00
NH3	141	47.3	41.6	166.3	310.6
NO2	152	25.8	23.0	63.5	88.9
R-SH	148	0.441	0.301	1.486	2.681
SO2	139	2.25	1.64	9.22	10.31

N - number of samples C_{avg} - average value C_{50} - median
C_{max} - maximum value C_{98} - 98th percentile

Seasonal differences between pollutant concentrations were tested by means of one-tailed *t*-test and the results are shown in Table 2. The existence of seasonal difference for H_2S concentrations was not expected, so the two-tail test was used.

4. Discussion

The H_2S levels were below recommended air quality value (C_{avg} 2 μg m^{-3}, C_{98} 5 μg m^{-3}) given by the Ordinance on Recommended and Limit Air Quality Values (Ordinance) based on the Croatian Law on Air Quality Protection. Winter concentrations were not significantly higher compared to those during summer.

Table II
Seasonal differences in pollutant levels

	t	P(T<=t)	
H_2S	0.414	0.6787	two-tail
NH_3	**-8.060**	**0.0000**	one-tail
NO_2	1.810	0.0357	one-tail
R-SH	**-4.313**	**0.0000**	one-tail
SO_2	**5.048**	**0.0000**	one-tail

Ammonia concentrations were higher than recommended (C_{avg} 30 μg m^{-3}, C_{98} 100 μg m^{-3}) but below the limit values (C_{avg} 70 μg m^{-3}, C_{98} 250 μg m^{-3}) as expected in rural environments and showed significant seasonal difference. It could be explained by seasonal difference in temperature and elevated formation and evaporation during summer. Average concentration of nitrogen dioxide was lower compared to the recommended value (C_{avg} 40 μg m^{-3}) and the 98th percentile was slightly higher (C_{98} 60 μg m^{-3}), but both values were well below the limit values (C_{avg} 60 μg m^{-3}, C_{98} 120 μg m^{-3}). Seasonal difference was slightly pronounced. Mercaptans concentrations were lower compared to the limit values (C_{avg} 1 μg m^{-3}, C_{98} 3 μg m^{-3}). Seasonal difference in concentration was significant, showing higher concentrations during summer, as a consequence of higher temperatures. Concentrations of SO_2 were much lower compared to the recommended values (C_{avg} 50 μg m^{-3}, C_{98} 125 μg m^{-3}). Although quite low compared to limit values, winter concentrations were significantly higher compared to those measured during summer. It could be concluded that SO_2 is a product of space heating and long-range transport from the city.

Obtained concentration data, together with meteorological data, were subjected to the factor analysis in order to define virtual variables which will describe their interdependence. Factors were based on principal component extraction and rotated by normalized varimax method. Three significant factors were extracted and their loadings, as well as the percent of the explained variance are presented in Table III. Factor loadings greater than 0.5 are considered significant. The total variance explained by three factors accounted 57.1%.

Table III
Factor loadings and explained variance

	Factor 1	**Factor 2**	**Factor 3**
H_2S	0.1473	0.0453	**0.6118**
NH_3	**0.8054**	-0.0673	0.0696
NO_2	0.1206	**0.6021**	0.1501
R-SH	**0.7032**	0.0591	0.1034
SO_2	-0.0492	0.3850	**0.6011**
T	**0.6309**	-0.1130	**-0.5251**
RH	-0.0880	**-0.6183**	**0.5964**
P	-0.3349	**0.7276**	0.1234
Prp.Totl	0.2125	0.1807	0.1775

The first factor describes the strong influence of elevated temperature to the ammonia and mercaptans formation and evaporation. The significant seasonal differences in concentrations of these pollutants lead to the same conclusion. The second factor could be described as the influence of good weather on NO_2 levels (high air pressure and low relative humidity). During such weather conditions there is no washout in the atmosphere so the pollutant accumulation and its long-range transport are quite possible. The third factor is quite complex and hard to explain. It describes well-known association between SO_2 and temperature. The strong bind of relative humidity and H_2S concentrations to this factor, although correlation between these two parameters is not significant, could partially be explained by the fact that the efficiency of the sampling method using impregnated filters is strongly dependent on air humidity.

Average pollutant concentrations and the overall contribution of pollution (average concentration * average time percentage) depending on wind direction are shown in Figures 1-2. In Figure 3. relative wind direction frequencies during the measuring period are presented. It could be seen that there are no elevated concentrations when wind blows from the direction of the waste dump, but that some relatively strong local source of mercaptans and hydrogen sulphide do exist in the SSE direction (village). Because of the predominant northeast wind direction, the contribution of all pollutants from the direction of the waste dump at the measuring site is significantly higher compared to other directions, but that does not necessarily mean that the pollutants originated from that source.

Average pollutant concentrations

Overall pollutant contribution

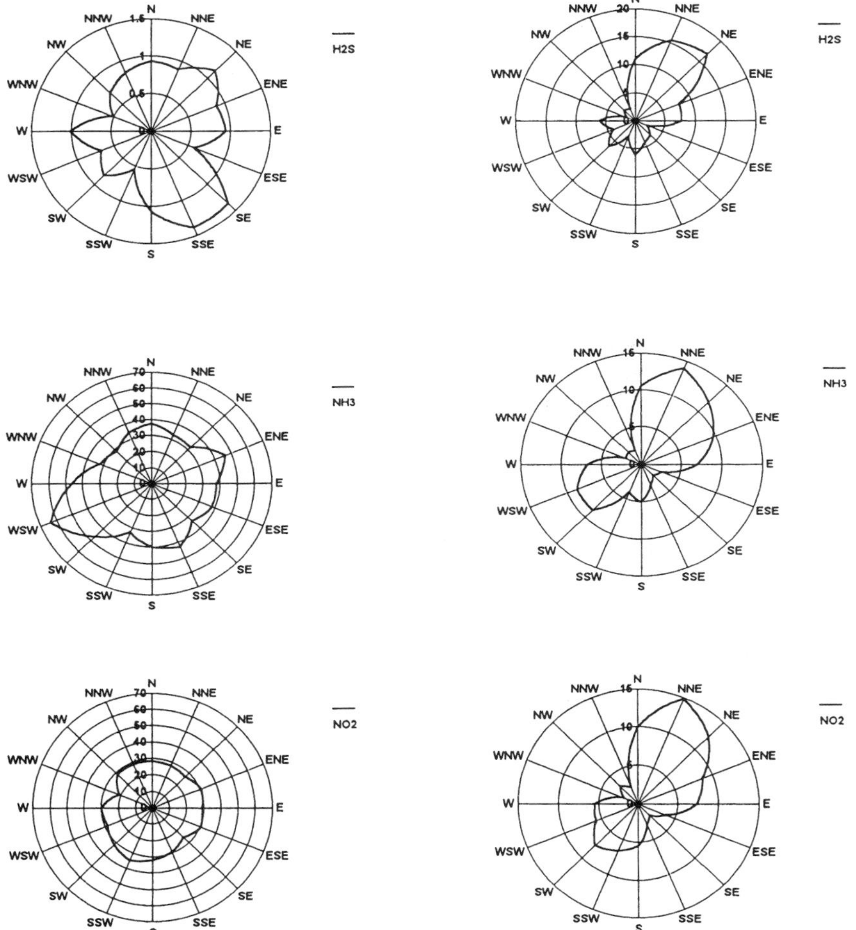

Figure 1. Average H_2S, NH_3 and NO_2 concentrations ($\mu g\ m^{-3}$) and average pollutant contribution in relation to the wind direction (%)

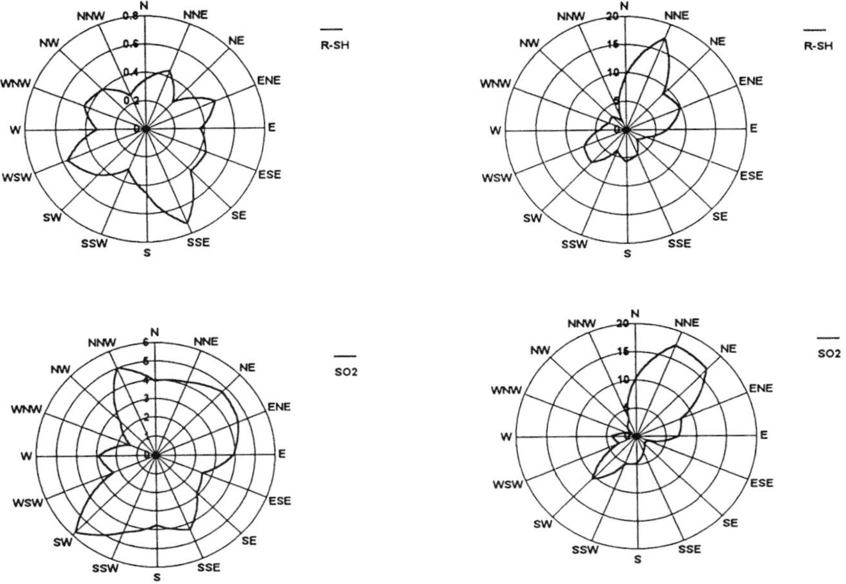

Figure 2. Average R-SH and SO_2 concentrations (µg m^{-3}) and average pollutant contribution in relation to the wind direction (%)

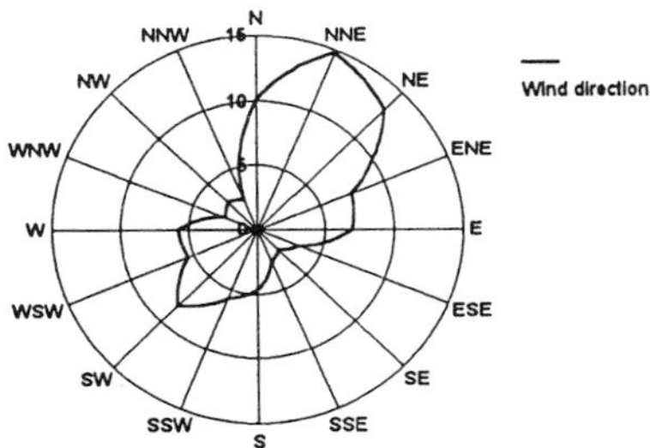

Figure 3. Frequencies of wind direction during the measuring period (%)

5. Conclusions

Only ammonia concentrations were higher compared to the proposed limit values, while concentrations of other pollutants were relatively low.

Good weather conditions are connected to NO_2 concentrations and its long-range transport, since there were no local sources recognized.

Ammonia and mercaptans concentrations show strong positive relationship with the air temperature, which result in elevated formation and evaporation.

It was recognized that the efficiency of the sampling method for hydrogen sulphide, using impregnated filters is strongly dependent on air humidity.

The existence of a local source of H_2S and R-SH in SSE direction was noticed. The overall contribution of this source is not significant since winds rarely blow from that direction.

Although gaseous pollutant concentrations were not elevated during northeast winds, because of the predominant northeast wind direction, the contribution of all pollutants from the direction of the waste dump at the measuring site is significantly higher compared to other directions, but that does not necessarily mean that the pollutants originated from that source.

References

ISO-4219, 1979, Air quality-Determination of gaseous sulphur compounds in ambient air-Sampling equipment.

ISO-4221, 1980, Air quality-Determination of mass concentration of sulphur dioxide in ambient air-Thorin spectrometric method.

ISO-6768, 1985, Air quality-Determination of the mass of nitrogen dioxide-Modified Griess-Saltzman method.

Moore, H.B.A., Helwig, H.L. and Grual, R.J.: 1960, A Spectrophotometric Method for the Determination of Mercaptans in Air, *Am. Ind. Hyg. Assoc. J.***21**, 466-468.

Vađić, V., Gentilizza, M., Hršak, J. and Fugaš, M.: 1980, Determination of Hydrogen Sulphide in the Air, *Staub-reinhalt Luft,* **40**, 73-75.

SAMPLING AND ANALYSIS OF ORGANIC COMPOUNDS IN DIESEL PARTICULATE MATTER

Eva LEOTZ-GARTZIANDIA[(1)], Véronique TATRY[(1)], Patrick CARLIER[(2)]
(1) INERIS (Institut National de l'Environnement Industriel et des Risques)
Parc Technologique Alata, B.P. 2, 60550 Verneuil-en-Halatte, France
(2) LISA (Laboratoire Inter-universitaire des Systèmes Atmosphériques)
Université Paris XII, 61 avenue du Général de Gaulle, 94010 Créteil cedex, France

Abstract. The fraction of atmospheric semi-volatile organic compounds (SVOC) is partitioned between the gaseous and particulate phases. Certain of these compounds eg. polynuclear aromatic hydrocarbons (PAH) and their derivatives have been shown to exhibit mutagenic or carcinogenic properties. Emissions from diesel engines are an important source of these contaminants. In a dilution chamber, we studied a diesel engine emissions. It is shown firstly, that the gaseous fraction is predominant (by up to 20 times) with respect to the particulate phase. Secondly, the polar compounds, neglected in the majority of previous studies, are the predominant species. A test campaign was carried out in Paris-Porte d'Auteuil which yielded similar results to the laboratory experiments.

Key words: diesel emissions, dilution chamber, atmospheric organic matter, gas/particle partitioning, PAH, oygenated PAH

1. Introduction

In urban air 50 to 80 % of fine particles come from traffic sources. In France, diesel engines account for 30 % of total traffic. In 1995 the French motor industry estimated that diesel vehicles were the source of 87 % of these particles. The contribution of gasoline vehicles without catalysed engines was 12 % and 1 % for the vehicles with catalysed engines (Société française de santé publique, 1996).

Diesel particles are very small, the average diameter is near 0.2 μm. These particles consist of solid carbonaceous soot particles (black carbon), that are associated with a complex mixture of organic compounds.

This paper presents the results of a laboratory study on diesel particles organic composition (gas and particulate phases) and a comparison with the results obtained in a yield campaign in Paris.

2. Diesel particles

The particle phase of diesel emissions consist of aggregates of spherical carbonaceous particles (about 0.2 μm in mass median aerodynamic diameter), onto which significant amounts of high-molecular-weight organic compounds are adsorbed when the hot engine exhaust cools down to ambient temperature

Environmental Monitoring and Assessment **65**: 155–163, 2000.

(H.E.I., 1995). The particles in diesel emissions are unique, because they have large surface areas allowing adsorption of organic compounds.

Organic compounds account, typically, for 10 to 40 % of the diesel particulate mass. These include high-molecular-weight hydrocarbons and semi-volatile hydrocarbons distributed between the gas and particle (adsorbed) phases (Kado et al., 1996; Simoneit, 1986; Westerholm et al., 1991) depending on physico-chemical properties; e.g. ambient temperature and humidity. This distribution between the two phases can occur during sampling, which leads to modification of the sample composition. Such modifications are called artefacts. The phase partitioning is important since it dictates the pollutants lifetime, hence their transport range and accumulation in the atmosphere. These facts are directly linked to health risk.

We can classify the organic compounds present in diesel particles into three groups:

- aliphatic hydrocarbons: alkanes, alkenes ...
- aromatic hydrocarbons: polycyclic aromatic hydrocarbons (PAH)
- polar hydrocarbons: oxygenated and nitrated PAH, dicarboxylics acids, alcanoic acids... (Allen et al., 1997)

Diesel particles have been classified by the " Centre International de Recherche sur le Cancer (CIRC) " and the US EPA (Environmental Protection Agency) as being " probably carcinogenic " in humans (Classe 2a of CIRC), and several hydrocarbons present in these particles are or are suspected to be carcinogenic and/or mutagenic (e.g. Benzo(a)pyrene (BaP), Benzo(a)anthracene (BaA), Dibezo(a,h)anthracene (DBahA), nitro-pyrene, fluorenone).

After inhalation these compounds can be absorbed by the organism and have the potential to induce mutations in humans. Therefore, they could conceivably play a role in both genotoxic and non-genotoxic carcinogenesis (H.E.I., 1995; Nielsen et al., 1996):

3. Objetives of this works

Our work was focused on three axes:

- the study of the effects of temperature in the vaporisation losses of organic compounds from filters during sampling (with a High Volume Sampler, HVS)
- characterisation of organic compounds in the diesel particles in a dilution tunnel, both in the particle and the gas phases and in different size fractions, with special emphasis on oxygenated PAH.
- the collection of atmospheric particles during a test campaign at Paris-Porte d'Auteuil in order to compare their composition with that of particles collected in the dilution tunnel experiments.

4. Experimental methods

4.1. CHOICE OF THE ORGANIC COMPOUNDS

They were chosen because they are known to be : molecular markers for diesel emissions, toxic or suspected toxic substances, photochemically active.

We chose:

- aliphatic hydrocarbons : alkanes containing 10 to 32 carbon atoms
- aromatic hydrocarbons : 17 HAP from EPA Method TO-13
- polar hydrocarbons : 1-napthaldehyde, fluorenone, anthraquinone, phenanthrene-9-carboxaldehyde

4.2. SAMPLING METHODS

- **High Volume Sampler (HVS)** : with this apparatus the particles are collected onto a quartz fibre filter and the volatile compounds are trapped by two cylindrical plugs of polyurethane foam (PUF) located behind the filter. The flow rate is quite 17 Nm^3/h
- **Impactor :** the impactor only permits the collection of samples of diesel particulate matter onto quartz fibre filters but in different size fractions. The flow rate was approximately 33 Nm^3/h

Four size ranges were selected:

Group A: > 4.2 μm
Group B: 4.2 to 1.3 μm
Group C: 1.3 to 0.4 μm
Group D: < 0.4 μm

After air sampling the PUF and filters were wrapped in aluminium foils and stored at 4 °C. They were always analysed within 48 h.

4.3. ANALYTICAL METHODS

At first the compounds were extracted from the filters and foams with CH_2Cl_2. The solvent was evaporated under a current of nitrogen. The dry extract was dissolved in n-hexane. To separate the compounds of the three different families, a silica gel alumina chromatographic column was used. The first fraction obtained with n-hexane contained the alkanes, the second fraction obtained with a mixture of CH_2Cl_2:n-hexane (20:80) contained the PAH, and the third fraction containing oxygenated PAH is obtained with a mixture of methanol:n-hexane (20:80).

Alkanes and oxygenated PAH are analysed by GC-FID and GC-MS and PAH are analysed by HPLC with fluorescence/UV detection and GC-MS.

Quantification was performed from the GC profiles using representative

Hydrocarbons as external standards. Analysis precision for the samples ranged from 5 to 15 %. Recovery rates of the analytical method and blanks were calculated for all the hydrocarbons studied and the results were corrected accordingly (Leotz-Gartziandia, 1998).

5. The dilution tunnel

A dilution tunnel was constructed such that it was possible to :

- to use a standard car
- to efficiently dilute the exhaust gases and to obtain an homogeneous mixture between exhaust gases and ambient air (the latter being used after filtration)
- to reproduce the chemical and physical phenomena (like reactions, and adsorption of the compounds on the particles)
- to monitor temperature, humidity and compounds like NO_x and CO_2
- to connect to the HVS and impactor

Stainless steel was used for the tunnel and quartz wool permitted thermal insulation.

The flow in the tunnel was optimised, for the different samplers to allow collection under isokinetic conditions, to 154 Nm^3/h and 126 Nm^3/h for the HVS and the impactor, respectively.

Blank runs were carried out on the air dilution with the two sets of apparatus. For oxygenated PAH and alkanes the results were under the detection limits. For PAH, the results for naphthalene (NAP), Fluoranthene(FL) and Phenanthrene (PHE) were between 1 and 6 ng/m^3 whilst for the others they were under the detection limits.

6. Results and discussion

6.1. VAPORIZATION LOSSES

The HVS filter was seeded with an internal standard containing all of the organic compounds under investigation. The PUF was placed behind the filter in the apparatus.

The HVS was connected to the dilution tunnel and purified air was drawn through the seeded filter for 4 hours at two different temperatures: 20 °C and 30 °C. 70 m^3 of air were sampled.

It is shown that the partition between the filter and the PUF is in agreement with the molecular weight of the compound and the increase of temperature.

The results observed for the compounds with the higher partition between the filter and PUF are shown in figure 1.

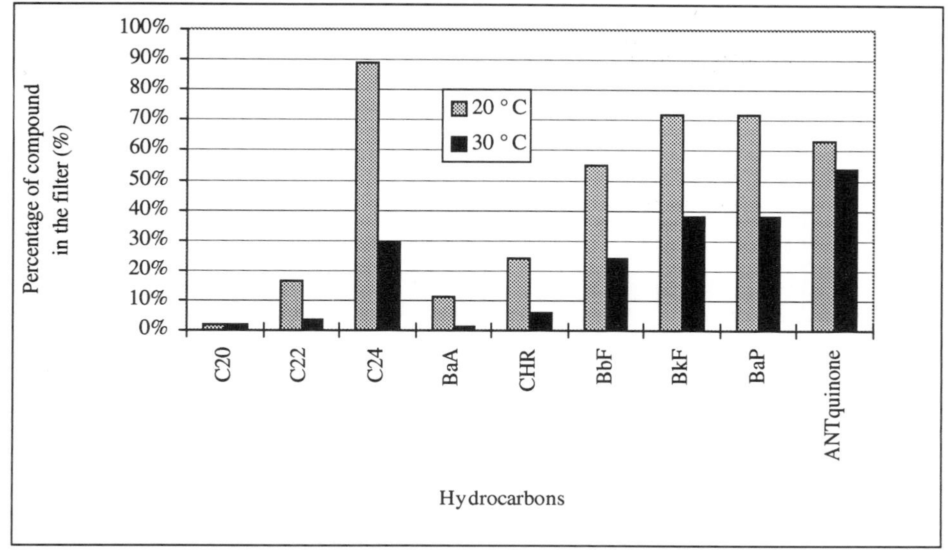

Figure 1. Results of the vaporisation losses for several hydrocarbons

We observed the importance of the temperature in the gas/particle partitioning for these compounds. A difference of only 10 °C provokes a loss of 60 % of C_{24} from the filter, for example.

In this study the partial sublimation of semi-volatile compounds during High Volume Sampling was observed. The importance of sampling the two phases (gaseous and particulate) was demonstrated by the quantitative sampling of these compounds a whole year.

6.2. CHARACTERIZATION OF ORGANIC COMPOSITION OF DIESEL PARTICLES

For this study The " Institut Français du Petrole, IFP " supplied a car and a reference fuel.

Real time determinations of temperature, humidity, carbon dioxide (CO_2) and oxides of nitrogen (NO_x) were carried out in the dilution tunnel.

Five minute samples were obtained at 30 °C with the HVS and the impactor and each experiment was carried out three times. A good repeatability of the experiments was observed.

All the organic compounds studied, were found in the diesel exhaust (gas and particulate), as follows :

In the particulate phase : alkanes > oxygenated PAH > PAH

In the gas phase : alkanes > PAH > oxygenated PAH

Compounds and profiles of alkanes and PAH observed in this study were similar to those reported in literature.

6.2.1. *HVS results*

The gaseous fraction is predominant with respect to the particulate phase from:
- up to 5 times more predominant for alkanes with15 to 18 carbon atoms (figure 2)
- up to 15 times more predominant for PAH between fluorene (FLN) and FL

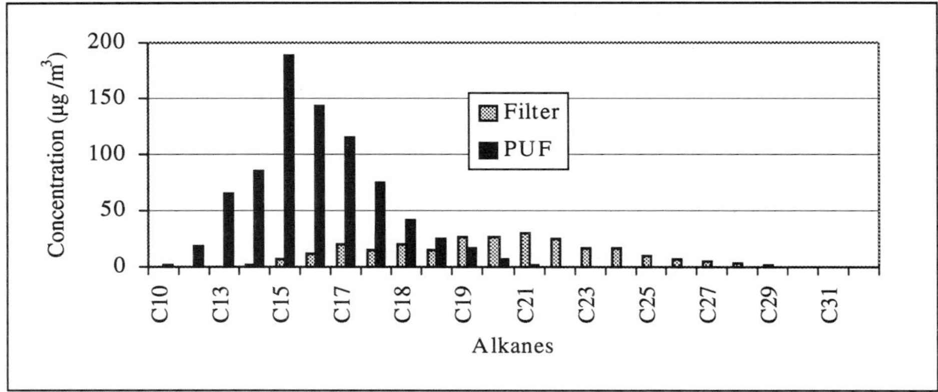

Figure 2. Concentration of the gaseous(PUF) and particulate(filter) phases of alkanes in diesel exhaust (the mean of three values).

The importance of sampling the gas phase is again demonstrated.

6.2.2. *Impactor results*

The results obtained with the impactor showed that all the studied compounds (alkanes, PAH and oxygenated PAH) are mainly in the very fine particulate fraction (particle size under 0.4 μm) (Figure 3)

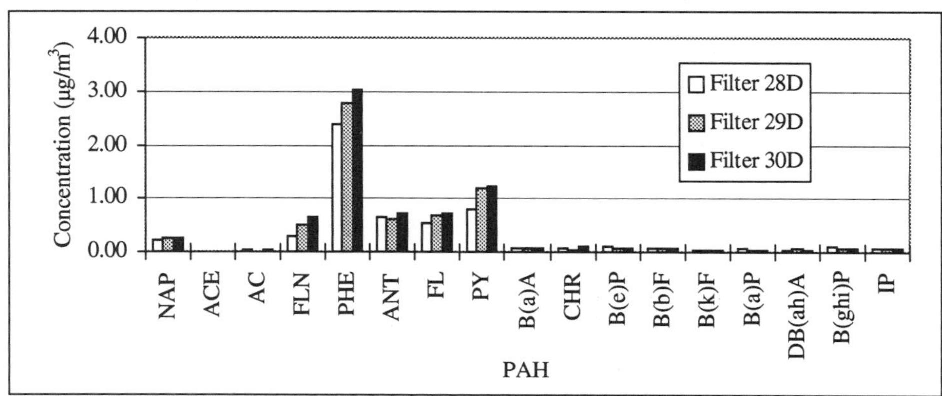

Figure 3. Concentration of PAH in the size group < 0.4 μm in diesel exhaust.

This fraction can reach the pulmonary alveoli, thus enhancing the risk of introducing the organic compounds, present in these particles, into the organism.

6.2.3. *Oxygenated PAH*

The results obtained showed that there is a greater proportion of oxygenated PAH than PAH in the particulate phase. For the compounds studied, the sum of concentrations of the 4 oxygenated PAH was 30 times greater than the sum of the concentrations of the 17 PAH This shows that the importance of these compounds has been underestimated as they have been neglected in the majority of previous studies. Very little toxicological data is available with respect to the evaluation of air quality.

7. Paris campaign

To complete the tunnel dilution tests, measurements were carried out near a motorway for a week (Paris-Porte d'Auteuil)

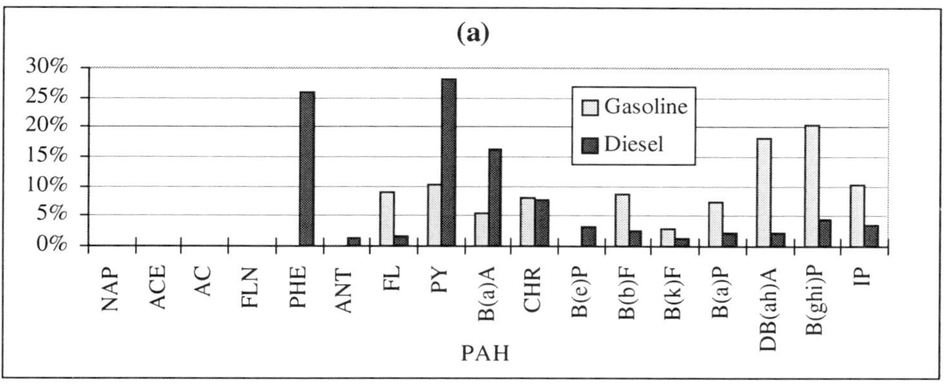

Figures 4. Comparison of diesel (results from the dilution tunnel) and gasoline (Miguel et al., 1998) PAH profils (a) with the results of Paris campaign (b).-

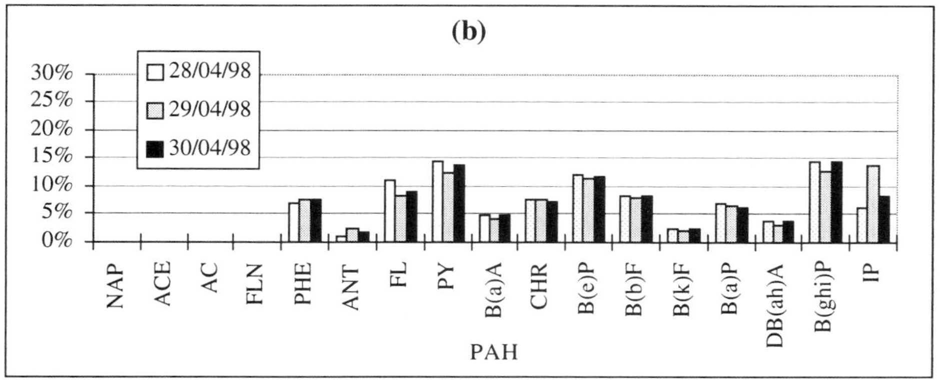

HVS was used to collect the gas and particle phases for alkanes, PAH and oxygenated PAH.

We observed the contribution of different sources of samples : vehicular emissions (diesel and gasoline) and natural emissions (measurements were carried out in spring near the “ Bois de Boulogne ”).

For the alkanes, we identified petroleum derivative indicators like prystane and phytane; for the polar fraction we identified stigmasterol. (a characteristic molecular marker emitted by plants).

The results showed similarities between the profiles of PAH in the dilution tunnel and during the test campaign, obviously the kind of traffic (diesel and gasoline) modifies the results (figure 4).

On the other hand we systematically observed:

- a predominance of the gaseous fraction with respect to the particulate phase:
- up to 5 times more predominant for alkanes with15 to 18 carbon atoms
- up to 10 to 20 times more predominant for PAH between FLN and FL
- higher of oxygenated PAH than PAH in the particulate phase, by a factor of almost 10.

It is important to know that some of the oxygenated PAH family are mutagenic, but current toxicological knowledge is poor.

8. Conclusion

The results showed the importance of the partition gas/particle in the sampling of semi-volatile organic compounds and the influence of temperature in the changes of this partition.The sampling of the gas phase is necessary.

We observed the greater fraction of oxygenated PAH in the particulate phase with respect to PAH : by a factor of almost 30 in the dilution tunnel and by a factor of almost 10 in ambient air This factor is certainly underestimated and a very little toxicological data is available, with respect to the evaluation of air quality.

Acknowledgement

This work was supported by French Ministry of Territorial Planning and Environment and by Ademe-Primequal Program, and carried out in collaboration with IFP (Institut Français du Pétrole). The authors would like to express their gratitude to Mrs. Nathalie Bocquet, Mrs. Marise Marlière, Mrs. Claudine Viley and Mr. Bruno Triart for their technical assistance.

References

Allen J.O., Dookeran N.M., Taghizadeh K., Lafleur A.L., Smith K.A. and Sarofim A.F.: 1997, Measurement of oxygenated polycyclic aromatic hydrocarbons associetad with a size-segregated urban aerosol. *Environ. Sci. Technol.,* **31**(7), 2064-2070.

Health Effects Institute (HEI).: 1995, Diesel exhaust : a critical analysis of emissions, exposure, and health effects. A special report of the institute's diesel working group.

Kado N.Y., Okamoto R.A., Kuzmick P.A., Rathbun C.J. and Hsieh D.P.H.: 1996, Integrated supercritical fluid extraction, biossay and chemical screening methods for analyzing vapor-phase compounds of an environmental complex mixture : Diesel exhaust. *Chemosphere*, **33**(3), 495-516.

Leotz-Gartziandia Eva.: 1998, Caractérisation chimique de la matière organique gazeuse et particulaire due aux moteurs Diesel, à l'émission et dans l'air ambiant, thèse de Doctorat en Sciences, Univerité Paris VII.

Miguel A-H., Kirchstetter T.W., Harley R.A. and Hering S.V. : 1998, On-road emissions of particulate polycyclic aromatic hydrocarbons and black carbon from gasoline and diesel vehicles. Environ. Sci. Technol., **32**(4), 450-455.

Nielsen T., Jorgensen H.E., Larsen J.C. and Poulsen M.: 1996, City air pollution of polycyclic aromatic and other mutagens : occurrence, sources and health effects. *The Science of the Total Environment*, **189/190**, 41-49.

Simoneit B.R.T.: 1986, Characterization of organic constituents in aerosol in relation to their origin and transport : A review. *Inter. J. Environ. Anal. Chem.*, **23**, 207-237.

Société française de santé publique.:1996, La pollution atmosphérique d'origine automobile et la santé publique. Bilan de 15 ans de recherche internationale Collection santé et société, N 4. ISBN 2-911489-02-0

Westerholm R.N., Almen J. Li H., Rannug J.U., Egeback K.-E. Et Gragg K.: 1991, Chemical and biological characterization of particulate-, semivolatile-, and gas-phase-associatited compounds in diluted heavy-duty Diesel exhaust : A comparison of three different semivolatile-phases samplers. *Environ. Sci. Technol.*, **25**(2), 332-338.

ELEMENTAL COMPOSITION OF URBAN AEROSOL COLLECTED IN FLORENCE, ITALY

FRANCO LUCARELLI[A], PIER ANDREA MANDÒ[A], SILVIA NAVA[A], MARINA VALERIO[A], PAOLO PRATI[B], ALESSANDRO ZUCCHIATTI[B]
[A] *Dipartimento di Fisica and I.N.F.N., Largo E.Fermi 2, 50125 Firenze (Italy)*
[B] *Dipartimento di Fisica and I.N.F.N., Via Dodecaneso 33, 16146 Genova (Italy)*

Abstract. An extensive investigation is in progress aiming at the characterisation of the air particulate composition in Florence. For our investigation, we use the external PIXE-PIGE beam facility of the I.N.F.N. Van de Graaff accelerator at the Physics Department of the Florence University. In order to gather information on both the longer- and shorter-time trends of the aerosol elemental composition, we are analysing both long temporal series (about 1 year) of 24-h Millipore filters collected by the health authorities in 3 different sites, and filters collected simultaneously in two of the above sites for about one month, by two streakers with one-hour resolution, providing size fractionation between particle size smaller than 2.5 μm and from 2.5 μm to 10 μm. The streakers sampling period includes two days during which the Municipality of Florence has banned the circulation of non-catalytic cars, due to the increase of NO_2 above the "recommended safety values". We present here the first obtained results for the sampling site located near a heavy traffic road.

Key words : Urban aerosol, streaker sampler, elemental composition, PIXE, factor analysis

1. Introduction.

Florence (400.000 inhabitants) is located in the central part of Italy about 80 Km east of the sea. The few big industries formerly present inside the town moved years ago from the centre; nevertheless the orographic configuration of the site (a closed basin) and its continental climate favour episodes with high atmospheric stability and heavy pollution [1] (in winter due to NO_2 and CO, in summer due to O_3). Therefore the urban atmosphere is monitored on-line with a public network of eight air quality measuring stations, sampling pollutant gases (NO, NO_2, CO, SO_2, O_3). Some of them measure the total suspended PM_{10} particulate (using Millipore filters exposed 24 hours to an air flux of 20 l/min, excluding particles larger than 10 μm) but do not include any aerosol composition determination. Recently we started an extensive investigation aiming at the characterisation of the air particulate elemental composition in Florence. In order to obtain information on both the long- and short-time trends of the aerosol composition, we are analysing both a long temporal series of the above quoted 24-h Millipore filters collected by the health authorities in 3 different sites (characterised by different urban settings) and filters collected simultaneously in two of the previous sites by two two-stage streakers with 1-

Environmental Monitoring and Assessment **65**: 165–173, 2000.

hour resolution. The data analysis is still in progress. We present here the first results obtained for the sampling site located close to a heavy traffic area.

2. Sampling and analysis

The public air quality measuring station we used is located close to a heavy traffic road (at about 5 m from the closest lane). This station is considered representative of heavy traffic area and it is among those used as a reference to check when the recommended safety concentration for CO are exceeded. The two-stage streaker sampler was located on the roof of the station (3 m above ground). The sampling campaign for the PM_{10} Millipore filters lasted from February 1997 till January 1998, the one with the streaker sampler lasted from January 21 till February 22 1998.

The two-stage streaker sampler (manufactured by P.I.X.E. International Corporation [2]) separates the fine (< 2.5 μm aerodynamic diameter) and the coarse (2.5-10 μm) modes of an aerosol. A pre-impactor stops particles with diameter >10 μm. A paraffin-coated 7.5 μm thick Kapton® foil is used as an impaction surface for coarse particles and a 0.4 μm pore-size Nuclepore® filter as a fine particle collector. The filter speed during sampling, the pumping orifice width and the beam size we use for the subsequent analysis are such that an overall resolution of about one hour is obtained on the elemental composition of air particulate.

The element concentrations in the aerosol samples were obtained by PIXE analysis. PIXE is extensively used for aerosol studies [3] because it is a fast, non-destructive, sensitive, multielemental technique. The samples were bombarded with 3 MeV protons from the I.N.F.N. Van de Graaff accelerator at the Physics Department of the University of Florence, with the external beam set-up [4]. For the streaker samples each spot corresponding to 1 hour of aerosol sampling was irradiated for 7 minutes with a beam intensity of 15 nA. Millipore filters were bombarded for about 10 minutes with an average beam intensity of 6 nA. X-rays were detected by two Si(Li) detectors located at different distances from the target and fitted with different filters to optimise the sensitivity to the widest possible range of Z. Helium flows into the volume in front of the smaller detector and all around the filter (we detected Na). PIXE spectra were fitted using the GUPIX software package [5]. Concentration uncertainties coming from PIXE analysis are generally around 5%; they are obviously higher when concentrations approach minimum detectable limits (about 10 ng/m^3 for elements from Na to V and 1 ng/m^3 -or below- for elements from Cr to Pb).

3. Results from the analysis of the streaker samples

The summary statistics for the four weeks measurement period are shown in Table I. Some elements (like S, K, Pb, Br, Ni) are present mainly in the fine fraction, other in the coarse one (Na). In figure 1 the average detected mass distribution measured for the two fractions is shown. In the fine stage S is the dominant element (around 50%). Lead and bromine concentrations, which are strongly correlated with traffic, account for about 3% and 0.7% of the mass.

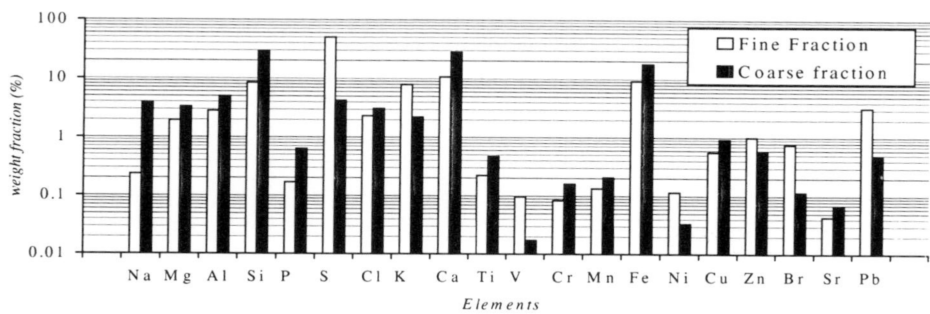

Figure 1. Average mass distribution for the two fractions.

Table I
Mean and maximum concentrations, sample standard deviation σ (μg/m3)

	mean	max	σ	mean	max	σ		mean	max	σ	mean	max	σ
	Fine fraction			*Coarse fraction*				*Fine fraction*			*Coarse fraction*		
Na	0.14	0.55	0.10	0.30	3.18	0.30	**V**	0.01	0.04	0.006	0.01	0.01	0.002
Mg	0.17	0.49	0.06	0.11	0.33	0.05	**Cr**	0.01	0.05	0.006	0.01	0.04	0.004
Al	0.20	0.70	0.12	0.16	0.73	0.11	**Mn**	0.01	0.03	0.006	0.01	0.04	0.005
Si	0.72	3.92	0.38	0.77	5.22	0.80	**Fe**	0.65	2.37	0.38	0.67	2.81	0.47
P	0.05	0.11	0.02	0.03	0.12	0.02	**Ni**	0.01	0.03	0.004	0.001	0.007	0.001
S	3.44	9.22	1.74	0.16	1.76	0.10	**Cu**	0.04	0.14	0.02	0.04	0.15	0.03
Cl	0.15	1.20	0.10	0.19	6.32	0.34	**Zn**	0.07	0.33	0.04	0.02	0.56	0.03
K	0.55	2.63	0.37	0.08	0.28	0.05	**Br**	0.05	0.18	0.03	0.01	0.03	0.004
Ca	0.72	3.19	0.53	0.96	6.15	0.68	**Sr**	0.01	0.02	0.003	0.004	0.03	0.002
Ti	0.03	0.06	0.01	0.02	0.09	0.01	**Pb**	0.24	0.79	0.14	0.02	0.12	0.01

In the coarse mode, Si and Ca are the dominant elements, representing together more than 50% of the measured mass. As an example of time sequence plots, Fig. 2 shows the Pb (in the fine fraction) and Na (in the coarse one) time patterns throughout the whole sampling period.In the coarse stage the pattern of

Na is similar to that of Cl: they are characterised by isolated peaks, as it would expect if they were due to marine aerosol transported occasionally by wind. Excluding an episode in last sampling week (with low Cl content) we obtained a correlation coefficient r=0.91 with a Cl/Na ratio 1.67 in good agreement with what expected for sodium chloride.

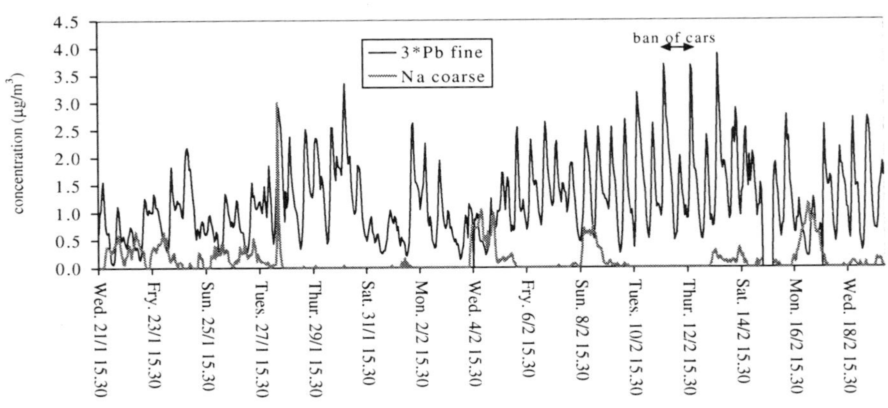

Figure 2. Sodium (coarse) and Lead (fine) concentrations during the whole sampling period.

In the fine fraction Pb and Br (not reported in fig.2) patterns exhibit correlated daily variations with 12 hours periodicity, connected to traffic, with peaks around 8.30 a.m. and 6.30 p.m. A similar time trend is shown by CO and NO concentrations obtained from gas measurement at the same sampling station. Pb and Br are highly correlated (r=0.95). The average Br/Pb ratio is 0.23, lower than the "ethyl ratio" but similar to those reported in literature for urban areas and to the values found in our previous work on Florence. In the coarse fraction Pb and Br are present in far lower concentrations but are again highly correlated (r=0.95) with an average ratio 0.23.

Because of the expectation of an increase of NO_2 concentration over the recommended safety values, the local authorities decided to ban the use of private non-catalytic cars for two days (February 11 e 12) between 8.30 a.m. till 6.30 p.m. Surprisingly there is no visible effect for Pb and Br which are exclusively emitted by the type of cars not allowed to circulate. A partial explanation is that ~20% of non catalytic cars is allowed to circulate for many reasons (commercial activities, emergencies etc.). Furthermore many people using scooters (allowed to circulate even in the restricted period) prefer to use leaded instead of unleaded gasoline. Finally it is probable that many people with non-catalytic cars did not respect the prohibition of circulation.

A multivariate analysis of the whole data set (like factor analysis or principal component analysis [6]) takes into account the correlations of all the variables simultaneously. The aim is to find a small number of factors which can describe most of the correlations of the variables. A large set of intercorrelated variables is replaced with a smaller number of indipendent ones. The new variables are derived from the original ones, and are simply linear combinations of those variables. The dimension of the measurement space is thus reduced, which simplifies the interpretation of the data and helps to identify and quantify the impact of the relevant sources on the receptor site. The VARIMAX rotated factor analysis was performed separately for the two fractions on the set of standardised (zero mean and unit variance) element concentration data using a commercial code [7]. The results are reported in Table II.

For the fine fraction four factors were obtained, representing 71% of data variability. The first factor identifies the "traffic" source through the presence of Br and Pb. It also contains crustal elements connected to dust resuspended by automotive transport. The second factor has high loadings for S, V, Ni and, to a lesser extent, also for K, Zn and Cl. This source could be connected to oil-combustion from domestic heating. The third factor, characterised by Al, Si, Ca, Ti and Sr, represents a soil dust source, separated from dust resuspended by cars. The fourth factor contains only Na and Mg and it is difficult to associate it with a specific source.

For the coarse fraction 3 factors were obtained, representing 80% of data variability. Factor 1 and factor 2 are the coarse fraction complement of factor 1 and 3, respectively, in the fine fraction and are connected to traffic and soil dust. The third factor is connected wind-transported sea-salt.

4. Results from 24 h sampling series

The 24-h sampling provide data useful for estimating average yearly and monthly concentration values and for air quality assessment according to the regulation in force. In figure 3 the concentrations in air of S and Pb are reported as an example. The values of the daily concentration of the different elements show strong variations due to the variability of both pollution sources and meteorological parameters, indeed a dependence on the characteristic of sampling period is observed (no working day, peculiar meteorological conditions etc). For example, in the first day of the year the increase of K, Sr and Ba concentrations is due to the use of fireworks. As in the 1-h resolution sampling, again Pb and Br, Na and Cl have very similar patterns.

The summary statistics for the whole year is reported in Table III. The concentrations of the various elements are very different, ranging from values of the order of some μg/m^3 (Si, S, Ca, Fe) to hundreds of ng/m^3 (Na, Mg, Al, Cl, K, Pb) and below.

Table II :
Varimax-rotated loadings obtained by factor analysis

	Fine fraction				Coarse fraction		
	Traffic	*Combustion*	*Soil*	*Factor 4*	*Traffic*	*Soil*	*Sea-salt*
Na	0.01	-0.07	-0.16	0.87	-0.10	0.21	0.95
Mg	0.22	0.04	0.25	0.8	0.18	0.43	0.72
Al	0.6	0.15	0.62	0.21	0.60	0.72	0.17
Si	0.35	-0.03	0.73	-0.17	0.66	0.47	0.00
S	0.09	0.82	0.03	-0.16	0.31	0.73	0.34
Cl	0.33	0.51	0.22	0.07	0.04	-0.02	0.89
K	0.49	0.54	0.24	-0.02	0.41	0.76	0.17
Ca	0.63	0.07	0.66	0.14	0.56	0.78	0.10
Ti	0.42	0.21	0.6	0.06	0.56	0.72	0.05
V	0.01	0.83	0.01	0.08			
Cr	0.54	0.17	0.05	0.1	0.78	0.34	0.04
Mn	0.77	0.18	0.37	0.1	0.79	0.58	0.05
Fe	0.87	0.15	0.35	0.03	0.88	0.42	0.01
Ni	0.48	0.7	0.18	-0.01	0.66	0.50	-0.10
Cu	0.87	0.19	0.27	0.02	0.90	0.33	-0.01
Zn	0.54	0.51	0.31	0.08	0.26	0.49	-0.05
Br	0.9	0.17	0.2	0.09	0.86	0.33	0.19
Sr	0.07	0.17	0.66	0.05	0.45	0.70	0.09
Pb	0.89	0.18	0.22	0.07	0.87	0.38	0.06
Variance	6.0	3.0	2.9	1.6	6.8	5.2	2.5
%variance	31.5	15.5	15.2	8.3	37.5	28.8	13.7

Some elements (As, Se, Rb, Cd, Sb) are usually below their MDL . Lead concentration is below the present (1.5 μg/m3, 24-h average) and future (0.5 μg/m3, 24-h average) exposure limits.

The data may be compared with those from our previous study (1988-1989) [4]. The sampling was made close to a heavy traffic road, but with a sampler with no size-fractionation nor any upper cut-point of the particle size but the data for elements mainly present in the fine fraction may be compared.

The concentrations for S, Pb, Br, Cu and Zn are much lower than those found 10 years ago, due to the use of unleaded gasoline since 90's, changes in motor oil additive, increased use of methane fuel for domestic heating.

The monthly average values show a more regular trend respect to daily values. Pb values show maximum values in winter and minimum in summer, as expected from traffic data. Also Si, like other crustal elements, has higher concentrations in winter. On the other hand, S, which is essentially a secondary aerosol component, has a different pattern, with higher values in summer.

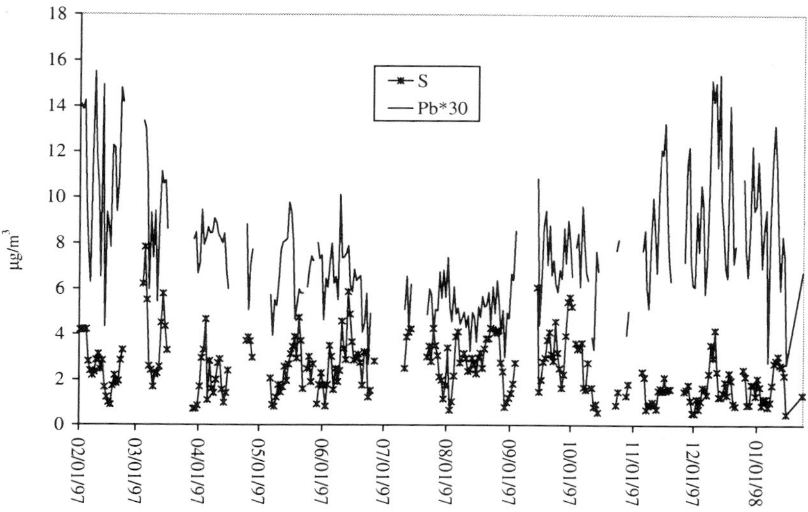

Figure 3. Daily concentration in air (μg/m³) for S and Pb.

Table III
Mean and maximum concentrations, sample standard deviation σ. *n* is the numbers of samples in which the element was above the minimum detectable limit.

Element	mean (μg/m³)	max (μg/m³)	σ (μg/m³)	*n*	Element	mean (μg/m³)	max (μg/m³)	σ (μg/m³)	*n*
Na	0.26	2.00	0.35	228	Fe	1.57	3.62	0.50	254
Mg	0.12	0.31	0.06	253	Ni	0.01	0.03	0.004	254
Al	0.36	1.14	0.18	254	Cu	0.08	0.19	0.03	254
Si	1.03	2.89	0.50	254	Zn	0.07	0.19	0.03	254
P	0.03	0.07	0.01	232	As	0.001	0.02	0.004	42
S	2.48	8.53	1.31	254	Se	0.001	0.01	0.001	96
Cl	0.24	2.92	0.46	254	Br	0.07	0.14	0.03	254
K	0.34	2.51	0.22	254	Sr	0.01	0.05	0.004	250
Ca	1.71	6.04	0.85	254	Zr	0.01	0.02	0.003	246
Ti	0.05	0.13	0.02	254	Sn	0.01	0.04	0.006	12
V	0.01	0.02	0.005	209	Sb	0.01	0.11	0.02	64
Cr	0.01	0.02	0.004	254	Ba	0.02	0.25	0.04	35
Mn	0.02	0.05	0.008	254	Pb	0.26	0.52	0.09	254

The average particulate matter composition in the four seasons is reported in fig. 4. Generally Si, S, Ca, Fe are in the range 1-10% of the total mass; Na, Mg, Al, Cl, K, Ti, Cu, Zn and Pb in the range 0.1-1%; the other elements <0.1%.

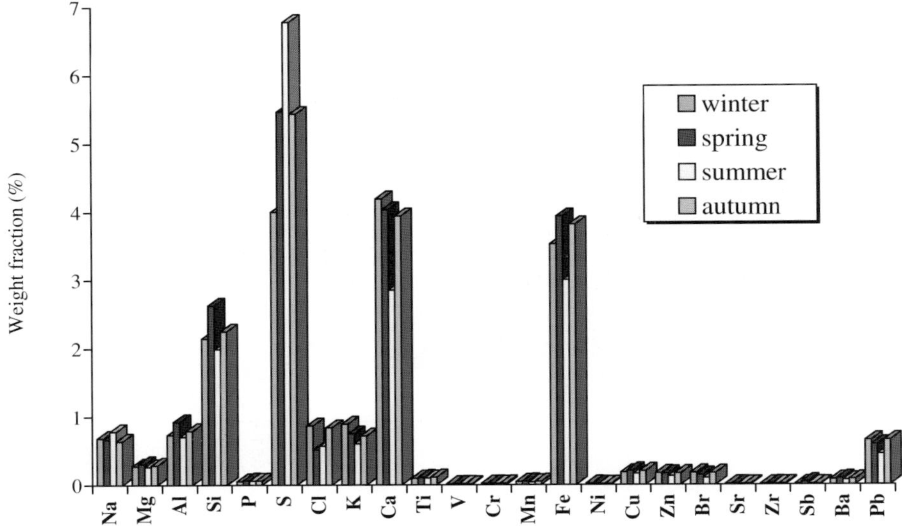

Figure 4. Average particulate matter composition with respect to the total PM10 mass.

The sum of the mass of the detected elements is typically 15-20% of the total PM_{10} mass (the analytical technique used does not detect H, N, C and O). Considering elements of crustal origin as oxides and S as ammonium sulphate we explain 40-50% of the total mass. Thus, about one half of the aerosol mass seems to be due mainly to carbon.

Factor analysis was performed on the set of standardised element concentration data and confirms the results obtained from the streaker data. Four factors were obtained, representing 83% of data variability. The first factor identifies the "traffic" source through the presence of Br and Pb. The second factor characterised by Al, Si, Ca, Ti and Sr, represents a soil dust source. The third factor is connected to sea-salt. The fourth factor (with high loadings for S and V) is connected to oil-combustion and/or domestic heating. A rough estimate for the main sources average contribution was obtained by multiple linear regression (MLR). The hypothesis is that the PM_{10} concentrations measured at the receptor site are controlled by n elements which are representative of n different sources : these n tracers elements are selected looking at the factor analysis results. For each factor obtained from the factor analysis, a tracer element was selected as an independent variable (Pb, Si, Na and S respectively) while PM_{10} mass was taken as dependent variable and MLR wad performed using a commercial code [7]. Vehicular traffic emission is the main particulate source. Resuspended soil dust and secondary sulphates give also an important contribution to total mass.

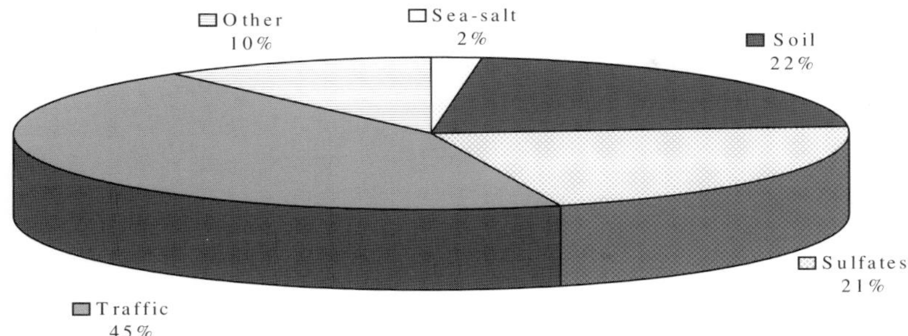

Figure 5. Average contribution of the main sources

5. Conclusions

This work represents a first effort to evaluate the aerosol composition in Florence. The data analysis is in progress. From the first results it is confirmed that traffic, because of both direct emission of exhaust and soil dust raising, is the main source of atmospheric pollution in Florence. The ban of non-catalytic cars seems to have had no significant effect on Pb and Br concentrations. Further work is to be done to correlate the results obtained so far with those from conventional gas measurements and with meteorological parameters.

References.

[1] Barbaro, A., Bazzani, M., Giovannini, F. and Nannini, P.: 1998, Episodi acuti di inquinamento atmosferico a Firenze, analisi dei dati e metodi di previsione, ARPAT- Cedif

[2] PIXE International Corp., P.O. Box 7744, Tallahassee, FL 32316, USA

[3] Proceedings of the Eighth International Conference on PIXE and its Analytical Applications, Padova, Italy, *Nuclear Instruments and Methods* **B109/110** (1996) 1.

[4] Del Carmine, P., Lucarelli, F., Mandò, P. A., Moscheni, G., Pecchioli, A.,and MacArthur, J. D.: 1990, PIXE measurements of air particulate composition in the urban area of Florence, Italy, *Nuclear Instruments and Methods* **B45**, 341-346

[5] Maxwell, J.A., Teesdale, W.J. and Campbell, J.L.: 1995, The Guelph PIXE software package II, *Nuclear Instruments and Methods* **B95**, 407-421

[6] Swietlicki, E., Puri, S., Hansson, H.C., Edner, H., *Atmos. Environ.* **30** (1996) 2795

[7] STATGRAPHICS PLUS @ Copyright 1995, Manugistic Inc., 2115 Jefferson Street, Rockville, Maryland 20852, USA

MONITORING OF PARTICLE MASS AND NUMBER IN URBAN AIR

N POMEROY, D WEBBER, C MURPHY

Chemical and Biological Defence Sector,
Defence Evaluation and Research Agency,
Porton Down, Salisbury, Wiltshire, SP40JQ, UK

Abstract. Measurements of aerosol particles in the air of an urban area in the UK have been made. Ambient air was sampled and the particulates measured after passing through a size selective PM_{10} inlet. Particle mass was measured using a Tapered Element Oscillating Microbalance (TEOM). Particle number and size distributions were obtained using an Electrical Aerosol Analyser (EAA) and an Aerodynamic Particle Sizer (APS). Measurements were also made of local meteorological parameters. Fine particle number concentrations were found to show better temporal agreement, including diurnal variation, with particle mass concentrations than the coarser particle number concentrations.

Key words : urban air, aerosol, particles, PM_{10}, air monitoring.

1. Introduction

Respirable particles contained in urban air derive from many diverse sources which are naturally occurring or anthropogenic in origin. Relative to people's exposure, these particle sources may be local or distant and of continuous, periodic or intermittent nature. Primary particles are directly emitted from such sources whereas secondary particles form indirectly as a result of chemical reactions of polluting gases in the air. In urban areas, the largest contribution of primary particles arise from the combustion products of road traffic and industry. Emissions inventories of sources in the UK reveal that most of the particulates in urban air arise from road traffic (~75%) with secondary particles, mostly nitrates and sulphates, and coarse particles such as resuspended "road-dust" making up the remainder (APEG, 1999).

The presence of these particulates in ambient urban air is increasingly implicated in the incidence of pollution related health effects (COMEAP, 1995; Anderson et al, 1996). The interactive effects of exposure to gas and particulate phases of pollutant aerosol, including potent aero-allergens, can lead to a synergism in the manifestation of adverse health effects.

Routine monitoring of airborne particulate matter present in the ambient atmosphere is usually achieved by the determination of particle mass in the PM_{10} size fraction (ie. particles having an aerodynamic diameter of less than 10 µm). This imparts a significant bias towards the larger coarse particles between 10 and 2.5µm which deposit in the upper respiratory tract. Recent

Environmental Monitoring and Assessment **65**: 175–180, 2000.

epidemiological studies demonstrate a stronger association of adverse health effects with the number concentration of fine particles in the $PM_{2.5}$ size fraction (ie. particles having an aerodynamic diameter of less than 2.5μm) which penetrate deeper into the pulmonary regions (Seaton et al, 1995). Similarly, some physiological studies suggest the involvement of even smaller, ultra-fine nano-metre sized particles (<100nm diametre) in cardio-vascular dysfunction (Oberdorster et al, 1995). It is well documented that these fine and ultra-fine particles arise mainly from anthropogenic combustion sources which, in urban areas, are predominantly motor vehicle exhaust emissions (QUARG, 1996). Whilst they are present in relatively high number concentrations, there is little information concerning their relationship to the aerosol mass. There are occasions when relatively short lived, fine particle episodes can contribute significantly to the real-time mass burden as far as people's exposure is concerned, but these are invariably masked by the time averaging effect of the mass monitor. Particle number concentration might therefore be a better characterisation metric than particle mass for epidemiological investigations.

This paper presents some preliminary results of a field sampling campaign to characterise the ambient aerosol and makes comparisons of particle mass and number time series data.

2. Materials and Methods

Ambient air was sampled continuously by a mobile laboratory using an omnidirectional PM_{10} sampling head (Graseby Andersen) operating at an air flow of 100 litres per minute. On-line measurements of the sampled aerosol were made using a Rupprecht and Patashnick tapered element oscillating micro-balance (TEOM) to determine particle mass. Coarse particles, generally defined as those upwards of 2μm forming the coarse particle mode of the atmospheric aerosol size distribution, were measured using a twin beam laser velocimeter [TSI Model 3310 aerodynamic particle sizer (APS)]. Fine particles within the particle accumulation mode having diameters of between 2μm and 100nm and ultra-fine particles in the nucleation particle mode of 100nm and smaller, were measured using an electrical mobility analyser [TSI Model 3030 electrical aerosol analyser (EAA)].

The sampling location was a site situated on the eastern periphery of the urban conurbation of Lichfield in Staffordshire, UK and approximately 200 m from the nearest road.

Data were collected continuously throughout the period 19-27 November 1996.

3. Results

Particle concentrations were derived from the size frequency distributions measured by the APS and EAA instruments. Total particles in each instrument's range, as described above, were plotted as time series of hourly averages along with the hourly average mass values derived by the TEOM.

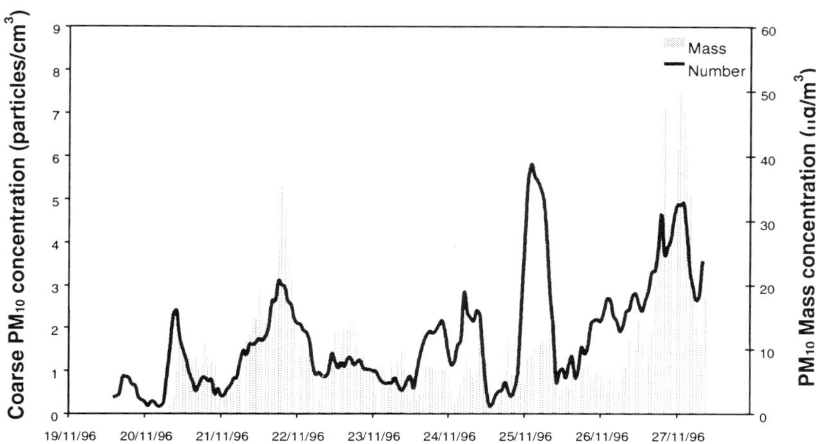

Figure 1. Temporal concentration profiles of particle mass and coarse particle number.

Figure 1 shows the temporal concentration profiles of particle mass and coarse particle number. The excursions of coarse particles between 24th and 26th November, whilst diverging from the mass trend, still maintain some degree of temporal agreement in that they are reflected in the mass profile. The wind direction for this period was generally north-westerly coming from the western Atlantic; however, the air-mass back trajectories prior to the mass peaks on the 21st and 27th show the involvement of a circulating European air-mass. A small amount of rain (~1.5mm) fell just after midday on the 24th and again on the 25th (~0.5mm). A possible explanation for these excursions is that the change in air mass origin may have also resulted in a change of particle composition having greater hygroscopicity such that there was greater water content in the particles when they were measured by the APS. However, whereas the particle stream entering the APS is at ambient temperature, the TEOM sample stream is heated to 50°C prior to interception by its filter to remove water (and the inevitable semi-volatile species).

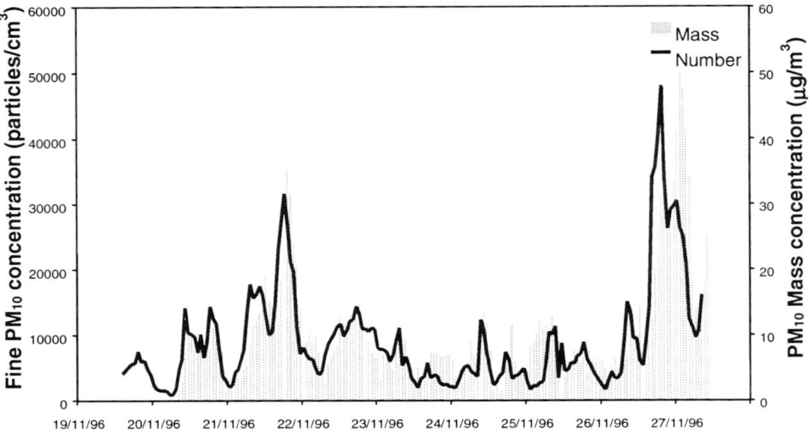

Figure 2. Temporal concentration profiles of particle mass and fine particle number.

Figure 2 shows the temporal concentration profiles of fine particles and particle mass from which can be seen a much better temporal agreement than with the coarse particles. Fine particles dominate the aerosol size distribution where almost all the total particle number concentrations are below 100 nm diameter, of which the majority are in the "ultrafine" range smaller than 50 nm. Whereas the much larger coarse particles make up less than 0.05% of the total particle number (Table 1), they account for most of the particulate mass. However, the temporal number profiles shown here demonstrate better agreement with the mass profile between fine particles than with the coarse particles. Surprisingly, excursions in coarse particle number concentrations are not matched by the mass concentrations as would be expected.

Date	PM_{10} (µg/m3)	Fine PM (#/cm3)	Coarse PM (#/cm3)	Coarse PM (% total #)
20/11/96	7.21	6389	0.86	0.01
21/11/96	13.78	14167	1.74	0.01
22/11/96	9.24	9270	1.33	0.01
23/11/96	5.91	5503	1.15	0.02
24/11/96	5.73	4687	1.27	0.03
25/11/96	7.58	5511	2.69	0.05
26/11/96	13.53	15186	2.84	0.02
27/11/96	28.74	19211	3.88	0.02

Table 1. Daily average mass and number concentrations.

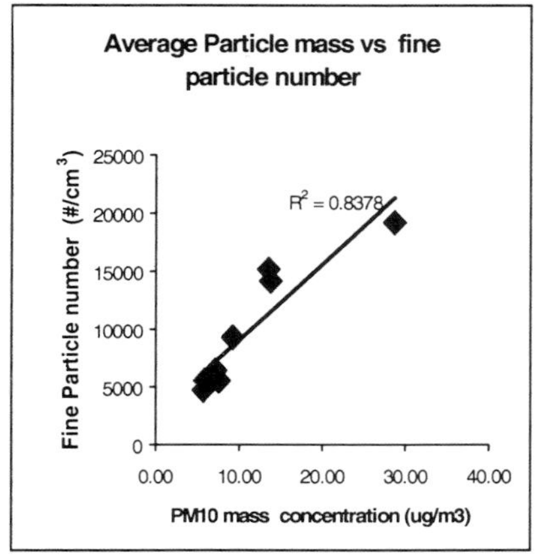

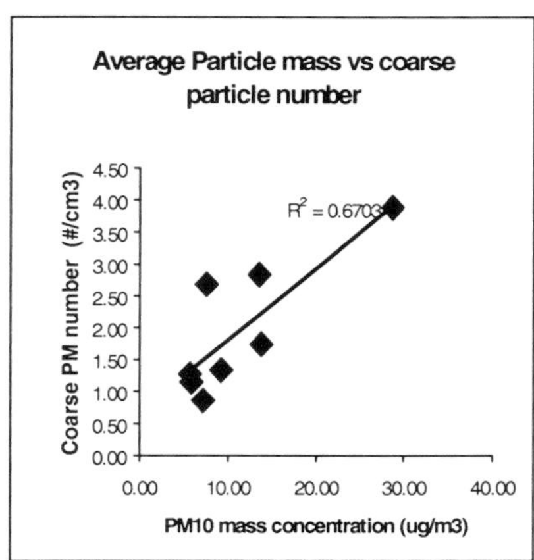

Figure 3. Correlation of daily data for PM10 mass and number concentrations.

The daily averages of both fine and coarse particle number concentrations together with particle mass are shown in Table 1. These values are used to make correlations which are plotted in Figure 3 from which it can be seen that the fine particle concentrations correlate better with the mass concentrations than the coarse particle concentrations. Whilst this finding was somewhat unexpected, it must be emphasised that the time series data represent only a very brief "snapshot". However, the data does show that, although derived by instruments operating at opposite ends of the respirable particle size spectrum, fine particle concentrations do appear to better reflect the mass concentrations. Longer term sampling campaigns are needed to obtain comparative data from different sites.

References

Anderson, H., de Leon, A., Bland J, Bower J, Strachan D 1996. Air Pollution and Daily Mortality in London:1987-92, *British Medical Journal*, 312, 665-669

APEG, 1999. Source apportionment of Airborne Particulate Matter in the United

Kingdom. Report of the Airborne Particles Expert Group, Department of the Environment, Transport and the Regions, UK.

COMEAP, 1995. Health Effects of Non-Biological Particles, Committee on the Medical Effects of Air Pollutants, Department of Health. HMSO, London.

Oberdorster G, Gelein R M, Ferin J, Weiss B. 1995. Association of particulate air pollution and acute mortality: Involvement of ultrafine particles? *Inhalation Toxicol* 7,111-124

QUARG, 1996. Airborne particulate matter in the United Kingdom. Third report of the Quality of Urban Air Review Group, Department of the Environment, UK.

Seaton, A., MacNee, W., Donaldson, K., Godden, D. 1995 Particulate air pollution and acute health effects. *The Lancet* 345,176-178.

APPLICATION OF CFD METHODS FOR MODELLING IN AIR POLLUTION PROBLEMS: POSSIBILITIES AND GAPS

ALEXANDER BAKLANOV,
Danish Meteorological Institute (DMI), Lyngbyvej 100, DK-2100 Copenhagen, Denmark, E-mail: alb@dmi.dk

Abstract. Different urban air pollution problems deal with complex structure of air flows and turbulence. For such problems the Computer Fluid Dynamics (CFD) methods become widely used. However, this approach despite a number of advantages has some problems. Experience of use of CFD tools for development of models and suggestions of their applications for a local scale air pollution over a complex terrain and stable stratification are discussed in this paper, including:

- Topography and complex geometry: choose of the co-ordinate system and computer grid;
- Turbulence closure for air pollution modelling: modified k-ε model for stable stratified ABL;
- Boundary conditions for vertical profiles of velocity for stable-stratified atmosphere;
- Effects of the radiation and thermal budget of inclined surfaces to dispersion of pollutants;
- Artificial sources of air dynamics and circulation.

Some examples of CFD applications for air pollution modelling for a flat terrain, mountainous area, mining open cast and indoor ventilation are discussed. Modified k-ε model for stably-stratified ABL is suggested. Due to the isotropic character of the k-ε model a combination of it in vertical with the sub-grid turbulence closure in horizontal can be more suitable for ABL. An effective scheme of boundary conditions for velocity profiles, based on the developed similarity theory for stable-stratified ABL, is suggested. Alongside with the common studies of atmospheric dispersion, the CFD methods have also demonstrated a good potential for studying anthropogenic and artificial-ventilation sources of air dynamic and circulation in local-scale processes of air pollution.

Key words: air pollution, Computer Fluid Dynamics (CFD) methods, environment modelling, atmospheric boundary layer (ABL), turbulence, stable-stratified atmosphere, complex terrain

1. Introduction

Different urban air pollution problems deal with complex turbulent air flows:

- Local-scale circulations: mountain areas, canyons and valleys, street canyons, mining open casts, industrial buildings and mining workings;
- Thermal effects: internal and stable-stratified (SBL) boundary layers, air circulations, catabatic and anabatic winds, nuclear and industrial accidents;
- Transport and diffusion of heavy gas or particles in atmosphere;
- Artificial and ventilation sources of air dynamics and circulation.

For such problems the Computer Fluid Dynamics (CFD) methods and tools become widely used. Thus, Svensson (1986) was one of the firsts who started to use commercial CFD tools for various air-pollution problems. Delaunay et al. (1996) studied gas dispersion from road tonnels in urban air by CFD and compared it with field data. Sini et al. (1996) included the thermal effects into CFD studies of the pollutant dispersion in urban street canyons. Murena et al. (1997) studied CO dispersion in urban street canyons by commercial CFD without comparisons with

Environmental Monitoring and Assessment **65**: 181–189, 2000.

experiments. Burman (1997) studied the influence of density of a heavy gas on its dispersion over complex terrain. Besides, several developed microscale 3-D numerical models for air dynamics and pollution (for overview see Schatzmann et al.,1997) are analogous to CFD tools. However, the CFD approach despite a number of advantages has many gaps and demands further development.

Experience of use of the CFD tools for development of models and their applications on example of a local scale air pollution over a complex terrain and for stable stratification are discussed in this paper. It includes a problem-oriented CFD software, developed by the author (Baklanov, 1988) and a general-oriented commercial CFD software, like the PHOENICS system of the CHAM company (PHOENICS, 1991), and briefly covers the following aspects:

- Topographical and complex geometry effects: choose of the co-ordinate system and computer grid;
- Turbulence closures for air pollution modelling: modified k-ε model for SBL;
- Boundary conditions for vertical profiles of velocity for SBL;
- Artificial sources of air dynamics and circulation;
- Effects of the energy budget of a surface to dispersion of pollutants.

2. Model description

Models of the ABL dynamics often use the hydrostatic approximation, without consideration atmospheric compressibility, in Boussinesq's approximation. Introducing potential temperature and Exner's function for pressure enables to omit small terms and to linearise non-linear terms in the equations of movement. However, for atmospheric currents in street canyons or over a complex terrain, especially for local scale, when the vertical and horizontal scales are comparable, tear-off lee and wave movements behind mountains or obstacles, in mountain valleys and cirques might play an important role. Hence, a system of equation in non-hydrostatic approximation was considered (Baklanov, 1988, 1995). Besides, in environmental problems, e.g. by modelling possible consequences after a hypothetical nuclear accident or fires, large thermal sources and temperature contrasts of hundred degrees might occur. For such cases, the approximation of free convection could be insufficient, and for the model of local atmospheric processes under influence of large thermal sources the fully compressible equations of hydrodynamics were used (Baklanov *et al.*, 1996).

To describe topographical and complex-geometry effects, 3 methods were tested:

1. A terrain-following "sigma z" vertical co-ordinate was employed for a complex terrain. Main advantage of the method is a possibility to increase resolution of the computer grid close to the surface. However, it performs poorly for slopes with more than 30 degrees of declination.

2. Method of fiction domains (Aloyan *et al.*, 1982) for a very complex terrain. The method is suitable for street canyons, local areas with strong topography and

indoor pollution. It can be used for any domain geometry, however has limited possibilities to increase grid resolution close to complex-form surface.
3. Body-fitted co-ordinates (BFC) with the finite volume numerical method (PHOENICS, 1991). It is universal, although sensitive to types of grid generation.

Different dispersion models of Eulerian and Lagrangian type can be used with CFD for air pollution applications. For testing and verification of the CFD methods for atmospheric pollution an Eulerian 3-D model of pollution transport was used (Baklanov *et al.*, 1996).

3. Turbulence closure models for air dynamics and pollution modelling

Correct turbulence closure is very important for estimation of turbulent diffusion, wind fields and mixing height h estimation for dispersion modelling. Different turbulence closure models were used: Smagorinsky-Lilli's model of sub-grid turbulence (Galperin & Orszag, 1993), modified k-ε model, k-l model and others.

The standard k-ε model of the turbulence closure (Rodi, 1987) is widely used in many applications of CFD, but has some problems for modelling of ABL dynamics. So, let us consider it in more details. One of the effects is that the model overestimates the turbulent length scale and eddy viscosity above the shear layer. Detering & Etling (1985) modified the k-ε model for neutral-stratified ABL by correction of the model constant $c'_{1\varepsilon}$ depending on the turbulent length scale and the characteristic scale for the ABL. However, for cases of the stable-stratified boundary layer the comparison with empirical data can give insufficient coincidence. We suggest a modified k-ε model for a stable stratification case by a formulation of the constant (a function) $c'_{1\varepsilon}$ of the k-ε model for SBL as following:

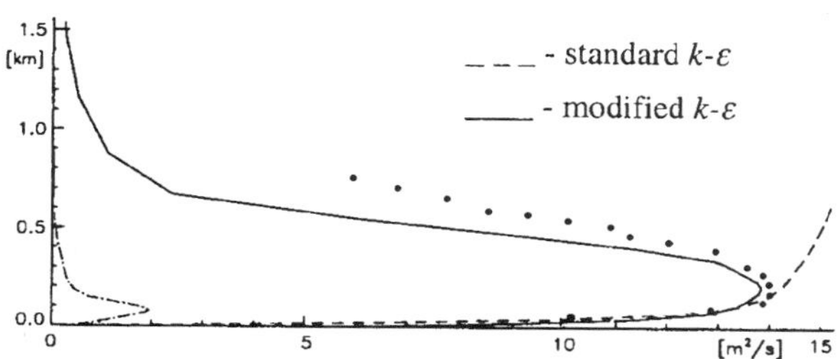

Fig. 1. The eddy viscosity for the neutral (line) and stable (dash dot) cases. • - measurement data (Lettau, 1962).

$$c'_{1\varepsilon} = c_{1\varepsilon} \frac{c_D k^{3/2} f^{1/2}}{\varepsilon d_h (u_* L)^{1/2}}, \qquad (1)$$

where k -turbulent kinetic energy, ε - its dissipation, f - Coriolis parameter, L - Monin-Obukhov's scale, u_* - the surface friction velocity, c_D and $c_{1\varepsilon}$ - constants in the standard k-ε model, and preliminary d_h=0.0009, but to be optimised by experimental data. Fig. 1 demonstrates the eddy profiles of the modified k-ε model for different ABL stratification and comparison with the Leipzig profile for near-neutral case (Lettau, 1967). Eddy profile by the standard k-ε model (dash line) shows its overestimation by factor of 10-12 over the surface layer and the mixing height, one of very important characteristics for dispersion modelling, is several times overestimated.

The *k-l* model shows good results for a flat terrain, but for complex flows determination of *l* becomes problematic. The sub-grid turbulence model (with $\tilde{k}$ = 0.26) gives a very good correlation with experimental data for complex flows (see Table 1). However, using it in LES mode for stable conditions demands a very high grid resolution and is expensive for applications. Therefore, modified k-ε models currently appear to be the most tested and applicable. Due to the isotropic character of the k-ε model using it in vertical coupled with the sub-grid approach in horizontal can be more suitable for ABL. However, for modelling of non-local turbulence effects and non-steady regimes the k-ε model has serious gaps.

Table 1. Comparison of the velocities of the numerical simulation and physical model measurements for a symmetric canyon (Kalabin *et al.*, 1990)

Distance, m	0.0	0.2	0.4	0.6	0.8	1.0	1.1
Numerical model	0.60	0.46	0.41	0.38	0.36	0.31	0.60
Hydromodel	0.60	0.47	0.40	0.38	0.32	0.25	0.60

4. Boundary conditions for vertical profiles of velocity for SBL

Since model results can be strongly dependent on the boundary conditions in any CFD model, various boundary conditions were set and their influence on modelling results was analysed (Baklanov *et al.*, 1996).

One of considerable problems of numerical local and meso-scale modelling of stable-stratified ABL is discrepancy of input and output boundary conditions for meteorological profiles, that take into account the Coriolis forces, surface layer structure etc. An effective scheme of boundary conditions for inlet velocity profiles, based on the similarity theory for SBL (Zilitinkevich *et al.* 1998), was applied:

$$u = \frac{u_*}{k}\left[\ln\frac{z}{z_{0u}} + b_1\zeta + b_2\Pi\zeta^2 + b_3\zeta^3\right],\quad v = -\frac{fh}{k^2}\left[-\zeta\ln\zeta + a_1\zeta + a_2\zeta^2 + a_3\zeta^3\right], \quad (2)$$

where, u and v - components parallel and perpendicular to the surface stress, z_{0u} - surface roughness length, k - von Karman constant, a_1=4δ^{-2}+$\tilde{\Pi}$, a_2=-3/2$\tilde{\Pi}$, a_3=1/3-4/3δ^{-2}+2/3$\tilde{\Pi}$, b_1=-3+Π, b_2=-3/2Π, b_3=2/3+2/3Π; $\Pi \equiv C_R\delta^2 + C_L h/L + C_N Nh/u_*$ and $\tilde{\Pi} \equiv \tilde{C}_R\delta^2 + \tilde{C}_L h/L + \tilde{C}_N Nh/u_*$ - composite stratification/rotation parameters, $\zeta \equiv z/h$ and $\delta \equiv fh/ku_*$ - the dimensionless height and rotation rate, N - Brunt-Väisälä frequency, h - ABL height. Preliminary comparison with experimental and LES data (Zilitinkevich *et al.*, 1998) gives the following constant values: $C_R = 7, \tilde{C}_R = 0$, $C_L = 4.5, \tilde{C}_L = -7$ $C_N = 0.4, \tilde{C}_N = -1.5$. Montavon (1998), based on a k-ε model for SBL, corrected the constants, however, due to the above-mentioned problems of k-ε models, they need further verification with measurement data.

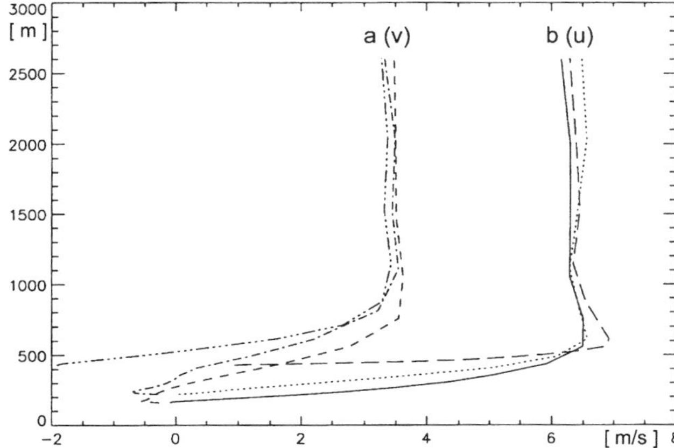

Fig. 2. The horizontal velocity profiles (a: v-component, b: *u*-component) at different distances: (dashed line (a) and line (b) - in front of the mountain, in the beginning of the area; dash -3 dots (a) and dashed line (b) - on the top of the mountain; dash dot (a) and dot line (b) - behind the mountain, in the centre of the area.

Fig. 2 shows a simulation of wind fields over a separate mountain (400 m) for the Monchegorsk experiment (Baklanov *et al.*, 1996). It presents vertical profile of the wind horizontal components *u* and v over different part of topographical surface. Already from the first site, close to the boundary, influence of Coriolis effect is clearly observed as negative values of v in the low layers. The wind profile over the mountain top has a characteristic intensifying the wind over the surface at the expanse of compressing the flow, and behind the mountain slope in the centre of the domain the surface velocity has some weakened in comparison with initial profile in the beginning of the domain, due to topographical resistance.

5. Effects of the radiation and thermal budget of a complex surface

For street canyons and areas with complex topography during a calm the radiation and thermal budget of a surface can strongly affect dispersion of pollutants, however most of the models don't pay enough attention to this. Let's determine the surface temperature from the equation of radiation and thermal budget at the boundary 'atmosphere-soil':

$$c_s \rho_c (k_s \frac{\partial T_s}{\partial n})_\delta - \rho c_p (\frac{\nu_t}{\sigma_{T\Theta}} \frac{\partial \theta}{\partial n})_\delta - \rho L_w (\frac{\nu_t}{\sigma_{T\Theta}} \frac{\partial q}{\partial n})_\delta = R_\delta + I_T \quad (3)$$

where the 1st term - soil heat flux at the surface δ; 2d term H_δ - sensible heat flux from soil to the atmosphere; 3d term- latent heat flux; R_δ - radiation budget of the surface; I_r - anthropogenic surface heat fluxes, c_s, ρ_s - specific heat and density of the soil; k_s - soil heat conduction coefficient, θ, q- air potential temperature and specific humidity, T_s - soil temperature, $\sigma_{T\Theta}$ -turbulent Prandtl number for heat.

The effect of large incidence angles of the slopes/walls in mountain/street canyons was accounted for (Baklanov *et al.*, 1996). Beside orientation and incidence (Fig. 3a), the radiation budget also depends on the fluxes of its various components. Radiation may be reflected from one slope/wall to another or from a horizontal to an inclined surface, etc. For the radiation budget of inclined surfaces we used the following equation: $R_\delta(x,t) = (S_\delta + D_\delta)(1 - A_\delta) - F_\delta$, where S_δ, D_δ - fluxes of direct and dissipated solar short-wave radiation onto the inclined surface; A_δ - albedo of the surface; F_δ - effective long-wave radiation flux onto the slope.

For the flux S_δ onto an arbitrary-oriented surface the following equation is used:

$$S_\delta = S_0 \frac{p_m}{r_c^2} \{ \cos\alpha(\sin\varphi \sin\delta_0 + \cos\varphi \cos\delta_0 \cos\Omega) + \sin\alpha[\cos\psi_n *$$

$$*(\sin\varphi \cos\delta_0 \cos\Omega - \cos\varphi \sin\delta_0) + \sin\psi_n \cos\delta_0 \sin\Omega \,]\,\}, \qquad (4)$$

where φ - latitude; Ω - solar hour angle; δ_0 - declination of the Sun at a given time, calculated from the real middle of the day, α - slope angle. Areas with illumination and shadow where determined at every time step after finding the boundary of the shadow at the surface. Assuming that D_δ and F_δ are isotropic, one can obtain the following relation for slopes: $D_\delta = D\cos^2\frac{\alpha}{2}$; $F_\delta = F\cos^2\frac{\alpha}{2}$, where D, F - fluxes of the dissipated and effective long-wave radiation for a horizontal surface. For calculation of incoming F^+ and outgoing F^- components F let's use the van Ulden & Holtslag (1985) correlation and the Stefan-Boltzmann law: $F^+ = \sigma c_1 T_r^6 + c_2 N$; $F^- = \sigma\varepsilon_* T_\delta^4$; where T_r - air temperature at a reference height, N - fraction of the sky, covered with clouds, c_1 and c_2 - empirical coefficients, ε_* - surface emissivity, σ - Stephan-Boltzman constant.

Calculations of the soil heat flux and the skin surface temperature for complex structure of the soil were done by using a complete set of equations of soil conductivity (Baklanov 1988) or by a more economical method, based on the force-restore equation for the surface skin temperature (Deardorf 1978). For the sensible heat flux from the soil to the atmosphere H_δ a parameterisation based on the Monin-Obukhov similarity theory was used. However, models with flux parametrisation in the surface layer based on the similarity theory can underestimate the turbulent heat flux for stable stratification and should be improved.

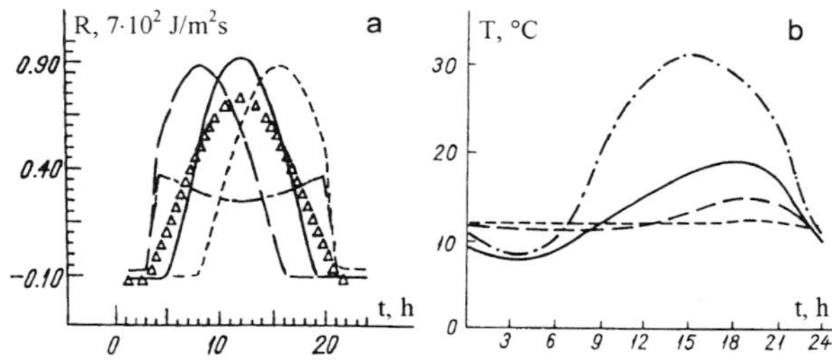

Fig.3. Daily alteration of thermal parameters on the different slopes: a - radiation budget of: __ -S, _._. - N, -- -E, _ _ - W slopes, ΔΔΔ -horizontal surface; b - temperature of: __- air, _._. - surface, _ _ - soil 0.25 m depth, --- - 0.5 m depth.

6. Artificial and ventilation sources of air dynamics and circulation

One of the methods to decrease air pollution concentrations in localised zones is artificial enhancement of the atmospheric ability for pollutant dispersion and carrying-out. Such a method, common for indoor ventilation and tunnels, in the last decades was used for larger scale open systems, such as open pits, street canyons and even city districts. But the efficiency of such a method is doubtful and demands further examination. CFD-based models of artificial impact (albedo shift, thermics, fan turbulent jets and so on) on the processes of ventilation in local zones for such objects are offered (Baklanov & Rigina, 1994). Due to the sub-grid scale of initial vent jets, grid nesting and analytical models were suggested for simulation of primary part of vent jets (Baklanov, 1988).

Figure 4 presents simulated fields of velocity and isolines of concentrations for artificial ventilation of the Saamsky open-cast (depth 230m) by 2 fans: horizontal (9.6 m/s) and vertical (7.5 m/s). The emission source is at the bottom. The ventilators are located to creat an air corridor: the horizontal fan sends polluted air to the zone of operation of the vertical one, wich throws out the pollution outside the open pit. In the case 'a' a recirculated zone has been formed, the pollution is diluted by effective intermixture of the air in the zone of jets operation. A particular carrying-out pollutant outside the open pit occurs.

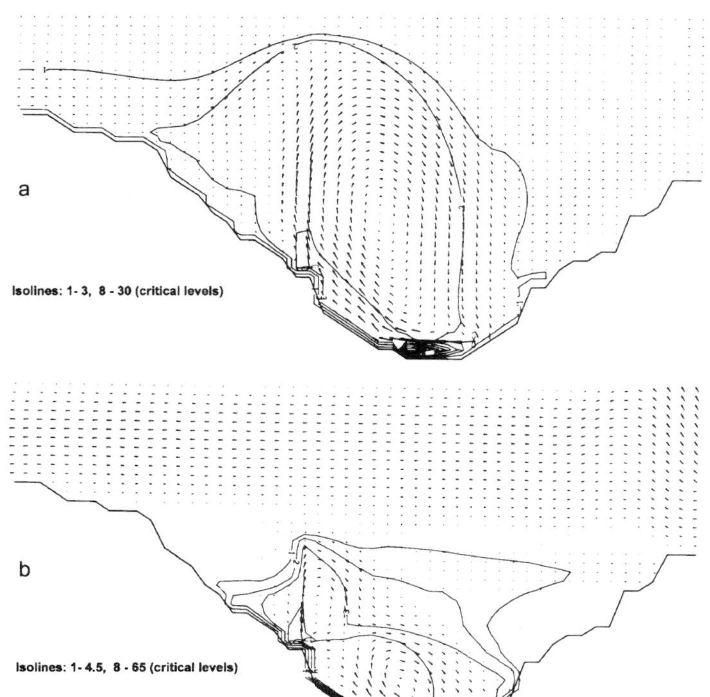

Fig. 4. Artificial ventilation of the Saamsky open pit by two fans with horizontal and vertical currents (vector field of velocity and isolines of concentration *C* in the vertical cross section):
a - neutral stratification, b - inversion γ=-0.02°C/m.

During stable stratification γ = -0.02°C/m (Fig. 4b) two recirculated zones are formed, where the pollutant is mixed, but the area of active circulation, the degree of opening and the range of the vertical current are sharply decreased, carrying-out the pollution from the open pit doesn't happen. Mean concentration in the working part of the open-pit is more than twice as high as in case 'a'. Exploitation of this ventilation scheme in the open-cast during a few years supported the modelling results: during inversion the ventilation scheme was ineffective and didn't provide carrying-out of pollutants.

7. Summary and Conclusion

The examples demonstrate the possibilities of CFD techniques in solving a broad range of air pollution problems. The CFD techniques are rather complex, but capable of handling situations where due to complex geometry, thermal effects or flow conditions other simpler methods fail. Large computer time necessary for CFD simulations makes fine resolution of the flow field difficult.

The standard k-ε model of the turbulence closure is widespread in many applications of CFD, but has to be corrected for modelling of ABL dynamics. Modified k-ε model is suggested for case of stable stratification.

The k-l model shows good results for a flat terrain, but for complex flows the determination of l becomes problematic. The sub-grid turbulence model gives a very good correlation with experimental data. However, using it in LES mode for stable conditions demands a very high grid resolution and is expensive for applications. Therefore, modified k-ε models currently appear to be the most verified and applicable. Due to the isotropic character of the k-ε model a combination of it in vertical with the sub-grid approach in horizontal can be more suitable for ABL. However, for modelling of non-local turbulence effects and non-steady regimes the k-ε model has some gaps.

Considerable problem is discrepancy of input and output boundary conditions for meteorological profiles taking into account the Coriolis forces, surface layer structure etc. An effective scheme of boundary conditions for velocity profiles, based on the developed similarity theory for stable-stratified ABL, is suggested.

For street canyons and areas with complex topography for calm situations the radiation and thermal budget of a surface can strongly affect dispersion of pollutants. Models with flux parametrisation in the surface layer based on Monin-Obukhov's similarity theory can underestimate the turbulent heat fluxes for stable stratification and should be improved.

CFD applications for air pollution modelling for mining open cast and indoor ventilation by artificial sources and turbulent jets are also tested. Alongside with the common studies of atmospheric dispersion and urban air pollution, the CFD methods have also demonstrated a good potential for studying anthropogenic and artificial-ventilation sources of air dynamic and circulation in air pollution problems.

Acknowledgement

This study was partly done at the Russian Academy of Science: Kola Science Centre and Computing Centre of Siberian Division and partly within a grant of the

Swedish Defence Research Establishment (FOA) in Umeå, and the author is very grateful to his colleagues Prof. S. Zilitinkevich, V. Penenko, A. Aloyan, Dr. E. Näslund, P.-E. Johansson, J. Burman and O. Rigina for fruitful collaboration.

References

Aloyan, A., Baklanov, A. & Penenko, V.: 1982, Fictitious regions in numerical simulation of quarry ventilation, *Soviet Meteorology and Hydrology*, **7**, pp.32-37.

Baklanov, A.: 1988, Numerical modelling in mine aerology, Apatity: KSC RAS, 200 p. (in Russian)

Baklanov, A.: 1995, Numerical modelling of atmosphere processes in mountain cirques and open pits, *Air Pollution III,* V.1: Theory and Simulation, Comp. Mech. Publ., pp.231-237.

Baklanov, A., Burman, J. & Näslund, E.: 1996, Numerical modelling of three-dimensional flow and pollution transport over complex terrain during stable stratification, Scientific report, FOA, Umea, Sweden (FOA-R-96-00376-4.5-SE), 40 p.

Baklanov, A. & Rigina, O.: 1994, Research of local zones atmosphere normalisation efficiency by artificial currents, *Air Pollution II*, V.1: Computer Simulation, Comp. Mech. Publ., 553-561.

Burman, J.: 1997, A study of the influence of topography and density on the dispersion in a gas cloud, Scientific report, FOA, Umea, Sweden (FOA-R-96-00304-4.5-SE), 27 p.

Deardorf, J.W.: 1978, Efficient prediction of ground surface temperature and moisture with inclusion of a layer of vegetation, *J. Geophys. Res.*, **83C**, 1889-1903.

Delaunay, D., Flori, J.P. & Sacré, C.: 1996, Numerical modelling of gas dispersion from road tunnels in urban environments: Comparison with field experiment data, 4th Workshop on Harmonization within Atmospheric Dispersion Modelling for Regulatory Purposes, Ostende, Belgique, 361-368.

Detering, H. W. & Etling, D.: 1985, Application of the E-ε turbulence model to the atmospheric boundary layer, *Boundary-Layer Meteorology,* **33** , 113-133

Galperin, B. and Orszag, S.A. (Editors): 1993, Large Eddy Simulation of Complex Engineering and Geophysical Flows, Cambridge University Press.

Kalabin, G., Baklanov, A. & Amosov, P.: 1990, Calculating the aerogas dynamics of chamber-like workings on the basis of mathematical modelling, *Sov. Min. Sci.*, Vol. **26**:1, 61-73.

Lettau, H.H.: 1962, Theoretical wind spirals in the boundary layer of a barotropic atmosphere. *Beitr. Phys. Atm.* 35, 195-212.

Montavon, C.: 1998, Simulation of atmospheric flows over complex terrain for wind power potential assessment, PhD thesis #1855, Lausanne, EPFL.

Murena, F., d'Allesandro, S. & Gioia, F.: 1997, CO dispersion in urban street canyons, *PHOENICS Journal*, v. **10**, n. 4, 407-415.

PHOENICS: 1991, The PHOENICS Reference Manual. CHAM Development team. UK.

Rodi, W.:1987, Examples of Calculation Methods for Flow and Mixing in Stratified Fluids, *Journal of Geophysical Research,* **92**, no C5, 5305-5328.

Schatzmann, M., Rafailidias, S., Britter, R. & Arend, M. (Editors): 1997: Database, monitoring and modelling of urban air pollution. Inventory of models and data sets. EC COST 615 Action. 109p.

Sini, J.-F., Anquentin, S. & Mestayer, P.G.: 1996, Pollutant dispersion and thermal effects in urban street canyons. *Atmos. Envir.*, **30**:15, pp. 2659-2677.

Svensson, U.: 1986, PHOENICS in environmental flows. A review of applications at SMHI. *Lecture Notes in Engineering* **18**, pp 87-96.

Van Ulden, A. and Holtslag, A.A.M.: 1985, Estimation of atmospheric boundary layer parameters for diffusion applications, *J. Clim. Appl. Meteor.*, **24**, pp. 1196-1207.

Zilitinkevich, S., P.-E. Johansson, D. V. Mironov & A. Baklanov: 1998, A similarity-theory model for wind profile and resistance law in stably stratified planetary boundary layers. *J. Wind Eng. Ind. Aerodyn.*, v. **74-76**, n. 1-3, pp. 209-218.

Environmental wind tunnel study on a municipal waste incinerator

Corti A.(*), Contini D.(*), Manfrida G.(*), Procino L.(**)

(*) Dipartimento di Energetica «Sergio Stecco» Univ. Degli Studi di Firenze - via S. Marta 3 50139 Firenze

(**) CRIACIV – Dipartimento di Ingegneria Civile Univ. Degli Studi di Firenze - via S. Marta 3 50139 Firenze

Abstract In this paper results concerning a diffusion experiment on a small scale model of a waste incinerator, carried out in a boundary layer wind tunnel, will be presented. At the beginning a description of the measurement system that has recently been placed in the wind tunnel will be given together with a study of repeatability of the measurements results. A comprehensive study of the flow mean characteristics at different position in the tunnel working section will be reported and at the end some vertical and horizontal concentration profiles results will be described and discussed.

Keywords: Wind tunnel, environmental impact, plume dispersal models, FID

1. Introduction

Recently the boundary-layer developed wind tunnel of the "Centro Interuniversitario di Aerodinamica delle Costruzioni ed Ingegneria del Vento" (C.R.I.A.C.I.V.) has been equipped with a system that allows tracer mean concentrations in gaseous samples to be measured. The systems is based on a Flame Ionisation Detector (FID) and allows 12 samples of gas, containing a known tracer, to be taken from different positions inside the tunnel and analysed on-line so that the average concentration can be evaluated with reference to previous calibrations of the FID unit [Contini 1998].

In the particular field of environmental studies of diffusion of emissions, small-scale experimental results are very useful for developing, improving or testing numerical codes, which up to now rely strongly on empirical parameters and/or on field data sets which are affected by large uncertainties; moreover, the analysis of the effects at ground level deriving from emissions of pollutants from man-made sources can be evaluated in spite of the traditional use of mathematical diffusional codes for environmental impact assessments.

In the present case, experimental results of a small scale model of a waste incinerator featuring two parallel stacks 60 m tall are presented (figure 1).

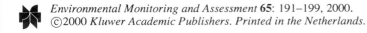

Environmental Monitoring and Assessment **65**: 191–199, 2000.

The CRIACIV environmental wind tunnel is not able to simulate temperature stratification; the presented results refer to effects determined with neutral atmospheric conditions. Within this limit, an accurate statistical study of the local meteorology was performed in order to simulate the most representative atmospheric conditions.

For the characterisation of diffusion phenomena acting on the plumes emitted from the stacks, measurements on the spatial distribution of the pollutant plume were made in order to establish its main parameters: effective height of plume, land reflection effect and vertical and horizontal dispersion parameters (σ_z and σ_y).

2. Case study

At a preliminary phase of the study, a Gaussian diffusion model (US-EPA ISC3), modified for the wind profile parameter (mean wind profile exponent), was used to evaluate approximately the dispersion effects of the main pollutants over the area at ground level. Consequently, an adequate scale (1:400) was selected for the wind tunnel measurements, considering both the available dimensions and the necessity of a good geometrical representation.

In table 1 a description of the main parameters together with their values in the full scale and in the small scale model is reported.

A standard buoyancy scaling relationship has been used to evaluate the correct flow and emission conditions in the small scale model [Obasaju & Robins 1998, Robins 1980]. The scaling relationship for the concentration C is based on the non-dimensional equation:

$$\frac{CU_{ref}H^2}{Q} \approx f\left(\frac{\rho_{gas}W^2}{\rho_a U_{ref}^2}, \frac{(\rho_{gas}-\rho_a)gH}{\rho_{gas}U_{ref}^2}\right) \tag{1}$$

which implies the conservation in the small scale model of the two parameters on the right side of the equation, describing respectively the plume momentum and the plume buoyancy. For a complete evaluation of the ground-level pollutant concentrations for different conditions of wind directions, different orientations of the model were studied using a rotating table. Average conditions for wind velocity were considered corresponding to a velocity at 10 meters of about 6 m/s equivalent to about 8.5 m/s at stack height. In equation (1) the non-dimensional concentration field is obtained normalising with the reference flow speed U_{ref} evaluated outside the boundary-layer and the boundary layer depth H.

Parameter description	Full scale	Small scale (1:400)
Stack height (h)	60 m	0.15 m
Stack internal diameter (D_i)	1.5 m	0.00375 m
Stack flow-rate (Q)	77000 Nm^3/h	1.96 l/min
Vertical exit speed of gas (W)	17.5 m/s	2.95 m/s
Mean wind profile exponent (α_p)	0.19	0.19
Mean wind speed at the stack height (U_{stack})	8.5 m/s	0.95 m/s
Reference flow speed U_{ref}		1.2 m/s
Ambient temperature (constant) (T_a)	298.16 °K	298.16 ° K
Emitted gas temperature (at the stack) (T_s)	398.16 °K	298.16 ° K
Density ratio ($\alpha=\rho_{gas}/\rho_a$)	0.692	0.307
Boundary-layer depth (H)	280 m	0.7 m

Table 1) Values of the parameters in the full scale and in the small scale model. Flow velocities in the small scale model are given with an accuracy of about ±10%

With the considered scale factor a maximum value of 280 m in full scale for boundary layer is available, with the present wind tunnel configuration.

3. Experimental set-up

A 0.7 m tall neutral boundary-layer was developed by using spires as vortex generator [Irwin 1980] at the tunnel inlet and roughness elements, with variable height from 10 to 50 mm (the shorter close to the model), covering the entire tunnel length (figure 2). Different layout of roughness panel and spire dimensions were tested in order to obtain a representative wind profile for a rural location.

With the final layout of the wind tunnel turbulence promoters, an average wind speed profile with a power-law exponent in good agreement with the values encountered at rural site was found.

Figure 1 – Small scale model placed inside the wind tunnel.

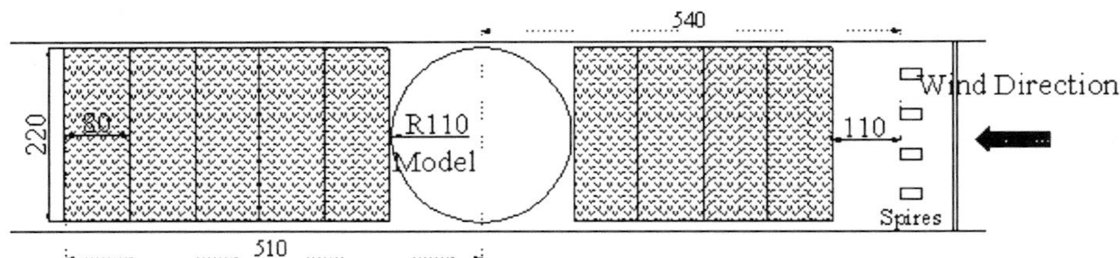

Figure 2 – Inside Wind Tunnel layout used for generating the neutral boundary-layer

Wind profile measurements were repeated at different distances from the stacks in order to check that the flow is completely developed and with small gradients in the mean property in the longitudinal direction; also the turbulence intensity distribution should be compatible to what can be expected for rural conditions, with sufficiently low values at the higher levels of the profile.

A range of variation on the wind power law exponent from 0.16 to 0.19 depending on the distance from the stack was obtained in the measurement (Figure 3; wind velocity is normalised with respect to the value in the external flow, that is , at a reference height of 1.4 m from the wind tunnel ground level).

The profile, fitted with a logarithmic law, has a frictional velocity u^*/U_{ref} = 0.058 and a roughness length z_o=0.84 mm, that is equivalent to 0.34 m in real scale, with a good approximation to rural standard conditions. The log profile fitted to the measurement data is also reported in figure 3.

Mean velocity and turbulent profiles were obtained using a hot wire anemometer with a averaging time of about 4.5 minutes at a sampling frequency of 500 Hz. Results show that the flow is not changing appreciably, close to the stack, in the longitudinal direction so that the boundary-layer is quite well developed in spite of the non standard position of the model only 6H from the vorticity generator, chosen in order to have a measurements zone of about 5m downwind of the stacks. At 1600m from the stacks some small changes in the boundary-layer are present probably as a consequence of the end of the fetch in the tunnel. This zone. in fact, is near the converging section of the tunnel leading to the fan. This is necessary to study ground level concentration (GLC) on a range of about 2km downwind of the stacks because there is a village placed approximately between 1.5 and 2.5 km from the stacks.

Simulations of plume dispersion were carried out by using a mixture of Helium and Ethylene (20% in volume) in order to have both a suitable tracer and the correct plume buoyancy.

Concentrations of tracer emitted from the stacks is measured for vertical and horizontal traverses based on a sampling valve unit with 12 inputs; air sampled from the measurement zone is continuously flushed by a vacuum pump; in this way the system can be represented as a continuos sampling, and the control of the sampling valve unit implies just a a few seconds delay for stabilising the input line to the FID detector. This last is really a gas chromatography unit, inclusive of complete sample conditioning, and is used to feed the FID detector which can reveal the presence of hydrocarbons such as ethylene. The gas cromatograph, used as a total hydrocarbon analyser, is based on a Flame Ionisation Detector calibrated by using a mixture of air and ethylene with a volumetric concentration of 53.5 ppm (the measurement range being between 10 and 200 ppm).

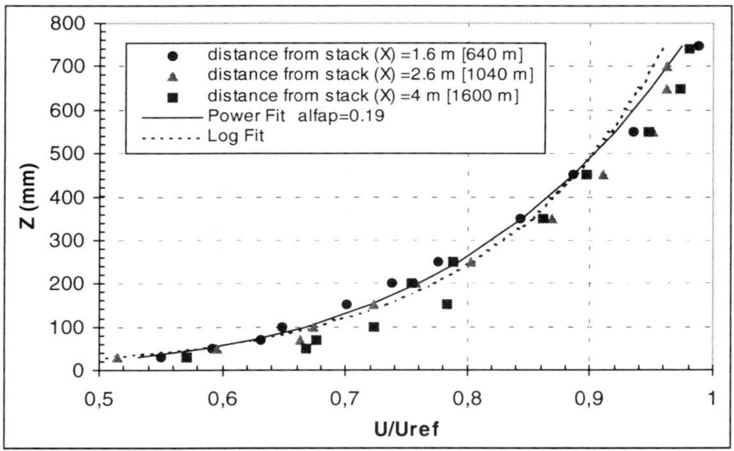

Fig. 3) Mean velocity profiles at different distances from stacks.

Due to the fact that the wind tunnel is operated in an industrial building, forming thus a closed system, and considering the time necessary for the execution of a single campaign of sampling and measurement, a background probe located up-flow with respect to the emission points (stacks) was used for the evaluation of the background concentration of ethylene. For this background tracer increment, a linear variation with time was considered, measuring its value as first and last sample during each single campaign.

A phase of the study was dedicated to the validation of the measurements results, in order to ascertain the correct behaviour of the sampling and analysis line from the probe to the analysis system (FID). This test was executed connecting directly each sampling probe to a precisely calibrated mixture of air and ethylene. The results obtained show a good uniformity among measurements obtained fluxing the mixtures across the sampling line and the reference value (dotted line) obtained by a direct connection to the FID unit (figure 5).

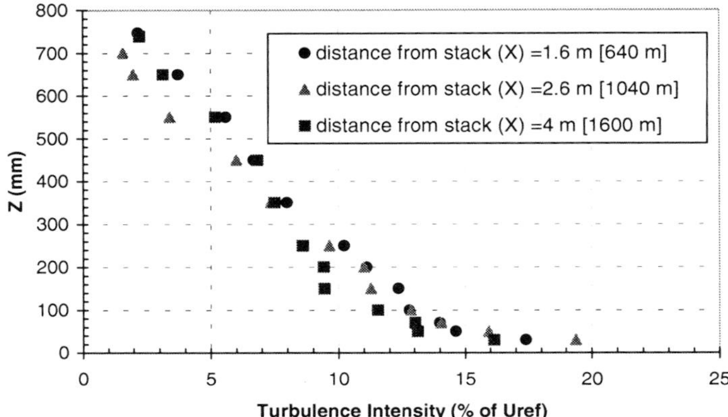

Fig. 4) Turbulence intensity profiles for different downwind distances from the stacks.

Because of the many variables involved in a diffusion experiment, each with its own random errors, it is difficult to give an analytical description of the experimental uncertainty. An estimate of this uncertainty can be obtained by repeating the same measurements in nominal identical conditions and looking at the difference in the results. Therefore some horizontal concentration profiles taken at full stack height have been repeated in order to estimate the repeatability.

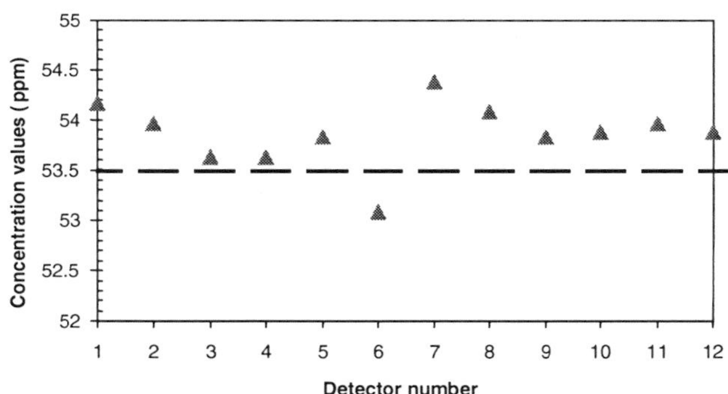

Fig. 5) Line sampling measuring of the calibration mixture. Horizontal dashed line represent the concentration of the calibration mixture to be measured

Results are shown in figure 6 where two concentration measurements profiles, each taken at X=640 and X=1040m, are shown; the average line fit and a band representing 15% of the local mean value is also shown.

All the measured points are within the 15% band (a part from two boundary points), which means that a repeatability of 15% can be reasonably expected, with a better precision close to the maximum concentration.

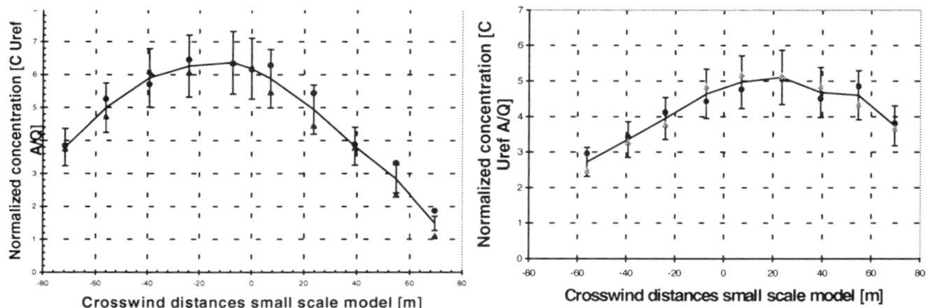

Fig. 6) Horizontal profile variation for X=640m (left) and X=1040 m(full scale),range of variation of 15% is shown.

4. Results

Measurements of vertical and horizontal concentration profiles were carried out at different distances downstream from the stacks. Different wind directions were analysed in order to evaluate different conditions of diffusion due to the presence of industrial buildings around the stacks. Here only one direction is reported, this is the most important because is the one in which the wind is blowing towards the inhabited village.

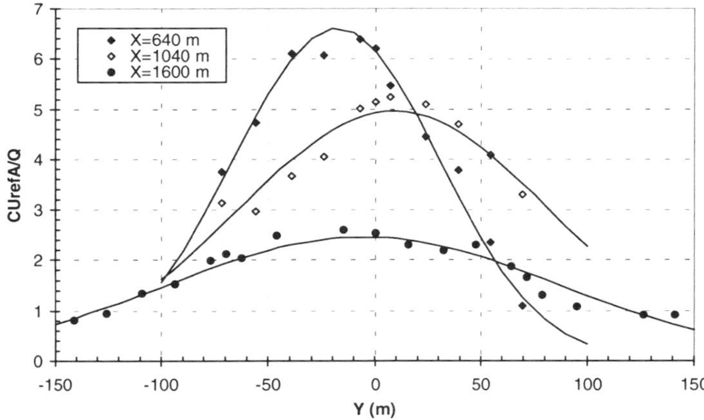

Fig.7) Variation of vertical concentration profiles with distance downwind from stacks measured at the plume centreline.

The vertical profile of concentrations in the plume was measured at three points downwind the sources at 1600, 2600 and 4000 mm (in small scale model dimensions) equivalent to approximately a range of measurement downwind from

640 to 1600 meters in real scale dimensions. Results are shown in figure 7 together with a fit with a Gaussian profiles (without the reflection terms). Measurements show a plume that is continuously spreading and the values of the σ_z evaluated by using the Gaussian fit are 48.13 m, 61.1 m and 79.31 m for X=640m, X=1040 and X=1600m respectively.

These values are relatively large with respect to the plume centreline height so therefore an important effect of ground reflection is present. The effect of terrain reflection is evident for the plume profile at 1600 meters real scale distance downwind the stacks. From data obtained in the present case, with an high wind velocity, plume raise is not so relevant.

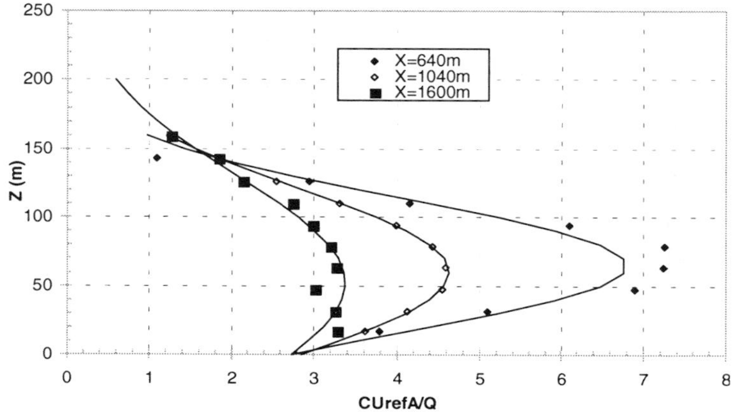

Fig. 8) Horizontal concentration profiles at three downwind distances (marks) with the results of the Gaussian fit (continuous lines).

Some examples of results for the horizontal concentration profiles (cross-wind) taken at full stack height are reported in figure 8 together with a Gaussian fit. In this case the measurements provide the following values for the plume lateral spread σ_y: 48.22m, 73m and 93.1m for X=640, X=1040m and X=1600m respectively. This values are consistent with the vertical spread and slightly larger as usually predicted in this situations by the analytical Gaussian codes. The centerline position obtained by the Gaussian fit is different for the three cross-wind profiles being –18m for X=640m, about 8m for X=1040m and about -5.2m for X=1600m. In the small scale model these values correspond to only about some tens of millimeters and they could indicate some minor local flow misalignment. The effects of the disturbances are also evident in the ground level concentration map reported in figure 9. This map has been obtained by five longitudinal lines of 11 detectors placed at Y=0, Y=±54m and Y=±126m. The results show the value of the maximum GLC and its position downwind of the stacks at about 850m.

Maximum value of nitrogen dioxide concentration was estimated to be 59.3 $\mu g/m^3$, well inside air quality regulation Italian limits (200 $\mu g/m^3$), considering an emission of NO_X equal to 400 mg/Nm^3 at the two stacks.

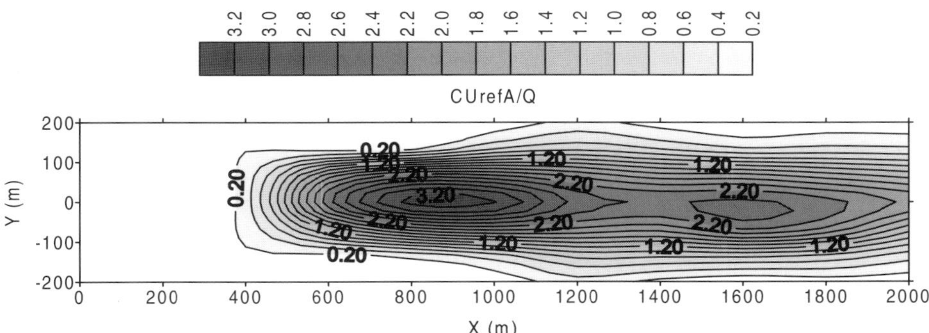

Fig. 9) Ground level concentration (GLC) map

5. Conclusions

The measurement should be integrated with a more extensive characterisation of the flow-field and some complete maps of the ground level concentration field, possibly eliminating some distortions in the flow. However the results presented show that the overall system, that is up to now a unique facility in Italy, is working quite well with a repeatability of the concentration profiles measurements of about 15%. The first series of results presented shows therefore the potential of the system allowing useful information for environmental impact assessment and mathematical simulation code validation.

6. References

Contini D., Ph. D. Thesis, Mechanical Engineering, Florence University, 1998.

Obasaju E. D., Robins A. G., Environmental Monitoring and Assessment 52, pp. 239-254, 1998.

Irwin. P. A., Journal of Wind Eng. and Ind. Aerod., 1981.

Robins A. G., Studies in Environmental Science n. 8, France 1980.

ON THE PERFORMANCE AND APPLICABILITY OF NONLINEAR TWO-EQUATION TURBULENCE MODELS FOR URBAN AIR QUALITY MODELLING

JAN EHRHARD and RAINER KUNZ
Institut für Technische Thermodynamik, Universität Karlsruhe (TH), D-76128 Karlsruhe, Germany
E-Mail: jan@ehrhard.de

NICOLAS MOUSSIOPOULOS
Laboratory of Heat Transfer and Environmental Engineering, Aristotle University Thessaloniki, GR-54006 Thessaloniki, Greece

Abstract. It is well known that the commonly used k-ϵ turbulence models yield inaccurate predictions for complex flow fields. One reason for this inaccuracy is the misrepresentation of Reynolds stress differences. Nonlinear turbulence models are capable to overcome this weakness while being not considerably more complex. However no comprehensive studies are known which analyze the performance of nonlinear turbulence models for three-dimensional flows around building-shaped structures. In the present study the predictions of the flow around a surface-mounted cube using three nonlinear two-equation turbulence models are discussed. The results are compared with predictions of the standard k-ϵ turbulence model and wind tunnel measurements. It is shown that the use of nonlinear turbulence models can be beneficial in predicting wind flows around buildings.

Key words: Urban atmosphere, numerical simulation, turbulence modelling, Boussinesq hypothesis, nonlinear stress-strain relation

1. Introduction

Turbulence has a significant influence on the transport of momentum, heat and pollutants. Therefore the numerical modelling of turbulence plays a crucial role in providing accurate microscale wind fields which are necessary to reliably predict transport and dispersion of pollutants in the vicinity of buildings.

Due to their robustness and efficiency two-equation turbulence models based on the linear Boussinesq hypothesis – like the well-known k-ϵ-model (Jones and Launder, 1972) – are most widely applied to wind engineering and air pollution problems. According to Boussinesq the effective turbulent shear stress is proportional to the strain rate. However, this linear stress-strain relation is violated in complex flows and therefore represents a major drawback. Turbulence models based on the Boussinesq hypothesis are unable to properly describe turbulent flows with body force effects arising from streamline curvature and Reynolds stress differences are not predicted correctly. Therefore their application to highly complex flows can give rise to considerable

Environmental Monitoring and Assessment **65**: 201–209, 2000.

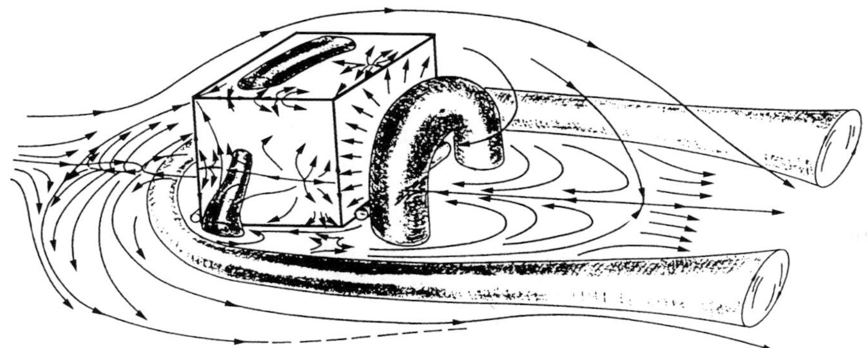

Figure 1. Flow pattern around surface-mounted cube (Martinuzzi and Tropea, 1993)

inaccuracies (see Ehrhard *et al.*, 1999). These perceptions have led to the development of nonlinear stress-strain relationships that transcend the Boussinesq hypothesis, but are not considerably more complex in structure (Speziale, 1987). Nonlinear two-equation turbulence models include nonlinear effects in the modelling of the Reynolds stresses within a two-equation framework. Therefore they are suited to improve the prediction of complex flow fields where turbulence becomes highly anisotropic.

Wind flow in the vicinity of buildings represents a formidable task for turbulence models due to severe pressure gradients, streamline curvature, separation and reattachment. Within the present study the performance of the standard k-ϵ turbulence model and three nonlinear two-equation turbulence models is compared with respect to the flow around a building-shaped structure. The accuracy of the investigated turbulence models is assessed by comparing the numerical predictions with results from wind tunnel experiments.

2. Validation test case

The flow around a surface mounted cube was employed for the present investigation since the cube is the most simple idealisation of a building. In spite of the simple geometry the corresponding flow is very complex with strong pressure gradients, streamline curvature and multiple, unsteady separation regions as shown in Figure 1.

Detailed flow visualisation studies and LDA measurements for the flow around a surface mounted cube were carried out by Martinuzzi and Tropea (1993). The cube was placed in a developed channel flow at a Reynolds number $Re_H = 40000$ based on the cube height H and the bulk velocity U_b. The height H of the cube was half of the channel height.

3. Model Description

The three-dimensional, microscale model MIMO, which was applied in the present study, is based on the finite-volume method for solving the Reynolds averaged conservation equations of mass, momentum and energy (Winkler, 1995). Additional transport equations for humidity, liquid water content and passive pollutants can be solved if required. A staggered grid arrangement is used and coordinate transformation is applied to allow non-equidistant mesh size in all three dimensions in order to achieve a high resolution near the ground and near obstacles.

To compute the unknown Reynolds stresses and turbulent fluxes of scalar quantities in the averaged conservation equations several linear and nonlinear turbulence models have been implemented. The numerical simulations in the present study were carried out using the standard k-ϵ turbulence model (Jones and Launder, 1972) [SKE] based on the linear stress-strain relation and the following turbulence models based on a nonlinear stress-strain relation: Craft *et al.* (1996) [CRA], Lien *et al.* (1996) [LIE] and Shih *et al.* (1995) [SHN]. Linear and quadratic terms for the computation of Reynolds stresses are contained in the model by Shih *et al.* (1995). The models by Craft *et al.* (1996) and Lien *et al.* (1996) further consider terms consisting of cubic products of the strain and vorticity tensors.

4. Grid and boundary conditions

The calculations were performed on a non-equidistant grid having $96 \times 48 \times 48$ grid points in the x-, y- and z-directions, respectively. This rather large number of grid points was chosen to ensure grid independent solutions and to meet the requirements for the wall boundary condition. Because of the symmetry only half of the channel width needed to be calculated. A side view of the computational domain is sketched in Figure 2.

At the inflow boundaries fully developed channel flow was prescribed and a zero gradient condition was applied at the outlet.

To enable the treatment of the near wall region the models by Craft *et al.* (1996) and Lien *et al.* (1996) include low-Reynolds-number terms to resolve the viscous sublayer. However the large bandwidth of length scales which is involved in typical atmospheric flow problems prohibits

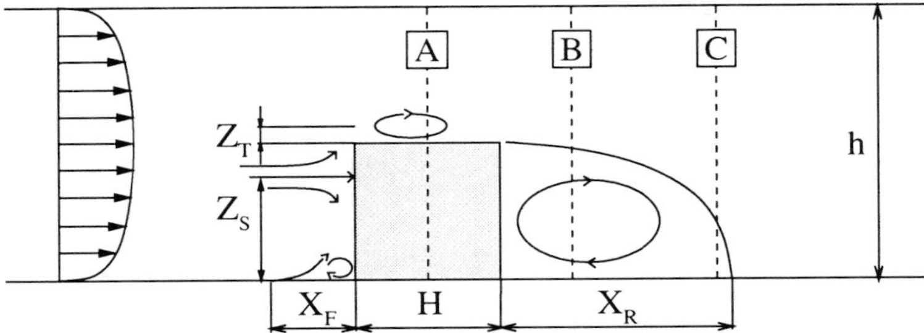

Figure 2. Side view of the investigated geometry

the resolution of the viscous sublayer. Therefore in the present study no-slip conditions were enforced by employing a two-zonal wall function procedure (Sondak and Pletcher, 1995). Wall functions relate the quantities at the grid points adjacent to the wall to the friction velocity at the wall. Low-Reynolds-number terms in the models by Craft *et al.* (1996) and Lien *et al.* (1996) have been neglected. In contrast to the logarithmic law the two-zonal wall function is valid in the viscous sublayer as well as in the log region. This leads to improved predictions especially near stagnation or reattachment points.

5. Results and discussion

In Figure 3 mean velocity profiles in the plane of symmetry for three different locations (see Figure 2) are shown. On top of the cube (position A) the recirculation bubble predicted by the nonlinear models is in good agreement with the experiment whereas for the standard $k-\epsilon$ model (SKE) the bubble is clearly too small. The latter is due to the well-known overprediction of turbulent kinetic energy near stagnation points. A high level of turbulent kinetic energy leads to an overprediction of turbulent diffusion and therefore counteracts separation. The nonlinear models do not overpredict turbulent kinetic energy and therefore they are able to accurately reproduce the recirculation bubble on top of the cube. Also in the region close to the wall, where anisotropies are strong, the predictions by the nonlinear models are superior.

In the wake of the cube (position B) the numerically predicted profiles show some distinct differences. The nonlinear models predict a strong gradient in the velocity component U which is also observed in the experiment. However the location of the vortex center is obviously not in line with the experiment. The standard $k-\epsilon$ model is not able to predict these large gradients and therefore the velocity profile is much more smooth. According to Lakehal and Rodi (1997) the major reason

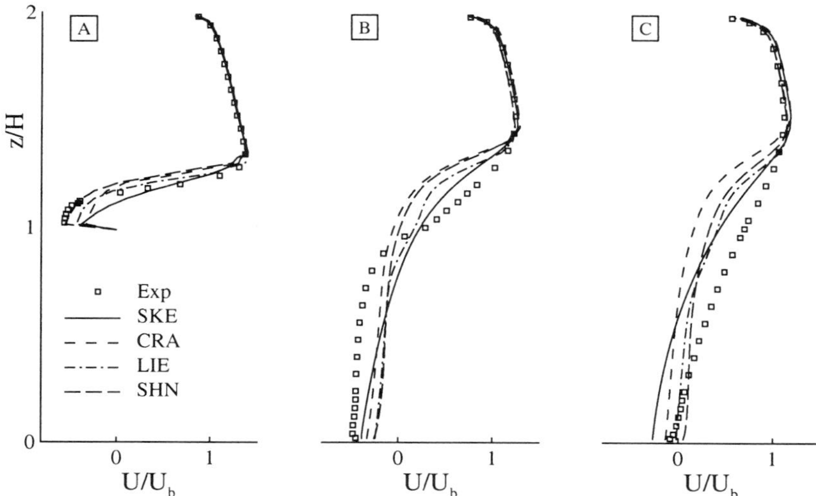

Figure 3. Mean velocity profiles in the plane of symmetry

for the discrepancy in the wake is vortex shedding which was observed in the experiments. Vortex shedding contributes to the momentum exchange but is of course not accounted for in steady calculations. However it is to note that due to the relatively low turbulence level of the approach flow vortex shedding is more significant in our validation test case than in atmospheric boundary layer conditions.

In the region of reattachment (position C) the nonlinear models still predict large gradients, although the flow field has relaxed notably. Nevertheless the profiles close to the wall indicate that the reattachment length is predicted well by the nonlinear models. At this location the standard $k - \epsilon$ still predicts a large negative component of U indicating that the location is still clearly inside the the recirculation zone.

A comparison of streamlines in the plane of symmetry is shown in Figure 4. On top are the streamlines calculated from the wind tunnel data (Exp.). All plots reveal the typical features of this kind of flow: a rather large recirculation region behind the cube and a small vortex right in front of the cube. In connection with the two-zonal wall function all models are also capable of predicting the separation bubble on top of the cube. However, the predicted height of this bubble is overestimated by the nonlinear SHN model whereas it is underestimated by the SKE model. It is to note that a smaller recirculation bubble should lead to a shorter leeward reattachment length. But in spite of the underestimation of the recirculation bubble the reattachment length X_R is still overestimated by the linear SKE model. This indicates that the turbulence level in the wake is severely underestimated by the SKE

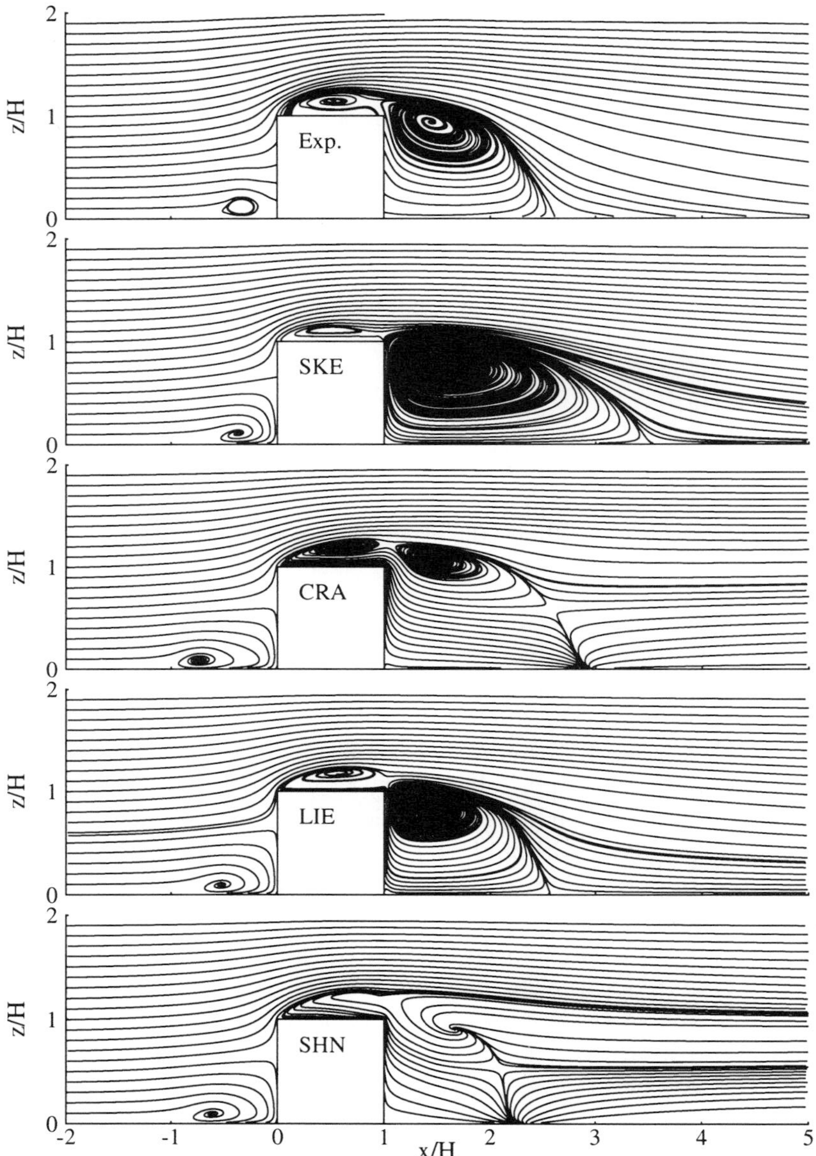

Figure 4. Streamlines in the plane of symmetry

model. The nonlinear models predict a shorter recirculation zone which fits better to the experimental data. However the results of the CRA model and the SHN model show a spurious saddle point in the wake. This is caused by an overpredicted level of turbulence which spoils the development of the vortex downstream of the obstacle and indicates

Table I. Investigated turbulence models and characteristic lengths of the flow field

Model	Identity	Z_S/H	Z_T/H	X_F/H	X_R/H
Jones and Launder (1972)	SKE	0.66	0.10	0.67	2.50
Craft *et al.* (1996)	CRA	0.67	0.21	1.03	1.94
Lien *et al.* (1996)	LIE	0.68	0.18	0.83	1.57
Shih *et al.* (1995)	SHN	0.69	0.23	0.93	1.25
Martinuzzi and Tropea (1993)	Exp	0.65	0.17	0.84	1.66

that these two models are not well suited for the kind of flows considered here. On the other hand, the nonlinear LIE model performs excellently and leads to a notably improved prediction with respect to the SKE model.

The characteristic lengths of the flow field are summed up in table I. The numerically predicted height of the forward stagnation point lies between 0.66 and 0.69 whereas the experimental value is 0.65, i.e. all turbulence models predict the height of the forward stagnation point well. The recirculation bubble on top of the cube is predicted best by the nonlinear LIE model. As mentioned before the SKE model underestimates this bubble. The LIE model predicts reattachment at X=1.57 whereas in the experiment the flow reattaches at 1.66. That means again a fairly good agreement for the nonlinear LIE model. Finally the windward separation length X_F is predicted best by the LIE model whereas it is underestimated by the SKE model.

The findings presented in this study are supported by computations which have been performed using additional test cases. For these results and a detailled analysis of the prediction of Reynolds stresses the reader is referred to Ehrhard (1999).

6. Conclusions

The present study showed that the predictions by the investigated models differ notably. This is due to the lack of universality of all two-equation turbulence models. Especially the nonlinear models have been calibrated mainly on simple shear flows and therefore their range of applicability has to be checked carefully before employing them.

The nonlinear model by Lien, Chen and Leschziner (LIE) seems to be an appropriate model for the type of flows which are of interest in urban air quality studies. The overall flow structure predicted by this model is in very close agreement with experimental results. But still there is a significant discrepancy between the numerical prediction and the experiment in the wake region. This is partly due to the calibration process mentioned above. No nonlinear model is known which has been calibrated and extensively tested against the type of flows considered in this study. Therefore it is obvious that there is potential to further improve the predictive capabilities of turbulence models based on a nonlinear stress-strain relation with respect to flows around buildings. A nonlinear turbulence model focused on this type of flow has currently been developed by one of the authors (Ehrhard, 1999). However the main reason for the discrepancy in the wake is vortex shedding. In order to overcome this weakness unsteady calculations are essential. Unfortunately unsteady calculations increase CPU time by at least an order of magnitude which prohibits their application to highly complex flows of engineering interest.

For the nonlinear turbulence models investigated in the present study the CPU time effort is only about 15 to 20% higher than for the linear $k - \epsilon$ models. Therefore appropriately calibrated nonlinear turbulence models represent a good compromise between numerical effort and predictive capabilities. Their use in the numerical prediction of urban air quality can indeed be beneficial.

References

Craft, T.J., Launder, B.E. and Suga, K.: 1996, Development and application of a cubic eddy-viscosity model of turbulence, *Int. J. Heat Fluid Flow* **17**, 108-115.

Ehrhard, J., Kunz, R. and Moussiopoulos, N.: 2000, A comparative study of two-equation turbulence models for simulating microscale wind flow, Accepted for publication in *Int. J. Environment and Pollution.*

Ehrhard, J.: 1999, *Untersuchung linearer und nichtlinearer Wirbelviskositätsmodelle zur Berechnung turbulenter Strömungen um Gebäude*, Fortschritt-Berichte VDI, Reihe 7, Nr. 367, VDI-Verlag, Düsseldorf, Germany.

Jones, W.P. and Launder, B.E.: 1972, The prediction of laminarization with a two-equation model of turbulence, *Int. J. Heat Mass Transfer* **15**, 301-314.

Lakehal, D. and Rodi, W.: 1997, Calculation of the flow past a surface-mounted cube with two-layer turbulence models, *J. Wind Eng. Ind. Aero.* **67&68**, 65-78

Lien, F.S., Chen, W.L. and Leschziner, M.A.: 1996, Low-Reynolds-number eddy-viscosity modelling based on non-linear stress-strain/vorticity relations, *Engineering Turbulence Modelling and Experiments 3*, ed. Rodi, W., Elsevier Science Publishing Co.

Martinuzzi, R. and Tropea, C.: 1993, The flow around surface-mounted, prismatic obstacles placed in a developed channel flow, *J. Fluids Eng.* **115**, 85-92.

Shih, T.-H., Zhu, J. and Lumley, J.L.: 1995, A new reynolds stress algebraic equation model, *Comput. Meth. Appl. Mech. Engrg.* **125**, 287-302.

Sondak, D.L. and Pletcher, R.H.: 1995, Application of wall functions to generalized nonorthogonal curvilinear coordinate systems, *AIAA J.* **33**, 33-41.

Speziale, C.G.: 1987, On nonlinear $k - l$ and $k - \epsilon$ models of turbulence, *J. Fluid Mech.* **178**, 459-475.

Winkler, C.: 1995, *Mathematische Modellierung der quellnahen Ausbreitung von Emissionen*, Fortschritt-Berichte VDI, Reihe 7, Nr. 268, VDI-Verlag, Düsseldorf, Germany.

Numerical Simulations of the Urban Roughness Sub-Layer : a First Attempt

E. Guilloteau* and P. G. Mestayer
Laboratoire de Mécanique des Fluides
UMR CNRS-Ecole Centrale de Nantes 6598
Ecole Centrale de Nantes, BP 92101
F-44321 Nantes Cedex 3, France

Abstract. Most of the mesoscale models use roughness parameters to characterise the ground and to compute the surface stress. As the experimental determinations of the urban roughness parameters are rare and not very reliable, a new methodology based on microscale numerical simulations is presented here and the first results from two-dimensional simulations with different roof shapes are analysed.

Firstly, it appears that the roof shape has a large influence : large difference in the Reynolds stress profile and in the roughness sub-layer thickness, enhancement of the exchanges at the roof level by buildings with attic. It also appears that the fetch necessary to obtain a constant flux layer is unrealistic compared to the real spatial homogeneity of quarters in European cities. Consequently, a new parameterisation of the urban ground-induced friction is to be developed without reference to the constant flux layer theory.

Keywords: Numerical simulation, Roughness sub-layer, Urban ground, Roughness parameters, Roof shape

1. Introduction

The transport and dispersion of atmospheric pollutants are mainly driven by two mechanisms : advection and diffusion. The former is related to the mean wind speed while the latter is related to the turbulence, and in the lower atmosphere both of them are affected by the ground obstacles (e.g. buildings, trees, streets...), which are known to have a large influence on the dynamic and thermodynamic structure of the lower atmosphere.

Studies of pollutant transport and dispersion at the scale of an urban agglomeration require numerical simulations over domains of some tens of kilometres. However, present computer capabilities do not allow in the same time calculations with a grid resolution sufficient to explicitly simulate flows at the scale of the urban obstacles. Because of this computational limitation, the real ground and its obstacles must be modelled by an "apparent" ground. This apparent ground is usually

* Present address : Institute of Hydromechanics, University of Karlsruhe, Kaiserstr. 12, D-76128 Karlsruhe, Germany

Environmental Monitoring and Assessment **65**: 211–219, 2000.

characterised by roughness parameters for the dynamics equations, and sensible and latent heat fluxes for the thermodynamics equations.

The atmospheric surface layer assumptions (horizontal homogeneity, stationarity, negligible influence of Coriolis force) yield the constant flux and, by integration, the logarithmic velocity profile where the roughness, representing the friction induced by the roughness sub-layer on the atmospheric flow, is characterised by two parameters : the displacement height h_d, which represents the height of a fictitious floor felt by the atmospheric boundary layer flow, and the roughness length z_0, which is the distance above the fictitious floor at which the extension of the logarithmic profile reaches zero.

These parameters often depend only on the surface element geometry; in the case of urban elements, the geometry may be described by two factors : the area density λ_P, which is the ratio of the built-up area to the total ground area, and the frontal density λ_F, which is the ratio of the frontal area (normal to the wind direction) to the total ground area.

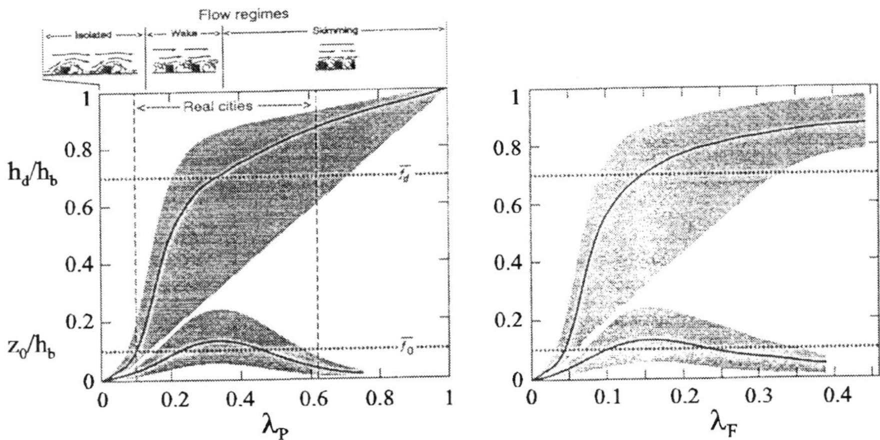

Figure 1. Roughness parameters normalised by building height, as a function of the area and the frontal densities (after Grimmond and Oke, 1999).

Figure 1 displays estimates of the roughness parameter variation with the area and frontal densities. Both parameters are normalised by the average building height h_b. The rule-of-thumb approximations of these parameters, $\overline{f_d} = 0.7$ and $\overline{f_0} = 0.1$, as well as the corresponding street canyon flow regimes, are also shown. The very large uncertainty of these roughness parameters can be seen. The general behaviour seems to be the following : the displacement height increases with the densities but never reaches the average building height, and the roughness length increases with densities until the skimming flow regime is attained, further it decreases with increasing densities.

A new method for determining the urban environment roughness parameters, based on microscale numerical simulation, is presented in the next two Sections and the first results stemming from two-dimensional simulations are analysed in Section 4.

2. Roughness Parameter Determination

Roughness parameters can be deduced from wind tunnel and field studies. However high quality data must fulfill several criteria : the terrain must be flat and homogeneous, the thermal stratification must be taken care of, the fetch must be larger than the average building height by at least two orders of magnitude, the measurement height must be above the roughness sub-layer and within the first tenth of the atmospheric boundary layer, the sensors must disturb the flow the least possible, appropriate measurement methods, in terms of sampling period, response characteristics... must be used, and the analysis method should rather lead to the independent determination of the friction velocity u_*, the roughness length z_0 and the displacement height h_d. On the other hand, turbulence-based approaches are preferable to wind-based estimates, and especially to the mean wind profile methods. As these criteria are difficult to be fulfilled all together, only a few reliable experimental determinations of the urban environment roughness parameters are available.

The criteria required for high quality measurements are not a heavy constraint for numerical simulations : flat terrain and thermal stratification are set, no measurement apparatus disturbs the flow, all measurement heights are accessible, provided a high enough calculation domain, the fetch and the spatial resolution are essentially dependent on the affordable computation cost and the "measurement quality" is mainly related to the turbulence closure model, to the wall function and to the care in grid resolution and boundary conditions.

3. Numerical Tools

The CFD code CHENSI version 4 (Guilloteau, 1999) solves steady state incompressible Reynolds momentum equations, with the finite volume method (Ferziger and Peric, 1996), on a non-uniform staggered Cartesian mesh (Harlow and Welch, 1965). The pressure-velocity coupling is solved by the artificial compressibility method of Viecelli (1969), the advection scheme is the hybrid scheme of Spalding (1972) and the turbulence closure model is the improved $k - \epsilon$ model of Chen and Kim (1987). The boundary condition used on building surfaces is the standard rough wall function.

The methodology used to determine roughness parameters is the numerical simulation of an internal boundary layer development, representing the perturbation of an upstream atmospheric surface layer by the presence of a large district characterised by a regular arrangement of stylised buildings. If the fetch is long enough, a sub-layer in equilibrium with this spatially homogeneous ground may be observed. The roughness parameters are then derived from the analysis of this sub-layer.

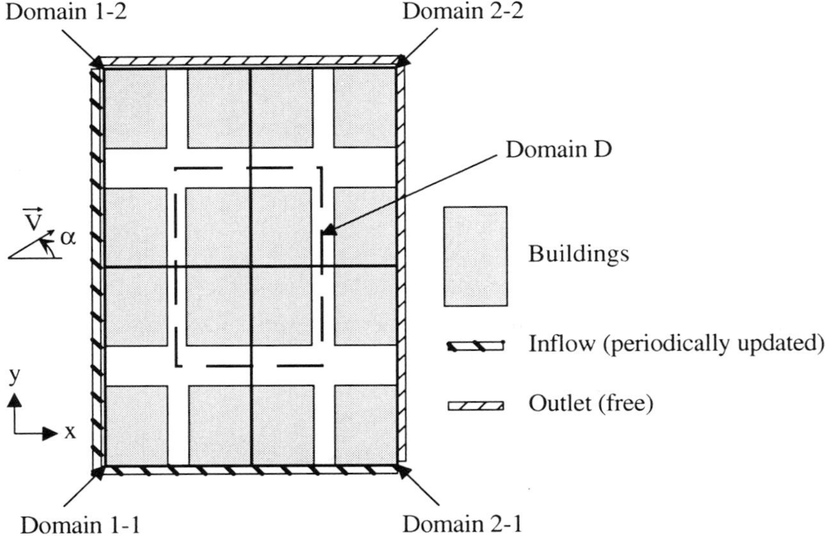

Figure 2. Fetch simulation and boundary conditions (seen from above).

In order to simulate the growth of the fetch, we consider a simulation domain twice as large as the minimum domain in each horizontal direction (Figure 2). The profiles are fixed at the inlet and free at the outlet. The wind and turbulence fields in the domain 1-1, 1-2 and 2-1 are periodically updated with the wind and turbulence fields of the domain 2-2. Thus, the fetch has grown by one domain size. This periodical update is done as many times as necessary to obtain a thick enough sub-layer in equilibrium. Thus, the inflow profiles are initially the upstream constant flux layer profiles : logarithmic profile for the horizontal velocity ($U = \kappa^{-1} u_{*0} ln((z - h_b)/z_{01})$), constant profile for the turbulent kinetic energy ($k = u_{*0}^2/\sqrt{C_\mu}$) and equilibrium state for the dissipation rate ($\epsilon = \kappa^{-1} u_{*0}^3/(z - h_b)$). At each update, the inflow profiles are replaced with the profiles previously computed at the edges of the domain 2-2.

The computational domain extends from the ground to a height of 2 km : it includes the canopy layer (below the roof level), the roughness sub-layer and the internal boundary layer, capped by the undisturbed upper initial atmospheric surface layer. The resulting vertical profiles of horizontal wind and turbulence are further horizontally averaged over the domain D in order to minimise the influence of lateral boundary conditions.

4. First Results

We studied a 2D configuration and considered three cases with two building shapes : buildings with flat roof and buildings with schematic attic, which can be regarded as a more realistic building shape (Figure 3). The frontal area density λ_F is the same for each of the three cases, and the area density λ_P is the same for cases 1 and 2 and larger than for case 3.

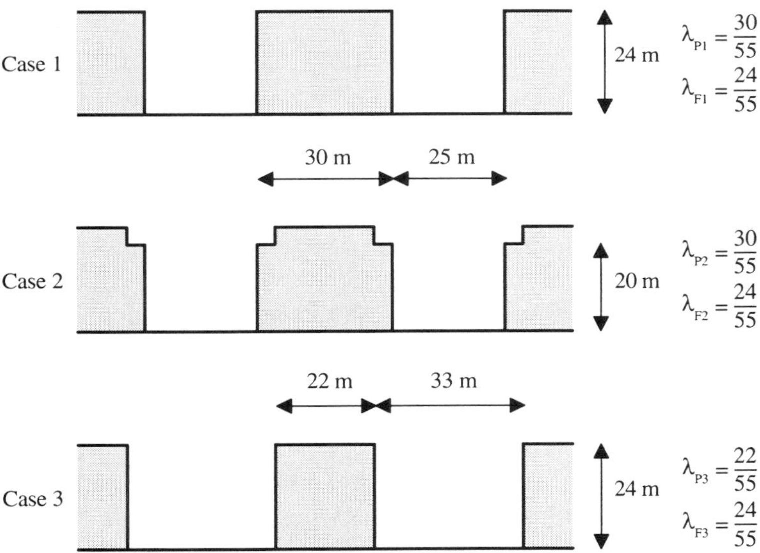

Figure 3. Studied case configurations.

First, we studied the internal boundary layer growth (Figure 4). We fitted the layer thickness with the power law $z_h = aX^b + h_0$, where X is the fetch, a and b are a function of z_{01} the upstream roughness length, and z_{02} the effective roughness length of the terrain; here z_{01} equals 0.01 m. The fitting parameter h_0 may have some similarity with the effective displacement height. The fitting uses the non-linear regression method of Marquardt-Levenberg (Press *et al.*, 1986).

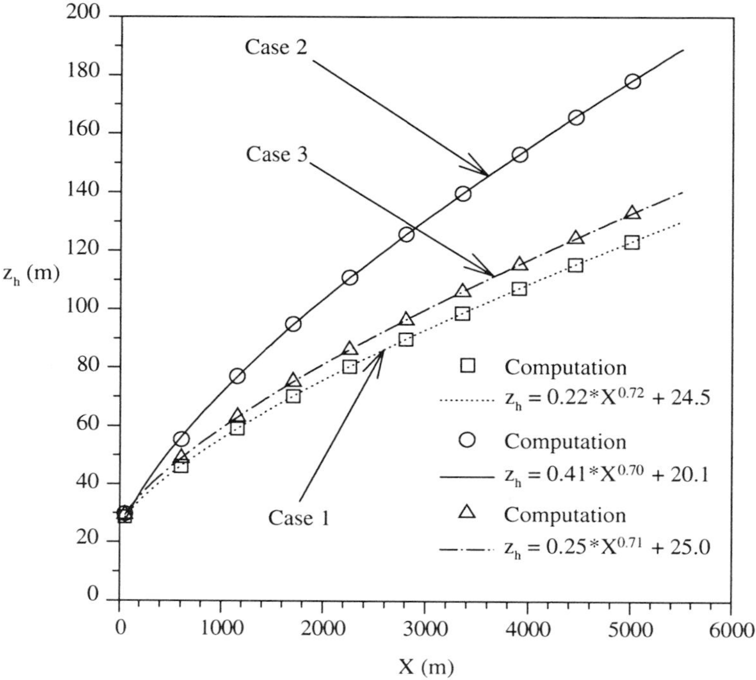

Figure 4. Internal boundary layer development for the three studied cases.

The fitting results show very little dispersion ($r^2 \simeq 0.99$). The power b seems on the order of 0.7, which is in close agreement with the results of Taesler and Karlsson (1981) for the growth of an urban boundary layer. With the exception of case 2 (with attic), the values of h_0 are larger than the building height, which may mean that h_0 is not a displacement height. Indeed, we expect from Figure 1 a displacement height value in the range of 0.8-0.9 h_b (i.e. 19-21.5 m) for these cases and, given the very good fitting, the difference between h_0 and h_d (about 20%) cannot be the only effect of the regression method. For cases 1 and 3, the parameter a is very similar but smaller than in case 2. This can be explained by a stronger flow disturbance due to the attic. In fact, although the presence of the attic does not change the flow regime (still a skimming flow), buildings with attic disturb the flow immediately above roof level more than buildings with flat roof, and case 2 cannot be regarded as an intermediate geometry between cases 1 and 3. The horizontal velocity at roof level is less accelerated in the case with attic than in the other two cases and the flow enters more within the street in this case than in the cases with flat roof. This is due to a larger leeward depression created by the attic. The presence of the attic thus enhances the exchanges at the roof level, which is also of importance for pollutant dispersion and atmosphere-street exchanges.

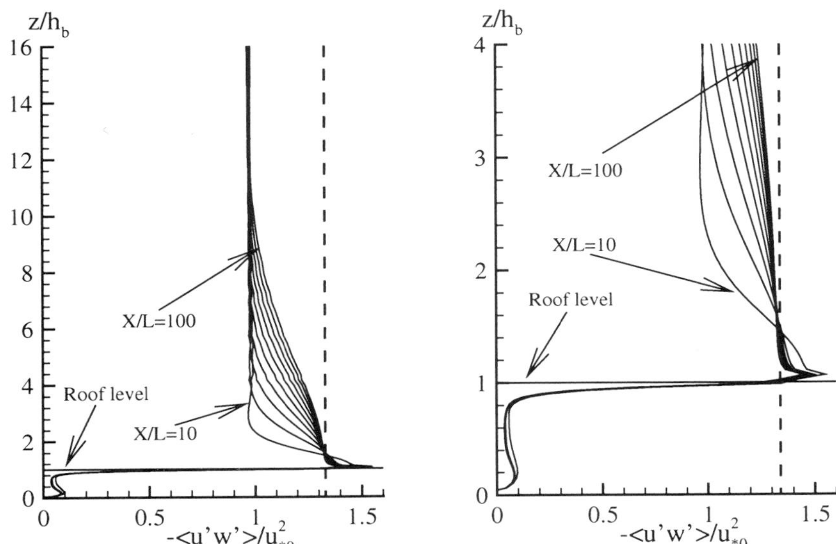

Figure 5. Evolution of the Reynolds stress with the fetch for case 1.

Figure 5 shows, for case 1, the evolution of the Reynolds stress vertical profile as a function of the fetch, from $10L$ to $100L$ where L is the length of the minimal domain, equal to 55 m in all three cases. The Reynolds stress $- < u'w' >$ is normalised by the upstream friction velocity u_{*0}. The altitude is normalised by the building height. The fact that the vertical profiles get closer to each other with increasing fetches means that the flow structure becomes less dependent on the initial perturbation, and spatial homogeneity is close to be reached, as required to deduce roughness parameters. The normalised Reynolds stress varies in the first tens of metres above the roof about linearly as a function of z, with a slope varying with the fetch. The normalised Reynolds stress seems to become constant and close to 1.34, but the zoomed picture (right-hand side part of the Figure) shows that the fetch is still not large enough since the constant flux layer is not developed enough to derive precise roughness parameters and to compare the displacement height with our fitting parameter h_0.

For cases 2 and 3, similar results are shown in Figure 6. A linear variation of the normalised Reynolds stress may be observed in the

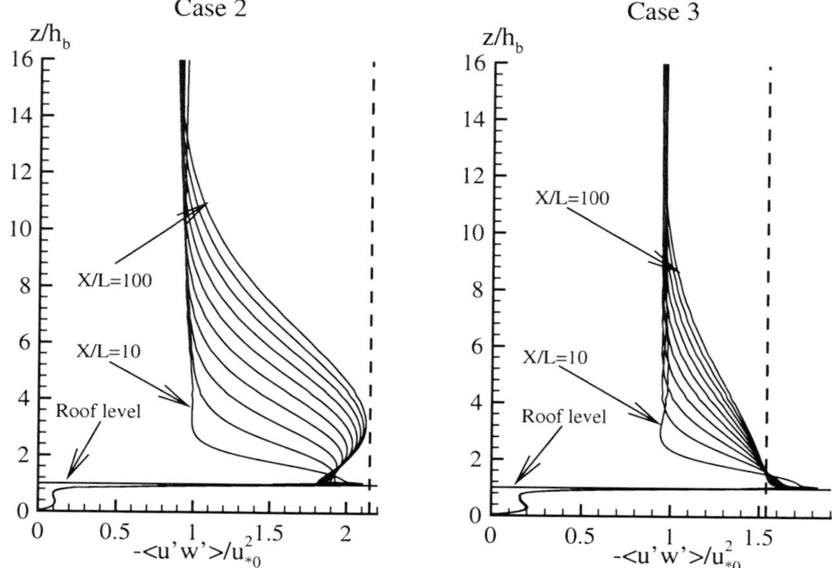

Figure 6. Evolution of the Reynolds stress with the fetch for cases 2 and 3.

first tens of metres above the roof. For the case with attic, the spatial homogeneity is far to be reached, and the vertical profiles do not really merge. However, we may extrapolate this behaviour and expect that, for a larger fetch, a constant flux layer will be observed with a higher value of the constant normalised Reynolds stress, probably about 2.15. The street geometry parameters of cases 1 and 2 are the same but the presence of the attic makes the canopy rougher and the roughness sub-layer thicker.

For case 3, a constant flux layer may be appearing, but a larger fetch would be required to derive roughness parameters. The value of the constant normalised Reynolds stress would be probably about 1.54; this means that the canopy of case 3 is rougher than that of case 1. This is in accordance with the general behaviour presented in Section 1 : in skimming flow regime, the roughness decreases with the increasing area density.

5. Conclusion

We showed the large influence of the roof shape on the roughness parameters and on pollutant dispersion. We also showed that the fetch required to observe a constant flux layer is very large and unrealistic for European city quarters, larger than 5.5 km in the cases studied here.

For realistic fetches usually smaller than 2 km, the flux is far from being constant. The vertical variation of the flux on the first tens of metres above roof level is close to linear, with a slope varying with the fetch. Therefore the constant flux layer theory cannot be applied and the classical wall function, parameterising the momentum exchange with the help of roughness parameters, should not be used in mesoscale model simulations. For this reason, new wall functions must be developed and the tools and methodologies presented here should contribute to this development.

Acknowledgements

This work has been done with the support of Electricité de France (grant #2M6321/AEE2154). The numerical calculations were made on the supercomputers of the Institut de Développement et de Recherche pour l'Informatique Scientifique (IDRIS) of the Centre National de la Recherche Scientifique (CNRS) in Orsay, France.

References

Chen, Y.S. and Kim, S.W. : 1987, *Computation of turbulent flows using an extended k-ϵ turbulence closure model*, Technical Report CR-179204, NASA

Ferziger, J.H. and Peric, M. : 1996, *Computational methods for fluid dynamics*, Springer-Verlag, New-York - Heidelberg - Berlin

Grimmond, C.S.B. and Oke, T.R. : 1999, 'Aerodynamic properties of urban areas derived from analysis of surface forms', *J. Appl. Meteorol.* **38** (9), pp. 1262–1292

Guilloteau, E. : 1999, *Modélisation des sols urbains pour les simulations de l'atmosphère aux échelles sub-méso*, Ph.D. Thesis, Université de Nantes - Ecole Centrale de Nantes (in French)

Harlow, F.H. and Welch, J.E. : 1965, 'Numerical calculation of time-dependent viscous incompressible flow of fluid with free surface', *Phys. Fluids* **8**, pp. 2182–2189

Press, W.H., Flannery, B.P., Teukolsky, S.A. and Vetterling, W.T. : 1986, *Numerical Recipes*, Cambridge University Press, Cambridge

Spalding, D.B. : 1972, 'A novel finite-difference formulation for differential expressions involving both first and second derivatives', *Int. J. Num. Methods Engng.* **4**, pp. 551–559

Taesler, R. and Karlsson, S. : 1981, *Power law estimates of the urban wind profile*, Reports No. 59, Met. Inst. Uppsala.

Viecelli, J.A. : 1969, 'A method for including arbitrary external boundaries in the MAC incompressible fluid computing technique', *J. Comput. Physics* **4**, pp. 543–551

PHYSICAL MODELING OF EMISSIONS FROM NATURALLY VENTILATED UNDERGROUND PARKING GARAGES

BERND LEITL and MICHAEL SCHATZMANN
Meteorological Institute of the University of Hamburg
Bundesstraße 55, D-20146 Hamburg, Germany
E-mail: leitl@dkrz.de

Abstract. The dispersion of pollutants from naturally ventilated underground parking garages has been studied in a boundary layer wind tunnel. Two idealized model setups have been analysed, one was simulating pollutant dispersion around an isolated rectangular building and one was representing dispersion in a finite array of idealized building blocks. Flow and dispersion close to modelled ground level emission sources was measured. The results illustrate the complexity of the flow around buildings and provide insight in pollutant transport from ground level sources located directly on building surfaces. As a result, areas critical with respect to high pollutant concentrations could be visualized. Particularly, the results show high concentration gradients on the surface of the buildings equipped with modelled emission sources. Inside the boundary layers on the building walls, a significant amount of pollutants is transported to upwind locations on the surface of the building. The paper documents the potential of physical modelling to be used for the simulation and measurement of dispersion close to emission sources and within complex building arrangements.

Key words: pollutant dispersion, car exhaust, underground parking garage, wind tunnel modelling, concentration measurements, flow measurements

1. Introduction

Predicting the dispersion of air pollution in urban areas is one of the most challenging problems environmental research is facing. Especially in urban areas, the dispersion of pollutants is dominated by a number of complex boundary conditions like different types of emission sources, constantly changing meteorological conditions and complex arrangements of structures. Specific air pollution problems arise from toxic or carcinogenic pollutants released for instance from parking garages. Since many of the big cities experience a general lack of sufficient parking spaces, an increasing number of new office and residential buildings is provided with underground parking garages to solve the permanent parking problems. In small building units, a natural ventilation of the parking area is often preferred to other solutions for economical reasons. Several building codes have been developed in Germany in order to ensure a safe air quality in parking spaces as well as to avoid health risks in closed parking areas. Carbon monoxide (CO) was chosen as leading pollutant for the assessment of air quality in underground garages because it can be measured easily. Besides, a reliable and permanent CO-monitoring can be used for controlling mechanical emergency ventilation systems in situations that are critical with respect to air quality and pollutant concentrations. However, newer studies give evidence that

Environmental Monitoring and Assessment **65**: 221–229, 2000.

there is a possibility of reaching or even exceeding the threshold values for other critical air pollutants inside and outside of the parking area though the CO-based air quality demands are satisfied inside a naturally ventilated parking garage. Benzene (C_6H_6) represents another air pollutant critical to human health, which is used for assessing the impact of the exhaust from underground parking garages on air quality outside the parking area. Benzene was selected as second leading component for two reasons. First, benzene was found to be a carcinogenic air pollutant, which requires low threshold values and a special assessment of exposure time to avoid risks for human health. Secondly, benzene is permanently emitted from underground garages. Driving vehicles release benzene as one component of the vehicles exhaust. Significant amounts of benzene are also released by parking vehicles because of evaporation of fuel and the effect of 'tank breathing'. In a recent study (TÜV Nord, 1996), maximum source strengths of up to 75 µg/s benzene per ventilation outlet of underground garages with natural ventilation were estimated. In a similar study by BAUCH (1993), average concentrations of 9.5 µg/m³ benzene with peak concentrations up to 18.1 µg/m³ and a 90th-perzentile of 15 µg/m³ were measured in the exhaust gas of an underground parking area. Considering typical mean background concentrations of 5 to 7 µg/m³ benzene in large cities and a threshold value of 10 µg/m³ benzene defined in Germany, it becomes evident that even small amounts of benzene emitted by naturally ventilated parking garages can cause exceedances of the threshold value given by environmental guidelines.

In order to study the flow and pollutant concentrations patterns close to the air outlets of naturally ventilated underground parking garages, a set of illustrative wind tunnel experiments has been set up in the BLASIUS wind tunnel of the Meteorological Institute at Hamburg University. The primary goal was to visualize critical emission situations near simplified building structures as well as on building surfaces. In addition, it was intended to provide a set of high quality experimental data, which can be used for testing the quality of numerical modelling, applied to a particular urban-type dispersion problem.

2. Experimental Setup

The experiments were carried out in a 16 m long, conventional type boundary layer wind tunnel with a test section 1.5 m wide and 1 m high. To obtain a sufficient geometrical resolution in flow and concentration measurements a model scale of 1:200 was selected. By means of a specific setup of turbulence generators (spires) and roughness elements on the wind tunnel floor the lower part of the atmospheric boundary layer (up to 100 m height in full scale) was modelled in the test section of the wind tunnel. Systematic boundary layer measurements and an iterative improvement of the spire - roughness configuration enabled even the constant shear layer found in field data to be modelled up to 100 m height full scale. Despite of the large model scale and the rather small cross-section of the tunnel, the parameters of the boundary layer flow agree well with field data for a suburban roughness (Figure 1). The height of the model boundary layer was exceeding the height of the model building by a factor of four. According to common guidelines for physical modelling this ensures that the wind tunnel results are not affected by the limited height of the model boundary layer and that similarity of the model results to field conditions can be achieved (VDI 3783).

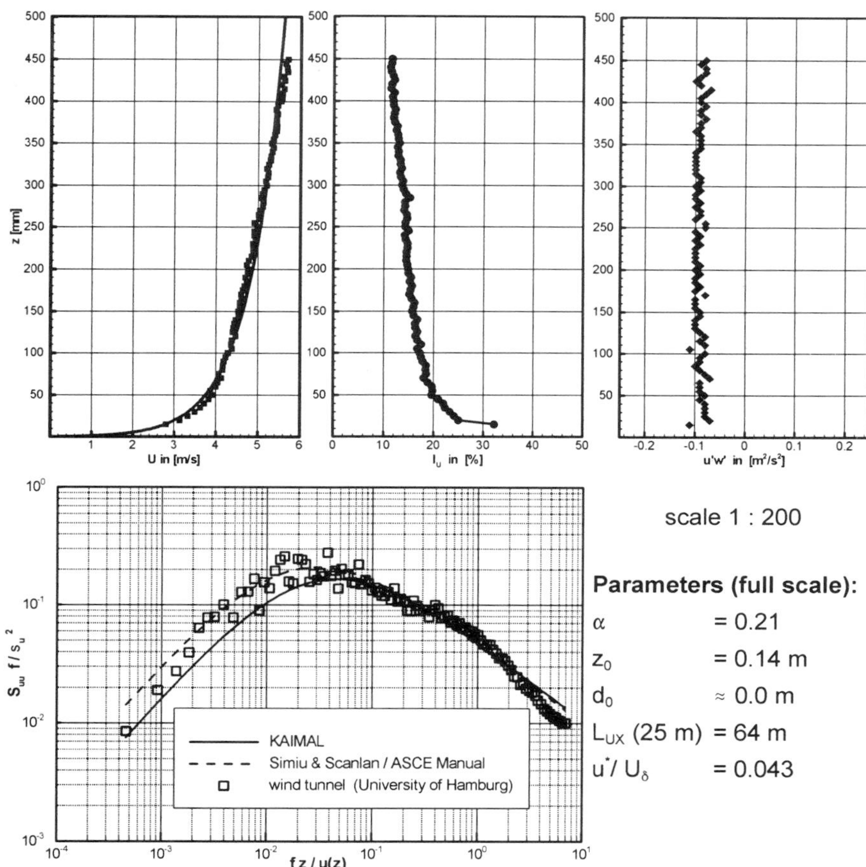

Figure 1. Atmospheric boundary layer flow modeled in the wind tunnel (mean wind profile, turbulence intensity profile, shear stress profile and characteristic parameters of the approaching flow of the partially modelled atmospheric boundary layer).

A 2D fibre-optic Laser-Doppler-Anemometer (~LDA, DANTEC®) with 500 mm focal distance was used for non-intrusive flow measurements. Due to the long focal distance, flow disturbances caused by the LDA probe head were not detectable even for near-wall measurements. For concentration measurements, a fast FID (flame ionisation detector, Cambustion Ltd.®) with a frequency resolution of up to 200 Hz and a mobile sampling head was used. Both, the LDA fibre-optic probe as well as the FastFID sampling unit were mounted on a computer controlled 3D traversing system to ensure sub-millimetre positioning accuracy and high efficiency of automated measurements at more than 1000 measurement points per test case. Besides mean values and statistical parameters of the measured quantities, time series of the flow velocity components and the concentration of a tracer were recorded and analysed off-line with respect to their fluctuation characteristics. Two source configurations with different complexity have been studied in detail. The first setup was consisting of an isolated rectangular building with four area sources attached on the lee side of the building (configuration A) close to the ground. As second configuration, a finite array of 27 rectangular buildings with exactly the same shape and dimension was investigated in the wind tunnel (configuration B). In configuration B four rows of model buildings with three buildings per row were mounted upstream of the source building and two rows of buildings were attached downstream (Figure 2).

3. Results

In Figures 3 and 4, measurements of two components of the velocity field are presented. The results illustrate, that the average flow pattern is symmetrical for winds perpendicular to the front side of the building(s). For configuration A, the characteristic flow structures like flow separation on the flat roof of the building as well as on both sides of the model could be measured. The typical horseshoe-vortex structure in front of the obstacle and the large recirculation area on the leeward side of the model building can be detected as well (Figure 3). The second setup (configuration B) did not show flow separation on top of the model buildings. In this configuration the separation zones on the sidewalls of the building were replaced by an almost uniform flow in the street canyons parallel to the wind. For the crosswind street canyon behind the source building an asymmetric recirculating flow pattern was detected in the vertical measurement plane which differs significantly from the almost centric flow pattern usually found in a long street canyon (Figure 4). The flow in configuration B was also found to be more unstable and intermittent than in configuration A. Long averaging times, corresponding to several hours in full scale, were required to capture even the aperiodic changes of the flow patterns observed during LASER light sheet visualization experiments and in order to get representative results.

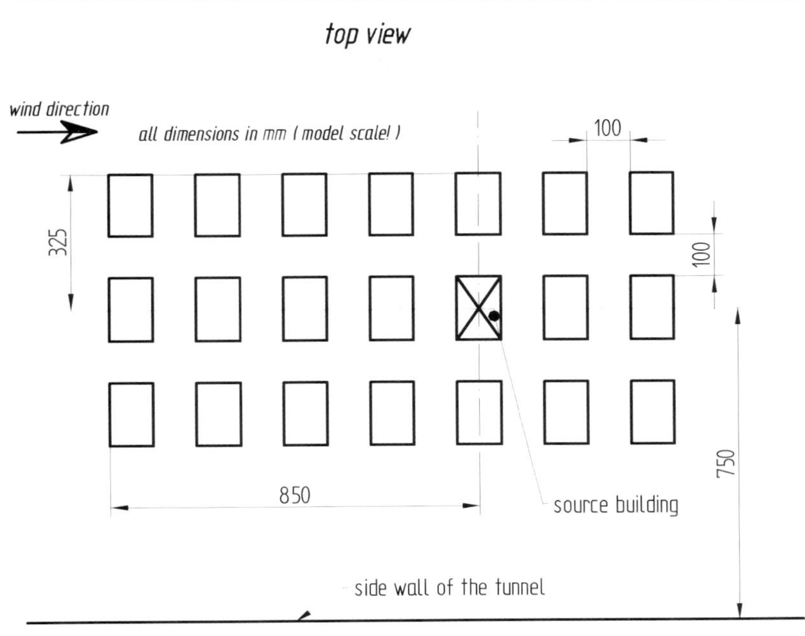

SOURCE BUILDING

SIDE VIEW

25 m
H
Z
X
20 m
0,8H

TOP VIEW

Y
X
30 m
1,2H

LEE SIDE BUILDING WALL

(location of emission sources)

11,25 m
0,45H
Z
Y
BUILDING
1 m
0,04H
3,75 m
0,15H

source element
A = 4,622 m²
3,4 m
R 2 m
0,08H
1,4 m
0,056H

University of Hamburg, Meteorological Institute

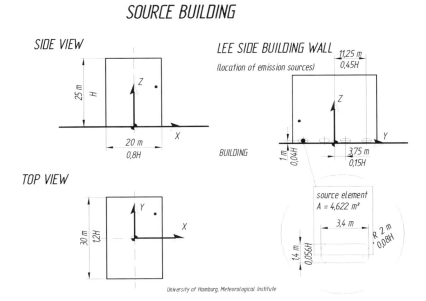

Figure 2. Experimental setup, building model and location of the emission sources used in the wind tunnel.

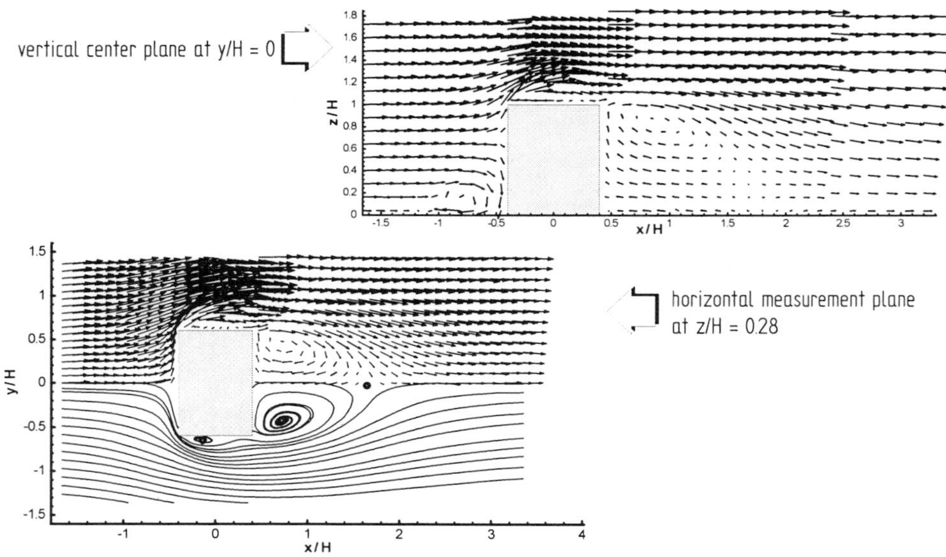

Figure 3. Illustrative results of 2D flow measurements – configuration A, isolated rectangular building (measured velocity vectors and streamtraces calculated from 2D measurements).

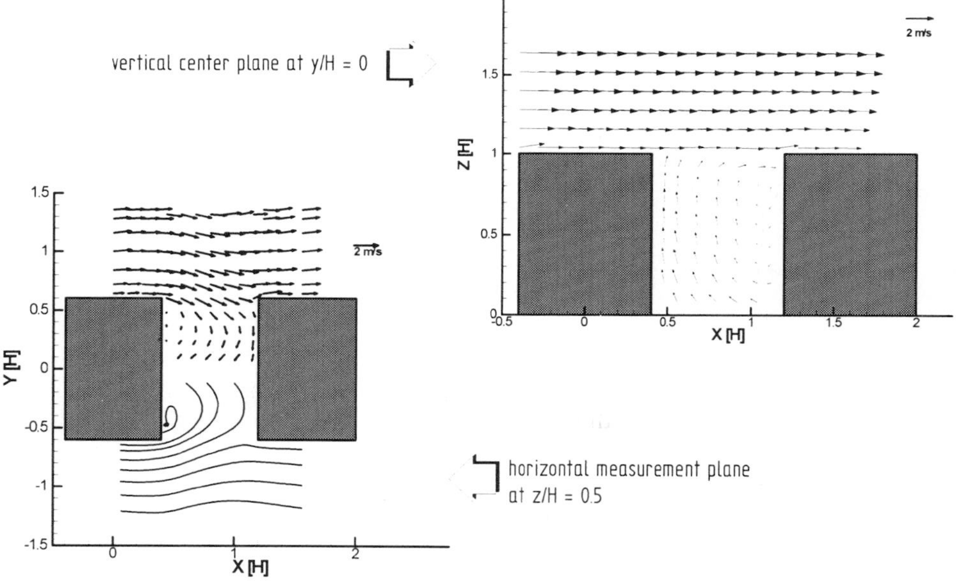

Figure 4. Illustrative results of 2D flow measurements – configuration B, array of buildings (measured velocity vectors and streamtraces calculated from 2D measurements).

Example concentration distributions are given in Figure 5 and 6. The measured concentrations have been converted into non-dimensional values of $K = (C_{measured}/C_{source}) \cdot u_{ref} \cdot H^2 / Q$ for comparison with similar results from numerical simulations and field measurements. $C_{measured}$ is the measured tracer concentration at a certain measurement point, C_{source} stands for the tracer concentration at the source, u_{ref} defines a reference wind speed, H is a reference height (building height) and Q indicates the emission source strength. Defining a maximum allowable concentration of a given air pollutant and a given source strength, safe distances can be derived directly from contour plots of measured K-values. The results of high-resolution concentration measurements indicated that close to the walls of the source building high concentration gradients can be found and a significant transport of pollutants along the surface of the building occurs.

A comparison with similar results from numerical simulations (TÜV Nord, 1996) shows that safe distances calculated from numerical simulations are constantly smaller than the values derived from our wind tunnel experiments. Obviously, the limited spatial resolution of discrete numerical models affects the predicted pollutant concentration especially in regions with strong concentration gradients found close to building surface. Considering, that apartment buildings and office complexes often have windows near to ventilation outlets of parking garages it becomes evident that numerical model results are likely to underestimate the air quality problems and health risks around ventilation outlets. Analysing the time series of measured concentrations it was found that the dispersion in structured arrays of buildings is characterized by low frequency, aperiodic changes of the concentration distribution. As a result, the instantaneous dispersion patterns substantially differ from what a time averaged concentration plot shows. Especially in the street canyons parallel to the wind (setup B) areas with rather low mean immission concentrations were characterized by very high short-term concentrations.

4. Conclusions

Based on results of two simplified test configurations the capability of wind tunnel modelling of emissions from underground parking garages has been demonstrated. It could be shown that physical modelling leads to reliable results with high resolution in space and time even in areas close to the emission sources. Effects like pollutant transport within the internal boundary layer on building surfaces could be resolved and quantified. The instantaneous character of flow and dispersion patterns was investigated by analysing high frequency time series of point-wise flow and dispersion measurements. A comparison with similar results from numerical simulation gives evidence that conventional numerical modelling is substantially underestimating concentration gradients close to the modelled emission sources. To avoid risks for human health, emissions

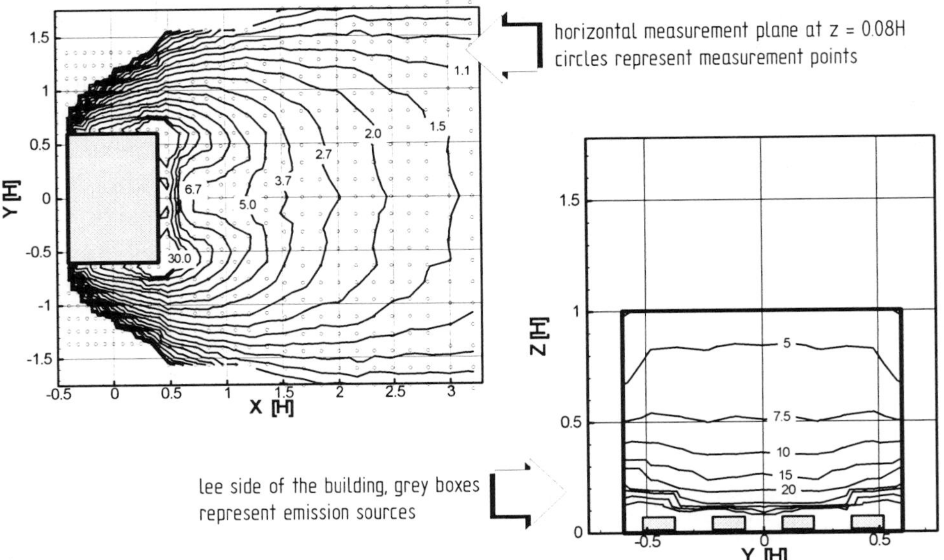

Figure 5. Illustrative results of concentration measurements – configuration A, isolated rectangular building (contours of non-dimensional concentrations K and location of the measurement points).

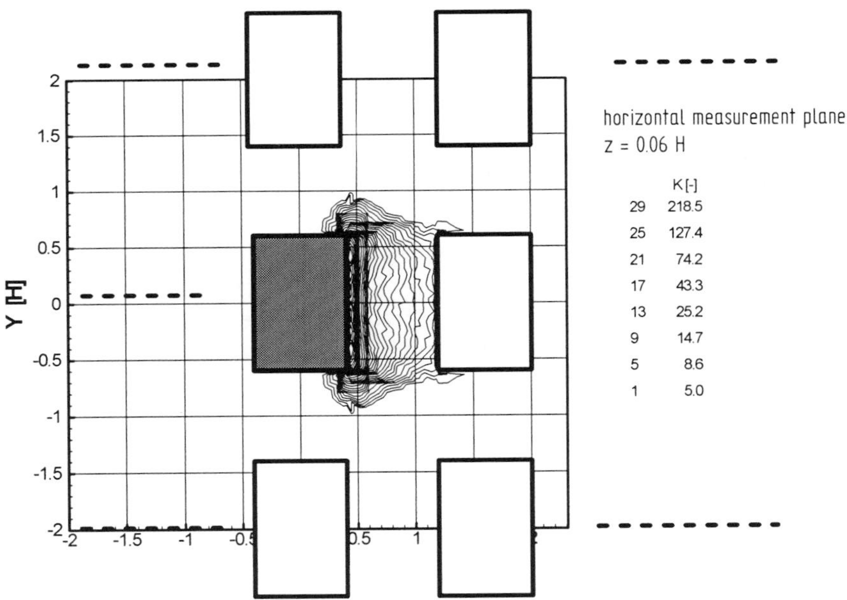

Figure 6. Illustrative results of concentration measurements – configuration B, array of rectangular buildings (contours of non-dimensional concentrations K).

from naturally ventilated underground parking garages should be investigated in more detail by means of systematic wind tunnel experiments and backing field measurements. In addition, the results of systematic wind tunnel measurements could be used for the development, improvement and validation of numerical dispersion models applied to urban-type dispersion problems.

Acknowledgment

The authors wish to express their appreciation for support from the German Federal Environmental Agency (grant 296 43 831).

References

B.A.U.CH.: 1993, Sachbericht zum Projekt "Belastung der Innenraumluft durch Emissionen aus Tiefgaragen, *Beratung und Analyse - Verein für Umweltchemie e.V., Wilsnacker Str. 15, D-10559 Berlin,* p. 47

Environ Corporation: 1984, Sensitivity Analysis of the California Department of Health Services Risk Assessment of Benzene, *Prepared for the American Petroleum Institute, Washington*

ARC MONOGRAPHS: 1982, Benzene and Annex. In: Some industrial chemicals and dyestuffs. *Lyon, International Agency for Research on Cancer, IARC Monographs on the Evaluation of the Carcinogenic Risk of Chemicals to Humans, Vol. 29*

Kaimal, J.C., Wyngaard, J.C., Izumi, Y., Cote, O.R.: 1971, Spectral characteristic of surface layer turbulence, *Quart. J. Roy. Meteorol. Soc., Vol. 98, No. 417, pp. 563 – 589*

Simiu, E., Scanlan, R. H.: 1986, Wind Effects on Structures – Part A: The Atmosphere, *John Wiley & Sons, Inc.*

TÜV Nord e.V.: 1996, Studie - Anforderungen an die Lüftung von Tiefgaragen unter dem Gesichtspunkt von Schadstoffimmissionen, *Technischer Überwachungs-Verein Nord e.V., Große Bahnstraße 31, D-22525 Hamburg,* p.70

VDI 3783: 1999, Physical modelling of flow and dispersion processes in the atmospheric boundary layer, *VDI, Environmental Meteorology, VDI/DIN Manual of the Commission on Air Pollution Prevention, VDI 3783 - 12 (Draft), p.22*

NEAR FIELD DISPERSION IN THE URBAN ENVIRONMENT - A HYDRAULIC FLUME STUDY

R.W. MACDONALD, B.J. COULSON and P.R. SLAWSON
Department of Mechanical Engineering, University of Waterloo, Waterloo, Ontario, Canada N2L 3G1
E-mail: rwmacdon@mecheng1.uwaterloo.ca

Abstract. The dispersion of material released from a point source immediately upwind of an obstacle array has been examined in a hydraulic flume with a low level of background turbulence. The main purpose of the experiments was to examine the interaction of the plume and the internal boundary layer (IBL) created over the obstacle array. The obstacle array consisted of 11 rows of cubes at 16% packing density in a staggered arrangement. Plume dispersion was measured using flow visualization with Rhodamine dye and also with a thermal tracer technique. During the experiments the source release height was varied between z = 0 and z = 4H, where H is the obstacle height. For the low-level releases, the upper boundary of the plume followed the growth of the IBL over the array. For higher level releases ($z/H \geq 2$) the rate of plume growth was much reduced until the point downstream where it descended into the IBL, after which it experienced the intense turbulent mixing within the array. This suggests that the urban lateral spread parameter σ_y should be a strong function of height in situations where the turbulence level in the ambient approach flow is low. These results highlight the importance of the ambient turbulence even in strongly obstacle-affected dispersion.

Key words: dispersion, turbulent boundary layer, physical modelling, urban pollution

1. Introduction

The ability to account for the effects of obstacles on plume dispersion ("obstacle-resolving" models) has been identified as one of the priority areas in urban air pollution modelling (SATURN, 1999). Existing regulatory models use a wide assortment of procedures for predicting near-field dispersion around built-up areas, sometimes with inconsistent results (Robins, 1999). Fundamental knowledge is still sparse in several areas such as the effect of upstream flow conditions, the interaction of a plume with the growing boundary layer over a new surface, the effect of source height and wind direction, etc. This has motivated more systematic studies of the problem, and in recent years there have been several wind tunnel studies using gas tracers to examine the dispersion behaviour around groups of buildings (*e.g.*, Davidson *et al.*, 1996; Hall *et al.*, 1998; Macdonald *et al.*, 1998; Robins *et al.*, 1998; Theurer *et al.*, 1996).

Because of the complexities of the near-field urban dispersion problem, most progress has been made using physical models in wind tunnels (Schatzmann *et al.*, 1999). As an alternative to wind tunnel modelling, the present study makes use of a water flume to examine the near-field dispersion in an obstacle array. A

Environmental Monitoring and Assessment **65**: 231–238, 2000.

water flume has several advantages over a wind tunnel, such as: a) the lower kinematic viscosity permits lower flow rates, thus slowing down the turbulent dispersion mechanisms; b) a simple dye tracer can be used for flow visualization; and, c) mineral particles present in the flow provide natural seeding for LDV measurements. In addition, the relative incompressibility of water means that buoyancy forces are often negligible for moderately heated plumes, and it is therefore possible to use heat as a scalar tracer.

With the exception of Robins *et al.* (1998), few of the above references considered the effect of source height. They were mainly interested in ground-level releases, which give the maximum influence of the obstacles. In the present study, the effect of source height for releases upstream of an obstacle array was examined. This is of obvious importance since the most intense point sources of pollutants are usually elevated, and it is important to know how much influence the surface obstacles will have on releases from different heights.

2. Experimental Facility and Methods

The hydraulic flume used in the experiments was 1.2 m wide by 0.9 m high with a total length of 12.2 m. The 2.4-m long test section was sufficiently large to accommodate an 8 x 11 array of H = 50-mm cube obstacles (representative of simple buildings at about 1:200 scale) with low blockage. The cube obstacles were arranged in a staggered array on the floor of the flume with a spacing between the obstacles of S/H = 1.5. The resulting surface coverage of the obstacles was 16%. The free-stream Reynolds number for these experiments, based on the cube height, was Re = 3500. Although the boundary-layer flow over the obstacle array was turbulent, the cubes were roughened with stipple-tone paint to ensure fully rough turbulent flow on their surfaces as well.

Because of the primary interest in obstacle-generated turbulence, the atmospheric boundary layer was not fully simulated in these experiments. The cubes were located in the natural boundary layer that formed on the floor of the flume and extended to a height of about 3H. The mean profile followed a 1/7-power law and the longitudinal turbulence intensity in the approach flow varied from about 7% at half-cube height to only 2% in the free stream. This low free-stream turbulence made it easier to discern the upper limit of the internal boundary layer over the cubes. The low approach-flow turbulence also emphasized the dispersion due to the cube-generated turbulence. On the other hand, the low levels of turbulence meant a lack of large-scale turbulent structures in the lateral direction and this had important consequences when comparing the dispersion results to other physical modelling facilities, as discussed below.

For flow visualization, concentrated Rhodamine dye was ejected from a source consisting of an L-shaped probe of 1.2-mm diameter. For the detailed measurements of the plume structure, a source tracer of heated water (25^0 C overheat) was used. The non-dimensional buoyancy flux of the resulting plume, based on the velocity at cube height (F_0/HU_H^3), was only 0.0034, too low for any significant plume rise in these experiments (Coulson, 1998). The temperatures in the dispersing plume were measured by an array of 19 thermistor probes, calibrated to give a precise determination of temperature to within ±0.02 ^{0}C. The probe array was traversed vertically over the depth of the plume at various stations downstream. A thermistor was also located inside the source tube to measure the reference temperature at the release point. The relative concentration at a particular point could then be expressed as $(T_i-T_a)/(T_s-T_a)$, where the subscripts *a* and *s* refer to the ambient and source temperatures and T_i is the temperature measured by the ith thermistor. Further experimental details (averaging times, sampling rates, *etc.*) can be found in Coulson's thesis (1998).

3. Results and Discussion

In this paper we present two types of results from the flume. In the first type of experiment, the source was located a distance 2H upstream of the array, while the source height was varied from z = 0.5H to 4H. For each setup, the Rhodamine dye plume was photographed up to 36 times in plan and elevation view at 5-second intervals. Using a reference grid on the test section wall, the mean visible plume outline could then be traced out from the resulting images. Figure 1 shows the average visible plume outlines for sources at z/H = 0.5, 1, 2, and 3. When the source was near ground level (z/H = 0.5), the lateral dispersion was greatly enhanced due to the plume-obstacle interactions. The plume width was quickly reduced with increasing source height, and even at z/H = 1 the plume was significantly narrower. At z/H = 3 the lateral dispersion was greatly reduced, and was similar to that of an unobstructed plume in the low turbulence free stream flow.

The vertical plume extent was also measured by flow visualisation. Figure 2 shows results for plumes released at heights of z/H = 0.5, 1, 2 and 3. The distance x/H at which the lower edge of the visible plume touched down increased greatly with source height, until at z/H = 3 the source did not descend into the array over the entire test section. When z/H = 3, the height of the plume centerline was approximately constant over the range of the test section. This suggests that the internal boundary layer over the array had not grown sufficiently to interact with the elevated plume. However, with the low level releases (z/H = 0.5 and z/H = 1.0 in Figure 2) the upper edge of the plume grew at the same rate as the internal boundary layer over the cubes (which was at a rate proportional to $(x/H)^{0.4}$ in this array).

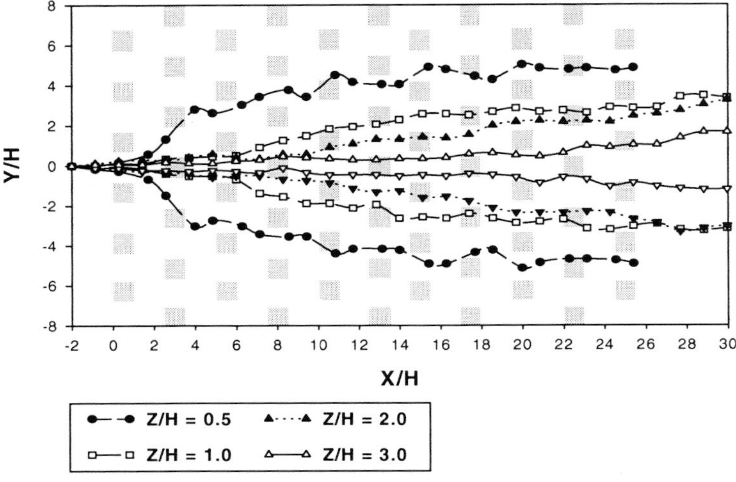

Figure 1. Plan view of mean plume outline for dye plumes released at various heights from the ground. All releases from 2H upstream of the obstacle array.

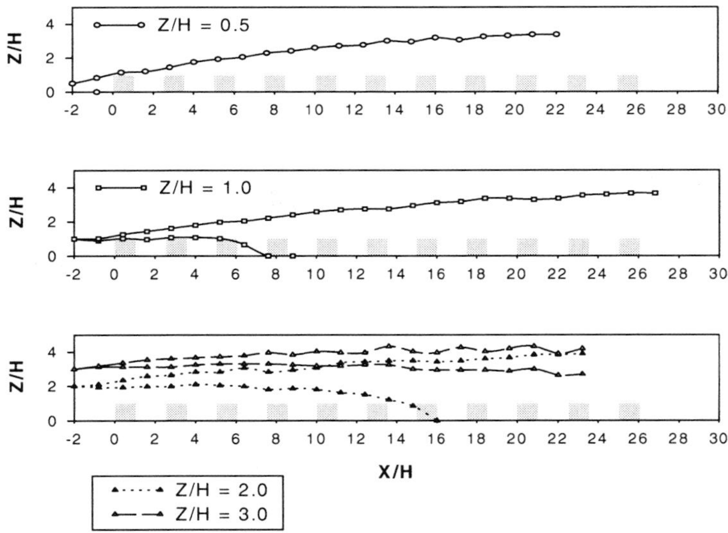

Figure 2. Elevation view of mean plume outline for dye plumes released at various heights from the ground. All releases from 2H upstream of the obstacle array.

INVESTIGATION OF BUILDING - INFLUENCED ATMOSPHERIC DISPERSION USING A DUAL SOURCE TECHNIQUE

ILIAS MAVROIDIS [1] and RICHARD F. GRIFFITHS [2]

Environmental Technology Centre, UMIST, P.O.Box 88, Manchester, M60 1QD, U.K.

E-mail: imavr@tee.gr [1], J.Rigby@umist.ac.uk [2]

Abstract. Dispersion of atmospheric contaminants in the vicinity of an isolated cubical model building was investigated in the field. A dual source/receptor technique was used in the experiments, which was proved to be very useful for the investigation of pollution dispersion. This experimental technique involved the simultaneous release of two different tracer gases from two different point sources, and the deployment of a FID (Flame Ionization Detector) co-located with a UVIC® (Ultra-Violet Ion Collector) detector. Both mean concentrations and concentration fluctuation statistics were examined. In this paper concentration fluctuation statistics are presented. The effect of the upwind source location on intermittency values and on the cumulative density function (cdf) is examined. The exact location of a source placed upwind of an obstacle has a very significant and complex effect on both mean concentrations and concentration fluctuations. As the lateral or vertical displacement between the two sources is increased, cross-correlation values between data taken simultaneously by two co-located detectors decrease rapidly.

Key words: Atmospheric dispersion, isolated obstacles, field experiments, dual source, concentration fluctuations, intermittency

1. Introduction

Increased concern over the problem of atmospheric pollution in built-up areas has highlighted the need for detailed investigations of atmospheric flow and dispersion of contaminants in the vicinity of buildings. Examination of flow and dispersion in the wake of isolated obstacles is very useful for evaluating doses due to routine and accidental releases of airborne radioactive effluents or hazardous materials of any kind, since those are usually released near large structures. Furthermore, information of the flow characteristics near an isolated obstacle may be used to describe the region close to a discharge in the development of an urban dispersion model. Hosker (1984) and Meroney (1982) give useful reviews of flow and dispersion near buildings. Hall *et al* (1996) present an up-to-date review of the literature on urban dispersion.

Most of the research on building-influenced dispersion is concerned with the examination of mean concentrations over a given time, typically minutes or hours in length. Although this description is adequate for many of the applications of dispersion around buildings, there are cases where the investigation of concentration fluctuations is also of importance. The ignition of a flammable gas, for example, depends on the concentration lying between the lower and upper flammability limits on a time scale corresponding to that of the ignition process. Concentration fluctuations are also important in the case of malodour nuisance, and in the case of

Environmental Monitoring and Assessment **65**: 239–247, 2000.

those toxic substances that have a damage function of the form $C^n t$, where n is an index typically between 1 and 3, C is the concentration and t is the exposure (Griffiths and Megson, 1984). Fackrell and Robins (1982) present a thorough investigation of concentration fluctuations. In their paper it was shown that most of the fluctuations are produced very near the source and that further experimentation is required for the investigation of concentration fluctuations. The authors suggested that the effect of instrumentation response should be examined and instrumentation should be further developed. Mylne and Mason (1991) studied concentration fluctuations in a dispersing plume of pollutant in the atmosphere and emphasized the need for experimental measurements of concentration fluctuations to provide statistical data and model verification. There are only a small number of reported field experiments investigating concentration fluctuations in the vicinity of obstacles. These include the early work of Jones and Griffiths (1984) and the experiments of Higson *et al* (1995), which examined the effect of atmospheric stability on concentration fluctuations in the vicinity of an isolated cube.

This paper presents the results of an experimental study investigating atmospheric dispersion around isolated obstacles in the field, examining mainly the fluctuating behaviour of a dispersing plume near an obstacle. A new experimental technique is deployed and the performance of the technique is discussed.

2. Experimental Details

The experiments were performed at Altcar Field Site on the northwest coast of England, which is a site with a fairly flat surrounding terrain. The atmospheric stability ranged from moderately unstable to neutral (Pasquill categories C and D). The concentration measurements were supported by meteorological data collected by an ultrasonic anemometer positioned upwind of the obstacle. Detailed meteorological measurements, which were conducted at the same site, gave a mean roughness length of approximately 12mm for the prevailing wind direction (Kourniotis, 1996). In the trials described in this paper a cubical model building positioned normal to the mean wind direction was used (H=1.15m). It should be noted that, since the model building represented real structures at a scale of approximately 1/10, the building scale and the boundary layer scale are not the same.

The dual source/receptor system technique that is used in the experiments was developed by Jones (1995). This experimental technique involves the simultaneous release of two different tracer gases from two point sources, and the deployment of two different types of co-located detectors. The co-located detectors used in the experiments described in this paper are the Flame Ionization Detector and the UVIC® (Ultra-Violet Ion Collector) detector. The tracer gases were chosen such that one is detected only by the UVIC® detector (ammonia) and the other only by the FID (propane). The UVIC® detector is a fast-response photo-ionization detector that provides a useful calibratable range from about 0.01 to 1000 ppm by volume and has a response time of about 0.02 seconds. The development of the UVIC® detector is described in Griffiths *et al* (1998). The FID detector is capable of detecting sub-ppm concentrations and has a response time of approximately 0.002 seconds. The response of the FID is almost linear to increasing carbon content of saturated hydrocarbons, such as propane.

The main experimental configuration (co-located sources) is presented in Figure 1. Two pairs of co-located detectors were used in the experiments, and were located 0.5H and 3.0H downwind of the rear face of the obstacle, while three more UVIC® detectors were located downwind of the rear face of the obstacle. All the detectors were located at a height of 0.5H. The two sources were co-located at the beginning of each experimental session at a location 2.0H upwind of the centre of the front face of the obstacle and at a height of 0.5H. Then the source was moved stepwise along one of the three axes and a separate trial was conducted for each ammonia source location.

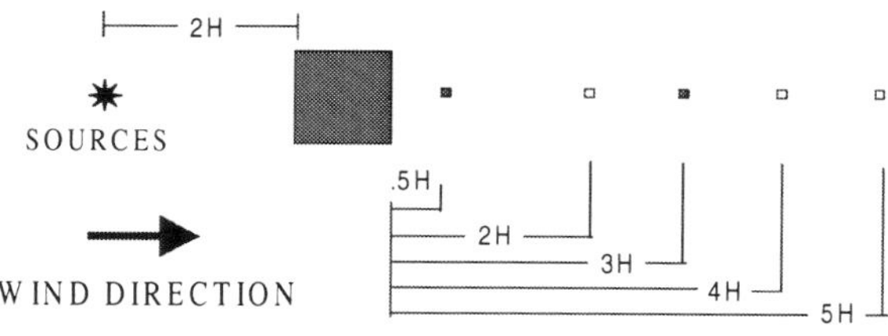

Figure 1. Plan view of the experimental configuration for a cube normal to the flow (co-located sources)

3. Results and Discussion

3.1 THE DUAL SOURCE TECHNIQUE

The application of the dual source technique in the field has been limited. A similar technique was used by Sawford *et al* (1985), who conducted experiments in which two different tracers were released, namely sulphur hexafluoride and phosphorus smoke, and two sensors were deployed. The sensors were co-located from 25m to 100m downwind of the sources and eighteen-second average concentrations were used to study concentration fluctuations. The dual source technique was used in the trials described here mainly to explore its applicability and its advantages for field investigations of atmospheric dispersion. The dual source technique was found to be very useful in many respects. The main advantages of the technique, as were observed during its deployment in the field trials are discussed below.

(i) It is an on-line method of detecting problems that may occur during the experiments. From the time series of data displayed on the PC monitor during data acquisition it was possible to see if something was wrong during an experiment, especially since the propane source was always at the same location and was used as a reference.

(ii) The time series of concentrations detected simultaneously by the two co-located instruments may be easily compared to examine the effects of the instrumentation on data acquired in the trials. Figure 2 shows sample time series of data in the case of co-located sources and detectors (the detectors are placed 0.5H downwind of the cube). The two time series are almost identical. Any differences in the magnitude or the characteristics of the time series are mainly attributed to differences in temporal and spatial resolution of the two types of instruments. Peak values are seen to be higher in the case of the FID, as would be expected given that the FID has a shorter response time.

As the ammonia source is displaced from the propane source the corresponding signals become less and less similar. Figure 3 shows the signals from a pair of co-located detectors placed 0.5H downwind of a cube, in the case of an ammonia source displaced 1.5H laterally from the centreline (and from the propane source). In this case the two time series are completely different, with the signal of the UVIC® detector being very intermittent, with one high peak which is attributed to a sudden change in the wind direction resulting in gas being entrained in the recirculation region. In the experiment presented in Figure 3, the wind direction changed suddenly to approximately +17 degrees from the normal to the cube at time t=220 s. This shift in the wind direction lasted approximately 3 seconds.

(iii) The fact that the propane source was always located at the same position allowed several repetitions of the same experiment for the different obstacles. As a result a statistically stable ensemble average for that particular propane source location (2.0H upwind of the obstacle, on the centreline and at a height of 0.5H) could be obtained. This reduced significantly the variability attributable to comparing time series obtained sequentially rather than simultaneously.

(iv) The effect that the displacement of the ammonia source has on the concentration patterns may be examined through the changes in the cross-correlation function between data recorded simultaneously by a pair of co-located FID and UVIC® detectors. Figure 4 shows results from the two pairs of co-located FID and UVIC® detectors, one placed 0.5H downwind and the other 3.0H downwind of the cube, in the case of lateral displacement of the ammonia source.

For co-located sources the cross-correlation value is higher than 0.9 for both pairs of detectors, which is indicative of the similarity of the two time series. At the other end of the experimental range examined here, when the ammonia source is displaced 2.0H from the centreline (and the propane source), the cross-correlation value is almost zero, as gas is probably not entrained in the near-wake for this source location. In general, cross-correlation values become smaller as the ammonia source is displaced further apart from the centreline, and become almost zero when the ammonia source is displaced 1.5H laterally from the centreline. The cross-correlation function has a similar behaviour when the ammonia source is vertically displaced.

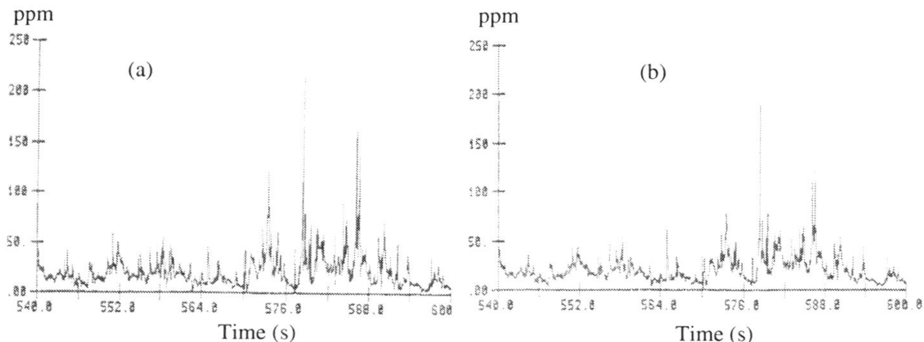

Figure 2. Data time series from a pair of co-located (a) FID and (b) UVIC® detectors, from the dual source trials around an isolated cube. Detectors are located 0.5H downwind of the cube. Sources are co-located 2.0H upwind, on the centreline and at a height of 0.5H.

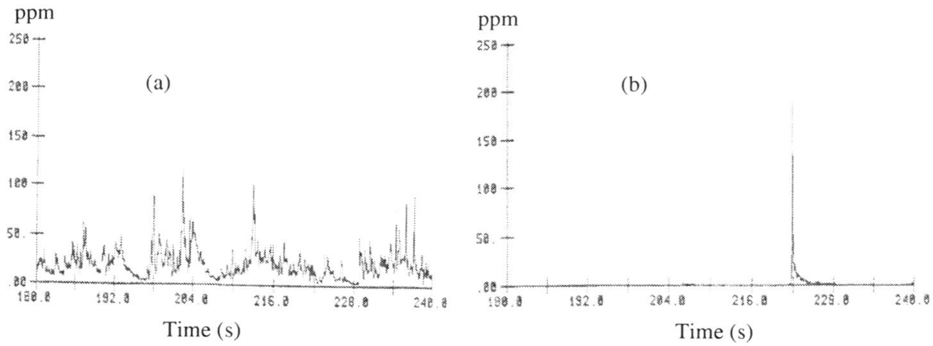

Figure 3. Data time series from a pair of co-located (a) FID and (b) UVIC® detectors, from the dual source trials around an isolated cube. Detectors are located 0.5H downwind of the cube. The ammonia source is displaced 1.5H laterally from the centreline.

It should be noted though that the dual source technique has some disadvantages, which mainly arise from the special requirements of the method. Such requirements are the use of two tracer gases, two source configurations and two different types of detectors. This involves some extra effort during the experiments and also increases the costs involved. Another important requirement is the deployment of two appropriate gases and the use of gases with higher purity.

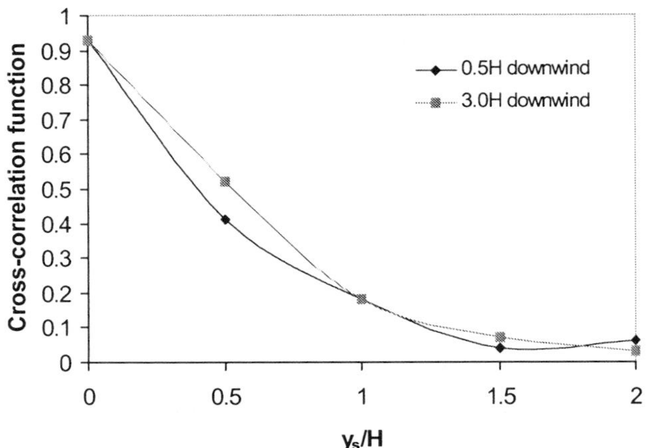

Figure 4. Cross-correlation values for lateral displacement of the ammonia source.

3.2 CONCENTRATION FLUCTUATIONS IN THE WAKE OF A CUBE NORMAL TO THE FLOW

The main statistical parameters examined in this section are: (a) the intermittency, which is defined here as the proportion of time for which concentration is at or below a threshold value, which in this case is nominally zero concentration, and (b) the cumulative distribution function (cdf) which gives the total number of concentration readings up to the maximum concentration observed, versus increasing concentration (normalized with the mean concentration).

As already described, in the beginning of each experimental session the source was located 2.0H upwind of the obstacle, on the centreline and at a height of 0.5H. Then the source was moved stepwise along one of the three axes and a separate trial was conducted for each ammonia source location. In the following sections the dependence of concentration fluctuation statistics on lateral and vertical source location are briefly discussed.

3.2.1 Dependence of Concentration Fluctuation Statistics on Lateral Source Location

Figure 5 shows the dependence of intermittency values on the lateral location of the ammonia source. Intermittency values are very low and almost constant for source displacements up to 0.5H from the centreline (which coincides with the lateral boundary of the obstacle), since for these source locations a large proportion of the plume is entrained in the near-wake. In the recirculation region (0.5H downwind) intermittency values are zero, which indicates not only that gas is entrained in this region, but also that the gas is well mixed (and not only intermittently present). Indeed, within this region of recirculating flow a very high degree of mixing is occurring. Intermittency values increase significantly when the source is laterally displaced more than 0.5H from the centreline. For a source displaced about 1.75H from the centreline intermittency is close to unity, which reflects the fact that gas is not entrained in the near-wake from this source location, except in the case of a sudden change in the wind direction (see Figure 3). As the source is displaced laterally more than 0.5H from the centreline mean concentrations reduce. The values of intermittency for the different source locations indicate that this variation of mean concentrations is to a great extent attributable to the variation of intermittency. A similar effect was observed by Mylne and Mason (1991) in a plume dispersing in open terrain. It should be noted that the conditional mean concentration (i.e. the mean concentration which excludes the periods for which concentration is at or below the threshold value) varies only slightly in general, but when intermittency is close to unity it may present peaks because of sudden episodes of increased concentration (e.g. when the wind direction changes for a very short period as shown in Figure 3).

The cdf curve for a source located on the centreline (and at a height of 0.5H) shows that the entrainment and mixing of gas in the recirculation region (Figure 6a) results in very low concentration fluctuation intensity values ($I=\sigma_c/C$), since the slope of the central part of the cdf curve is very steep. In the recirculation region (Figure 6a) the peak-to-mean ratio (c'_{max}/C, where c'_{max} is the maximum instantaneous concentration observed and C is the average concentration over the 3-minute period) is very low. Further downwind (Figure 6b) the gas is not as well mixed, and as a result the slope of the cdf curve is slightly less steep. In the case of a source displaced 1.0H from the centreline (at a height of 0.5H) the cdf curve is very different. The values of intermittency - indicated by the point where the cdf curve crosses the y-axis - are higher than 0.7. Thus gas is very intermittently entrained in the near-wake, and this results in bursts of gas, which increase the concentration fluctuation intensity.

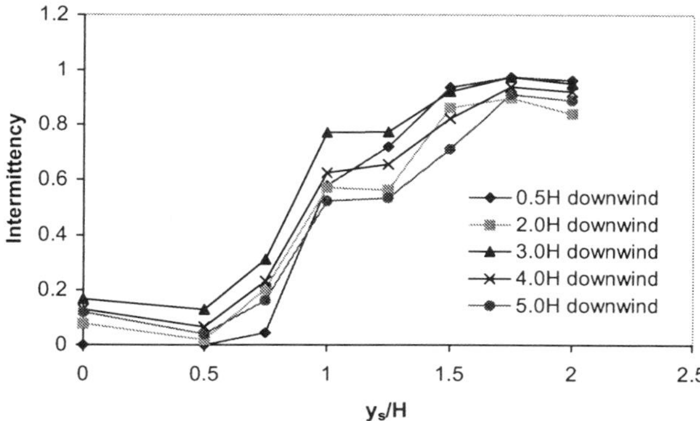

Figure 5. Dependence of intermittency in the wake of a cube normal to the flow on lateral source location.

3.2.2 Dependence of Concentration Fluctuation Statistics on Vertical Source Location

For source heights up to the cube height, intermittency values are zero in the recirculation region, indicating a very good level of mixing. Intermittency values 2.0H to 5.0H downwind of the cube range between 0 and 0.2. For source heights more than 1.0H intermittency increases continuously, since gas is very intermittently entrained in the wake for these source heights, until it reaches a value of unity for a source height of 1.75H. The cdf curve for a source height of 1.5H suggests that in the recirculation region (Figure 6a) intermittency is approximately 0.5, which is still a much lower value than that expected if the cube was not present, since neither the downwash nor the mixing due to the cube would occur. Further downwind, where the mixing effect of the cube is much reduced (Figure 6b), intermittency is close to unity. This stresses the effect that the high degree of mixing occurring in the recirculation region has on mean concentrations. The slope of the cdf curve is much less steep, especially for the detector located outside the recirculation region (3.0H downwind), implying that the intensity of concentration fluctuations has increased significantly for this source height.

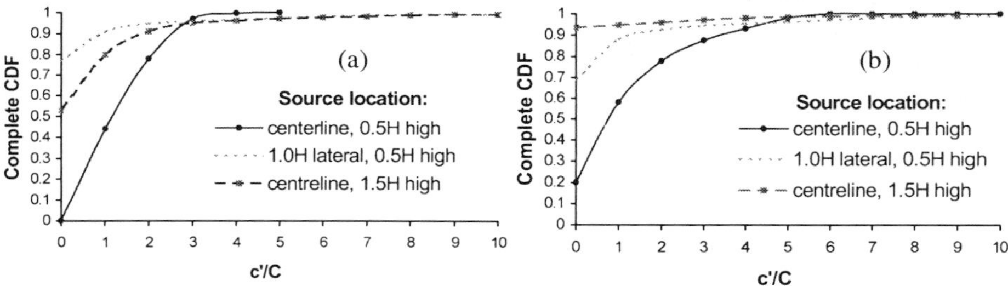

Figure 6. Cdf curves for concentrations measured (a) 0.5H downwind and (b) 3.0H downwind of a cube normal to the flow for three different source locations.

4. Summary and Conclusions

Atmospheric dispersion around isolated obstacles was investigated in the field using a dual source/receptor system technique. This technique proved to be very useful for the investigation of pollutant dispersion problems, since it allows comparison of data collected from various source locations (ammonia) with data simultaneously collected from a fixed source location (propane). Cross-correlation between data simultaneously collected from two co-located detectors in the same trial is examined. For co-located sources cross-correlation values are higher than 0.9 for both pairs of co-located detectors and as the displacement between the two sources is increased, cross-correlation values decrease rapidly. The dual source technique is very useful in providing data for statistical analysis and modelling purposes. The possibilities offered by this technique have not yet been fully explored, and the technique should be further investigated and refined.

Concentration fluctuation results indicate that as the source is displaced from the centreline it is in general the case that intermittency values increase. This accounts, at least partly, for the subsequent reduction of mean concentrations. As the cdf curves show, concentration fluctuation intensity values increase as well. The above results imply that, even in the recirculation region, investigation of concentration fluctuations is important since for a source displaced further away from the centreline gas entrainment is very intermittent and thus sudden peaks in concentrations occur, which are followed by long periods of zero concentrations.

Acknowledgements

This work has been carried out under contract to the Atmospheric Dispersion Modelling Group at DERA Porton Down with funding provided by the UK MoD (agreement 2044/15/CBDE). Thanks are due to Dr. C.D.Jones for many helpful discussions.

References

Fackrell, J.E. and Robins, A.G.: 1982, Concentration fluctuations and fluxes in plumes from point sources in a turbulent boundary layer, *J. Fluid Mech.* **117**, 1-26.

Griffiths, R.F. and Megson, L.C.: 1984, The effect of uncertainties in human toxic response on hazard range estimation for ammonia and chlorine, *Atmospheric Environment* **18**, 1195-1206.

Griffiths, R.F., Mavroidis, I. and Jones, C.D.: 1998, Development of a fast-response portable PID: model of the instrument response and validation tests in air, *Measurement Science and Technology* **9,** 1369-1379.

Hall, D.J., Spanton, A.M., Macdonald, R. and Walker, S.: 1996, A review of requirements for a simple urban dispersion model, BRE Client Report No. 77/96, BRE, Watford, U.K.

Higson, H.L., Griffiths, R.F., Jones, C.D. and Biltoft, C.: 1995, Effect of atmospheric stability on concentration fluctuations and wake retention times for dispersion in the vicinity of an isolated building, *Environmetrics* **6**, 571-581.

Hosker, R.P. Jr.: 1984, Flow and diffusion near obstacles, *Atmospheric Science And Power Production*, (Ed. Randerson, D.), U.S. Department of Energy, 241-326.

Jones, C.D.: 1995, *personal communication*

Jones, C.D. and Griffiths, R.F.: 1984, Full-scale experiments on dispersion around an isolated building using an ionised air tracer technique with a very short averaging time, *Atmospheric Environment* **18**, 903-916.

Kourniotis, S.: 1996, Characterization of the atmospheric boundary layer approaching the Altcar Field Site, MSc Dissertation, UMIST, Manchester, U.K.

Meroney, R.N.: 1982, Turbulent diffusion near buildings, *Engineering Meteorology*, (Ed. Plate, E.), Elsevier, 481-525.

Mylne, K.R. and Mason, P.J.: 1991, Concentration fluctuation measurements in a dispersing plume at a range of up to 1000m, *Quart.J.Roy.Met.Soc.* **117**, 177-206.

Sawford, B.L., Frost, C.C. and Allen, T.C.: 1985, Atmospheric boundary-layer measurements on concentration statistics from isolated and multiple sources, *Boundary Layer Met.* **31**, 249-286.

DISPERSION IN URBAN ENVIRONMENTS
Comparison of field measurements with wind tunnel results

MICHAEL SCHATZMANN*, BERND LEITL, JOACHIM LIEDTKE
Meteorological Institute, University of Hamburg
Bundesstraße 55, D-20146 Hamburg, Germany
(*author for correspondence, E-mail: schatzmann@dkrz.de)

Abstract: A wind tunnel study was performed to determine the dispersion characteristics of vehicle exhaust gases within the urban canopy layer. The results were compared with those from a field monitoring station located in a street canyon with heavy traffic load. The agreement found was fair. In the second part of the paper it is shown how wind tunnel data can be utilized to supplement and thereby enhance the value of field data for model validation purposes. Uncertainty ranges were quantified which are inherent to mean concentration values measured in urban streets.

Key words: pollutant dispersion, car exhaust gas, street canyon, wind tunnel measurements, field experiments, model validation

1. Introduction

European legislation requires the installation of monitoring stations in city areas with heavy traffic loads. Field data sets covering the emission and pollution situation over sufficiently long periods (1 year or more) are now available.

At Hamburg University, a monitoring site located in Hanover/Germany was selected and replicated in a boundary layer wind tunnel experiment. The objectives of this work were (1) to show the degree of agreement achievable in small-scale physical model experiments, (2) to investigate the reasons for possible deviations between laboratory and field data, (3) to help interpret specific features both data sets exhibit, and (4) to provide data sets for model validation purposes.

2. Experimental Setup

The experiments were carried out in the multi-layer wind tunnel of Hamburg University. The length, width and height of the working section of this tunnel are 8.7 m x 2.3 m x 1.0 m. The tunnel was operated in a neutral stratification mode. Details of the tunnel characteristics are given in Schatzmann et al., 1995.

In order to obtain sufficient spatial resolution for the subsequent model experiments, a boundary layer in the scale of 1:200 was established. Only the lowest about 100 m of the field boundary layer were properly represented in the tunnel. The boundary layer characteristics of the approach flow were in

Environmental Monitoring and Assessment **65**: 249–257, 2000.

fair agreement with those of a typical urban boundary layer. It had a mean vertical velocity profile exponent of n = 0.28, a roughness length of z_o . 1 m, a vertical turbulence intensity distribution according to ESDU (1974) and a shear velocity of $u_* = 0.055\ U_4$ (with U_4 being the free stream velocity). Also the lateral length and time scales were in agreement with reality (Pascheke et al., 1999).

Fig. 1 shows a view on the wind tunnel model of the Hanover site. The location of the monitoring station which is operated by the State Environment Agency of Lower-Saxony (NLÖ, 1994 and 1995) in the center of the model area at the pedestrian walkway in a four-lane street canyon with a load of about 30.000 vehicles/day. Based on automated traffic counts and information on the composition of the German vehicle fleet, reasonably good estimates of pollutant emission rates were available. The above-roof wind and background concentrations were also measured.

In the wind tunnel, the four traffic lanes were represented by four line sources. Special care was taken to achieve uniform and nearly momentum

Fig. 1: View on a section of the wind tunnel model of the Hanover site. The position of the monitoring station is the center of the figure.

free emissions, thereby using the experience accumulated in previous experiments (Meroney et al., 1996). In the final design of the source, hundreds of needles from syringes with an internal diameter of 0.2 mm and a length of 42 mm were mounted next to each other (Fig. 2). The line sources had a length of about 200 m (full scale).

Velocity measurements in the wind tunnel were made at the same (above-roof) position as in the field and in a reference height 100 m above ground . A standard hot wire anemometer was used.

In order not to disturb the flow within the street canyon, the concentration detector (based on the flame ionisation principle) was built into one of the model buildings. Only the tiny suction pipe reached into the canyon and ended at the position of the air intake of the corresponding field monitoring station (1.5 m above ground).

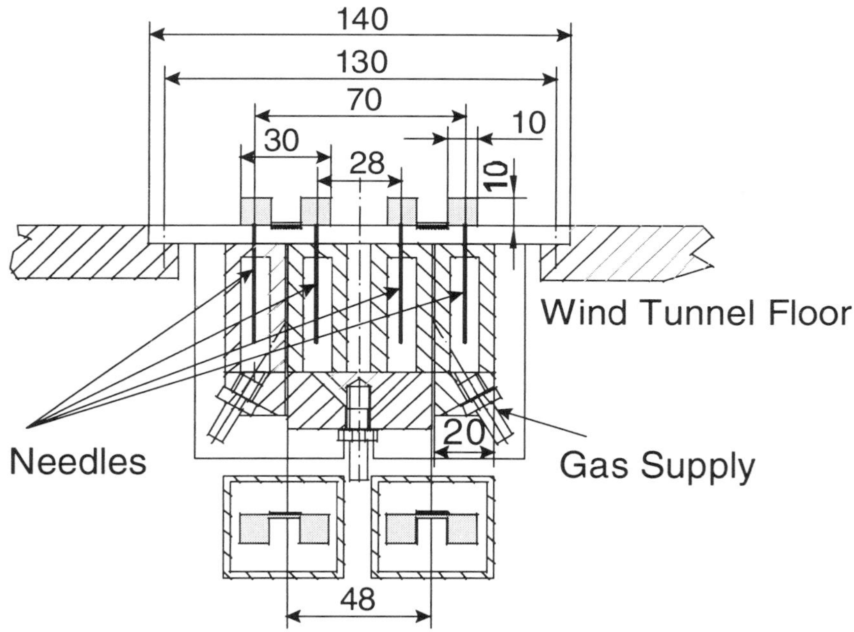

Fig. 2: Design of the line source.

3. Results

In order to make the results from the field and from the wind tunnel comparable with each other, NO_x concentrations observed in the field over a period of one year (1994) were grouped according to the wind direction (10° steps) and brought into the non-dimensional form $c^* = C \cong u_{ref} \cong H/(Q/L)$, where C is the time mean value of the measured concentration (30 min average, in excess above background), u_{ref} is a reference velocity, taken at a height of 100 m, H is a characteristic length (the average height of the surrounding buildings, H = 20 m) and (Q/L) is the source strength of the line source.

This presentation is commonly chosen. It reflects the fact that the measured concentration C should be linearly dependent on the source strength and inversely proportional to the wind velocity. Other factors like, e.g., ambient stratification are regarded to be of secondary importance for in-canopy layer dispersion processes (although they produce some scatter).

The determination of the emission rate per unit length Q/L is usually somewhat of a problem during field measurements. To obtain an estimate as reliable as possible, automated traffic counters were used at the Hanover site which registered not only the number of vehicles per time interval but discriminated also between passenger cars and trucks. In combination with knowledge about the composition of the German vehicle fleet in the year 1994, the prevailing driving pattern alongside the monitoring station and corresponding emission factors for specific groups of vehicles, Schädler (1996) computed Q/L-values which we simply took over.

Fig. 3 shows the results. Presented are non-dimensionalized concentrations c^* as a function of wind direction (wind from north corresponds to 0°, see Fig. 4) at the position of the monitoring station inside the street canyon. The curve marked with circles represents the wind tunnel data and those with squares the corresponding field data. The difference between the two field curves results from different emission models used in the computation of Q/L. The curve with open squares is based on Schädlers (1996) report. For the corresponding curve with closed squares factors according to the newer emission model MOBILEV (Skrzipczyk, 1997) were used. As can be seen, the uncertainty in field emission rates accounts for differences in c^*-values of about 40%. Whether this value is a realistic measure of the true uncertainty remains unknown.

The agreement between field and wind tunnel data is generally fair, with the exception of wind directions around 280°. The reason is not fully understood, but it was noticed that for this wind direction sector the monitoring station is located in a zone with large concentration gradients. This means small probe positioning errors have large effects. To show this, we removed the concentration sensor from its original but geometrically complex position (see Fig. 5), moved it into a comparable section of the street canyon about 80 m (full scale) to the north and repeated the measurements. The c^*-values from the wind tunnel are now significantly lower for winds from 280° to 300° but remain about the same for the wind directions in the range of 250° to 270° (Fig. 6).

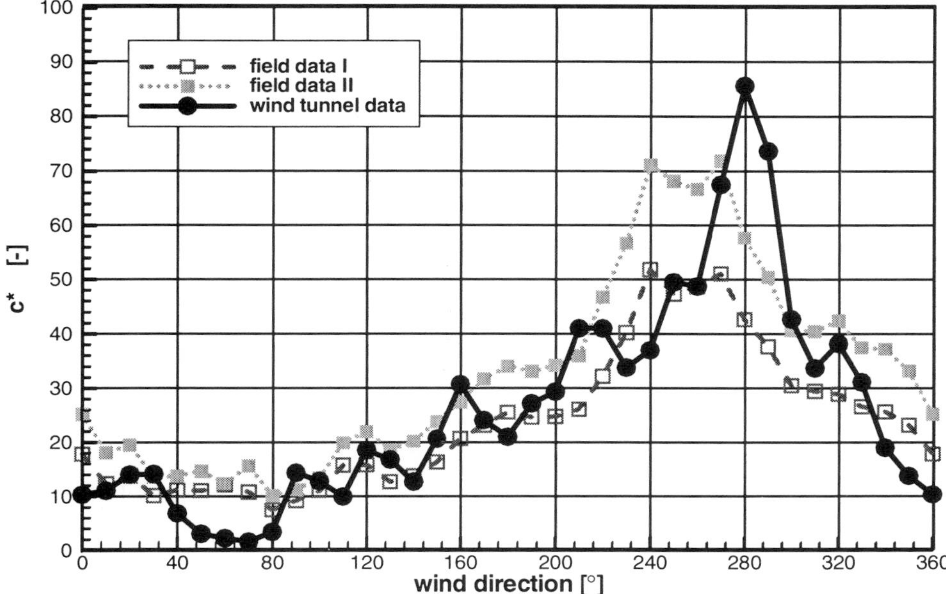

Fig. 3: Comparison of results from field and wind tunnel measurements. For explanations see text.

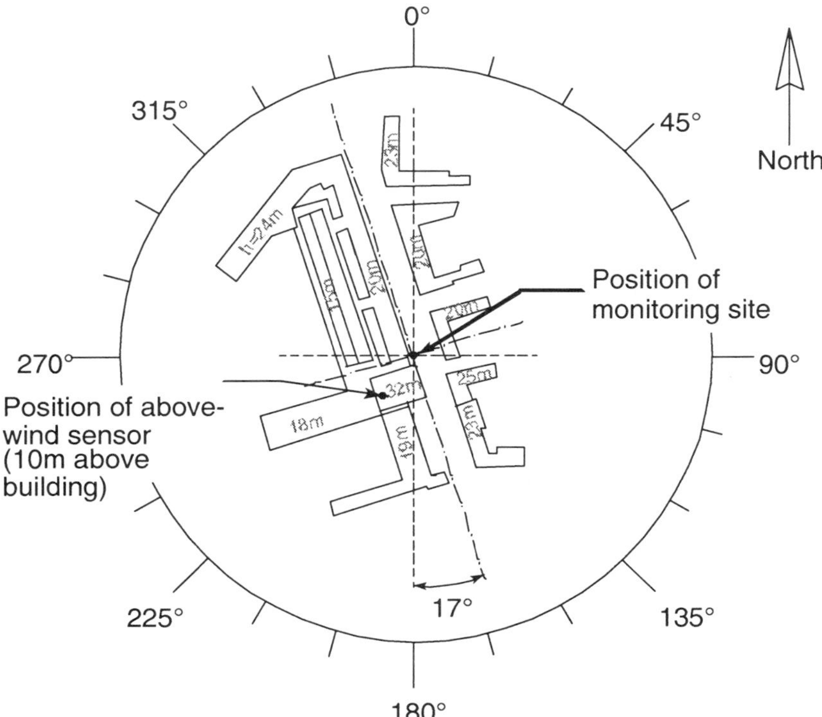

Fig. 4: Top view on the monitoring site and indication of wind directions.

Other factors which might have an effect on c* obtained in the field are vehicle-induced turbulence or unsymmetric heating of the canyon walls. Both must have an effect under very low ambient wind conditions which were not especially considered here. First results of ongoing research (Kovar-Panskus et al., 1999) give reason to believe that vehicle-induced turbulence should be of rather limited importance as far passenger cars are concerned. With respect to trucks this statement might not hold but we are not yet able to give a final answer. Provided there is wind, Monin Obukhov theory implies that thermal effects should be rather small.

Wind tunnel models do not only have a geometrical scale, they also have a time scale. When we use the same wind velocity in the tunnel as in the field, all processes in our 1:200 model are 200 times faster than in the field. Half-hourly mean concentrations as they were determined in the field correspond to 9 s averages in the wind tunnel. Although in the wind tunnel all boundary conditions were carefully controlled and kept constant over the duration of an experiment, we needed several minutes of averaging time to obtain mean conentration values repeatable within ± 5 %.

The concentrations in the wind tunnel were measured utilizing a fast flame inonisation detector with a frequency responce of approximately 400 Hz. This high resolution in time enabled us to collect time series which subsequently were averaged over intervals of 9 s only. Fig. 7 shows the result. The error bars shown indicate the largest and smallest 9s-average concentration found in the time series. They should be attached to the field data curves in Fig. 3. They indicate the variation of mean concentrations due to the fact that averaging times of 30 min (full scale) are simply too short to provide steady state results. To simply increase the averaging intervals in the field is not advisable since the ambient conditions are subject to the diurnal cycle and change constantly.

These findings fully corroborate the suspicion raised in a previous paper (Schatzmann et al., 1997) that mean concentrations determined in field experiments within the urban canopy layer exhibit a large inherent statistical variability which is linked to the unsteadyness of the flow field.

4. Conclusions

The present study shows that the interpretation of data measured at urban street monitoring stations is not an easy task. There is a large inherent uncertainty in such data which is caused by the continuous change of the atmospheric boundary layer due to the diurnal cycle. Before these data are utilized to validate numerical models they should be analyzed and enhanced by appropriate boundary layer wind tunnel experiments.

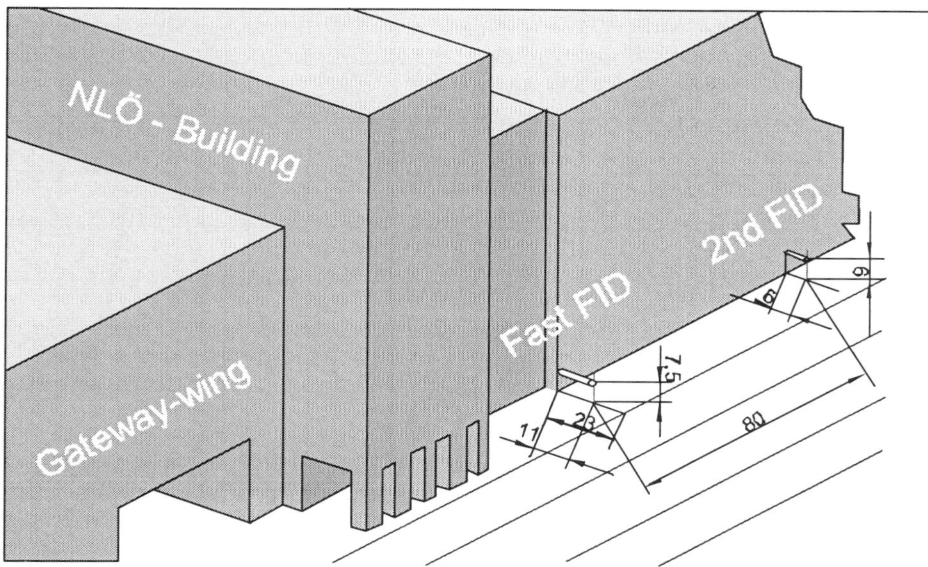

Fig. 5: Sketch of the site with indication of the monitoring position and the position of the 2nd flame ionisation detector.

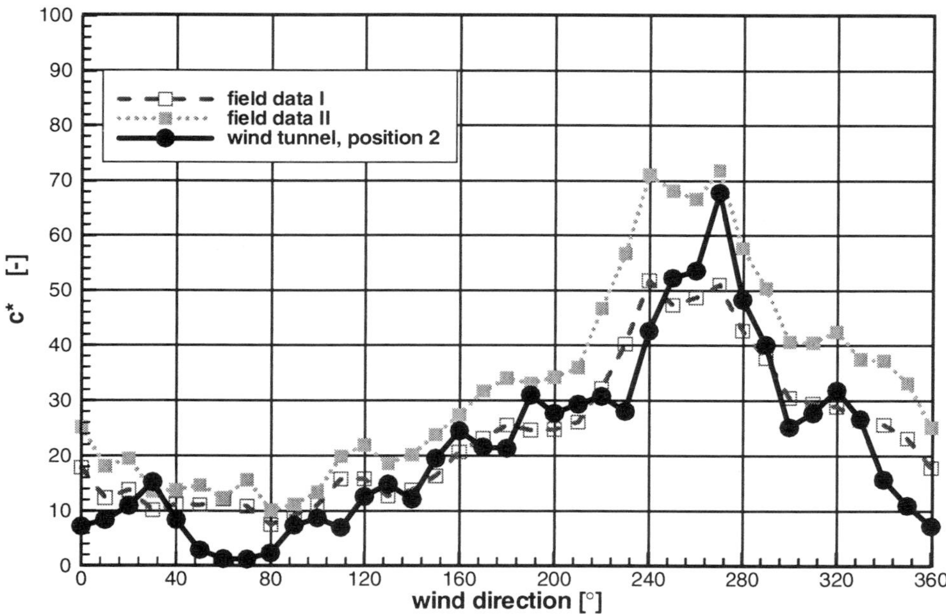

Fig. 6: Comparison of the field data with those from the 2nd measurement position in a geometrically less complex environment.

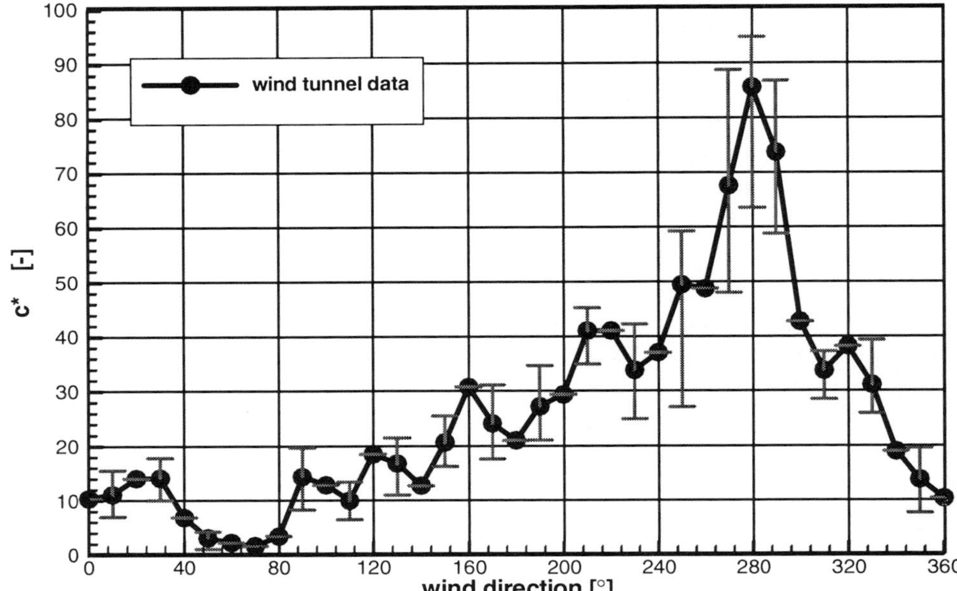

Fig. 7: Time mean values of concentration and the expected variation of the field data due to insufficiently long averaging intervals.

Such laboratory experiments do not provide only error bars. They can also be used to determine suitable positions at which the monitoring stations should be located in order to achieve results as representative as possible.

Acknowledgements

The authors are grateful for financial support from PEF (Projekt Europäisches Forschungszentrum für Maßnahmen zur Luftreinhaltung, Forschungszentrum Karlsruhe), and UBA (German Federal Environmental Agency Berlin).

We would like to thank W.J. Müller (NLÖ Hannover) and G. Schädler (Ing.-Büro Lohmeyer, Karlsruhe) for providing us with data and other helpful information.

References

ESDU: 1974, Characteristics of atmospheric turbulence near ground. Part II: Single point data for strong winds (neutral atmosphere). Item 1674031. Engineering Science Data Unit, London, U.K.

Kovar-Panskus, A., Leitl, B., and Schatzmann, M. (1999) Modelling of car-induced turbulence in a boundary layer wind tunnel. Proceedings, EUROMECH Colloquium 391, Prague, Sept. 13-15.

Meroney, R.N., Pavageau, M., Rafailidis, S., and Schatzmann, M.: 1996, Study of line source characteristics for 2-d-physical modelling of pollutant dispersion in street canyons. Journ. Wind Eng. and Ind. Aerodyn., 62, pp. 37-56.

NLÖ: 1994, Lufthygienisches Überwachungssystem Niedersachsen - Luftschadstoffbelastung in Straßenschluchten. Bericht, Niedersächsisches Landesamt für Ökologie, Göttinger Str. 14, D-30449 Hannover, ISSN 0945-4187 (in German).

NLÖ: 1995, Lufthygienisches Überwachungssystem Niedersachsen - Standortbeschreibung der LÜN-Stationen. Bericht, Niedersächsisches Landesamt für Ökologie, Göttinger Str. 14, D-30449 Hannover, ISSN 0945 4187 (in German).

Pascheke, F., Leitl, B., and Schatzmann, M.: 1999, Physical modelling of boundary layer characteristics for in-canopy layer dispersion processes. Proceedings, EUROMECH Colloquium 391, Prag, Sept. 13-15.

Schatzmann, M., Donat, J., Hendel, S., and Krishan, G.: 1995, Design of a low-cost stratified boundary layer wind tunnel. Journ. Wind Eng. and Ind. Aerodyn., 54/55, pp. 483-491.

Schatzmann, M., Rafailidis, S., and Pavageau, M.: 1997, Some remarks on the validation of small-scale dispersion models with field and laboratory data. Journ. Wind. Eng. and Ind. Aerodyn., 67/68, pp. 885-893.

Schädler, G., Bächlin, W., Lohmeyer, A., van Wees, T: 1996, Vergleich und Bewertung derzeit verfügbarer mikroskaliger Strömungs- und Ausbreitungsmodelle. Projekt Europäisches Forschungszentrum für Maßnahmen zur Luftreinhaltung. Bericht FZKA-PEF 138, Forschungszentrum Karlsruhe, ISSN 0948 535X (in German).

A SIMPLE MODEL FOR URBAN BACKGROUND POLLUTION

RUWIM BERKOWICZ
National Environmental Research Institute, Department of Atmospheric Environment, Frederiksborgvej 399, DK-4000, Roskilde, Denmark
E-mail: rb@dmu.dk

Abstract. A simple urban background pollution model is presented. Contributions from the individual area sources, subdivided into a grid net of a resolution of 2km x 2km, are integrated along the wind direction path assuming linear dispersion with the distance to the receptor point. Horizontal dispersion is accounted for by averaging the calculated concentrations over a certain, wind speed dependent wind direction sector, centred on the average wind direction. Formation of the nitrogen dioxide due to oxidation of nitrogen monoxide by ozone is calculated using a simple chemical model based on assumption of a photochemical equilibrium on the time scale of the pollution transport across the city area. The rate of entrainment of fresh rural ozone is governed by this time scale. The model is suitable for calculations of urban background when the dominating source is the road traffic. For this source the emissions take place at ground level, and a good approximation is to treat the emissions as area sources, but with an initial vertical dispersion determined by the height of the buildings.

Key words: urban background, pollution models, traffic pollution, model validation

1. Introduction

The highest pollution concentrations are usually found in street canyons with intense traffic. The main contribution to the pollution levels is attributed here to the direct emissions from the street traffic while the background contribution constitutes only a small portion. The situation is, however, different concerning the secondary pollutants, such as the nitrogen dioxide (NO_2). Concentrations of NO_2, even in a very trafficked street, depend strongly on the background concentrations and on the availability of ozone (O_3). The main purpose of the urban background model presented here is just to provide a quick and simple method for estimation of background concentrations for street canyon calculations. Examples of model validation on data collected in Copenhagen, Denmark, are shown. The model is, however, still under development and the results should be considered as preliminary only.

Environmental Monitoring and Assessment **65**: 259–267, 2000.

2. Description of the model

Although many sources can contribute to the urban background, the dominating source for such pollutants as NO_x, CO and many hydrocarbons is often the road traffic. Emissions from road traffic take place at ground level, and a good approximation is to treat these emissions as area sources.

An emission survey for Copenhagen was constructed by the Danish Road Directorate (Bendtsen and Reiff, 1996) subdivided into a grid net with a resolution of 2km x 2km and covering most of the Great Copenhagen. For each grid, a typical diurnal variation was provided too. The average daily NO_x emissions (in $kg.grid^{-1}.day^{-1}$) are shown in Figure 1.

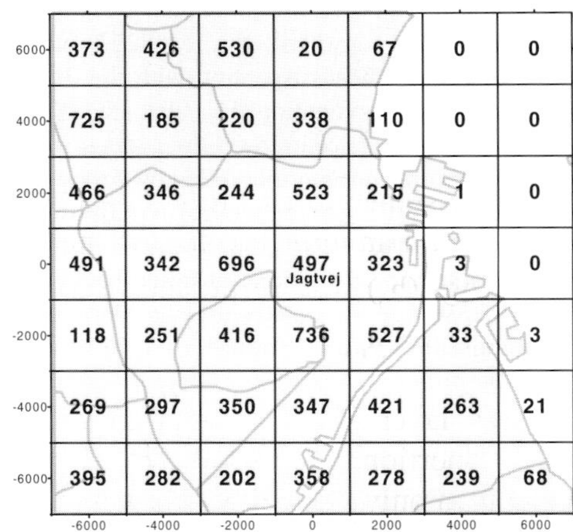

Figure.1. Average daily emissions of NO_x in Copenhagen (in $kg.grid^{-1}.day^{-1}$).

2.1 THE MODEL CONCEPT

The physical concept applied in the model is schematically depicted in Figure 2. The basic equations are also shown in this figure.

A number of simplifications and approximation was necessary in order to make the model be dependent on routinely available data only.

- Concentrations at a receptor point (calculation point) are calculated by numerical integration of contributions from each of the individual area sources along a path determined by the actual wind direction (Eq. (1)). The step size in the numerical integration is 50 m. The emission density, Q, is the value at the distance x from the receptor point. The rural background concentration is added to the contributions from the urban area.

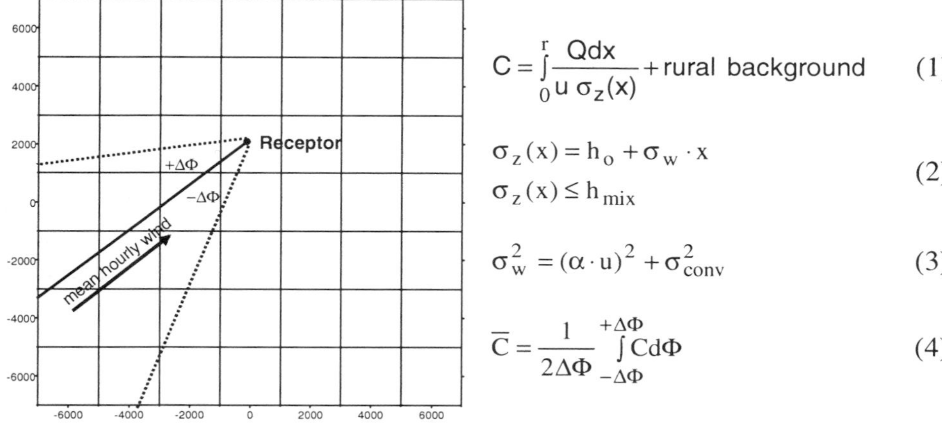

Figure 2 Schematic illustration of the model concept. The basic model equations are shown too.

- The vertical dispersion is assumed to be proportional with the distance to the receptor-point (Eq. (2)) but with an initial vertical dispersion determined by the average height of the buildings (h_o). The vertical dispersion is eventually limited by the height of the mixing layer, h_{mix}.

- The height of the mixing layer is not a routinely available parameter, it is therefore taken to be proportional to wind speed. This is a very crude approximation and in fact, can only be justified for neutral conditions. However, as shown in the subsequent discussion of the model and the results, the height of the mixing layer has only a small influence on the calculated concentrations.

- The vertical turbulence (Eq. (2)), which governs the vertical dispersion, is assumed to be composed of two parts: the mechanical turbulence, taken proportional to wind speed (α=0.1), and the convective turbulence, given by the free convection velocity scale.

$$\sigma_{conv} = \left(h_o \frac{g}{T} \overline{w'T'} \right)^{1/3}$$

The sensible heat flux, $\overline{w'T'}$, is estimated using the so-called Resistance Method, originally developed for the Danish point source dispersion model OML (Olesen and Brown, 1988). The height at which the convective velocity scale is calculated is set equal to the average height of the buildings h_o.

- Horizontal dispersion is accounted for by averaging the calculated concentrations over a certain, wind speed dependent wind direction sector centred on the average wind direction (Eq. (4)). The averaging is performed by numerical integration. A similar procedure is used in the Danish Operational Street Pollution Model (OSPM) (Berkowicz, 1999)

The basic principles of the model are not new. A similar concept was already proposed by Hanna (1971). He has additionally demonstrated that predictions from a simple model based on summation of contributions from individual area sources provides as good results as computations with more sophisticated models. It was also demonstrated in this work that the most important parameters governing the pollution concentrations in an urban environment are the wind speed and emission strength of the local area source.

2.2 FORMATION OF NO_2

Formation of the nitrogen dioxide due to oxidation of NO by ozone is calculated using a simple chemical model based on assumption of a photochemical equilibrium on the time scale of the pollution transport across the city area. The rate of entrainment of fresh rural ozone is governed by this time scale. The only reaction scheme accounted for is

$$NO + O_3 \Leftrightarrow NO_2 + O_2$$

This procedure is similar to that applied in OSPM (Hertel and Berkowicz, 1989; Berkowicz, 1999). It has also been used in connection with a method for prediction of high NO_2 concentrations in Danish cities (Hertel and Berkowicz, 1990). The parameters governing NO_2 concentrations are: the urban background concentrations of NO_x and the rural background concentrations of NO_2 and O_3. The rural background concentrations are presently derived from measured concentrations at a rural monitoring station, Lille Valby, located at about 25 km West of Copenhagen.

3. Results

The calculated annual concentrations of NO_x and NO_2 for the year 1995 are shown in Figure. 3.

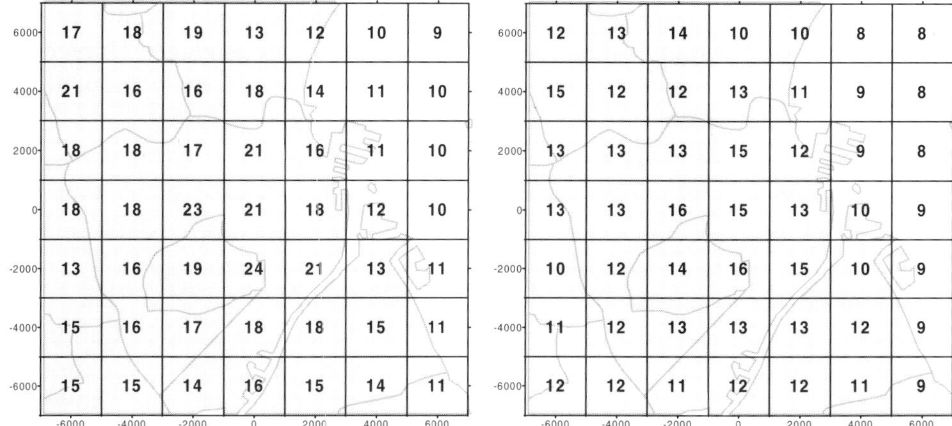

Figure. 3. Calculated annual concentrations of NO_x (left) and NO_2 (right) in Copenhagen (in ppb).

The variation of both NO_x and NO_2 concentrations across the city is quite modest. The highest concentrations are found in the central part of the city and they decline slightly towards the city borders. As expected, the variations in the concentrations of the secondary pollutant, NO_2, are smaller than in the case of the primarily emitted NO_x. Background measurements at a location within the city centre can thus be considered as being representative for an area of several kilometres around this location.

4. Comparison with measurements and Discussion

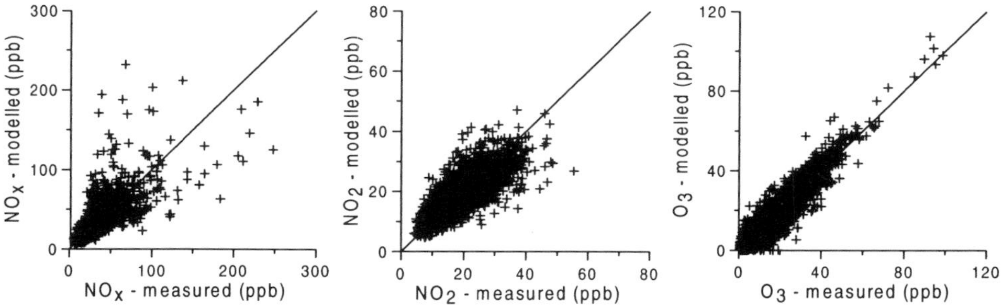

Figure 4. Comparison of measured and calculated hourly mean values of NO_x, NO_2 and O_3 for the urban background station in Copenhagen. Only daytime hours are shown.

Comparison of measured and calculated hourly mean values of NO_x, NO_2 and O_3 is shown in Figure 4, but for daytime hours only (8 to 18). Measurements are from the permanent urban background station in Copenhagen (Kemp *et al.*, 1998) located on the roof of a building belonging to the Copenhagen University (0,0 co-ordinates in Figure 1). The meteorological data used for computations are taken from a 10m mast situated on the same building.

The scatter depicted in Figure 4 is large but the correlation is reasonable. The agreement with measurements for NO_2 and O_3 is considerable better than for NO_x. Taking into account that the background concentrations of these two pollutants have the largest impact on the street level concentrations, the results can be considered as satisfactorily.

The average diurnal variation of the calculated and measured concentrations is shown in Figure. 5. The rural background concentrations from the monitoring station in Lille Valby, which are used in the model calculations, are shown as well. The overall agreement is good, however, the model slightly overpredicts the concentrations during the afternoon rush hours and underpredicts during the night hours. This is presumably due to the shortcomings in the applied diurnal variation of the emissions.

The rural background contribution is remarkable large. For NO_2 it constitutes almost 50%. The decline in urban ozone concentrations compared to the rural values compares very well with the increase in the NO_2 concentrations. This proves that even on the urban scale the main production mechanism of NO_2 is the NO oxidation by ozone (Palmgren *et al.,* 1996).

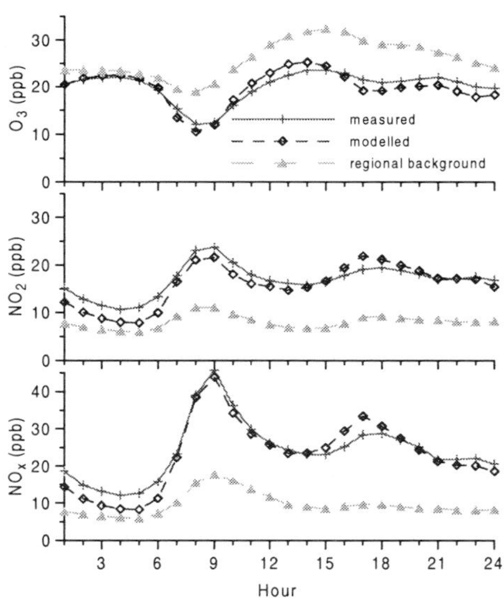

Figure 5. Average diurnal variation of the measured and modelled concentrations. The rural background concentrations are shown too.

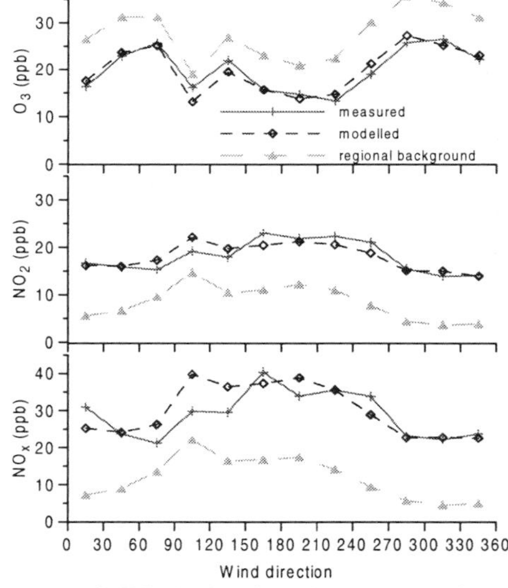

Figure.6. Wind direction dependence of the annually averaged measured and modelled concentrations. Only daytime hours are selected for this comparison.

The influence of the rural background concentrations on the model results is shown in Figure 6, where the measured and modelled annually averaged concentrations, but only over the daytime period, are plotted as a function of the wind direction. The agreement is reasonable good but it is evident that the concentrations of NO_x and NO_2 are overestimated for easterly wind directions. This is presumably due to overestimation of the regional background when the rural monitoring station is under influence of the urban plume from Copenhagen. The concentrations measured at the rural station are clearly highest for the easterly winds.

4.1 SENSITIVITY ANALYSIS

The model was tested with different procedures for generation of the input parameters. The accuracy of the prediction of the heat flux had only a marginal effect on the model results. Actually, practically the same results were achieved assuming that the heat flux constitutes a constant fraction of the global radiation. Global radiation is easily measured by routine instruments or can be calculated from cloud cover data. The reason for this weak dependence on the heat flux is that the mechanical turbulence is the most dominating dispersion factor in urban areas. This conclusion is, at least, true for Danish cities but might not be justified for urban areas in Southern Europe, where the convective situations are more frequent.

The model results were also insensitive to parameterisation of the mixing height. This behaviour is, however, easy to explain taking into account the results presented in Figure 7. In this figure are shown calculated annually averaged concentrations for daytime hours as a function of the radius of the emission area field. It is seen that only small additional contributions to the calculated concentrations arise from emissions that take place at a distance larger than about 3 km from the receptor point. The main contribution comes thus from the nearby sources. For such short distances the dispersion of pollutants is normally not affected by the mixing height and this explains the relatively weak influence of the mixing height on the modelled concentrations. A similar conclusion was also reached in the paper by Hanna (1971). Again, this statement might not be true for urban areas that are much larger than Copenhagen, or in situations when strong, ground based inversions with a shallow mixing layer are occurring frequently.

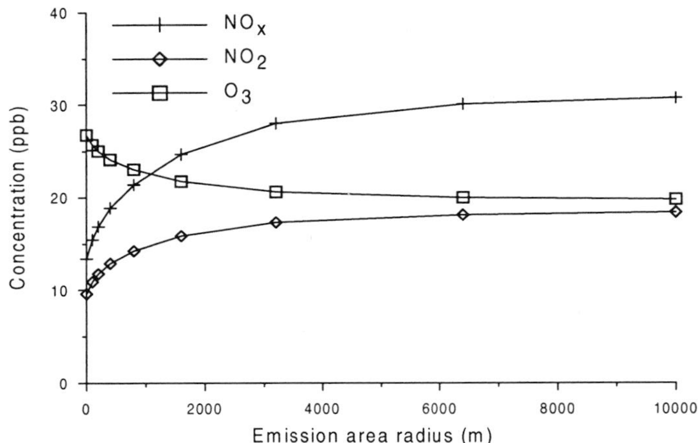

Figure 7. Dependence of the calculated annually averaged concentrations on the radius of the area of the emission field.

The rural background contribution is an important factor. The detailed analysis of model results reveals that for most of the situations with significant overprediction of the observed concentrations the contribution from the rural background was exceptionally high. As mentioned previously, the rural background measurements used in this study were not always representative for the modelling area.

5. Conclusions

The Urban Background Dispersion Model is still under development, but the preliminary results presented here are promising. It can be concluded that the simple O_3 - NO chemistry that is applied here for modelling of NO_2 formation is adequate for urban areas similar to Copenhagen This concept must, however, be tested on other urban areas where the climatic conditions can be significantly different.

It is planned to couple the model with a numerical whether prediction model, which will supply the necessary meteorological data, incl. heat fluxes and mixing heights (Brandt *et al.,* 1999). The regional background data will be provided by a large-scale transport model, thereby problems with using a particular rural monitoring station will be avoided. After this extension it is the intention to use the model together with OSPM for forecasting air pollution on street scale and also for general information on traffic pollution levels in Danish cities.

Acknowledgements

The experimental data used in this study were provided from the Danish National Air Quality Monitoring Programme and the assistance of Drs. Finn Palmgren and Kåre Kemp is highly appreciated. The Copenhagen emission inventory was performed by the Danish Road Directorate in the framework of the Danish National Research Programme on Traffic Pollution in Urban Areas.

References

Bendtsen, H. and Reiff, L.: 1996, Byområders trafikskabte luftforurening - En overslagsmetode til emissionskortlægning (with English Summary), Rapport nr. 43, Vejdirektoratet, Copenhagen, Denmark, pp127.

Berkowicz, R.: 1999, OSPM - A parameterised street pollution model, 2nd International Conference on Urban Air Quality, Madrid, 3-5 march 1999.

Brandt, J., Christensen J.H. and Zlatev, Z.: 1999, Operational air pollution forecast modelling using THOR, *Physics and Chemistry of the Earth*, To appear.

Hanna, S.R.: 1971, A simple method of calculating dispersion from urban area sources, *J.Air Poll. Control Ass.,* **21**, 774-777.

Hertel, O. and Berkowicz, R.: 1989, Modelling NO_2 concentrations in a street canyon, *DMU Luft* **A-131**, pp31.

Hertel, O. and Berkowicz, R.: 1990, Beregning af NO_2-koncentrationer i byområder i forbindelse med varsling af forureningsepisoder (with English Summary). *DMU Luft* **A-139**, pp43.

Kemp, K, Palmgren, F. and Manscher, O.H.: 1998, The Danish Air Quality Monitoring Programme. Annual Report for 1997, NERI Technical Report No. 245, pp59.

Olesen, H.R. and Brown, N.: 1988, The OML meteorological preprocessor - a software package for preparation of meteorological data for dispersion models, *MST Luft* **A122**.

Palmgren, F., Berkowicz, R., Hertel, O. and Vignati, E.: 1996, Effects of reduction of NO_x on NO_2 levels in urban streets, *Sci. Total Environ.*, **189/190**, 409-415.

MODEL ASSESSMENT OF AIR-POLLUTION IN PRAGUE

JOSEF BRECHLER
Dept. of Meteorology and Env.Protection, Faculty of Math. and Phys., Charles University, V Holešovi kách 2, 180 00 Prague 8, the Czech Republic. E-mail:josef.brechler@mff.cuni.cz

Abstract. Results of relatively simple gaussian dispersion model are presented. This model, developed in the beginning of this decade, is suitable mainly for determination of spatial distributions of annual mean concentrations of such kinds of air-pollution as sulphur dioxide (SO_2), mixture of nitrogen oxides (NO_X) or suspended particulate matter (SPM) from all types of emission sources located within the city – point sources, line sources and area sources. Model has been used in years 1994, 1996 and 1998 to assess the air-pollution distributions of the above mentioned kinds in the Prague area and the development of the air quality state. It contributed to the decision making process when possible impact of some changes in traffic system or in system of domestic heating, for example, has been investigated. In this contribution, behaviour of sulphur dioxide (SO_2) and mixture of nitrogen oxides (NO_X) ground concentration distributions are analyzed with respect to changing emission situation in Prague in recent years. Results show that SO_2 ground concentration level decreases mainly due to changes in local heating methods and type of fuel used. Different trend can be seen in NO_X ground concentrations thanks to rapid increase in the traffic density.

Key words: urban air-pollution, dispersion model, model assessment, ground concentration distribution.

1. Introduction

There is a network of monitors in Prague that measure ground concentration of various kinds of air-pollution. This overall network, in fact, consists of two different networks. One is run by the Czech Hydrometeorological Institute (CHMI) and is equipped with automatic monitors and the other is run by the hygienic service of the Prague city. It is necessary to say that, firstly, both networks are equipped with different types of monitors and thus the span of measured species differs and, secondly, different goals were aimed when these networks were constructed. The CHMI network, which consists of 13 automatic monitors, has been constructed with the aim to give the most representative information about ground concentrations of air-pollution in the most extensive neighbourhood of each monitor. Hygienic service network, which consists of 11 monitors, follows the aim to give information about air quality just in the location of a monitor and its close neighbourhood; these monitors have been located approximately 2 meters above the bottom of the street canyons in the highly populated areas, for example. Remaining 3 more monitors run by CHMI (not automatic monitors) or another institution than hygienic service have been used for some specific measurements. Such a network of 27 monitors could

Environmental Monitoring and Assessment **65**: 269–276, 2000.

seem relatively dense enough to give a correct information about spatial distribution of air-pollution concentrations within the area of Prague. But, as it has been already mentioned, there is the inconsistency between those two networks that create the overall Prague monitor network. Another problem consists in the fact that Prague terrain is of a relatively complicated shape, which also affects the spatial distribution of ground concentration of air-pollution. With these facts in mind the municipal authorities supported the development of a model tool and its further use in the assessment of air-pollution distribution within the city area and/or in the decision making process when some changes in emission situation were supposed and it was necessary to judge the impact of these changes.

When the use of air-quality model was planned there were two possible approaches. The former has consisted in using relatively simple gaussian dispersion model where some modifications had to be accepted. But in conditions that prevail in the area of Prague (relatively complicated orography) and with the meteorological data that have been available (yearly mean wind roses computed by mesoscale model, details are mentioned elsewhere) this type of model is suitable mainly for computation of long term values like annual or seasonal mean concentrations. The latter has been based on development of more sophisticated model that would be able to describe the instantaneous distribution of ground concentrations but the development of a such model would take much longer time and the result of this development was not too sure in a relatively short time and, moreover, there were available neither corresponding high resolution fields of input meteorological data (wind field, temperature field) nor a tool, that could be used for computing them routinely. Due to these difficulties, the first approach has been chosen. Another reason that resulted in the choice of gaussian type of model consisted in the fact that only yearly mean values of emissions from point and area sources have been available.

In this contribution we would like to present that also such a simple type of model can give quite important information about air pollution state on annual mean level and when used in subsequent years a development in air quality state can be examined. Model can also be used in decision making process if reliable emission data are available.

2. Model Application

A gaussian dispersion model, described in Brechler *et al.* (1997), has been used for the assessment of annual mean ground concentration values of suspended particulate matter (SPM), sulphur dioxide (SO_2), a mixture of nitrogen oxides (NO_X) and carbon monoxide (CO) originating from all possible kinds of emission sources located within the city area and its neighbourhood. All the sources can be divided in point, line and area sources. The first group involves all point sources located within the area of Prague city together with the most important sources from surrounding districts and the most powerful point sources located in the area of the Czech Republic (stacks of thermal power plants). These sources are indicated with their positions, stack heights, emission rates of the above mentioned air-pollution and so-called thermal outputs, which are parameters needed in computation of plume rises, details will be given elsewhere. Next group of emission sources consists of so-called traffic sources. This group involves both line sources - parametrizations of street and road parts - and area sources – cross-roads, petrol stations, parking-sites, railway and bus stations. The third group of emission sources consists of local domestic heating sources (a great number of small chimneys) which are still very important in Prague even on annual mean concentration scale. Domestic heating have been parametrized as area sources with emission rates equal to the sums of individual emission rates of single chimneys located within each area source square divided by the area of that square, with the heights of area sources equal to average heights of buildings at each square and with thermal outputs equal to average thermal outputs of chimneys located in these areas.

As there are many powerful point emission sources outside Prague area (industrial plants and thermal power plants) that can affect values of ground concentrations within the city area, these sources have been involved into emission data set. Impact of the rest of emission sources located outside Prague area and impact of emissions from surrounding states has been involved via so-called transfers (= long range transport). Each transfer depends only on wind direction and has it been equal to the value of concentration (contribution from unknown sources) that had to be added to the computed value of concentration from the known sources.

A thermal output, that is proportional to thermal energy of exhausted emissions, has been used in plume rise computation. This approach was chosen because originally there were no data in the Czech emission inventories available for using the usual Brigg´s formulae - see, for example, Boubel *et al.* (1994). An instantaneous plume rise had to be computed according to formula of Lukas et al, see Berlyand (1991), for example. A modified method for gradual plume rise computation that uses data involved in the Czech emission inventory has been developed but not published yet.

Meteorological data consist of mixing layer heights for different vertical stability categories and a set of yearly mean wind roses covering the whole model domain. As the shape of Prague terrain is relatively complicated it was decided to use following approach. The whole model area has been divided into smaller areas and each area has been attributed with one representative yearly mean wind rose. These wind roses has been computed by the mesoscale model of the Institute of Atmospheric Physics of the Czech Academy of Sciences, Svoboda and Štekl (1994). Each emission source located in this smaller area has been attributed to this representative wind rose and also all receptor points located in this area have been attributed to that. This approach allows to involve the impact of terrain together with the others standard methods used in gaussian dispersion models in description of air-pollution transport and spreading process and thus to assess a ground concentration spatial distribution in a better way. Prague area has been divided into 142 smaller areas for which the same number of characteristic wind roses has been constructed. If, for example, there is northerly wind on 850 hPa level the low level wind directions in different parts of Prague can differ from this upper level wind direction and they can differ one from another due to orographic irregularities. The described approach consisted in involving these orographic impacts in a more realistic way than if only one wind rose would have been used. But, on the other hand, this approach is possible to use only in case of long term characteristics (annual or seasonal means) if the corresponding long term wind roses are available. Results for instantaneous values can be obtained but they are only of qualitative nature.

An emission sources input file involves data about approximately 90 powerful point emission sources that are located outside Prague (thermal power plants, heating plants and major industrial plants with stacks higher than 150m in the area of the Czech Republic and main point sources from surrounding districts), approximately 750 point emission sources located in the area of Prague, little bit less than 1900 line sources, about 270 cross-roads, circa 130 petrol stations, parking-sites, railways and bus stations etc. and more than 2140 squares that parametrize domestic heating sources (only squares with non-zero emission rates were taken into account with each square side equal to 500 m). The whole area has been covered with regular grid mesh of more than 8300 receptor points in which concentrations and other characteristics have been computed. But, if necessary, any arbitrarily chosen new receptor point can be added to those above mentioned regularly spaced points.

Outside the Czech Republic the well-known Pasquill stability categories (see Seinfeld, 1986, for example) have been used in air-pollution modelling but in the Czech Republic so-called Bubník – Koldovský classification was recommended by the Czech Ministry of Environment in the past. This is also a reason for using the latter in the presented model. The Bubník – Koldovský classification is based only on the value of vertical gradient of temperature and splits all possible conditions of vertical temperature stability into 5 categories spanning from very stable stratification (so-called strong inversion category - temperature increases vertically more than 1.6 K / 100m) to unstable one (so-called convective category - vertical decrease of temperature is greater than 0.8 K / 100 m). There are two categories for inversion conditions (the above mentioned strong inversion and a weak inversion), a category equal to isothermal condition, one corresponding to neutral stratification and the above mentioned convective category.

The gaussian dispersion model has been used mainly for computation of annual means of ground concentrations of SPM, SO_2, NO_X and CO and values that can be derived from them. It is necessary to say that it would be possible to use this model also for computation of some instantaneous characteristics (e.g. analogies of 30' mean values of ground concentrations). But mainly meteorological data are given as long term means that are homogeneous over the whole computational domain but short term concentrations are strongly affected with real atmospheric conditions at a given place and time that are far enough from the homogeneous conditions. Thus the use of this model has been rejected for computation of something like instantaneous values of concentrations especially in such complicated terrain conditions that can be met in Prague. Deformations of meteorological fields (flow field, temperature field, stability field) due to the impact of terrain irregularities are very important factors in concentration distributions and they are not well described in simple gaussian dispersion models.

3. Results

There are several output files containing results. For the great majority of results the table output format has been chosen in order that they can be used without changes in geographic information system (GIS) software. The most important output contains spatial distributions of annual mean values of ground concentrations of SPM, SO_2, NO_X and CO (in $\mu g\ m^{-3}$). From another output file it is possible to show the distribution of concentrations of any above mentioned types of air-pollution with respect to wind directions at each receptor point. Shares of emission sources groups (point sources in Prague, traffic sources, domestic heating sources, for example) in per cents are given in another output files and this type of output is useful when impact of the only one group of emission sources (e.g. traffic) has to be investigated. Similarly the impact of individual sources (both in $\mu g\ m^{-3}$ and in per cent) can be found in other output files. To limit the number of sources contained in these files only sources with contributions higher than 1 per cent of computed overall ground concentration value at a given receptor point are involved.

This model has been used in assessment of annual mean ground concentration distributions of the above mentioned species for the emission data from 1994, 1996 and 1998. Some limitations can be found in the results – computed values of SPM concentrations, for example, are equal to about 50 per cent of measured values of SPM, because only so called primary emissions of this type of pollution are taken into account but monitors measure also so-called secondary emissions. The secondary emissions mean once deposited and again taken up particulates from the surface. These secondary emissions are not parametrized in the this model and they depend on such factors as quality of soil surface, contend of moisture in the uppermost soil layer, vegetation, wind speed and intensity of turbulence, for example. Computed values of CO are also low in comparison with measurement because the impact of natural background has not been involved. Results of annual mean concentrations of SO_2 (in $\mu g\ m^{-3}$) for emission data from 1996 and 1998 are presented in Figure 1 and 2.

From the results showed it is evident a strong improvement of the air quality concerning SO_2 pollution between 1996 and 1998. It corresponds to measures that have been accepted in previous years. For example, in many residential flats local heating facilities has been changed and now heat is produced centrally in heating plants where natural gas is used, for example. Brown coal has not been so frequently burned yet as a source of heat in areas, where the local heating facilities are still used. The highest level of annual mean ground concentrations of SO_2 can still be found in central parts of Prague where the density of local heating still used is the highest but the hygienic limits (= 60 $\mu g.m^{-3}$) is not exceed. In Figures 1 and 2 origin of co-ordinates (in meters) correspond to the bottom of the Wenceslas square and rectangular around the origin shows the historic centre of Prague.

In NO_X concentration distribution the situation is different from that of SO_2. The central part of Prague is relatively highly polluted with this type of pollution and increasing trend was modelled and has been measured, too. The reason consists in several factors. Firstly, in 1996 and 1998 there were still a lot of cars that did not use catalysts and cars that were old and in a bad technical state. Secondly, the increase of car number has been very fast and this number is several times higher than it used to be. Thirdly, the consequence of the situation mentioned in previous item is that infrastructure has not been prepared for such a dramatic change in traffic density. There were no ring around Prague and all cars, including lorries, passing Prague from east to west or from north to south had to go through its centre. All this reasons result in increase of the level of NO_X ground concentration within the city.

4. Concluding remarks

The model presented here has been used not only for assessment of air-quality state in 1994, 1996 and 1998 but also some possible future emission scenarios have been modelled. These scenarios dealt with possible changes in the Prague traffic system, changes in the system of heat supply to residential flats etc. It is an advantage of the model approach that when having reliable emission data corresponding to future situation it is possible to model the impact of that

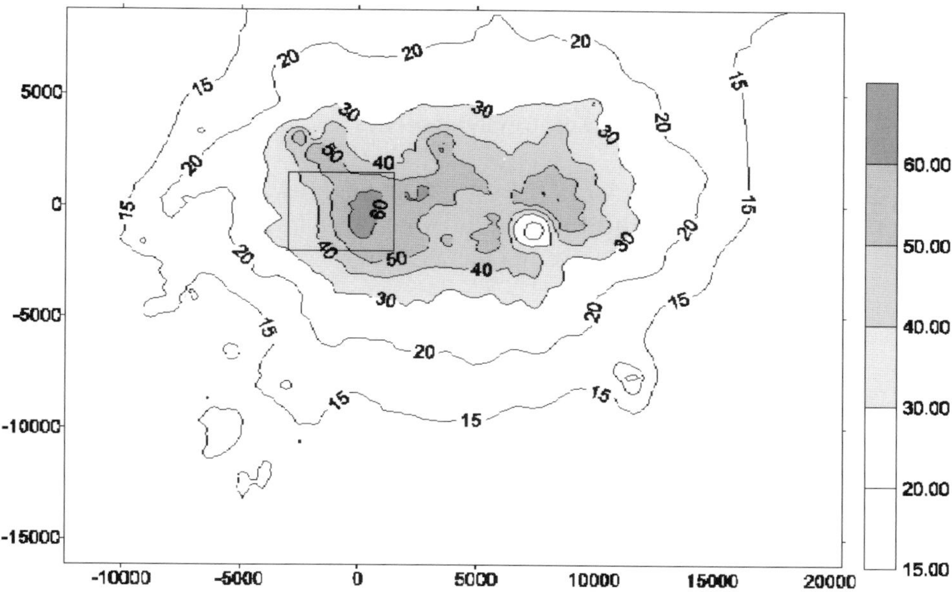

Figure 1. Annual mean ground concentration of SO_2 for 1996 emission data (in μg.m^{-3}).

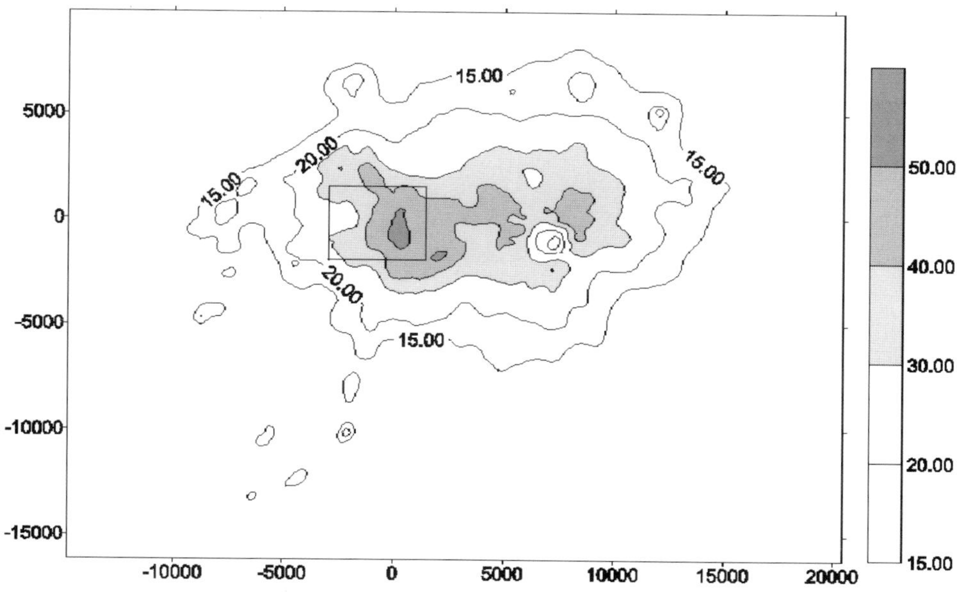

Figure 2. Annual mean ground concentration of SO_2 for 1998 emission data (in μg.m^{-3}).

situation on air quality and thus to help in decision making process.

Acknowledgements

Author wishes to thank to Ing. V. Píša, the head of ATEM company for his support in this activity, and to Institute of Municipal Informatics of the City of Prague for the overall support given to this task.

References

Berlyand, M. E.: 1991, Prediction and Regulation of Air Pollution, Kluwer Academic Publishers, Dordrecht.

Boubel, R. W., Fox, D. L., Turner, D. B. and Stern, A. C.: 1994, Fundamentals of Air Pollution, Academic Press, London.

Brechler, J., Píša, V., Pretel, J.: 1997, Air pollution state modelling, *Meteorological Bulletin* **50**(4), 110 – 116 (in Czech).

Seinfeld, J. H.: 1986, Atmospheric Chemistry and Physics of Air Pollution, A Willey Interscience Publication, New York.

Svoboda, J. and Štekl, J.: 1994, Mesoscale modelling of a flow modification caused by orography, *Meteorol. Zeitschrift, N.F.* **3**, 233 - 241.

FORECASTING AIR QUALITY PARAMETERS USING HYBRID NEURAL NETWORK MODELLING

MIKKO KOLEHMAINEN*, HANNU MARTIKAINEN, TERI HILTUNEN AND JUHANI RUUSKANEN,
University of Kuopio, Department of Environmental Sciences, P.O.Box 1627, FIN-70211 Kuopio, Finland
*email: mikko.kolehmainen@uku.fi

Abstract. Urban air pollution has emerged as an acute problem in recent years because of its detrimental effects on health and living conditions. The research presented here aims at attaining a better understanding of phenomena associated with atmospheric pollution, and in particular with aerosol particles. The specific goal was to develop a form of air quality modelling which can forecast urban air quality for the next day using airborne pollutant, meteorological and timing variables.

Hourly airborne pollutant and meteorological averages collected during the years 1995-1997 were analysed in order to identify air quality episodes having typical and the most probable combinations of air pollutant and meteorological variables. This modelling was done using the Self-Organising Map (SOM) algorithm, Sammon's mapping and fuzzy distance metrics. The clusters of data that were found were characterised by statistics. Several overlapping Multi-Layer Perceptron (MLP) models were then applied to the clustered data, each of which represented one pollution episode. The actual levels for individual pollutants could then be calculated using a combination of the MLP models which were appropriate in that situation.

The analysis phase of the modelling gave clear and intuitive results regarding air quality in the area where the data had been collected. The resulting forecast showed that the modelling of gaseous pollutants is more reliable than that of the particles.

Key words : air quality, forecasting, neural networks, Self-Organizing Map, Sammon's mapping

1. Introduction

Urban air pollution has emerged as the most acute problem in recent years because of its detrimental effects on health and living conditions. To prevent any further decline in air quality, scientific planning of analytical methods and pollution control is required. Within this framework it is necessary; (i) to analyse and specify all pollution sources and their contribution to air quality, (ii) to study the various factors which cause the pollution phenomenon, and (iii) to develop tools for reducing pollution by introducing alternatives to existing practices.

Air quality phenomena have traditionally been modelled using physical reality as the starting point, and this information has then been coded into differential equations, for example. Methods of Computational Intelligence offer a

Environmental Monitoring and Assessment **65**: 277–286, 2000.

completely different approach, in which the model is constructed entirely using measured data. Computational Intelligence is a fairly new discipline which is based on the hypothesis that reasoning can be realised using computation (Poole et al., 1998). The methods used in computational intelligence include a number of forms of computation, the best known of which are neurocomputing and fuzzy logic. Neurocomputing is based on principles that have been discovered by investigating the brain and its structures (Haykin, 1994), while fuzzy logic is based on fuzzy set theory, which extends traditional bivalent logic into continuous group membership with truth values between 0 and 1 (Zimmermann, 1991). Neural networks are used as a tool in most of these applications. As examples, neural models for ozone concentrations have been constructed (Yi and Prytok, 1996 ; Comrie, 1997) and a model that predicts hourly NO_x and NO_2 concentrations (Gardner and Dorling, 1999). Most of the work has focused on comparing feed-forward neural networks (especially Multi-Layer Perceptrons) with traditional methods such as the ARIMA model and linear regression. The results show in general that the neural models perform as well as these methods or better (Garder and Dorling, 1998).

There has been some criticism for this "black box" modelling, however, on the grounds that it offers too little support for understanding the physical phenomena that are being considered. The research presented here therefore aims at attaining a better understanding of phenomena associated with atmospheric pollution, meteorological parameters, and in particular, aerosol particles. The aim is to show how a set of neural and related methods can be used (i) to understand better the relationship among airborne pollutant and meteorological variables, (ii) to use expert knowledge to support the modelling, and (iii) to create a model which is able to forecast concentrations of airborne pollutant variables.

2. Materials and Methods

2.1 SELF-ORGANIZING MAPS (SOM)

One of the best known unsupervised neural learning algorithms is the Self-Organizing Map (SOM) (Kohonen, 1995), the aim of which is to find prototype vectors that can represent the input data set and at the same time to achieve a continuous mapping from the input space to a lattice. This lattice can be an easily visualised 2-dimensional map, for example.

The weight vectors of the SOM are first initialised to random values. With each training pattern the winning neurone (Best-Matching Unit, BMU) is first found by comparing the input (measured) and weight vectors of the neurones by

Euclidean distance metrics. The weights of the winning neurone (BMU) and its neighbours are then moved towards the input vector according to a learning rate factor which decreases monotonically towards the end of learning. This unsupervised learning process is summarised in Figure 1 using a 4 by 4 Self-Organizing Map.

A variation of the SOM, called tree-structured SOM, was used in this work (Koikkalainen, 1994). The software implementation consists of several SOMs that are organised hierarchically in several layers in a pyramid-like fashion. The number of neurones at a larger level is four times the number at the previous level. Visual inspection of the measurement data is directed at one level at a time, however, and the results are comparable to those achieved by "standard" SOM. The level that was selected here for visualisation consists of 1024 neurones.

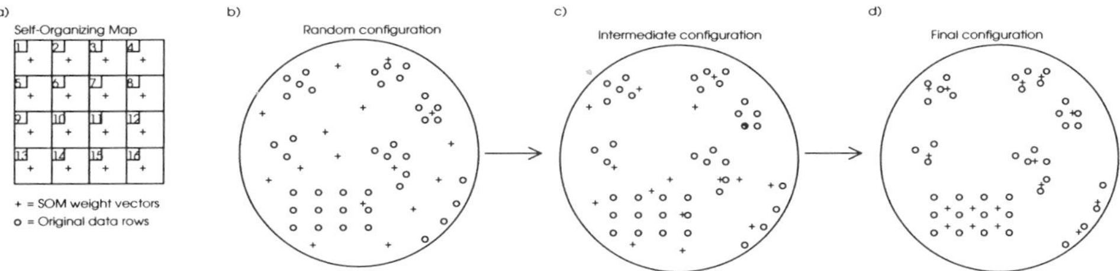

Figure 1. Visualisation of the SOM learning process. a) a Self-Organizing Map with 16 neurons. b) Weight vectors (+) are first initialized to random values. c) The weight vectors move towards the final configuration through intermediate configurations. d) In the final configuration the weight vectors represent a number of original data rows.

2.2 SAMMON'S MAPPING

Sammon's mapping is an iterative method based on a gradient search (Sammon Jr, 1969). The aim is to map points in n-dimensional space usually into 2 dimensions. The algorithm finds the locations in the target space so that as much as possible of the original structure of the measurement vectors in the n-dimensional space is conserved. The numerical calculation is more time-consuming than the SOM algorithm, however, which can be a problem with a massive data set. On the other hand, it is able to represent the relative distances between vectors in a measurement space and is thus useful for determining the shape of clusters and the relative distances between them. It is therefore of benefit to combine these two algorithms. Sammon's mapping is thus applied to the stage where the SOM algorithm has already achieved a substantial data reduction by replacing the original data vectors with a smaller number of representative prototype vectors.

2.3 MULTI-LAYER PERCEPTRON AND THE BACK-PROPAGATION ALGORITHM

Multi-Layer Perceptrons, which represent one type of feed-forward neural network, consist of processing elements and connections (Hecht-Nielsen, 1991). The processing elements, usually called neurones, are arranged in layers. There are three kinds of layers: input layers, hidden layers and output layers. An input layer serves as a buffer that distributes input signals to the hidden layer. Each unit in the hidden layer sums its input, processes it with a transfer function and distributes the result to the next layer, which is usually the output layer. The units in the output layer compute their output in a similar manner.

The most common supervised learning algorithm is the back-propagation (BP) algorithm, also called the generalised delta rule (Haykin, 1994). It is a gradient descent algorithm that is normally used to train the MLP network. For an introduction and overview of MLP applications in the atmospheric sciences, see Garder and Dorling (1998).

2.4 GROUP MEMBERSHIP EVALUATION USING FUZZY LOGIC

The goal of group membership evaluation is to represent data lines (measurements) in respect to clustered phenomena found in the data set. The process is initiated by finding the kernels of the groups. This can be done using the SOM algorithm and Sammon‘s mapping, for example. The kernels are areas where a certain phenomenon or property is known to be true with certainty. Other areas can now be evaluated using the kernels as reference points. The membership of each data line with respect to the group kernels is described using terms known from fuzzy logic (Zimmermann, 1991).

2.5 STATISTICAL PERFORMANCE INDICATORS

Selected statistical indicators were used to describe numerically the goodness of the estimates (Willmott, 1982). With the neural networks, one of the most common indicators is the Root Mean Square Error (RMSE). It can be calculated according to Equation (1).

$$RMSE = \left(\frac{1}{N} \sum_{i=1}^{N} [P_i - O_i]^2 \right)^{\frac{1}{2}} \qquad (1)$$

where N = number of data points, O_i = observed data point and P_i = predicted data point.

A relative measure of error called Index of Agreement (d) can be calculated according to equation (2). It is a dimensionless measure that is bounded into range 0..1.

$$d = 1 - \frac{\sum_{i=1}^{N} (P_i - O_i)^2}{\sum_{i=1}^{N} (|P_i'| - |O_i'|)^2} \qquad (2)$$

where $P_i' = P_i - \overline{O}$ and $O_i' = O_i - \overline{O}$.

3. Results and Discussion

3.1 PRE-PROCESSING OF DATA

The variables for the modelling were first selected. The meteorological variables used were wind speed, wind direction and temperature. The time information for each hourly data line made it possible to construct time dependence for airborne pollutant and meteorological variables. The airborne pollutant variables selected for modelling were NO, NO_2, CO and PM_{10}. As the tools used for neural modelling do not accept missing values, data lines having missing values were simply omitted. Furthermore, variables for wind direction, hour and month were each transformed into two auxiliary variables using the sine and cosine functions. This enabled the neural algorithms to work properly despite of the discontinuities in the original variables. Finally, the data set was split into a training set (90 %) and a test set (10 %).

3.2 SELF-ORGANIZING MAP OF AIR QUALITY

The data applied to the city of Kuopio during the years 1995-1997, and the analysis was performed using the Self-Organising Map (SOM) algorithm (Kohonen, 1995), which is able to visualise the resulting mapping from n dimensions to a 2-dimensional grid (Figure 2). The first phase of this process consisted of finding the kernels of the pollution episodes. The locations of the episodes selected are the neurones in the SOM map where the pollutant averages reach their highest values locally. In the next phase, fuzzy membership values were calculated for each data row and for each episode. These new variables described the distance of each data row from the prototypes (kernels) of each episode. Finally, the neurones were collected into several, possibly partially overlapping clusters, each representing one episode. This was done by calculating average values for the memberships of each neurone and comparing these with a pre-set limit value. Thus, the neurones where the average membership for a certain episode was higher that the limit were marked as belonging to the cluster (group of neurones) representing that episode in a fuzzy sense (Zimmermann, 1991).

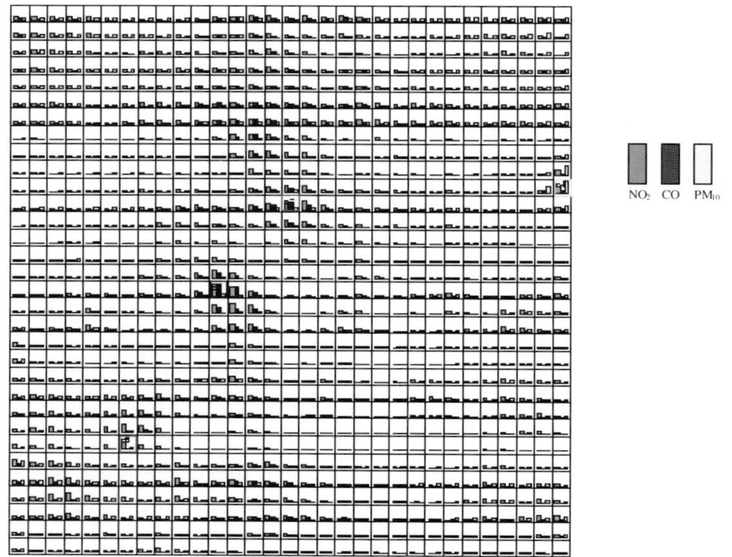

Figure 2. Self-Organizing Map of air quality

3.3 SAMMON'S MAPPING OF EPISODES

The clustered data were visualised using Sammon's Mapping algorithm (Sammon Jr, 1969), which is another way of mapping n-dimensional data into two dimensions (Figure 3). This enabled qualitative evaluation of the resulting clusters in two ways. Firstly, the consistency of each cluster could be observed, and secondly, the overlapping of clusters was detectable. By adjusting the limit value for membership, it was thus possible to find a suitable value for it by iteration.

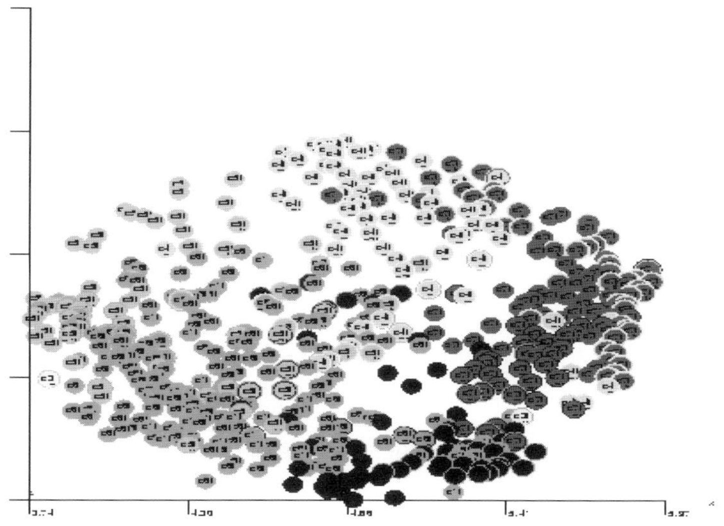

Figure 3. Sammon's mapping of episodes

3.4 DESCRIPTION OF THE EPISODES

In order to identify the clusters (episodes), statistics for the episodes were calculated and the mean values (median for time variables) are given in Table 1. By inspecting the distribution of values for each cluster, it was possible to characterise each episode in terms of time of year, time of day, wind speed and direction and composition of the set of the pollutants. Each episode was then assigned a description (Table 2), which made it possible to convert a forecast into an oral description at a later stage.

3.5 LOCAL MODELING USING MLPS

At the next stage, local numerical models were constructed for the episodes. The training data for the Multi-Layer Perceptron (MLP) modelling were first selected from the original data rows taking the same membership limit value as was used for clustering the neurones. Next, a MLP model was constructed for each episode and for each pollutant variable using the back-propagation (BP) algorithm.

Table 1
Descriptions of the episodes C1-C7 by mean values (month and hour by median)

Variable	C1	C2	C3	C4	C5	C6	C7
NO_2	65.9	91.3	82.7	48.7	65.1	65.4	59.7
CO	1.1	4.5	3.2	1.0	1.6	0.3	2.2
PM_{10}	79.9	20.7	96.2	128.3	87.5	31.0	23.1
Temperature	-4.4	-18.7	-7.9	-7.9	-7.2	18.7	-15.6
Wind sp.	2.2	1.4	1.5	1.1	1.3	2.6	1.4
Wind dir.	309	273	94	332	38	315	302
Hour(med)	21	17	21	7	21	22	11
Month(med)	3	12	3	4	3	8	1

NO_2 and PM_{10} expressed in $\mu g/m^3$, CO in mg/m^3, temperature in °C, wind speed in m/s and wind direction in degrees.

Table 2
Characterisation of the episodes by textual descriptions

Episode	Description
C1	Springtime evening, low wind from north-west, PM_{10} and NO_2 high
C2	Winter afternoon, very low wind (inversion), NO_2 and CO high
C3	Springtime evening, very low wind, all pollutants high
C4	Springtime morning, very low wind, PM_{10} and NO_2 high
C5	Springtime evening, very low wind, all pollutants elevated
C6	Late evening in late summer, low wind from north-west, NO_2 high
C7	Winter morning, very low wind, NO_2 and CO high

3.6 USING AND TESTING THE HYBRID MODEL

The hybrid neural network model (see Figure 4) was tested using the data set which had been put aside for this purpose in the pre-processing stage. The first part consisted of finding the Best Matching Unit (BMU) neurone in the SOM. This calculation consisted of using the meteorological and time variables of the test data line to find the closest neurone in the SOM. An average membership value could then be determined for each episode and the actual levels for individual pollutants calculated using a combination of the MLP models which were appropriate in that situation. Estimates from several submodels were combined by weighting the output of each MLP by the membership value and summing the results into one estimate.

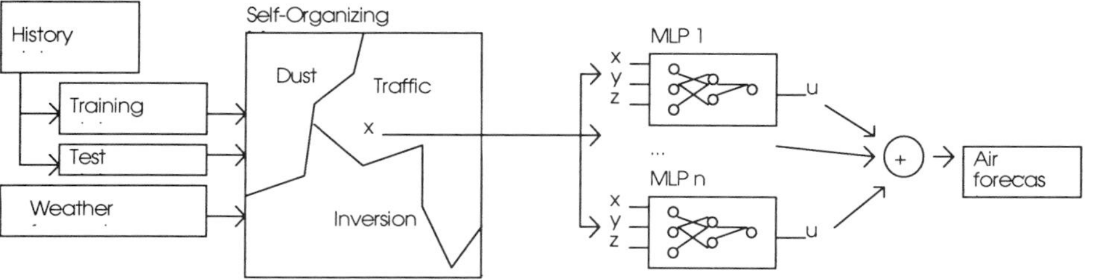

Figure 4. General principle of the hybrid neural network

3.7 EVALUATING THE GOODNESS OF THE MODEL

The goodness of the estimates was evaluated by using general statistical indicators, RMSE and the Index of Agreement (d) for three of the pollutant variables (Table 3). It can be seen that the accuracy of the model was higher for the gaseous pollutant variables than for the particles.

Table 3
Goodness of the model by statistical indicators

Variable	Min	Max	Mean	Median	RMSE	d
NO_2	0	123.0	15.2	11	12.2	0.66
CO	0	5.7	0.2	0.1	0.3	0.63
PM_{10}	0	159.0	12.3	9	11.1	0.47

NO_2 and PM_{10} expressed in $\mu g/m^3$ and CO in mg/m^3

4. Summary and Conclusion

The data mining phase of the modelling gives clear and intuitive results regarding air quality in the district where the data had been collected, and also supports the inclusion of expert knowledge in the model. The forecast results with the test data for good air quality were only moderate, which is probably due to the small amount of data available on episodes since the air quality in Kuopio, Finland, is generally good most of the time, with only a few bad pollution episodes per year. The forecast results also showed that the modelling of gaseous pollutants is more reliable than that of particles.

By inspecting the distribution of variables of the episodes using data mining, however, it can be deduced that the discriminative power for local models could be improved by using more descriptive meteorological variables, such as atmospheric stability and the height of the atmospheric boundary layer. Another important aspect that should be taken into consideration is the persistence of the weather, and thus also of air quality phenomena. This can be done by time-series forecasting, which will be the most important development aspect of the model for future work.

Acknowledgements

This research was funded by Tekes (the Technology Development Centre, Finland) and the University of Kuopio. More information at the web addresses:
http://erin.math.jyu.fi (NDA software for neural computing)
http://www.visipoint.fi (Visual Data software for data mining)

References

Comrie A. C.: 1997, Comparing neural networks and regression models for ozone forecasting, Journal of Air and Waste Management Assiciation **47**, 653-663

Gardner M. W., Dorling S. R.: 1998, Artificial neural networks (the multi-layer perceptron) - a review of applications in the atmospheric sciences, Atmospheric Environment **32**, 2627-2636

Gardner M. W., Dorling S. R. : 1999, Neural network modelling and prediction of hourly NO_x and NO_2 concentrations in urban air in London, Atmospheric Environment **33**, 709-719

Haykin S.: 1994, Neural Networks: A Comprehensive Foundation, Prentice Hall, New Jersey

Hecht-Nielsen R.: 1991, Neurocomputing, Reprinted with corrections, Addison-Wesley Publishing Company, Inc.

Kohonen T.: 1995, Self-Organizing Maps, Springer-Verlag, Berlin Heidelberg, Germany

Koikkalainen P.: 1994, Progress with the tree-structured self-organizing map, ECAI'94. Proceedings of the 11th European Conference on Artificial Intelligence, Aug 1994, Editor Cohn, A., Wiley&Sons, 211-215.

Poole D., Macworth A., Goebel R.: 1998, Computational Intelligence, A Logical Approach, Oxford University Press

Sammon Jr J.W.: 1969, A nonlinear mapping for data structure analysis, IEEE Transactions on Computers, **C-18**(5), 401-409

Willmott C.: 1982, Some comments on the evaluation of the model performance, Bulletin American Meteorological Society **63**, 1309-1313

Yi J., Prybutok V. R.: 1996, A neural network model forecasting for prediction of daily maximum ozone concentration in an industrialized urban area, Environmental Pollution **92**, 349-357

Zimmermann H.-J.: 1991, Fuzzy set theory and its applications, Second Edition, Kluwer Academic Publishers

THE SENSITIVITY OF MESOSCALE CHEMISTRY TRANSPORT MODEL RESULTS TO BOUNDARY VALUES

LENZ*, C.-J.; MÜLLER, F. and SCHLÜNZEN, K. H.
Meteorological Institute, University of Hamburg, Geomatikum, Bundesstr. 55, 20146 Hamburg, Germany
E-mail: claus-juergen.lenz@dkrz.de

Abstract. Using a high resolution meteorology-chemistry transport model, simulations were performed to estimate the sensitivity of the model results to nesting. The model results are compared with airplane measurements made during the TRACT field measuring campaign in September, 1992. For the meteorological part of the model the performance is enhanced using one-way nesting in a larger scale model, if the quality of the large scale driving data is sufficient. The sensitivity of the NO_x concentration results with respect to nesting of chemical quantities is rather low due to the poor quality of the forcing data. A correct description of the emission rates and the meteorological conditions may be more important. For ozone, the best results can be achieved with either no nesting or a meteorological and chemical nested model simulation, which is again a result of the poor quality of the forcing data.

Keywords: chemistry transport model, boundary values, nesting method, ozone concentration

1. Introduction

High resolution numerical chemistry transport models are generally used to calculate concentration fields of trace gases, their formation and reduction, as well as their interaction in model areas covering a characteristic size of about 50 x 50 km^2 to 300 x 300 km^2 with a typical horizontal grid resolution between 1 km and 10 km. Due to this limited model area, the concentrations of trace gases in the inner model domain depend not only on local emissions but also on emissions outside of the model domain. Advective and turbulent processes are the link between the trace gas concentrations in the considered model area and its surroundings. Therefore, at the model boundaries realistic trace gas concentrations, as well as meteorological data such as wind, temperature and humidity, must be supplied in most meteorological conditions to calculate a proper flux of trace gases into the model area.
Measurements of chemical trace gases and meteorological quantities are usually not available within a sufficient spatial and temporal resolution to drive a chemistry transport model. To receive concentrations and meteorological data at the lateral boundaries as realistic as possible, it is necessary to nest the high-resolution model into a larger scale model using suitable nesting methods.

---

present affiliation: Max-Planck Institute for Meteorology, Bundesstr. 55, 20146 Hamburg, Germany

Environmental Monitoring and Assessment **65**: 287–295, 2000.

In the present paper the sensitivity of the results of a high-resolution chemistry transport model with respect to nesting the model into a larger scale model is shown. The influence of nesting the meteorological or the chemical variables or both may be of different importance for a realistic forecast of the meteorological and/or chemical concentration fields. If the forecast accuracy would not differ much by nesting only the chemical or the meteorological variables, computing time and storage could be saved.
The model results shown in this publication are based only on one simulation period. Hence, the drawn conclusions could actually only be applied for the mentioned period. To generalize the results, the model study should be extended to some additional simulation periods.

2. Used models and input data

The sensitivity study with respect to nesting a high resolution chemistry transport model into a larger scale model was performed with the mesoscale transport and fluid model METRAS (Schlünzen, 1990; Schlünzen et al., 1996) and the mesoscale chemistry transport model MECTM, which have been developed at the University of Hamburg. METRAS is a three-dimensional non-hydrostatic model based on the conservation laws for mass, momentum, heat, and humidity. The advection/diffusion part of MECTM is the same as in METRAS, and the chemical reaction mechanism used is a modified version of the RADM2 mechanism (Stockwell et al., 1990). Both models, METRAS and MECTM, are running with the same spatial resolution.
METRAS and MECTM have been nested into the larger scale models MM5 (Grell et al., 1993) and CTM2 (Hass, 1991) from the EURAD model system (Ebel et al., 1997) (see Figure 1). For the prognostic meteorological variables, the method of nudging has been used (Schlünzen et al., 1996). MECTM is nested into time dependent boundary conditions interpolated from the trace gas concentrations of CTM2 (Niemeier, 1997; Müller et al., 2000).
The model study has been performed for the second measuring campaign of the TRACT experiment which took place in the bordering area of southwestern Germany, northeastern France and the northern part of Switzerland (Figure 2) on September 16, 1992 (Zimmermann, 1995). Airplane measurements of different meteorological and chemical quantities were taken along two flight patterns in the model area. These measurements have been used to compare the model results and to assess the importance of model nesting for selected quantities.

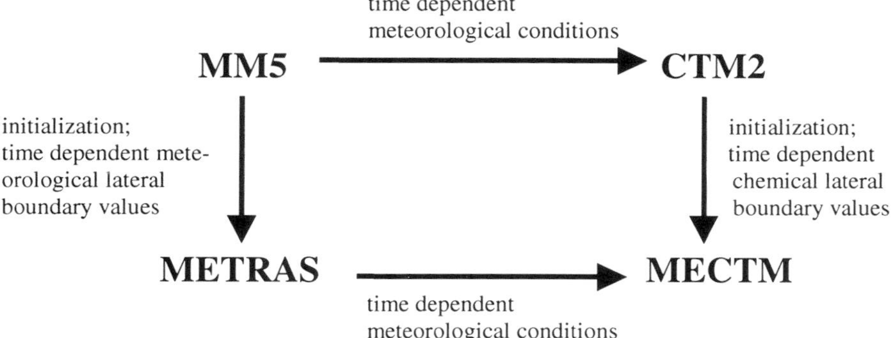

Figure 1: Nesting scheme of METRAS and MECTM in the larger scale models MM5 and CTM2

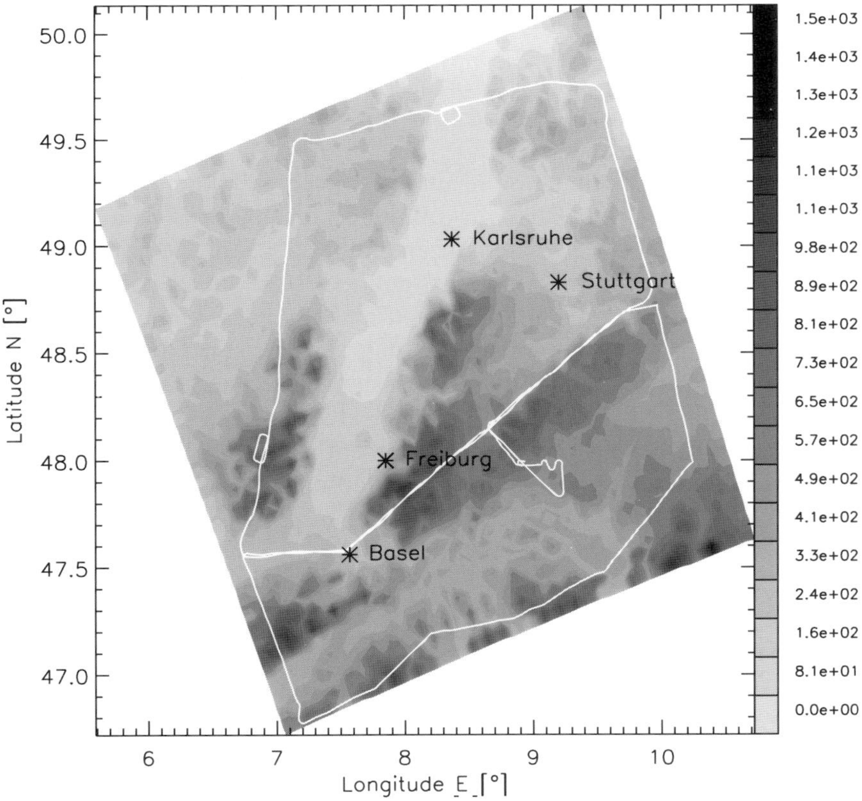

Figure 2: Orographical height [m] of the simulation area and flight patterns

3. Results and discussion

To determine the sensitivity of the model system METRAS/MECTM to model nesting, four model simulations have been performed:

- full nesting: nesting of meteorological and chemical variables,
- nesting of meteorology, but chemical variables are not nested,
- meteorology not nested, but chemical variables are nested,
- no nesting: neither meteorological nor chemical variables are nested.

This results in two different meteorology cases and four different chemistry cases.

In Figure 3, the agreement of the simulated and measured potential temperature (Fig. 3a and 3b) and specific humidity (Fig. 3c and 3d) is shown for the flight pattern in the northern part (Fig. 3a, 3c) and in the southern part (Fig. 3b, 3d) of the model area. The abscissa gives the difference of the model results and the measurements. The ordinate shows the relative cumulative frequency as the percentage of model results less than the difference value. The thick line represents the forcing data interpolated from the EURAD results to the METRAS grid. These data are used for the nesting of METRAS. The dashed and dotted lines represent the nested and the non-nested simulation for the meteorological variables respectively.

About 60 % of the nested simulation results for the potential temperature and of the corresponding forcing data have a difference to measured data below the measurement reliability of 1.5 K (Schlünzen et al., 2000) for the northern flight track. The value is 36 % for the southern flight track. The strong coupling of the forcing data and the higher resolution nested model can be seen. The use of a high resolution model does not enhance the quality of the simulation results.

In the unnested simulation (dotted line) the agreement of measured temperatures and the corresponding METRAS data is much lower than in the nested case in both parts of the model area. The disagreement between the simulated and measured temperature can be explained by the absence of large-scale advection of warmer air in the non-nested METRAS simulation which is included in the forcing data and in the nested METRAS run.

For the specific humidity (Figures 3c, 3d) the agreement of the nested METRAS results and the forcing data is again similar. 40 % of the data have a difference below the measurement reliability of about 0.9 g/kg (Schlünzen et al. 2000) for the northern flight track, 30 % for the southern flight track. However, larger differences are less frequent in the nested results compared to the forcing data. In the case of non-nested meteorological variables (dotted line) the agreement is much higher than in the nested case: About 58 % of the model results in the southern part of the model area and 68 % in the northern part differ less than

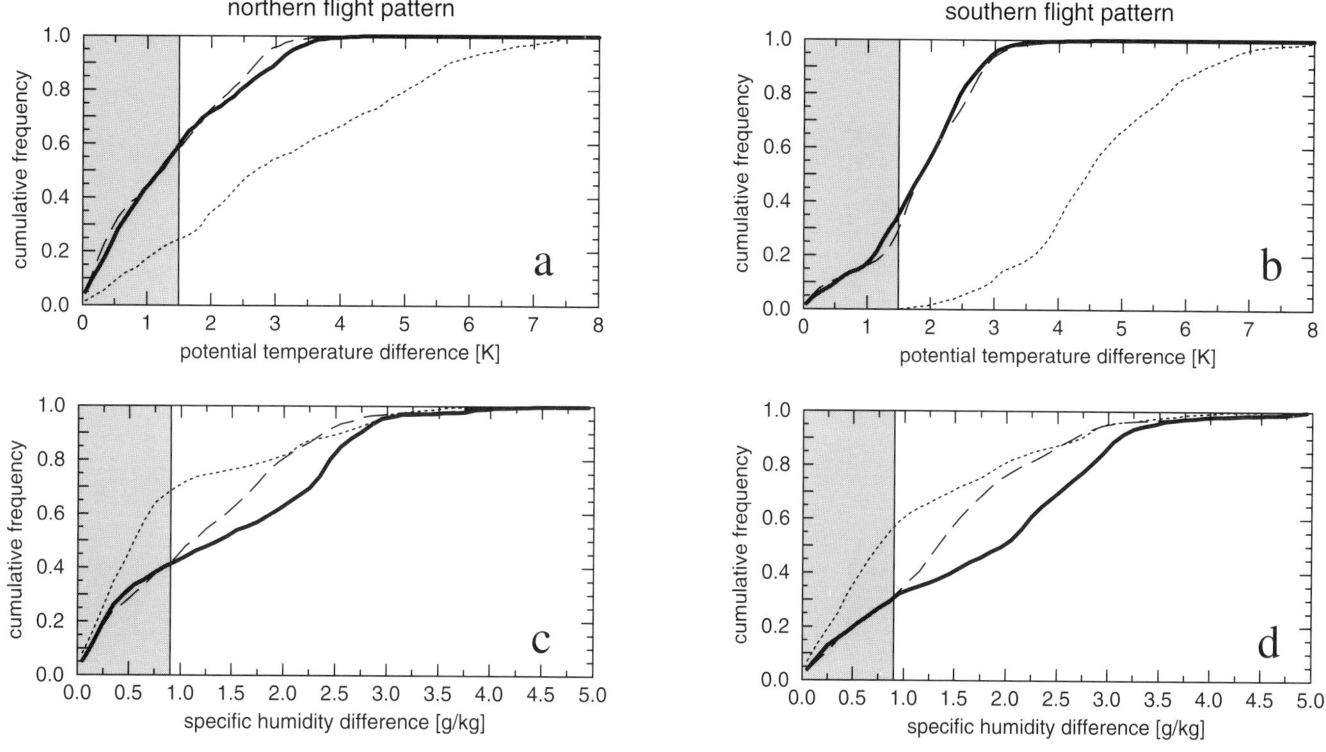

Figure 3: Cumulative distribution function for the difference of the simulated and measured potential temperature (Fig. a and b) and specific humidity (Fig. c and d) for the flight patterns in the northern part (Fig. a and c) and in the southern part (Fig. b and d) of the model area. The solid thick line represents the forcing data, the dashed and dotted lines the results of the nested and non-nested METRAS simulation respectively. Differences below 1.5 K or 0.9 g/kg are below the measurement reliability of temperature or specific humidity (Schlünzen et al. 2000).

0.9 g/kg from the measurements. For the humidity, the nesting reduces the agreement due to the quality of the forcing data. For the METRAS simulations, a tendency to correct the forcing data to higher humidity values and hence into the direction of the measurements can be recognized. This correction, however, cannot fully compensate the nesting data. Nesting can cause additional errors in the model results if the quality of the forcing data is not significant. In these circumstances, a model run without nesting can yield more reliable results than a nested simulation using poor forcing data.

In Figure 4 the agreement of the simulated and measured concentrations of nitrogen-oxides (NO_x) (Fig. 4a and 4b) and ozone (O_3) (Fig. 4c and 4d) are shown for the flight patterns in the northern (Fig. 4a and 4c) and in the southern part (Fig. 4b and 4d) of the model area.

The expected data reliability is about 1.6 ppbv of the measured values for NO_x and 2 ppbv for O_3 respectively (Schlünzen et al., 2000). In the northern part of the model area all high resolution model results of NO_x and O_3 are at least of similar quality with respect to the measurements as the forcing data. The same is true for the southern flight pattern; only the NO_x results for the two cases with non-nested meteorology agree less than the forcing data. For the northern flight track about 64 % (forcing data, fully nested case) to 71 % (partly nested cases and fully non-nested case) of the NO_x data and 36 % to 41 % of the O_3 data agree within the mentioned accuracy with the measurements. For the southern flight track the agreement is similar for NO_x (63 %) and higher for O_3 (48 %) for the nested meteorology. From these results the conclusion could be drawn that a proper nesting of trace gas concentrations is less important than a nesting of the meteorological fields. In the considered model study, the NO_x and O_3 concentrations seem to be less dependent on the flow across lateral boundaries than on the local processes, such as emission rates, and a more realistic description of the meteorological conditions in the surroundings of the sources. However, in the present case study, the NO_x boundary values provided by the forcing data are too low compared to the measurements and the NO_x flux across the model boundaries into the inner model domain is too small as well. Due to the chemical reactions the O_3 concentration is influenced strongly by the NO_x concentration and consequently it could not be expected that the O_3 concentration fields are simulated well if the NO_x concentration results are too low. For calculating reliable concentrations, the meteorology has to be nested. To improve the results for concentrations, these have to be nested too and the boundary values have to be of high quality. For proper model results the concentration fields have to be consistent with the meteorological fields.

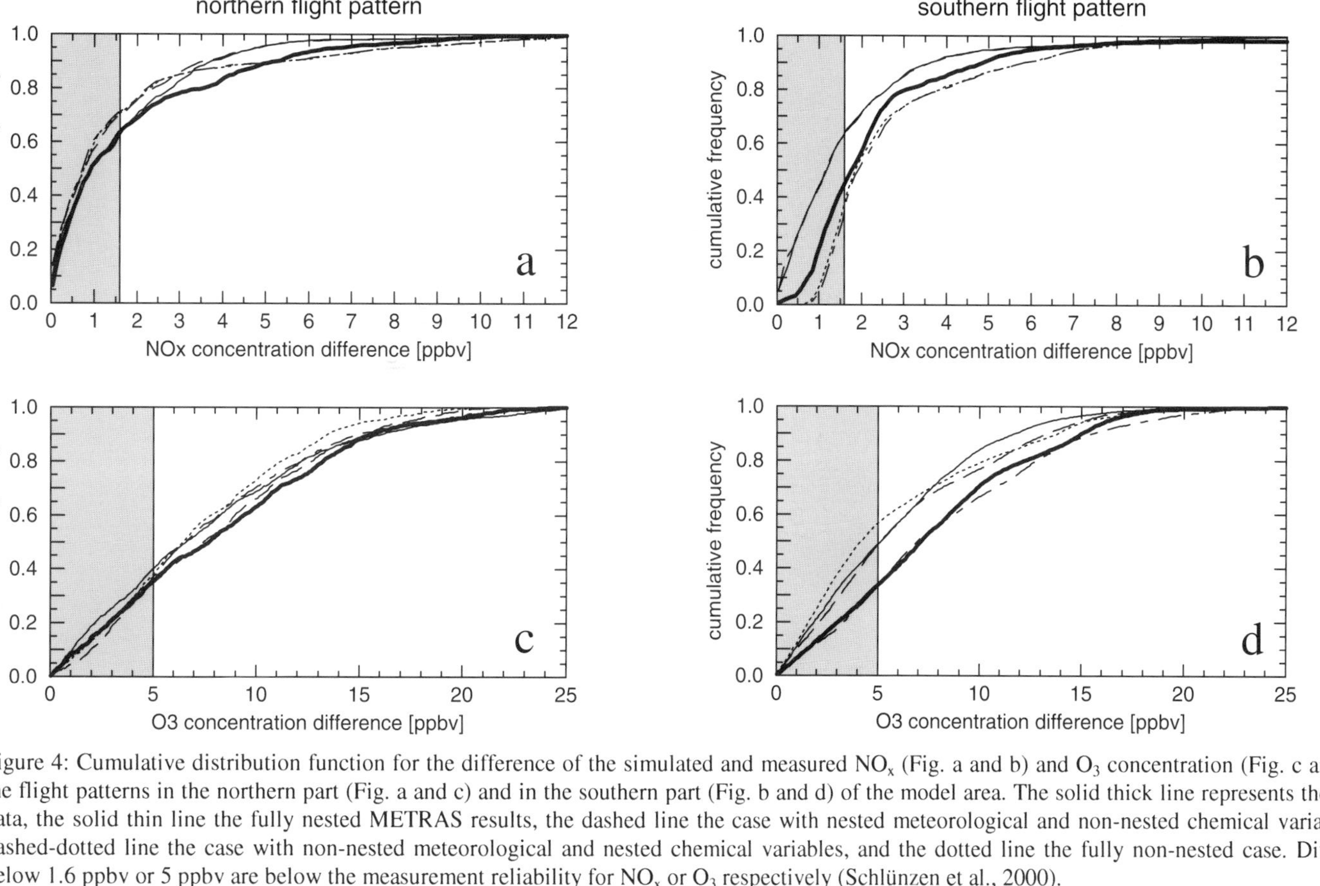

Figure 4: Cumulative distribution function for the difference of the simulated and measured NO_x (Fig. a and b) and O_3 concentration (Fig. c and d) for the flight patterns in the northern part (Fig. a and c) and in the southern part (Fig. b and d) of the model area. The solid thick line represents the forcing data, the solid thin line the fully nested METRAS results, the dashed line the case with nested meteorological and non-nested chemical variables, the dashed-dotted line the case with non-nested meteorological and nested chemical variables, and the dotted line the fully non-nested case. Differences below 1.6 ppbv or 5 ppbv are below the measurement reliability for NO_x or O_3 respectively (Schlünzen et al., 2000).

4. Conclusions

In general, the nesting of the meteorological part of METRAS into a larger scale model enhances the precision of the forecast of the meteorological variables. However, the forcing data adapted from the larger scale model must be of good quality, because the nested model METRAS can only partly correct deficiencies included in the forcing data.

In the considered case the forecast of NO_x and O_3 concentrations is more dependent upon a correct description of the meteorological boundary values than on the concentration fluxes across the lateral boundaries. The simulated NO_x concentrations are rather insensitive to the nesting of trace gas concentrations, which is probably a result of the poor performance of the forcing data. For the prediction of O_3 concentrations in the considered case, the nesting of meteorology has to be accompanied by the nesting of chemical variables. Further sensitivity studies, as well as simulations for other simulation periods, will be made to confirm the current results.

Acknowledgements

The contents of this work is in the responsibility of the authors. We thank the EURAD group in Köln for providing the MM5/CTM2 data, Klaus Nester and Andreas Wenzel for providing the emission, land use and measuring data and evaluation programs. Further we thank Laura Klein for suggestions on the manuscript. This work has been supported by the Federal Ministry for Science, Education, Research and Technology (BMBF) under Grant Number 07TFS10/LT1-C3.

References

Ebel, A., Elbern, H., Feldmann, H., Jakobs, H. J., Kessler, C., Memmesheimer, M., Oberreuter, A. and Piekorz, G.: 1997, Air Pollution Studies with the EURAD Model System (3): EURAD - European Air Pollution Dispersion Model System, Mitteilungen aus dem Institut für Geophysik und Meteorologie der Universität zu Köln, Heft 120.

Grell, A. G., Dudhia, J. and Stauffer, D. R.: 1993, A Description of the Fifth-Generation PENN STATE/NCAR MESOSCALE MODEL (MM5), NCAR Technical Note 398+IA. National Center for Atmospheric Research, Boulder, Colorado, USA.

Hass, H.: 1991, Description of the EURAD Chemistry-Transport-Model Version 2 (CTM2), Mitteilungen aus dem Institut für Geophysik und Meteorologie der Universität zu Köln, Heft 83.

Müller, F., Schlünzen, K. H. und Schatzmann, M.: 2000, Test of numerical solvers for chemical reaction mechanisms in 3D-air quality models, in print by *Environmental Modelling and Software*.

Niemeier, U.: 1997, Chemische Umsetzungen in einem hochauflösenden mesoskaligen Modell - Bestimmung geeigneter Randwerte und Modellanwendungen, Berichte aus dem Zentrum für Meeres- und Klimaforschung, Reihe A, 28. Zentrum für Meeres- und Klimaforschung der Universität Hamburg, Meteorologisches Institut.

Schlünzen, K. H.: 1990, Numerical Studies on the Inland Penetration of Sea Breeze Fronts at a Coastline with Tidally Flooded Mudflats, *Beitr. Phys. Atmosph.*, **63**, 243-256.

Schlünzen, K. H., Bigalke, K., Lenz, C.-J., Lüpkes, C., Niemeier, U. and von Salzen, K.: 1996, Concept and Realization of the Mesoscale Transport and Fluid Model 'METRAS', METRAS Technical Report, 5, Meteorological Institute, University of Hamburg, Bundesstr. 55, D-20146 Hamburg, Germany.

Schlünzen, K. H., Schaller, E., Ebel, A.: 2000, An evaluation strategy for mesoscale atmospheric chemistry transport models, in preparation for *Atmos. Environ.*

Stockwell, W. R., Middelton, P. and Chang, J. S.: 1990, The Second Generation Regional Acid Deposition Model Chemical Mechanism for Regional Air Quality Modeling, *J. Geophys. Res.*, **85, D10**, 16343-16367.

Zimmermann, H.: 1995, Field Phase Report of the TRACT Field Measurement Campaign, EUROTRAC International Scientific Secretary, Garmisch-Partenkirchen, Germany.

USE OF NEURAL NET MODELS TO FORECAST ATMOSPHERIC POLLUTION.

A. PELLICCIONI, U.POLI
Italian Institute for Occupational Safety and Prevention (ISPESL).
Environmental Department, Via Urbana 167, 00184 Rome, Italy.

Abstract. The forecast of the CO and NO_2 concentration levels has been obtained by the 3-layer Perceptron Neural Network with Error Backpropagation learning rule.
This study shows the 3-layer Perceptron performances in relationship with the choice of the activation functions parameters (depending on the statistical values of the input/output variables).
A first simulation set using data at 1 hour before has been utilized to forecast CO levels. A second simulation set with data at 12-24 hours before has been used to forecast NO_2 levels.
The Neural Net's performance appears to be very good both for the parameters activation function optimisation and the variables choice.

Key words: Atmospheric Pollution, Multilayer Perceptron, Neural Net.

1 Introduction

An accurate forecasting of pollutants levels in an urban area sometimes may be difficult to predict because of the presence of too many factors involved such as micrometeorological phenomena, different pollutant sources, atmospheric reactions of the chemical compounds (Bardeschi and others, 1991).

Altogether these factors give a non linear correlation structure of data, difficult to be captured through regression models.

Recently the Neural Net (NN), particularly the Multilayer Perceptron (MLP), has appeared to be a very efficient tool in locating the correlation structure of a data set, using only a supervised learning from a suitable training set (Faussett, 1994).

In this study a generalization of a 3-Layer Perceptron with error backpropagation learning rule (EPB) has been applied to forecast CO and NO_2 concentrations in Rome's urban city centre.

The aim of this study is the application of a new approach able to optimize the performance of the MLP, by linking the parameters of the activation function to statistical analysis of input data.

The following approach permits to eliminate the learning difficulties of the NN, typical of forecasting problems, and to optimise the network performance in reproducing experimental data trend.

Environmental Monitoring and Assessment **65**: 297-304, 2000.

2 Remarks on Perceptron models

The most frequent neural network architecture to forecast time series is the multilayer perceptron with error-backpropagation supervised learning rule. This architecture is commonly dictated by the following considerations:
a) the backpropagation algorithm is very easy to implement through computer programs;
b) there is a mathematical proof granting for the ability of a 3-layer perceptron with non linear activation functions to approximate whatever function of input variables, with unlimited precision, provided suitable assumptions be satisfied (Baum, 1988).

These considerations, however, do not help to find a fast convergence of the learning process or to choose the training set criterion in such a way to obtain a good performance related to the generalization step. Experience accumulated by different forecasting problems has shown that the convergence is very slow and there is often no convergence at all. The network convergence is hardly verified when the following standard activation function is assumed:

$$F(P) = 1/[1 + \exp(-P)] \tag{1}$$

where P is the total input signal received or the activation (Rojas, 1996). This happens because the function (1) does not contain any adjustable parameter and therefore it becomes very difficult to avoid undesired behaviours. A typical example is a situation where the values of hidden-output connection weights, at a given time step have the activation potential of all output units very high in the absolute value (typically greater than 1). In such a case the output of the function (1) is practically insensitive to variations of activation potential, hence of connection weights, because it always stays near 0 or 1. This implies that learning process, based only on connection weight variations, becomes inefficient and there is a little possibility to converge.

To solve this problem, the function (1) can be parameterized as follows:

$$F(x) = A/\{1 + \exp[-K(x - S)]\} \tag{2}$$

where A is the amplitude, K is the slope and S is the threshold. However, in order to have an efficient learning process, the values of parameters A, K and S must be chosen carefully. Generally, these parameters must grant an high sensitivity of the output values to the derivatives of the activation potentials within a given range. In Fig 1 and Fig. 2 the activation function and his derivatives are respectively shown for three different choices of the slope K (0.5; 1.0; 1.5), having assumed A=1 and S=0. It is important to note that the maximum sensitivity of the sigmoid function depends on the P values that are

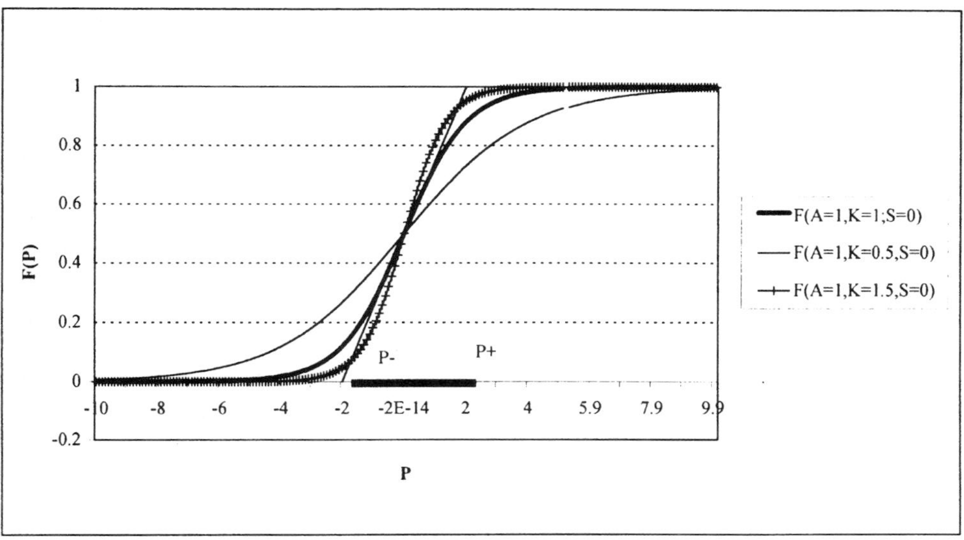

Figure 1. Trend of the Sigmoid Activation Function

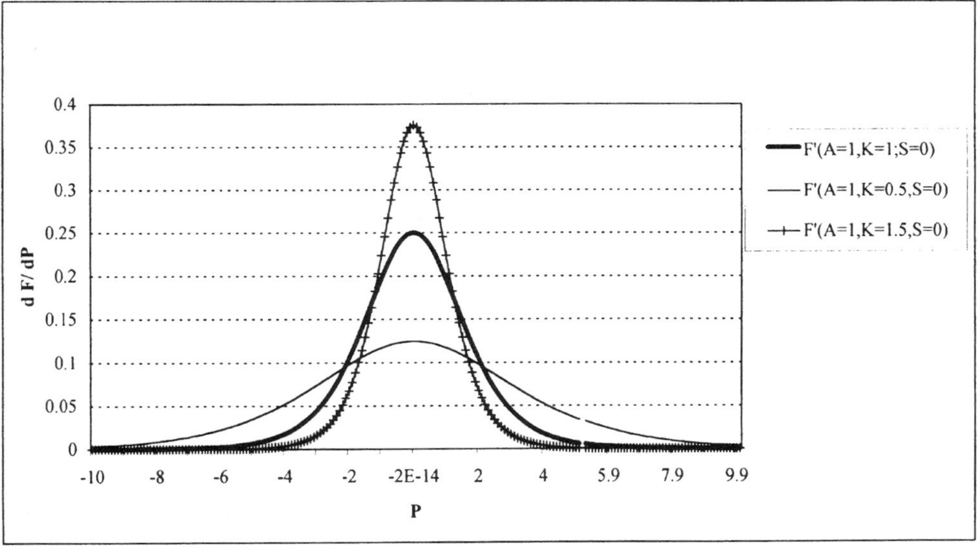

Fig 2. Trend of the derivative of the Sigmoid Activation Function

linked to the input data of each neuron. In particular, the range of maximum sensibility of the sigmoid (ΔP) can be obtained by geometric considerations from the following formulation:

$$\Delta P = P_{+} - P_{-} = \frac{2}{K} - \left(-\frac{2}{K} \right) = \frac{4}{K} \tag{3}$$

where K is the slope. Every input data that are in the ΔP range is projected in the range 0÷A.

The formula (3) can be used to estimate the sigmoid parameter ($K_{Optimum}$) for each layer (*e.g.* the hidden layer). In fact, estimating the maximum value of P through the input data (P_{input}) and assuming $P_{-} = - P_{+}$, it is possible to optimize the slope, $K_{Optimum}$, so that to maximize the efficiency of the neuron, using the formulation:

$$K_{Optimum} = \frac{4}{\Delta P} = \frac{2}{P_{input}} \tag{4}$$

This observation leads us to consider that if the slope K of the neuron has been fixed *a priori* (as in (1)), no training with a determined input data can be verified.

Still there is another important consideration to be analysed: in the MLP the input potential for the output layer is also determined by the activation function coming from the hidden layer.

All the observation mentioned above can be presented in the scheme below as a proposed optimisation of the activation function parameters for a symmetric *P* distribution:

	Hidden Layer	Output Layer
K(Hidden/Output)	$2/P_{Input}$	$2/A_H$
A(Hidden/Output)	A_H	P_{Output}

where A_H is the threshold of the hidden layer.

Therefore, concerning the optimisation of a single neuron, and given the estimate of the maximum values for the activation potential for the input layer (P_{Input}) and the output layer (P_{Output}), the only parameter to be determined is the threshold A_H.

Furthermore, the following questions can be pointed out:

- If the sigmoid parameters have been chosen accordingly to values reported in the above table, the best performance can always be obtained. Consequently, it is possible to clarify in an objective manner when the net is inefficient **due to** the data structure.
- For a single neuron, only one parameter (the threshold) has to be chosen and not four, when no objective criteria are adopted.
- The standard choice of the parameters (A=1, K=1, and S=0) for both neurons of all layers never optimize the performances of the MLP. This last assumption

derives from the consideration that the slope of the output layer is strongly linked to the threshold value of the input layer and that if $A_H=1$ has been assumed, the slope of the output layer could be 2 (instead of 1).

Unfortunately, such a criterion cannot be formulated into an algorithmic form. Therefore, in this paper the aim was to find the most convenient parameter values through an adjustment procedure based on the following criteria:

1. The high-sensitivity range of activation potential received by output units must overlap in a significant way with the typical range of effective activation potential produced by hidden units;
2. When the criterion 1. is satisfied, the slope of the activation function (2) must optimise the sensitivity to variations of the activation potential, *i.e.* enhancing it in such a way as to capture the information contained in the data but avoiding a too high sensitivity which could also capture noise;
3. When criterion 2. is satisfied the amplitude has to be chosen in such a way as to control the variability range of the output layer.

Implementing these criteria, we are forced to adopt different parameter values for each layer. In this paper, there have been reported two applications of the implementation of this strategy in forecasting the pollution levels of CO and NO_2 in an urban area. It must be noted that the use of the above mentioned criteria can lead to a more objective choice of the network parameters.

3 Results and discussion

The training set concerns measurements of atmospheric pollutants (CO and NO_2) taken at fixed-site monitoring stations of the *Presidio Multizonale di Prevenzione* network placed in the urban city centre of Rome.

The pollutant concentrations of the training data set have been normalised by the maximum value of the related pollutant.

In order to obtain typical situations (linked to the diurnal/nocturnal cycle of the anthropogenic emissions and to the daily micrometeorological conditions), the overall data set has been classified by means of Kohonen Net as non hierarchical method. Therefore, a classification of mean situations has been obtained (connected to more frequent emission conditions, as turbulence) as well as acute episodes (few cases of very high concentration levels).

The first simulation set has been designed to forecast CO levels at time t, starting from data taken at time (t-1). The adopted 3-layer Perceptron has 5 input units, 2 hidden units and 1 output unit giving the CO concentration at time t.

The EBP with zero momentum has been used, whereas the learning rate (the typical parameter to update the weights) has been set to 0.2 in the first thousand steps, and to 0.1 in the remaining steps. The training set contains 100 input/output pairs taken from clustered data relative to three months.

The analysis of both autocorrelation function (*i.e.* correlation between the CO concentration at time t and the CO concentration at time $(t\text{-}L)$ with (L=1, 2, ...24) and crosscorrelation function (*i.e.* correlation between the CO concentration at time t and the concentration of other pollutants at time $(t\text{-}L)$) have been computed in order to choose a suitable temporal step.

When lag L equal to 1, the highest autocorrelation and crosscorrelation coefficients occur. Meaningful correlations are obtained at lags 12 and 24, thus corresponding to the diurnal/nocturnal cycle and respectively to the periodicity of emissions.

The 5-dimensional input vectors contains CO, NO_X, NO and NO_2 levels at time $(t\text{-}1)$ and the hour at time $(t\text{-}1)$.

The results have shown that NN's performance is strongly dependent upon the choice of A, K and S: using the standard activation function F(x) for both hidden and output layer the correlation between the input data and the data foreseen by the network is equal to -0.02, while a correlation equal to 0.82 is obtained if the (2) is adopted with A=2, K=0.5 and S=1.5 for the hidden layer, and A=2, K=2 and S=0.5 for the output layer. The correlation improves because the above parameters are consistent with the statistical distribution of data used in the training set.

Fig.3 shows the comparison between the CO measured and the normalised CO reproduced by NN’s choices related to 15 days on April 1997. It’s worth nothing that the usual choice of the activation function never gives good results. On the contrary, in this case the best choice gives good performances except for higher peaks concentrations.

The second simulation set has been used to forecast the NO_2 concentrations, starting from data taken at time $(t\text{-}12)$ and $(t\text{-}24)$. The network has 14 input units: CO$(t\text{-}12)$, CO$(t\text{-}24)$, $NO_X(t\text{-}12)$, $NO_X(t\text{-}24)$, NO$(t\text{-}12)$, NO$(t\text{-}24)$, $NO_2(t\text{-}12)$, $NO_2(t\text{-}24)$, RH$(t\text{-}12)$, RH$(t\text{-}24)$, T$(t\text{-}12)$, T$(t\text{-}24)$ and hours at time $(t\text{-}12)$ and $(t\text{-}24)$, where RH is the relative humidity and T the temperature. Also in this case, the temporal step of input training data has been chosen taking into account the correlation of NO_2 with the input variables at different lags.

The output layer of the adopted 3-layer Perceptron has one unit, giving the NO_2 level at time t, and 3 hidden units. The best NN’s performance has been obtained through a suitable choice of A, K and S, *i.e.* A=2, K=1 and S=0.5 for the hidden layer, and A=1, K=1 and S=0 for the output layer. The correlation coefficient between experimental and foreseen data has been 0.73. The results have shown the importance to determine the most suitable variables, that are able to describe the correlation structure of the system.

A traditional autoregressive model (Wei, 1990) was applied to the same training data set in order to compare the network performance in forecasting NO_2 concentration. A correlation of 0.62 has been obtained. This result confirms the higher performance of NN to forecast pollution levels.

Fig.4 shows the comparison between the normalised NO_2as measured and reproduced by NN for the same period. While the trend is well approximated,

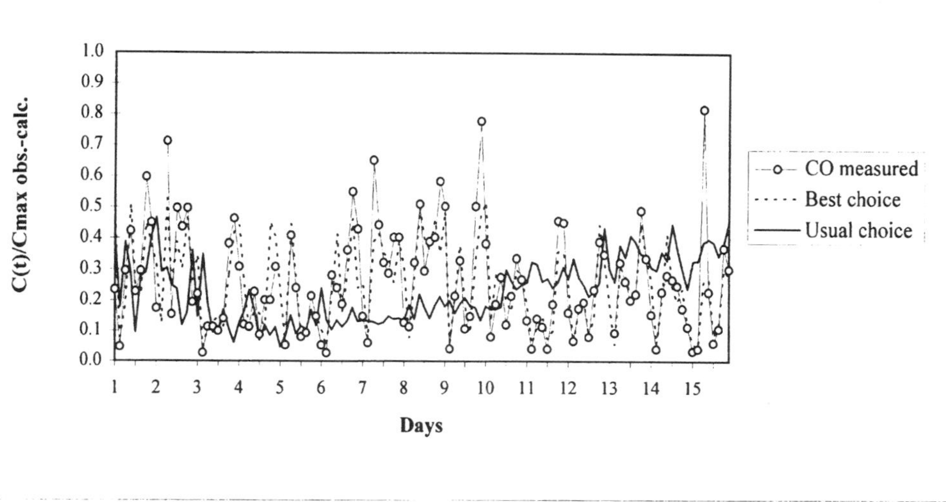

Figure 3. Temporal trend of Normalized CO reproduced by NN.

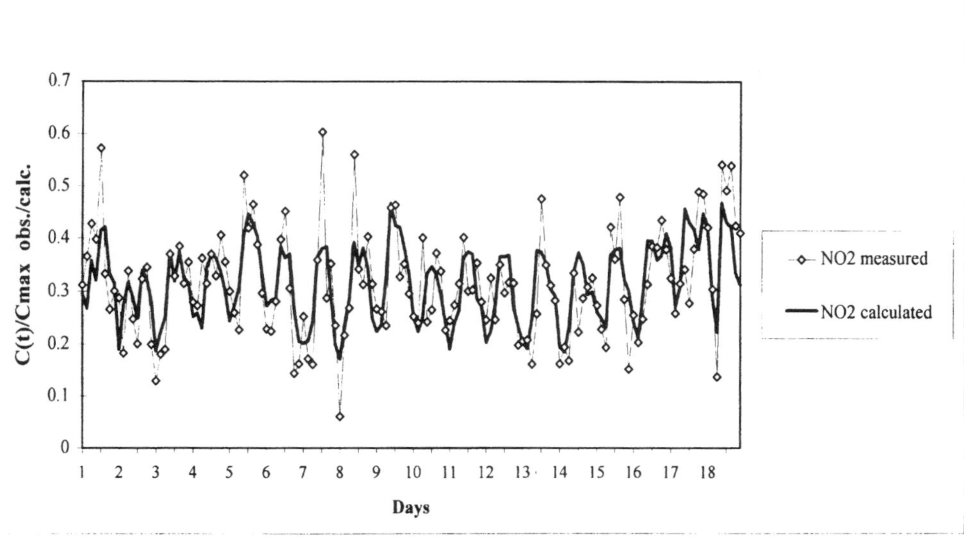

Figure 4. Temporal trend of Normalized NO_2 reproduced by NN.

the model presents some difficulty in predicting lower and higher NO_2 concentration.

4 Conclusions

The NN's performance seems to be very good and anyway better than the results given by commonly used parametric models. However, the good NN's performance is essentially related to a careful choice of the $F_{A,K,S}$ parameters. This implies that, besides a good selection of the training set through a Kohonen classification, a preliminary study of input data is always needed in order to:
- determine the statistical distribution of the input/output data set;
- remove not very meaningful data from the training set;
- choice a suitable normalization rule;
- compute the usual statistical correlation indexes among the variables.

In this study particular values of activation function parameters that optimise NN's forecasting ability are derived from the statistical analysis of input-output variables.

Acknowledgments

The Authors thank the *Presidio Multizonale di Prevenzione* of the Province of Rome for the cooperation in the data set provision.

References

Baum E.B., 1988, On the Capabilities of Multilayer Perceptrons, *Journal of Complexity*, 4, 193-215

Bardeschi A, Colucci A, Giannelle V, Cagnetti M, Tamponi M and Tebaldi G: 1991, Analysis of Impact on Air Quality of Motor Vehicle Traffic in the Milan Urban Area, *Atmospheric Environment,* 3, 415-428.

Faussett L: 1994, Fundamentals of Neural Networks. Architectures, Algorithms, and Applications, Prentice Hall International Editions.

Hornik K, Stinchcombe M and White H: 1989, Multilayer feedforward networks are universal approximations, *Networks*, 2, 359-366.

Rojas R., 1996, Neural Networkks: A systematic Introduction, Springer-Verlag, Berlin Heidelberg

Wei W S: 1990, Time series Analysis Univariate and Multivariate methods, Addison, Wesley Redwood City.

THE OZONE FINE STRUCTURE MODEL: MODEL CONCEPT AND OPTIONS

PETER SAHM, NICOLAS MOUSSIOPOULOS and JANN JANSSEN
Laboratory of Heat Transfer and Environmental Engineering
Aristotle University Thessaloniki, Box # 483, GR-540 06 Thessaloniki, Greece
E-mail:peter@aix.meng.auth.gr

Abstract. The Ozone Fine Structure Model (OFIS) model allows an adequate description of urban photochemistry at a very low computational effort. Thus, it may be used for statistical analyses of urban scale ozone levels, including the assessment of control strategies and exposure estimates. Recent model extensions allow taking properly into account emission inhomogeneities in the surroundings of the considered city and local circulation systems. The performance of both the standard and the extended OFIS model versions is illustrated by its application to Stuttgart and Athens. The results reveal that the model extensions allow for more versatility in inhomogeneous situations.

Key words: urban air-pollution, dispersion, ozone strategy, OFIS model.

1. Introduction

Exposure to air pollution represents a serious problem in many cities all over the world. Despite the achievements in the reduction of "traditional" emissions (SO_2 and particulate matter), the majority of European cities still exceed short-term WHO air quality guidelines. Nowadays, photochemical air pollution causes most of the concern, and in this context attention is focused on ozone - as the most prominent photo-oxidant - and nitrogen dioxide (NO_2).

Sophisticated mesoscale chemical transport model systems are suitable for calculating the diurnal variation of ozone patterns in a conurbation (Moussiopoulos et al., 1997). Model simulations of this type are indispensable, for example, to identify "hot spots" or to assess maximum concentration levels in a city. Such model systems normally consist of a nonhydrostatic prognostic mesoscale meteorological model and a 3D photochemical dispersion model. Due to the complexity of the model system the simulation period is usually restricted to individual episodes lasting a couple of days.

Effect-based research in Europe suggests using a long-term criterion as the no-damage and health-effect threshold (e.g. Kärenlampi and Skärby, 1998). As a consequence, integrated assessment models should also be able to address the long-term ozone exposure over a multi-month period. The computational complexity of sophisticated mesoscale chemical transport models and their need of

Environmental Monitoring and Assessment **65**: 305–312, 2000.

disaggregated emission input data makes it impossible to use them directly within the framework of an integrated assessment model concept. In order to overcome this gap, 'reduced-form' models, using statistical methods to summarise the reaction of a more sophisticated model concept have to be utilised.

As an example for such a simpler approach, the Ozone Fine Structure Model (OFIS) model allows an adequate description of urban photochemistry at a very low computational effort: This newly developed Eulerian model is capable of calculating ozone levels due to transport and photochemical transformation within an urban plume. Moreover, it can be used for obtaining exceedance statistics of ozone threshold values based on meteorological data and regional background concentrations for the period considered.

The subsequent section of this paper deals with the model concept of OFIS including several new options to account for more complex situations. The performance of OFIS is then assessed by its application to Stuttgart and Athens.

2. Model concept

2.1 STANDARD MODEL VERSION

The OFIS model was derived from well-tested full 3D models, and hence it retains all elements necessary to achieve a realistic statistical evaluation of urban scale ozone levels. The conceptual basis of OFIS is a coupled 1D-2D approach:

- Background boundary layer concentrations are calculated with a three-layer box model representing the local-to-regional conditions in the surroundings of the city considered. This model uses at input non-urban emission rates, meteorological data and regional scale model results for pollutant concentrations (e.g. EMEP model results (Simpson, 1993; 1995)).
- Pollutant transport and transformation downwind of the city is calculated with a three-layer multibox model representing a substantially refined version of MARS-1D (Moussiopoulos, 1990). It is assumed that one wind direction prevails for a specific day.

The distinction of three individual layers of time depending thickness allows adequately describing the dynamics of the atmospheric boundary layer: A shallow layer adjacent to the ground is used for simulating dry deposition and other sub-grid phenomena. This layer practically corresponds to the surface layer. The upper limit of the second layer coincides with the mixing height. The latter is described by Deardorff's prognostic equation (Deardorff, 1974). The top of the upper layer, which serves as a reservoir layer, is set at 3000 m, chosen so that it is well above the maximum mixing layer height. This methodology is identical to the one adopted in the multilayer model MUSE (Sahm et al, 1997) which is another simplified version of MARS (a few layers instead of 'normal' discretisation in the vertical direction; semi-implicit solver instead of a fully implicit one). A 1D version of the non-hydrostatic mesoscale model MEMO (Moussiopoulos, 1995) is utilised to calculate the vertical profiles of temperature, mean wind speed and turbulent exchange coefficient for both the city surroundings and the urban plume assuming Monin-Obukhov similarity at the lower boundary.

The mathematical analysis is based on the coupled, two-dimensional advection-diffusion equations for the ensemble averaged quantities of reactive species. The equations are solved by operator splitting according to the method of lines, that is by solving the advection dominated terms separately from the diffusion dominated ones (in vertical direction) and the chemical reaction terms. The concentration trends due to advection, vertical diffusion and entrainment are then treated as source terms in the chemical reaction equation system. The latter is solved in OFIS with a backward difference solution procedure, i.e. by applying the Gauss-Seidel iteration scheme (Kessler, 1995). The model uses a variable time step with an upper limit for the integration time increment (e.g. of 300 seconds).

Due to the modular structure of OFIS, chemical transformations can be treated by any suitable chemical reaction mechanism, the default being the EMEP MSC-W chemistry which has been described in detail in Simpson et al. (1993) and Simpson (1995).

The dry deposition process for both the city surroundings and the urban plume is calculated following the resistance model concept (Sahm, 1998).

The results of the standard version of OFIS were the basis of expertise provided to the European Commission (Moussiopoulos et al., 1998). The standard version was also used for the assessment of urban scale ozone levels for the needs of the EU State of the Environment Report 1998 (EEA, 1998).

2.2 EXTENDED MODEL VERSION TO ACCOUNT FOR INHOMOGENEOUS SITUATIONS

2.2.1 *Considering emission inhomogeneities in the city surroundings*

Assuming that the background boundary layer concentrations are independent of the wind direction is most likely erroneous in densely populated areas as high emitters (like another city or industrial areas) could well be located upwind of the considered urban area and thus affect air quality in the latter. By extending the domain of the three-layer multibox model well upwind of the city centre, large emitters can now be taken into account in the vicinity of the urban area depending on the prevailing wind direction of the respective day.

2.2.2 *Accounting for local circulation systems*

The assumption of one prevailing wind direction for a specific day does not allow to resolve local circulation systems. The extended version of OFIS accounts for a local circulation system such as the sea-breeze in coastal areas by inversing the wind direction of the urban plume in the lower two layers (i.e. up to the mixed layer) in the afternoon hours of days with weak synoptic forcing and off-shore wind direction. The inversion is simulated by interpolating the wind speed from positive values to negative values between two subsequent hours.

3. Case specifications

Simulations of the ozone levels in the area around Stuttgart and Athens were performed with both the standard and the extended versions of OFIS for each day between 1 April and 30 September 1990. For this period, the EMEP model provided regional background concentrations at a spatial resolution of 150×150 km^2 and a temporal resolution of 6 hours. Meteorological data as used by the EMEP model have been available at the same resolution. Emission data for Stuttgart and Athens, specifically the diurnal variation of the emission rates for urban, suburban and rural areas at a temporal resolution of 1 hour were derived from CORINAIR data.

Background boundary layer concentrations for both cities were computed with the three-layer box model embedded in OFIS for a domain of 150×150 km^2 which was assumed to be rural area. For each day in the period considered, pollutant transport and transformation downwind of the two cities were calculated in 5 km steps assuming the wind direction valid for the respective day. Dry deposition was accounted for by using a three-resistance model approach. Biogenic emissions were taken into account for rural areas.

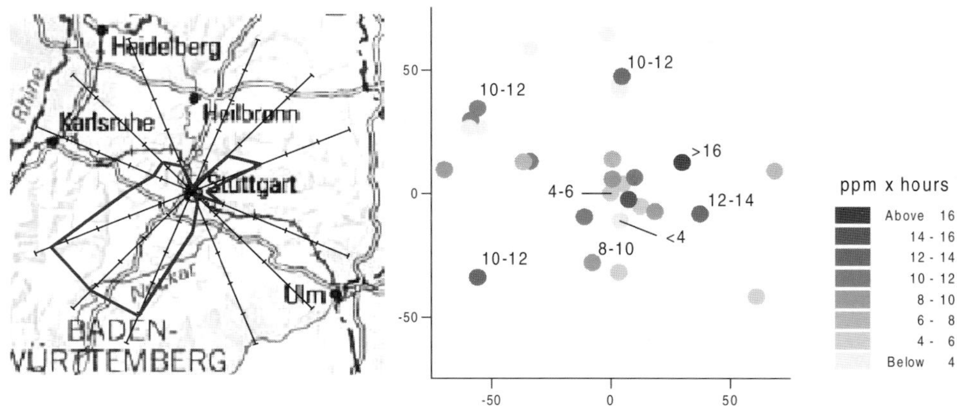

Figure 1. Simulation domain and wind direction statistics for Stuttgart as derived from EMEP model input (left) and AOT60 values derived from the measurements in Stuttgart for the period 1 April - 30 September 1990 (right).

As shown in Figure 1, prevailing wind directions at surface level in Stuttgart are those from W-SW. Also, winds from NE appear to occur rather frequently in the area. Figure 2 reveals that according to the regional scale information prevailing wind directions at surface level in Athens would be those from N-NW (white wind rose). If sea breeze effects are taken into account, winds from S-SE are found to occur most frequently (black wind rose valid for afternoon hours). In this context, days with sea breeze are defined as those with a ground level wind speed below 3 m/s and an offshore wind direction in the early morning.

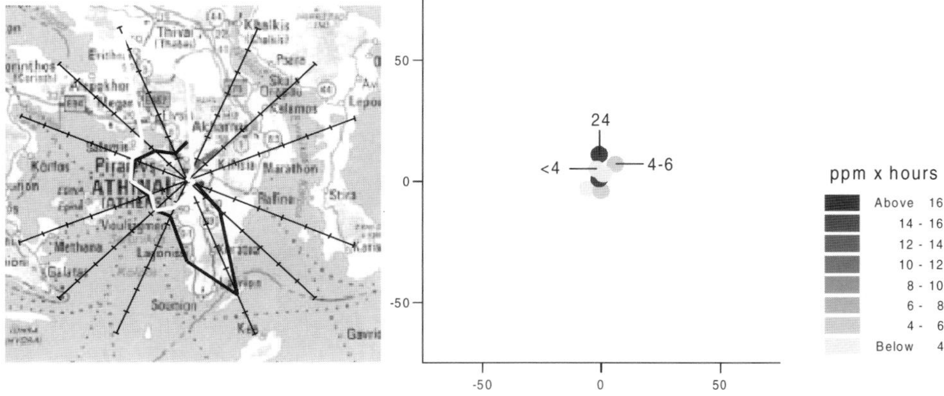

Figure 2. Left: Simulation domain and wind direction statistics for Athens as derived from EMEP model input (white) and assuming the wind direction inversion for days with sea breeze (black). Right: AOT60 values derived from the measurements in Athens for the period 1 April - 30 September 1990.

4. Results

The results are presented in terms of AOT60, the accumulated ozone exposure above a threshold value of 60 ppb (AOT60) which is frequently used as a pragmatic tool to quantify the exceedance of the WHO air quality guideline for protection of human health. Although the AOT60 value is based on the threshold approach and thus a very sensitive statistic, the fact that AOT60 integrates the exceedances gives a longer time-perspective to the results. Furthermore, it is much simpler to use and understand than the complex dose-response functions favoured in the U.S. (e.g. sigmoidal functions).

4.1 STUTTGART

In Figure 3 AOT60 values calculated with OFIS for the period 1 April – 30 September 1990 are shown for both the standard (left) and the extended version (right) of OFIS. According to the OFIS results for AOT60 in the surroundings of Stuttgart, highest AOT60 levels are reached to the NE of Stuttgart, i.e. the area which happens to be most of the time downwind of the city. Values ranging up to 16 ppm×hours agree satisfactorily with available measurements (cf. Figure 1) in spite of the fact that the situation in the individual measuring stations may be influenced by a variety of local influences not considered in OFIS. Taking into account emissions of neighbouring cities leads to lower AOT60 values in the city and higher AOT60 values in the urban plume. This corresponds to the observations.

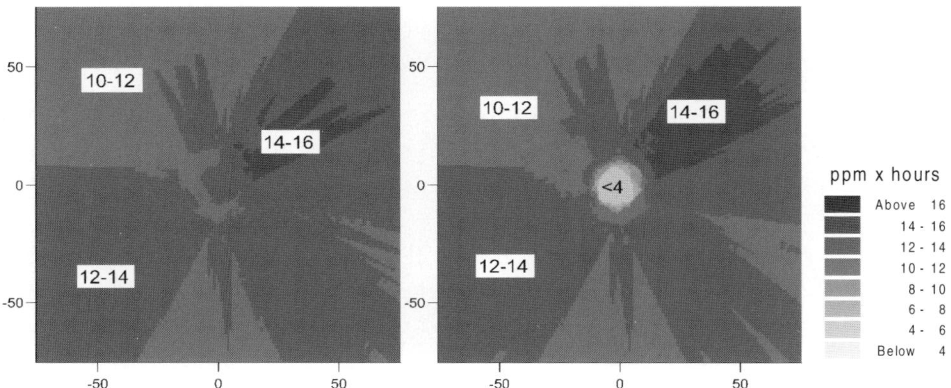

Figure 3. AOT60 values calculated with OFIS for Stuttgart. Standard (left) and extended version (right) of OFIS to account for emission inhomogeneities in the city surroundings.

4.2 ATHENS

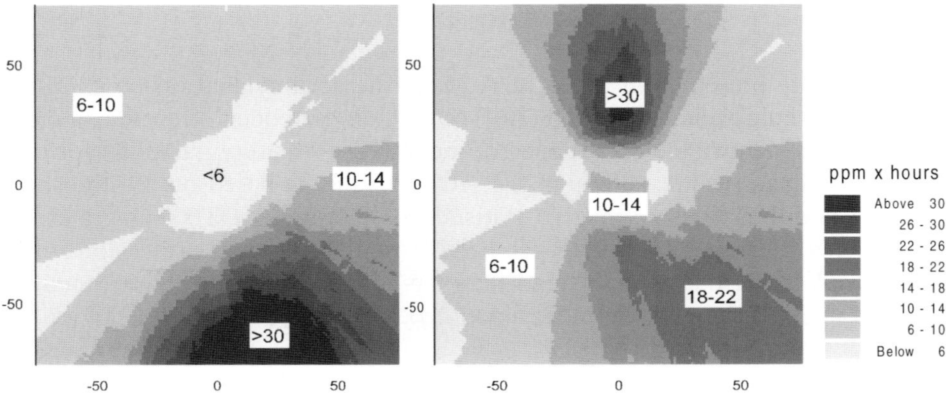

Figure 4. AOT60 values calculated with OFIS for Athens. Standard (left) and extended version (right) of OFIS to account for sea breeze effects.

In Figure 4 AOT60 values calculated with OFIS for the period 1 April – 30 September 1990 are shown for both the standard (left) and the extended version (right) of OFIS. The standard model version fails to predict the location of highest AOT60 values. Taking into account sea breeze effects leads to higher AOT60 values in the city, as polluted air masses are now re-advected over the city with the sea breeze. In agreement with available measurements, highest AOT60 values are now reached in the northern suburbs of Athens.

5. Conclusions

The OFIS model was derived from well-tested full 3D models. Therefore, it retains all elements necessary to achieve a realistic statistical evaluation of urban scale ozone levels. The model applications demonstrate its suitability for assessing ozone exposure at the urban scale. The presented extensions of the model were found to improve its performance in inhomogeneous situations (emission inhomogeneities in the city surroundings or local circulation systems).

References

Deardorff, J.W.: 1974, Three-dimensional numerical study of the height and mean structure of the heated planetary boundary layer, *Bound.-Layer Meterol.*, **7**, 81-106.

EEA: 1998, European State of the Environment Report, European Environment Agency, Copenhagen.

Kärenlampi, L. and Skärby, L. (eds.): 1998, Critical levels for Ozone in Europe: Testing and Finalizing the Concepts, UN/ECE Workshop Report, University of Kuopio, Department of Ecology and Environmental Science, Finland.

Kessler, Ch.: 1995, Entwicklung eines effizienten Lösungsverfahrens zur Beschreibung der Ausbreitung und chemischen Umwandlung reaktiver Luftschadstoffe, Verlag Shaker, Aachen, 148 pp.

Moussiopoulos, N., Sahm, P., Tourlou, P.M., Friedrich, R., Wickert, B., Reis, S. and Simpson, D.: 1998, Technical Expertise in the Context of the Commission's Communication on an Ozone Strategy , Final report to European Commission DGXI.

Moussiopoulos, N.: 1990, Influence of power plant emissions and industrial emissions on the leeward ozone levels, *Atmos. Environ.* **24A**, 1451-1460.

Moussiopoulos, N.: 1995, The EUMAC Zooming Model, A tool for local-to-regional air quality studies, Meteorol. *Atmos. Phys.*, **57**, 115-133.

Moussiopoulos, N., Berge, E., Bohler, T., Leeuw, F., Gronskei, K-E., Mylona, S. and Tombrou, M.: 1997, Ambient Air Quality, Pollutant Dispersion and Transport Models, European Topic Centre on Air Quality, Topic Report 19, European Environment Agency, EU Publications, Copenhagen, 94 pp., (URL: http://eea.eu.int/frdocu.htm).

Sahm, P.: 1998, Kopplung eines nicht-hydrostatischen prognostischen Grenz-schicht-modells und eines mesoskaligen Ausbreitungsmodells für reaktive Stoffe, Fortschr.-Ber. VDI, Reihe 15, Nr. 199, 170 pp.

Sahm, P., Kirchner, F. and Moussiopoulos, N.: 1997, Development and Validation of the Multilayer Model MUSE - The Impact of the Chemical Reaction Mechanism on Air Quality Predictions, Proceedings of the 22nd NATO/CCMS International Technical Meeting on Air Pollution Modelling and its Application, Clermont-Ferrand, France, June 2-6, 1997.

Simpson, D.: 1993, Photochemical model calculations over Europe for two extended summer periods: 1985 and 1989, *Atmos. Environ.,* **27A**, 921-943.

Simpson, D.: 1995, Biogenic emissions in Europe 2: Implications for ozone control strategies, *J. Geophys. Res.,* **100 D11**, 22891-22906.

GREY BOX AND COMPONENT MODELS TO FORECAST OZONE EPISODES: A COMPARISON STUDY

UWE SCHLINK and MARIALUISA VOLTA [1]
Human Exposure Res. and Epidemiology, UFZ, POBox 2, 04301 Leipzig, Germany
[1] *D.E.A., Università degli Studi di Brescia, Via Branze, 38, 25123 Brescia, Italy*

Abstract. For the purpose of short-term forecasting of high ozone concentration episodes stochastic models have been suggested and developed in the literature. The present paper compares the quality of forecasts produced by a grey box and a component time-series model. The summer ozone patterns for three European urban areas (two continental and one mediterranean) are processed. By means of forecast performance indices according to EC and WHO guidelines, the following features of the models could be found: The grey box model is highly adaptive and produces forecasts with low error variance that increases with the time horizon of forecast. The component model is more 'stiff' that results in a higher forecast-error variance and poorer adaption in detail. The forecast horizon, however, could be enlarged with this model. The accuracy of predicting threshold exceedance is similar for both models. This can be understood from the assumption of a cyclical time development of ozone that was made for both models.

Keywords: ozone, air pollution management, time-series model, cyclostationarity, Kalman filter, short-term forecasting

1. Introduction

To protect the population from adverse health effects early information and warnings of high ozone concentration are to be given in good time. Methods for the short-term prediction of atmospheric pollution have been developed based on very different approaches. The most popular methods are based on neuronal nets or stochastic models. The purpose of this paper is to compare the performance of two stochastic techniques: grey box and component time-series models are used to forecast exceedances of ozone threshold values. The ozone time-series of two consecutive summers of three European urban areas (two continental and one mediterranean) are used for calculations. A comparison of results, in terms of forecast performance indices according to the European Community directive 92/72/EEC and following the WHO guidelines, is presented and discussed.

2. Models

From the damped decay of its autocorrelation function it can be realised that the ozone time-series is autocorrelated, and therefore autoregressive models

Environmental Monitoring and Assessment **65**: 313–321, 2000.

could be applied. Autoregressive and moving average models (so called ARMA models) have a long tradition in air pollution modelling and forecasting. The problem of modelling ozone data, however, arises from its non-stationarity. To solve this problem two different approaches have been given among others in the literature. Their formalisms are very different and, therefore, the quality of their predictions was compared here.

Grey box stochastic models are extended autoregressive moving average models. They take into account the non-stationarity and non-linearity of the process. Therefore, unlike autoregressive models, they use parameters that are time dependent and vary in predefined classes (Finzi *et al.*, 1982,1984). The general formulation for these models is the following:

$$c(t) = \sum_{i=1}^{p} a_i s(t)c(t-i) + \sum_{j=1}^{r} b_j s(t)u(t-k_j) + \sum_{h=1}^{g} c_h s(t)\varepsilon(t-h) + \varepsilon(t) \qquad (1)$$

where $c(t)$ is the ozone concentration at time t, $s(t)$ is a properly defined category at time t, $\varepsilon(t)$ is a non-linear stationary white noise, $u(t)$ is an exogeneous input at time t and $k_j = 0,1,..$ is the lag time of a significant causal correlation between input $u(t)$ and output $c(t)$.

The model parameters can assume different sets of values, according to different categories $s(t)$, in order to better simulate non-stationary behaviour of the output variable $c(t)$. A particular kind of non-stationarity is the cyclostationarity, occurring whenever the process has an underlying periodic component (for instance, daily). In this case, the parameter sets vary from time to time following a periodic function.

While $u(t)$ is an explicit external input, there is also an implicit input to the structure of the model by the parameters a_i, b_i, c_i. The latter makes it possible to insert supplementary information about the physics of the phenomenon. This is the main reason to use the term 'grey box' in contrast to the so called 'black box' models, that do not take into respect additional knowledge about the physics of the process.

The grey-box models have been identified and estimated from the summer data June - August 1996 in order to subsequently forecast the summer 1997 period. For example, for Brescia we found the following grey box model:

$$c(t) = a_1 s(t)c(t-1) + a_2 s(t)c(t-2) + a_0 s(t) + \varepsilon(t) \qquad (2)$$

with the hourly ozone concentration $c(t)$. The autoregressive parameters a_1, a_2, a_0 describe the association of the present ozone concentration to the ozone values observed one and two hours before. Such relationship is able to model not only the persistency of the pollutants concentration, but also periodic changes (Slutzky, 1927; Yule, 1927). The association between successive observations, however, may not be constant during the course of the day. Considering the atmospheric conditions we can imagine, for instance, periods of higher atmospheric stability (early morning hours), periods without ozone formation (night), periods that show very probably ozone increase (morning to noon), and so on. One can identify such periods considering the course of the ozone concentration and its variance of an average day that is typical for summer conditions. Every day there is a cyclical return of such characteristic periods and, thus, it appears to be more appropriate to use different sets or categories of parameters a_1, a_2, a_0 for different periods. This way the knowledge about the physics of the ozone formation process is incorporated into the grey box model.

The second model that was created in this study is a component time-series model. Time series of air pollution exhibit characteristic frequencies that can be used for forecast purposes (Schlink *et al.*, 1997). That means, the component model utilises knowledge about regularly appearing atmospherical processes as well as the grey box model. It may utilise not only the diurnal cycle, but also cyclic behaviour due to local circulation systems (e.g. low level jet, Schlink *et al.*, 1999) in result of local orographic conditions or due to the heat island effect. The component model is also able to describe a change in the trend of ozone data that may be caused by transport during particular weather situations.

Practically, characteristic frequencies are identified by help of a spectral analysis. So we found that in the basic data the cycles of 24 h, 12 h and 6 h are predominant with changing intensity. These periods shape because the photochemical development of tropospherical ozone requires both the presence of precursors and sun radiation. In addition, the diffusion ability within the Atmospheric Boundary Layer increases during the course of the morning, and may finally become larger than the rate of ozone formation. The cycles with a period below 24 h may also be attributed to local phenomena, such as local wind circulation (e.g. Schlink *et al.*, 1999). The component model (3) contains the following components: the low-frequency ('trend') part $T(t)$, 12 h and 6 h seasonal parts $S_{12}(t)$, $S_6(t)$, and the residue $R(t)$.

$$\log c(t) = T(t) + S_{12}(t) + S_6(t) + R(t) \quad (3)$$

In order to estimate the components, a Kalman filter was used (Ng *et al.*, 1990). The Kalman filter acts as a low-pass for the trend and as a band-pass for the cyclical parts of the ozone time-series. Both the cut-off frequency and the bandwidth can be tuned by so called "hyperparameters", that are the ratios of the variances of the noise terms in the state-space model to the observation noise (Harvey *et al.*, 1992). The hyperparameters define the smoothness of the trend and the cyclical components in the output of the filter. The division of the ozone time-series into different components may be interpreted as a separation of processes running at separate time scales. Changes in ozone concentration at separate time scales may be linked to different atmospheric processes. For instance, the trend component describes long-term processes, such as transport of ozone. The diurnal component displays the formation of a daily ozone pattern due to the cycle of sun radiation. Shortest-term variations of the ozone time-series (<3hrs.) may be caused by changes in traffic or local turbulences. They are smoothed out by the Kalman filter, as they are considered to be at random (μ_t).

Because of the orthogonality of Fourier components, such a division in the spectral domain (ideally) yields independent parts. This property is very important for the interpretation and further investigation of the isolated components. This independence of the components enables each component to be forecasted on the basis of its own characteristics. For example, the trend is found to be a process with no clear direction and, therefore, was modelled using the random walk model. At each time point it is unknown whether the predicted value for the trend will be greater or smaller than the last observation. Hence, it is best to assume that the trend value does not change – which is precisely the forecast of the random walk model. Similarly, forecasting the amplitude of the seasonal components S_{12} and S_6 is performed using the integrated random walk model, that produces a constant linear change of the future values of the amplitude. Finally, the whole forecast is a superposition of the predicted components.

This forecasting algorithm has a number of advantages. It only needs a minimal amount of input data, namely only ozone observations of the previous day. The algorithm is easy to perform using the Kalman filter in the prediction-correction form (cf. Young *et al.*, 1991). Thus the forecasting procedure could be implemented on the computer that collects air quality monitoring data and can deliver new real-time predictions at each sampling interval.

Utilizing the explained filter algorithm 12-hour-forecasts were made for the summer 1997 period for each day based on the data of the previous day. Based on ozone time-series of the recent year the components were identified by help of spectral analysis and the hyperparameters were adjusted. This procedure was in analogy to the identification procedure for the grey-box model.

3. Data

The models have been identified, estimated and validated on summer ozone time-series for 1996 and 1997 of three differently polluted urban areas (Fig. 1): Brescia is located at the borderline between the Alps and the Po valley, in the most densely populated and industrialized region in Italy; Leipzig is in the plain industrial region of Sachsen and Sachsen/ Anhalt in Germany; and Catania is situated at the east coast of the Sicilia island (cf. also Finzi, 1998).

Figure 1. Geographical location of the three measuring regions under study to compare the prognostic models.

4. Results

The performance to forecast periods of 3 and 6 hrs was compared for the studied models in the following indices: mean value of errors E[e(t)] between forecast and observations, error variance σ_e , not explained variance σ_e^2/σ_y^2 and correlation between forecast and future observations (Corr). The percentage of true predictions (I1) and false alarms (I2) of braking a concentration threshold was calculated additionally. For Leipzig the thresholds have been choosen lower because of the lower ozone level that was observed generally in this city. In case of only few missing values (e.g. one or two hours for calibration) the Kalman filter can interpolate the data series. On the other hand, if the gaps are large the period of the gap and the following day could not be used by the Kalman filter because of a missing basis for prediction. For that reason the frequency of threshold breaks is different for the two models in Table I and II. For Catania the predictions of both models show the highest correlation to future observations due to the regular ozone profile in this city. Figure 2 displays the characteristic ozone pattern of Catania together with the prediction calculated with the component model.

In contrast, in Leipzig the ozone formation is strongly influenced by local traffic emissions, but not by regularly acting transport systems like in Catania or Brescia. Thus, the correlation between forecast and observation is the lowest for Leipzig (Table I).

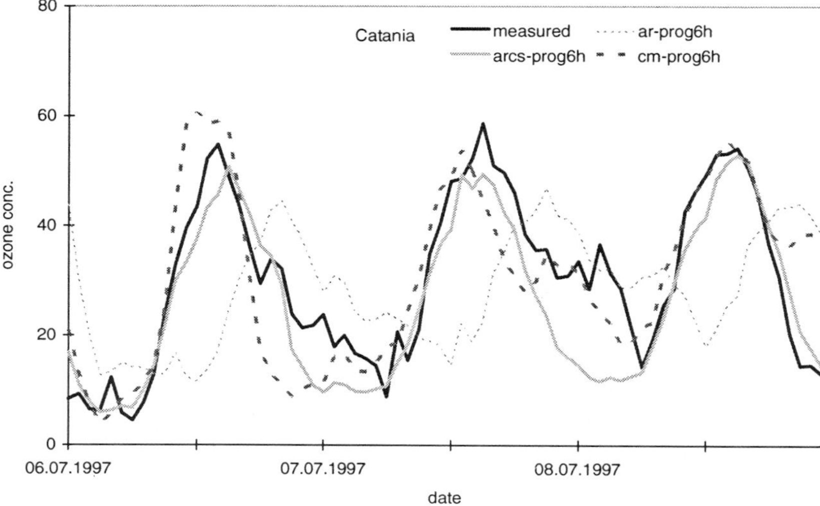

Figure 2. Comparison of the measured ozone time series (in ppb) and the 6 hrs. predictions calculated with the autoregressive (ar), grey box (arcs) and component (cm) models for Catania, July 6-8, 1997.

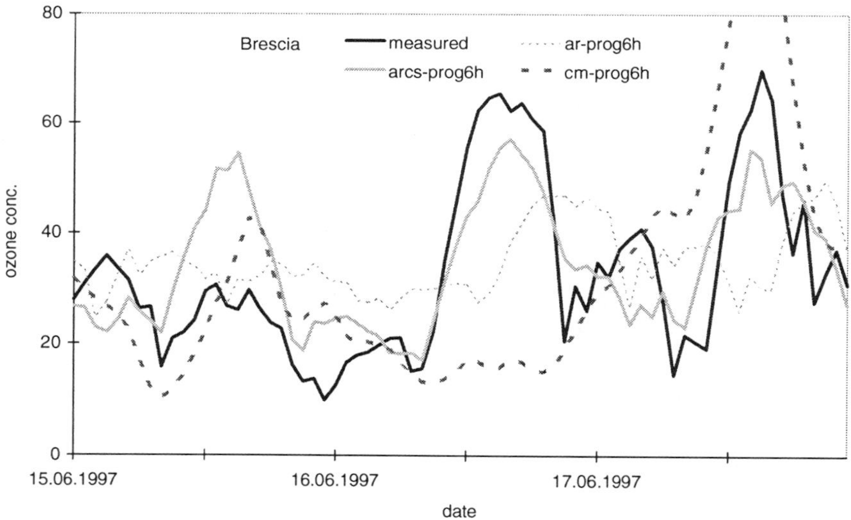

Figure 3. Comparison of the measured ozone time series (in ppb) and the 6 hrs. predictions calculated with the autoregressive (ar), grey box (arcs) and component (cm) models for Brescia, June 15-17, 1997.

The very high adaptivity of the grey box model to short-term variations (cf. Figure 2) can be realized from the low error variance (Table I). On the other hand, the component model is more 'stiff'. This property produces higher error variance (Table II) and poorer adaption in detail (Figures 2 - 4). The accuracy of predicting threshold exceedance is similar for both models. Unfortunately, an increased percentage of true predictions (I1) is generally associated to an increased percentage of false alarms (I2).

To explain the similar forecasting performance of grey box and component models we argue that both approaches are based on the assumption of a cyclical time development of ozone. Interestingly, the additional utilization of exogeneous inputs $u(t)$ in the grey-box model, like temperature or NO_x - concentration did not improve the forecasting performance essentially.

TABLE I:
Forecast performance indices of the grey box models.

Brescia, ARCS(2), stationary intervals:1-6,7-8,9-10,11-15,16-20,21-24 hrs										
Forecast	E[e(t)]	σe	σ_e^2 / σ_v^2	Corr	50 ppb:	I1	I2	70 ppb	I1	I2
3 hrs	0,798	11,52	0,534	0,687	N° of exc.	78,1	25	n° of exc.	52,9	5,4
6 hrs	1,682	12,56	0,634	0,617	443	74,9	26,1	34	38,2	4,4
Catania, ARCS(1), stationary intervals: 1-8,9-16,17-21,22-24 hrs										
Forecast	E[e(t)]	σe	σ_e^2 / σ_v^2	Corr	50 ppb:	I1	I2	70 ppb:	I1	I2
3 hrs	1,260	10,13	0,256	0,864	N° of exc.	88,3	14,5	n° of exc.	50,0	3,3
6 hrs	2,352	11,83	0,349	0,808	240	84,6	16,8	30	3,3	0,5
Leipzig, ARCS(2), stationary intervals: 1-7,8-13,14-18,19-22,23-24 hrs										
Forecast	E[e(t)]	σe	σ_e^2 / σ_v^2	Corr	45 ppb:	I1	I2	65 ppb:	I1	I2
3 hrs	-4,782	13,12	0,529	0,693	N° of exc.	79,4	35,7	n° of exc.	51,2	8,7
6 hrs	-7,728	16,53	0,840	0,499	286	77,6	55	41	48,8	16,8

5. Conclusions

Short-term ozone forecasting belongs to the most important aspects of urban air quality management. The development of new methods to predict the break of limit values is currently under research. In this context, the present study applies different techniques of statistical time-series modelling to forecast the course of hourly ozone measurements. It demonstrates that both grey box and component time series models may be used for European regions that are very different reffering to their geographical, orographical an emission conditions. Our work may be considered as one of the first steps to compare statistical models in their performance. Due to the use of the European Community directive 92/72/EEC the results may be valid for the application to air quality management systems under European conditions. The statistical models we compare refer to data that are measured directly in the region where poeple live (urban area). That means the forecasts are made for the relevant exposure of the population.

From the very different techniques of both approaches one could expect to get a different forecast performance. Indeed there are minor differences in the quality of predictions in various situations. However, surprisingly, the overall predictive power of both approaches proved to be similar. This finding confirms that both models utilise the same physical phenomena, though they are techically different. Both approaches are based on characteristic pattern in the ozone time-series and this way they utilise phenomena like ozone transport and cyclic change of meteorological and emission conditions.

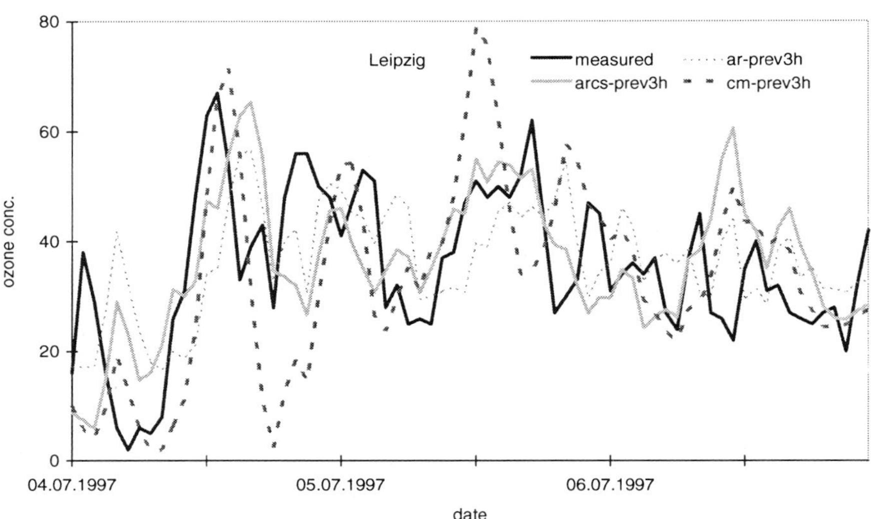

Figure 4. Comparison of the measured ozone time series (in ppb) and the 3 hrs. predictions calculated with the autoregressive (ar), grey box (arcs) and component (cm) models for Leipzig, July 4-6, 1997.

TABLE II

Forecast performance indices of the component model for Brescia, Leipzig and Catania [hyperparameters: low-pass (0.07), band-pass 12 hrs (0.00001), band-pass 6 hrs (0.001)].

Brescia										
Forecast	E[e(t)]	σe	σ_e^2/σ_v^2	Corr	50 ppb:	I1	I2	70 ppb:	I1	I2
3 hrs	0,336	13,33	0,638	0,665	n° of exc.	88,9	38,6	n° of exc.	80,7	11,1
6 hrs	0,557	15,38	0,849	0,559	559	86,8	47,2	57	75,4	13,7
Catania										
Forecast	E[e(t)]	σe	σ_e^2/σ_v^2	Corr	50 ppb:	I1	I2	70 ppb:	I1	I2
3 hrs	0,419	10,83	0,374	0,807	n° of exc.	79,6	9	n° of exc.	60	3,6
6 hrs	1,140	12,40	0,490	0,728	240	68,7	10	30	23,3	2,7
Leipzig										
Forecast	E[e(t)]	σe	σ_e^2/σ_v^2	Corr	45 ppb:	I1	I2	65 ppb:	I1	I2
3 hrs	0,050	15,07	0,703	0,633	n° of exc.	80,2	31,5	n° of exc.	48,6	9
6 hrs	-0,389	18,58	1,069	0,452	262	76,7	41,8	37	32,4	13,9

Changes in precursor concentrations, like NO_x, due to local turbulences or changed intensity of traffic may modify ozone concentration as well, but only at shortest time scales. Such variations were smoothed out by both modelling approaches as they are considered to be not essential for the prediction of episodes. In contrast, one can suppose that the prediction of the latter may be improved by taking into consideration the development of weather systems. Further research may clarify the association between synoptical situations and the structure of the statistical model.

Acknowledgements

The authors are grateful to Prof. Giovanna Finzi for helpful discussions, Mr. D. Bergoli for assistance in calculations and Dr. Oliveri (Comune di Catania), Dr. Goffi (Provincia di Brescia) and Mr. Gräfe (Sächsisches Landesamt für Umwelt und Geologie, Dresden) who provided the data sets for the study. This research has been partially supported by Italian MURST and Provincia di Brescia (I). The research was done in the frame of the SATURN project while the first author was visiting the University of Brescia and he gratefully acknowledges the financial support of the UFZ-Centre for Environmental Research, Leipzig-Halle Ltd. (D).

References

Finzi, G. and Tebaldi, G.: 1982, 'A mathematical model for air pollution forecast and alarm in an urban area', *Atmospheric Environment* **16**, 2055.

Finzi, G., Bonelli, P., and Bacci, G.: 1984, 'A stochastic model of surface wind speed for air quality control purposes', *Journal of Climate and Applied Meteorology* **23**, 1354.

Finzi, G., Volta, M., Nucifora, A., and Nunnari, G.: 1998, 'Real time ozone forecast: a comparizon between neural network and grey box models', *International ICSC/IFAC Symposium on Neural Computation*, Wien, Sept. 23-25.

Harvey, A.C.: 1992, 'Forecasting, Structural Time Series Models and the Kalman Filter', Cambridge University Press, Cambridge, 1992.

Kendall, M. and Ord, J.K.: 1993, 'Time Series', Edward Arnold.

Ng, C.N., and Young, P.C.: 1990, 'Recursive Estimation and Forecasting of Non-stationary Time Series', *Journal of Forecasting* **9**, 173.

Schlink, U., Herbarth, O., and Tetzlaff, G.: 1997, 'A component time-series model for SO_2 data: forecasting, interpretation and modification', *Atmospheric Environment* **31(9)**, 1285.

Schlink, U., Herbarth, O., Richter, M., Rehwagen, M., Puliafito, J.L., Puliafito, E., Puliafito, C., Guerreiro, P., Quéro, J.L., and Behler, J.C.: 1999, 'Ozone-Monitoring in Mendoza, Argentina: initial results', *Journal of the Air & Waste Management Association* **49**, 82.

Slutzky E.: 1927, 'The summation of random causes as the source of cyclic processes' (Russian), *Problems of Economic Conditions* **3**, 1; English translation in *Econometrica* (1937), **5**, 105.

Young, P.C., Ng, C.N., Lane, K., and Parker, D.: 1991, 'Recursive forecasting, smoothing and seasonal adjustment of non-stationary environmental data', *Journal of Forecasting* **10**, 57.

Yule, G.U.: 1927, 'On a method of investigating periodicities in disturbed series, with special reference to Wölfer's sunspot numbers', *Phil. Trans.* **A226**, 267.

OSPM - A PARAMETERISED STREET POLLUTION MODEL

RUWIM BERKOWICZ
National Environmental Research Institute, Department of Atmospheric Environment, Frederiksborgvej 399, DK-4000, Roskilde, Denmark
E-mail: rb@dmu.dk

Abstract. For many practical applications, as e.g. in support of air pollution management, numerical models based on solution of the basic flow and dispersion equations are still too complex. Alternative are models that are basically parameterised semi-empirical models making use of a priori assumptions about the flow and dispersion conditions. However, these models must, be thoroughly tested and their performance and limitations carefully documented. The Danish Operational Street Pollution Model (OSPM) belongs to this category of parameterised models. In the OSPM, concentrations of exhaust gases are calculated using a combination of a plume model for the direct contribution and a box model for the recirculating part of the pollutants in the street. Parameterisation of flow and dispersion conditions in street canyons was deduced from extensive analysis of experimental data and model tests. Results of these tests were used to further improve the model performance, especially with regard to different street configurations and a variety of meteorological conditions.

Key words: street pollution models, OSPM, parameterised models, street-canyon, traffic pollution, model validation

1. Introduction

Significant increase in computer power makes it now possible to use advanced 3D numerical models for air pollution studies. However, for many practical applications, as e.g. in support of air pollution management, the numerical models based on solution of the basic flow and dispersion equations are still too complex. We must also bear in mind that the quality of input data, such as e.g. emissions, is often not sufficient in order to justify application of the very complex numerical tools. Alternative are models that are basically parameterised semi-empirical models making use of a priori assumptions about the flow and dispersion conditions. These models must, however, be thoroughly tested and their performance and limitations carefully documented. The Danish Operational Street Pollution Model (OSPM), belongs to this category of parameterised models. Parameterisation of flow and dispersion conditions in street canyons was deduced from extensive analysis of experimental data and model tests. The basic features of the model are described in this paper and a number of examples of evaluation studies are presented. The problem of uncertainties in model results is also addressed.

Environmental Monitoring and Assessment **65**: 323–331, 2000.

2. Description of the model

The most characteristic feature of the street canyon wind flow is the formation of a wind vortex, so that the direction of the wind at street level is opposite to the flow above roof level, as illustrated in Figure 1.

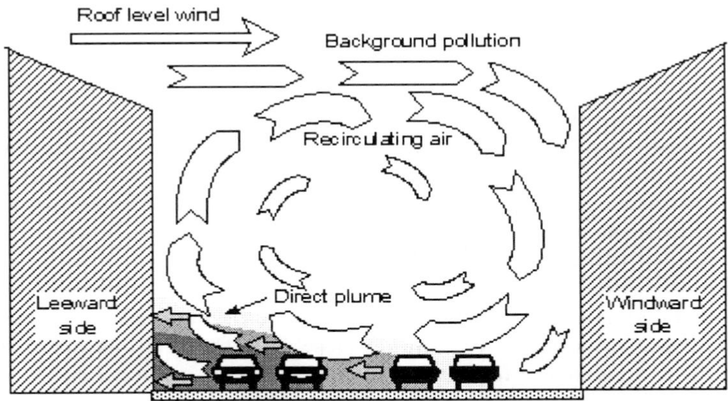

Figure 1. Schematic illustration of flow and dispersion conditions in street canyons.

These special flow conditions leads to a situation in which the pollutants emitted from traffic in the street are primarily transported towards the upwind building (leeward side) while the downwind side is primarily exposed to background pollution and pollution that has recirculated in the street.

In one of the earliest street pollution models, the STREET model by Johnson *et al.* (1973) this phenomena was described using two different expressions for the leeward and the windward sides of the street. The model predicts thus that the concentrations on the leeward side of the street are higher than on the windward side. These are the most essential features of pollutant dispersion in street canyons and therefore the STREET model, with some minor modifications is still widely used, especially for engineering applications. The more detailed features of pollution dispersion in street canyons can, however, not be described by such a simplified model as STREET. An innovative approach was introduced by Yamartino and Wiegand (1986) in their Canyon Plume-Box Model (CPBM). In this model the concentrations are calculated combining a plume model for the direct impact of vehicle emitted pollutants with a box model that enables computation of the additional impact due to pollutants recirculated within the street by the vortex flow. OSPM is based on similar principles (Hertel and Berkowicz, 1989a; Berkowicz *et al.*, 1997; Berkowicz, 1998).

2.1 FLOW AND DISPERSION

The main assumptions incorporated in the model are:

- The vortex is formed in the street whenever a wind component perpendicular to the street axis exists. The length of the vortex, calculated along the wind direction, is 2 times the upwind building height. For roof level wind speeds below 2 m/s, the length of the vortex decreases gradually with the wind speed. The buildings along the street may have different heights. Because the length of the vortex is assumed to be proportional to the upwind building height, this length will depend on the wind direction as well.

- The upwind receptor (leeward side) receives contribution from the traffic emissions within the area occupied by the vortex (the recirculation zone), the recirculated pollution and a portion of the emissions from outside of the vortex area. This portion is modelled by a wind speed and wind direction dependent factor, R, varying continuously from zero to one as wind direction changes from 0 to 90 degrees with respect to the street axis. The wind speed dependence is such that R approaches unity for vanishing wind speeds regardless of the wind direction.

- The downwind receptor (windward side) mainly receives contributions from the recirculated component, and if the vortex does not occupy the whole street, the traffic emissions from outside of the recirculation zone also contribute.

- As the wind speed approaches zero or is parallel to the street, concentrations on both sides of the street become equal.

- The direct contribution from within the recirculation zone is calculated with a plume model assuming linear dispersion of pollutants with the distance. Traffic emissions are assumed to be homogeneously distributed across the street.

- The recirculation part is described by a box model and concentrations are computed assuming equality of the incoming and outcoming pollution flux. The incoming flux is set equal to the traffic emission while the outcoming flux is governed by the turbulence at the top of the street.

An important feature of OSPM is modelling of the turbulence in the street. The turbulence in the street is assumed to be composed of two parts: a part dependent on wind speed (ambient turbulence) and a part due to traffic induced turbulence. The last dominates when the wind speed is low. Because the highest concentrations usually occur in near calm wind situations, appropriate modelling of these conditions is crucial (Kastner-Klein *et al.*, 1998; Ketzel *et al.*, 1999).

Wind meandering affects the dependence of the street concentrations on wind direction. In OSPM the wind meandering is taken into account by averaging the calculated concentrations over a wind speed dependent wind direction range. At low wind speeds, the averaging angle is large and the calculated concentrations are practically not dependent on the mean wind direction.

2.2 MODELLING OF NO_2 FORMATION IN STREETS

Transport and dispersion processes are not the only factors determining relationships between emission sources and ambient concentrations. Chemistry plays a crucial role in transformation of pollutants resulting in degradation of some species and formation of others. Considering transport of pollution on a larger scale, when the transport times involved are of the order of hours or even days, hundreds of chemical reactions must often be considered in order to account for the chemical composition of the air. The situation is quite different when dealing with processes in street canyons. Due to the very short distances between the sources and receptors, only the fastest chemical reactions can have any significant influence on the transformation processes in the street canyon air. It means that most of the pollutants emitted from traffic can be considered as inert components for which chemical transformations inside the street canyon are unimportant. On these time scales, such inert compounds are carbon monoxide (CO) and many hydrocarbons, which actually constitute the main composition of car exhaust gases. The situation is however different for nitrogen oxide gases which actually are the compounds most often considered in connection with impact of traffic pollution on human health. The main nitrogen oxides are nitrogen monoxide (NO) and nitrogen dioxide (NO_2), the sum of which is denoted as NO_x. Regarding health effects, NO is considered to be harmless, at least at concentrations expected in urban air. On the contrary, NO_2 can have severe adverse health effects on humans. Only a small portion of NO_x gases emitted by motor vehicles is in the form of NO_2, the main part being NO. The presence of NO_2 in ambient air is mainly due to the subsequent chemical oxidation of NO by ozone (O_3). Under sunlight conditions, photodissociation of NO_2 leads to partial reproduction of NO and O_3.

$$NO + O_3 \Leftrightarrow NO_2 + O_2 \tag{1}$$

The time scales characterising these reactions are of the order of tens of seconds, thus comparable with residence time of pollutants in a street canyon. Consequently, the chemical transformations and exchange of street canyon air with the ambient air are of importance for NO_2 formation. Modelling of these processes is implemented in OSPM (Hertel and Berkowicz, 1989b; Palmgren *et al.*, 1996) utilising an analytical solution of the kinetics of the reaction scheme (1). Exchange with the background air is modelled taking into account the residence time of pollutants in the street. Urban background O_3 and NO_2 play important role in NO_2 formation in the street air.

3. Results

OSPM has been extensively tested on field data (Hertel and Berkowicz, 1989a; Berkowicz *et al.*, 1997; Berkowicz, 1998). Results of these tests were used to further improve the model performance, especially with regard to different street configurations and a variety of meteorological conditions.

In this paper some recent model test results are shown using data from a monitoring stations in the street Albanigade in Odense, Denmark and Schildhorstrasse in Berlin, Germany. Albanigade is a street with somewhat irregular building architecture and only the area very close to the monitoring station can be considered as a street-canyon. Schildhornstrasse is a street with a very intense traffic (about 50000 vehicles per day) and is in a direct prolongation of a major highway.

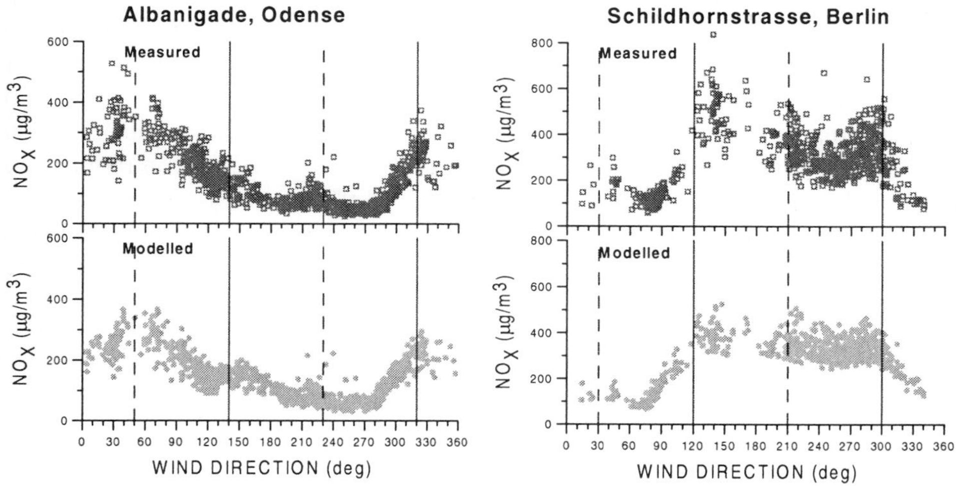

Figure 2. Wind direction dependence of measured and modelled concentrations. Only daytime hours and observations with wind speed between 4 and 7 m/s are shown.

In Figure.2 is shown a plot of measured and modelled NO_x concentrations as function of wind direction for those two streets. To reduce the variation in concentrations that is due to variation in the emissions, only daytime hours (from 7 to 18) are shown. Additionally, only cases with wind speed between 4 and 7 m/s are selected. Wind directions parallel and perpendicular with the street are indicated by continuous and dashed lines, respectively. It is seen that the model reproduces the wind direction dependence quite well, although not all the structure is resolved.

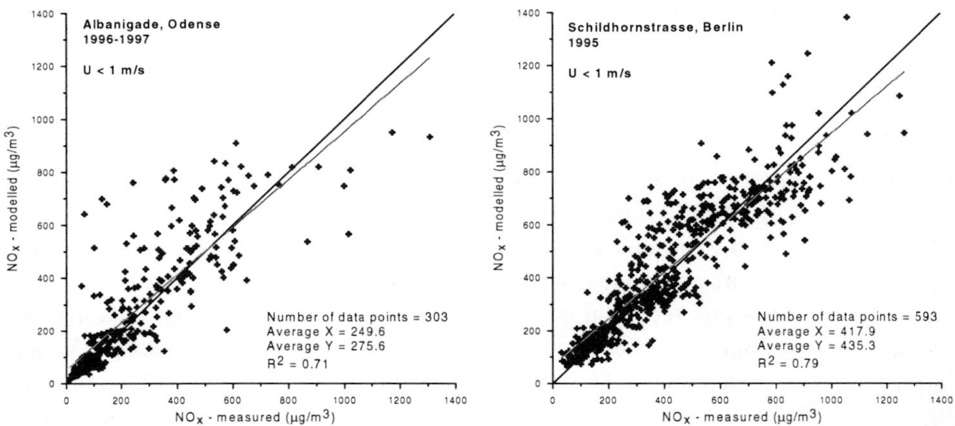

Figure 3. Comparison of measured and modelled concentrations in the case of wind speed <1 m/s

In Figure.3 results are shown for all the cases when wind speed is < 1 m/s. This is an important situation, when the traffic induced turbulence is believed to dominate dispersion of traffic emissions (Kastner-Klein *et al.*, 1998). The scatter of data shown in this figure is large, indicating that the modelling of traffic induced turbulence still needs to be improved or other dispersion mechanisms important at low wind speeds should be explored. However, an essential feature of the shown results is that the highest concentrations are reasonable well reproduced by the model. The coefficient of prediction is quite good for both streets (R^2=0.71 and 0.79) and the average of the modelled concentrations is close to the observed (the values are shown in the corresponding figures).

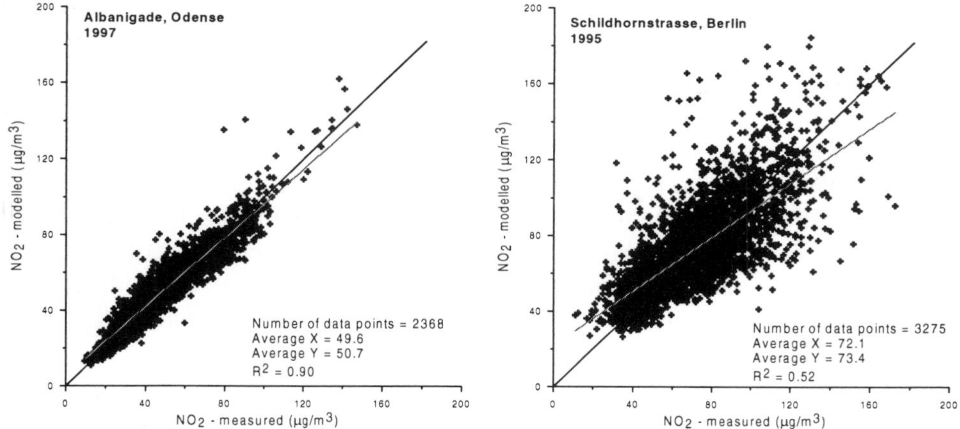

Figure 4. Comparison of measured and modelled concentrations of NO_2. Daytime hours only.

Comparison of measured and modelled NO_2 concentrations is shown in Fig. 4 (daytime hours only). The agreement is very good for Albanigade (R^2=0.90) but the scatter is much larger for Schildhornstrasse (R^2=0.52). A possible explanation of the large scatter apparent in the data from Schildhornstrasse might be that the urban background ozone concentrations used in the calculations are estimated as an average of measurements from 3 monitoring stations, all located in the suburbs of Berlin. These data may not always be representative for the actual urban ozone background. The sensitivity of street NO_2 concentrations to the urban background O_3 concentrations is well documented (Palmgren *et al.*, 1996).

An example of a recent application OSPM for a street in Helsinki, Finland is presented in Kukkonen *et al.* (1999).

4. Discussion and Conclusions

Comparisons between calculated and measured concentrations showed that a simple parameterised model like OSPM can successfully be applied for prediction of street traffic pollution. An important feature is the ability to predict pollution concentrations at low wind speed conditions, which normally lead to the most severe pollution episodes. Modelling these situations is often more complicated in advanced CFD models and is not easy to simulate in wind-tunnels. Simplifications applied in parameterisation of OSPM lead, however, to some deficiencies in the ability to resolve the fine structure in the concentration dependence on wind direction. Here, the CFD models might be superior. It is also obvious that parameterised models cannot provide insight into new phenomena. Physically well-founded CFD models are required for such studies.

Currently OSPM is compared with results from more advanced numerical models with the aim to improve the physical concept of the model parameterisation (Ketzel *et al.*, 1999). This study is conducted in the framework of the European research network on Optimisation of Modelling Methods for Traffic Pollution in Streets (TRAPOS). One of the conclusions from this study is that advanced numerical models still not adequately describe several important phenomena and empirical relationships might be required. This is particularly valid for modelling low wind speed conditions, when the traffic induced turbulence is of a great importance.

The quality of the predictions of NO_2 concentrations depends highly on the availability of representative urban background concentrations for NO_2 as well as O_3. Good results obtained for the location in the street Albanigade show that the implemented method is sufficiently accurate for prediction of formation of NO_2 in streets. In the case of Schildhornstrasse, the ozone background concentrations were estimated from measurements made at a remote location and this resulted in a much larger scatter than in the case of Albanigade. A simple model for estimation of urban background concentrations is presented in Berkowicz (1999). This model can be used in the case when measurements are not available.

It should also be noted that OSPM was successfully applied for analyses of measurements from routine monitoring stations with the aim to estimate the trends in traffic emissions (Palmgren *et al.*, 1999). Due to very modest computer time requirements, analyses of data from a long (several years) time period can be undertaken in few minutes.

Acknowledgements

Development and testing of OSPM was only possible due to availability of extensive measurements from Air Quality Monitoring Networks. Data from Albanigade were provided from the Danish National Air Quality Monitoring Programme and the assistance of Drs. Finn Palmgren and Kåre Kemp is highly appreciated. Data from Schildhornstrasse are kindly provided by Peter Lenschow from the Senatsverwaltung für Stadtentwicklung, Umweltschutz und Technologie, Berlin, Germany. Application of OSPM on these data is a part of the European Commission's project Auto Oil II supervised by Dr. Andreas Skouloudis from the Joint Research Centre, Ispra, Italy. Permission to use these data for this study is highly acknowledged.

References

Berkowicz, R., Hertel, O., Sørensen, N.N. and Michelsen, J.A.: 1997, Modelling air pollution from traffic in urban areas, In R.J. Perkins and S.E. Belcher (eds), Flow and Dispersion Through Groups of Obstacles, pp 121-141, Clarendon Press, Oxford, pp 121-141.

Berkowicz, R.: 1998, Street Scale Models, In J. Fenger, O. Hertel, and F. Palmgren (eds.), Urban Air Pollution - European Aspects, Kluwer Academic Publishers, pp. 223-251.

Berkowicz, R.: 1999, A simple model for urban background pollution, 2nd International Conference on Urban Air Quality, Madrid, 3-5 March 1999.

Hertel, O. and Berkowicz, R.: 1989a, Modelling pollution from traffic in a street canyon. Evaluation of data and model development, *DMU Luft* **A-129**, pp77.

Hertel, O. and Berkowicz, R.: 1989b, Modelling NO_2 concentrations in a street canyon, *DMU Luft* **A-131**, pp31.

Johnson, W.B., Ludwig, F.L., Dabbert, W.F. and Allen, R.J.: 1973, An urban diffusion simulation model for carbon monoxide, *JAPCA*, **23**, 490-498.

Kastner-Klein, P., Berkowicz, R. and Plate, E.J.: 1998, Modelling of vehicle induced turbulence in air pollution studies for streets, In Proceedings of 5th International Conference on Harmonisation within Atmospheric Dispersion Modelling for Regulatory Purposes, 18-21 May 1998, Rhodes, Greece (to be published in . *Int. J. Environment and Pollution*)

Ketzel, M., Berkowicz, R. and Lohmeyer, A.: 1999, Comparison of numerical street dispersion models with results from wind tunnel and field measurements, 2nd International Conference on Urban Air Quality, Madrid, 3-5 March 1999.

Kukkonen, J., Valkonen, E., Koskentalo, T., Walden, J., Karppinen, A. and Berkowicz, R.: 1999, Measurements and modelling of air pollution in a street canyon in Helsinki, 2nd International Conference on Urban Air Quality, Madrid, 3-5 March 1999.

Palmgren, F., Berkowicz, R., Hertel, O. and Vignati, E.: 1996, Effects of reduction of NO_x on NO_2 levels in urban streets, *Sci. Total Environ.*, **189/190**, 409-415.

Palmgren, F., Berkowicz, R. Ziv, A. and Hertel, O.: 1999, Actual car fleet emissions estimated from urban air quality measurements and street pollution models, *The Science of the Total Environment*, **235**, 101-109.

Yamartino, R.J. and Wiegand, G.: 1986, Development and evaluation of simple models for flow, turbulence and pollutant concentration fields within an urban street canyon, *Atmospheric Environment*, **20**, 2137-2156.

AN INTEGRATED APPROACH TO STREET CANYON POLLUTION MODELLING

PAUL S. ADDISON, JOHN I. CURRIE, DAVID J. LOW & JOANNA M. McCANN
Transport Research Institute, Napier University, Edinburgh.

Abstract. An integrated method for the prediction of the spatial pollution distribution within a street canyon directly from a microscopic traffic simulation model is outlined. The traffic simulation package Paramics is used to model the flow of vehicles in realistic traffic conditions on a real road network. This produces details of the amount of pollutant produced by each vehicle at any given time. The authors calculate the dispersion of the pollutant using a particle tracking diffusion method which is superimposed on a known velocity and turbulence field. This paper shows how these individual components may be integrated to provide a practical street canyon pollution model. The resulting street canyon pollution model provides isoconcentrations of pollutant within the road topography.

Key Words: street canyon, dispersion, pollution concentration, traffic microsimulation

1. Introduction

Microscopic car simulation models provide a realistic measure of traffic flow on a network and allow for the type and number of vehicles to be varied. Each vehicle is autonomous and follows a set of mathematical rules according to its type (Low and Addison 1998a,b; Addison, 1997). The simulation model determines each vehicle's car-following and overtaking behaviour and includes a statistical model of driver awareness and aggressiveness. In this way each individual vehicle responds to the geometry of the road network, and the presence of other road vehicles in a realistic fashion. The emissions data these models provide can be collected for individual sections of roads and used as initial input to dispersion models. A street canyon dispersion model is currently under development by the authors to run in conjunction with a microscopic traffic simulation model and provide isoconcentrations of pollutant within the road topography. Dispersion of the pollutant is calculated using a particle tracking diffusion method which is superimposed on a known velocity and turbulence field. Particle tracking methods (Addison et al, 1997; Addison et al, 1998) are extremely popular for modelling the Lagrangian dispersion characteristics of pollutants in a variety of fluid flow fields due to their flexibility and ease of use. The modelling of the production, emission and subsequent fate of the pollutants within a street canyon is extremely difficult due to a variety of factors. A full discussion of the relative merits of existing models is outwith the scope of this paper. For more information the reader is referred to Addison et al (1999) and the extensive references therein.)

Environmental Monitoring and Assessment **65**: 333–342, 2000.

In developing an integrated model of street canyon pollution consideration must be given to an appropriate choice of the following:

1 Microscopic traffic simulation model.
2 Vehicle emissions model.
3 Canyon velocity field: advection model.
4 Canyon velocity field: turbulence model.
5 Vehicle wake model to describe both the traffic induced turbulence and associated initial mixing of the pollutants behind the vehicle.
6 Atmospheric model including solar flux and mean wind speeds.
7 Dispersion model, e.g. Fickian, non-Fickian, particle tracking, solution of an advection-diffusion equation.
8 Canyon geometry model, due to the computational cost associated with refining the geometric grid.

The authors believe that in order to produce a reasonably accurate, site specific simulation of the concentration profiles within street canyons a global approach must be taken which integrates on-site measurements with computational models of the canyon. This paper details the initial attempts at producing a fully integrated approach to street canyon pollution modelling: from the microscopic simulation of traffic to the prediction of isoconcentrations of pollution within the canyon.

2. Pollution Data from 'Paramics'

The Paramics traffic simulation model, developed over the past six years by SIAS Ltd and Quadstone Ltd, is used here. For each vehicle travelling on a real road network it determines the instantaneous emission rate of carbon monoxide, carbon dioxide, hydrocarbons, oxides of nitrogen and particulate matter. This calculation is performed at every simulation time step and the amount of pollutant emitted along a link (a section of road between two of the points, or nodes, used by Paramics to define the network geometry) determined by summing the contribution from every vehicle on that link. Paramics allows the user to specify the length of time for which a record is produced. For each link Paramics details its length, the number of vehicles that travelled along it, the amount of fuel used by these vehicles, and the total amount of each pollutant emitted. (Table 1 shows a typical record). The emission rate is defined as a function of instantaneous speed, instantaneous acceleration, vehicle type (non-catalyst petrol, catalyst petrol, diesel), and engine size (less than 1400cc, between 1400cc and 2000cc, greater than 2000cc) and is based on data supplied by TRL.

In this study Paramics was used to simulate traffic flow around the Fountainbridge area of Edinburgh (see Figure 1) and the emissions from cars and light goods vehicles recorded over a two hour period. Table 1 shows data collected along one road of the network: Detectors 101 to 107 are on various links along this road and measure emissions from traffic on the westbound carriageway. Detectors 201 to 207 relate to the eastbound carriageway. The statistics from detector 101 (indicated by an arrow in figure 1) were used to calculate the average amount of carbon monoxide emitted per vehicle per metre length of road and this amount was used to represent an instantaneous point source release in the canyon dispersion model.

Table I

Typical Pollution Output from Paramics

Detector	Link (2 nodes)	Length (m)	Carbon Monoxide (mg)	Vehicle Count
101	**25b–106**	**90.82**	**2203551.000**	**1829**
102	76–23	93.52	1826275.500	1820
103	112–113	53.85	1248582.250	1558
104	22–105	59.64	1256641.125	1282
105	105–21	128.22	1421548.750	1292
106	21–20	59.94	1011953.250	1274
107	14–15	100.00	1816631.750	1185

Link-pollution after 2 hours using the basic 2 by 2 pollution matrix. Both light and heavy vehicles are on the network but only light vehicles are counted, and their pollution detected.
Note: Emission used in this paper shown in bold.

3. The Canyon Velocity Field

The wind flow within the street canyon being modelled is clearly important in determining the distribution of pollutant in the canyon. However, for the purposes of demonstrating our integrated approach it is sufficient, at this stage, to employ a rather naïve canyon velocity model. This model is the combination of both the mean velocity field (used for advection) and the field of turbulent velocity fluctuations (used for dispersion). Details of these component fields used are given below.

3.1 The Pollutant Transport Model

The dispersion of the pollutant is simulated using a Lagrangian stochastic particle model proposed by Thomson (1985). The mass of pollutant is represented by a collection of particles each defining a specific mass. Each particle is assumed to move independently of the others and its trajectory is described by

$$x_{n+1} = x_n + \left[U(x_n) + s_n\right]\Delta t \qquad [1]$$

At each time step the particle assumes the local mean wind speed, $U(x_n)$, and an additional velocity, s_n, representing turbulent velocity fluctuations. The velocity fluctuation s_n is given by the following three equations, where the first equation describes the fluctuation as if the turbulence were homogeneous, whilst the other two deal with any inhomogeneous behaviour. The latter two equations are applied alternately.

$$s_n^i = s_n'^i - \Delta t T^{ij} s_n'^j + \mu_n^i \qquad [2]$$

$$s_n'^i = s_{n-1}^i + \Delta t\left[\frac{\partial \sigma^{ij}}{\partial x^j} + \left(2\frac{\sigma^{ij}}{\rho}\right)\frac{\partial \rho}{\partial x^j}\right] \qquad [3]$$

$$\chi^{ij}(x_n)s_n'^j = \chi^{ij}(x_{n-1})s_{n-1}^j \qquad [4]$$

μ_n is a random variable drawn from a Gaussian distribution, σ^{ij} is the ijth component from the velocity covariance tensor, T^{ij} is the ijth component from the inverse matrix of local Lagrangian timescales.

The following conditions apply to the particle tracking model used:

1 - Particles are assumed to undergo elastic collisions with the canyon walls and ground.
2 - No deposition is assumed.
3 - Once a particle has attained a height greater than the canyon (in this case 40m) it is deemed to have 'escaped' and does not re-enter the canyon.
4 - Turbulence is assumed to be homogeneous.
5 - No chemical reactions are assumed to take place between pollutants.

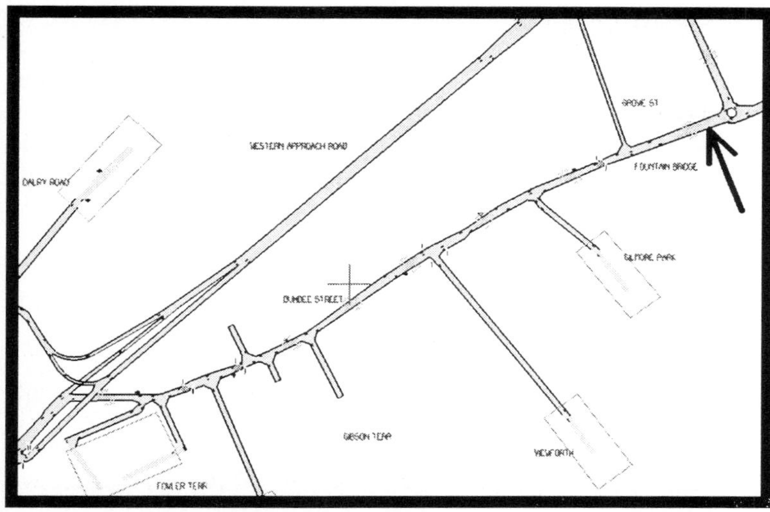

Figure 1: Paramics Model Area (The arrow indicates the location of the pollution monitoring site for the results presented herein.)

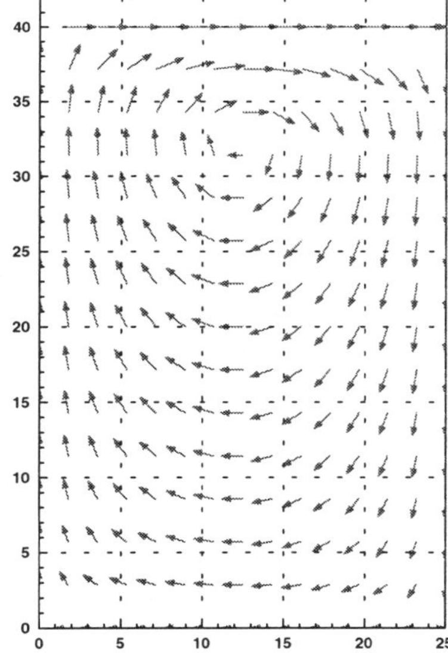

Figure 2: Mean velocity field in a 40m by 25m canyon as simulated by the Hotchkiss and Harlow model with a roof top wind speed of 5m/s

3.2 The Advective Velocity Field

A simple two dimensional model produced by Hotchkiss and Harlow (1973) is used herein to calculate the mean wind field in a canyon. The model describes a single vortex entirely contained within the canyon which although an over simplification defines the essential mechanism for dispersion. The velocity components are dependent on the canyon dimensions: height, *H*; width, *B;* and roof level wind speed, u_0.

$$u = \frac{A}{k}\left(e^{zy}(1+zy) - \beta e^{-kz}(1-kz)\right)\sin(kx) \qquad [5]$$

$$v = -Ay\left(e^{kz} - \beta e^{-kz}\right)\cos(kx) \qquad [6]$$

where x/z are the horizontal/vertical coordinates respectively and

$$k = \frac{\pi}{B} \quad \beta = e^{-2kH} \quad A = \frac{ku_0}{1-\beta}$$

The model does not allow for complexities such as non-flat roofs or asymmetrical canyons but Yamartino and Wiegand (1986), on comparing the model with a complex two-dimensional model, concluded that that the *H-H* model was satisfactory. Further, Yamartino and Wiegand (1986) give details of extending the model to take into account flow along the canyon employing a simple logarithmic velocity profile.

3.3 The Turbulent Velocity Field

The ***covariance tensor*** $\sigma(x,t)$ is a tensor whose components are the covariances of the turbulent velocities at (x,z). Yamartino and Wiegand (1986) suggest an empirically determined formula for the diagonal components:

$$d_i = \left(\sigma^{ii}\right)^{1/2} = f^i(x,z)\left[A_m^i\left(s_r^2 + \alpha^{i2} v_r^2\right)^{1/2} + \left(A_c^i + A_h h\right)\right], \quad i = u, v, w \qquad [7]$$

$$s_r^2 = u_r^2 + v_r^2 \qquad [8]$$

$$h = S + \frac{Nae_a}{B} \qquad [9]$$

where u_r, and v_r are the mean wind components measured at a reference height r. A_m and α characterise mechanical turbulence induced by the mean wind field; A_c characterises night time turbulence under the assumption of no traffic; A_h describes the turbulence induced by solar radiation and vehicle generated heat flux. The function $f(x,z)$ describes the spatial variability of the turbulence field. Herein, again for simplicity, we have used Yamartino and Wiegand's values for the constants.

The ***time scale tensor*** T^{ij} used is that suggested by Lanzani and Tamponi (1995) who propose the following form for the diagonal components of the inverse time scale tensor:

$$\frac{1}{T^{xx}} = \frac{kB}{d_u^*} \quad \frac{1}{T^{yy}} = \frac{k\min(B,H)}{d_v^*} \quad \frac{1}{T^{zz}} = \frac{kH}{d_w^*}$$

where d_u*, d_v*, and d_w* are the mean values of d_u, d_v, and d_w, averaged over the canyon.

4. Results

Figure 3 shows the initial results from a point source emission of pollutant within the canyon. The pollutant is released at a height of 0.5m in the centre of the canyon, and the subsequent isoconcentration contours are plotted through times 5 to 200 seconds since release. The two dimensional cross-sectional distribution pattern of the plume through time can clearly be seen. The advective velocity field of figure 2 is used. Noticeable from the plots is both the dispersion of the pollutant and clockwise mean advection of the cloud through time. A distinct canyon effect is shown: that is, wind at roof-top level, blowing from left-to-right, generates a recirculation vortex which moves the street-level pollution from right-to-left, and produces higher concentrations on the leeward side of the canyon.

5. Conclusions

The authors have outlined the basis for an integrated street canyon pollution model, incorporating a variety of component modules. A very simple example has been provided, in which pollutant emissions taken from a microscopic traffic simulation model have been used to produce time series of the pollutant isoconcentrations within a street canyon.

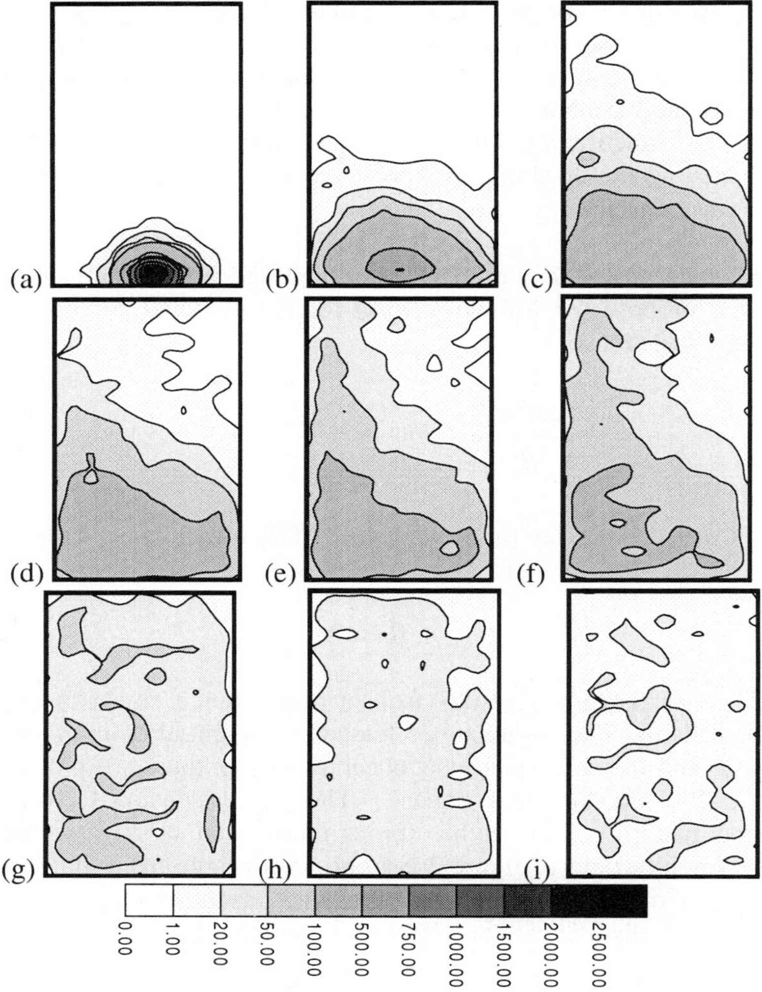

Figure 3: Development through time of the instantaneous release of 13.27mg of carbon monoxide in a 40 by 25 metre canyon. Concentration contours given in mg/m^3×1000 at times (a) 5, (b) 10, (c) 15, (d) 20, (e) 25, (f) 30, (g) 60, (h)100, and (i) 150 seconds after release. Note: Nonlinear scaling of concentration contour levels.

The authors have shown that it is possible to combine a variety of modular components to produce an integrated model. The modular approach described herein forms the basis of future work by the authors. The integrated model described can be extended to incorporate complex canyon geometries, non-stationary velocity fields (including inhomogenous turbulence), sophisticated wake models and dispersion models. Further work will concentrate on the exact form of these component modules. The integrated model needs to be developed together with roadside data which (significantly) includes spatial velocity measurements. The authors plan to undertake such a roadside study to measure simultaneously the traffic flows, wind velocities and pollution concentrations within a street canyon in order to calibrate their integrated model.

Acknowledgements

We acknowledge the assistance of SIAS Ltd, Edinburgh for providing the Paramics software used in the study. The Paramics data set used in this paper is the property of Scottish and Newcastle plc. The work was partly funded by EPSRC grant GR/L36086.

References

Addison P. S. (1997), EPSRC Grant GR/L3608 , '*Investigation of a Separation-Distance Centred Non-Linear Car-Following Model*'.

Addison P. S., Qu B., Nisbet A. & Pender G. (1997), A non-Fickian particle-tracking diffusion model based on fractional Brownian motion, *Int. J. Numerical Methods in Fluids*, **25**, pp 1373 - 1384.

Addison P. S., Qu B., Ndumu A. S. and Pyrah I. (1998), A particle tracking model for non-Fickian subsurface diffusion, *Math. Geology*, **16**, pp 127 - 151.

Addison P.S., McCann J.M., Low D.J. and Currie J.I., (1999) 'An Integrated Approach to Modelling Traffic Pollution in the Urban Environment', *Traffic Engineering and Control*. (In Press.)

Hotchkiss R. S. & Harlow. H. (1973), 'Air pollution in street canyons', *Rep. EPA-R4-73-029, U.S. Env. Protection Agency*, Washington D.C., United States.

Low D. J. & Addison P. S. (1998a), The Complex Dynamical Behaviour of Congested Road Traffic, *3rd IMA Conf. on Mathematics in Transport Planning and Control*, Cardiff Wales, April 1-3.

Low D. J. & Addison P. S. (1998b), A Nonlinear Temporal Headway Model of Traffic Dynamics, *Nonlinear Dynamics*, **16**, pp 127 - 151.

Lanzani G. & Tamponi M. (1995), A microscale Lagrangian particle model for the dispersion of primary pollutants in a street canyon, *Atmospheric Environment*, **29**, (23), pp 3465 - 3475.

Thomson D. J. (1985), A random walk model of dispersion in turbulent flows and its application to dispersion in a valley, *Q.J.R. Met. Soc.,* **112**, pp511 - 530

Yamarito R. J. & Wiegand G. (1986), Development and evaluation of simple models for the flow, turbulence and pollutant concentration fields within an urban street canyon, *Atmospheric Environment,* **20** (11), pp 2137 - 21562

DISPERSION OF POLLUTANTS IN STREET CANYON UNDER TRAFFIC INDUCED FLOW AND TURBULENCE

MIROSLAV JICHA, JIRI POSPISIL and JAROSLAV KATOLICKY
Technical University of Brno, Faculty of Mechanical Engineering, Institute of Power Engineering, Department of Thermodynamics & Environmental Engineering, Technicka 2, 61669 Brno, Czech republic,
Email: jicha@ktermo41.fme.vutbr.cz

Abstract: A 3-D Eulerian-Lagrangian approach to moving vehicles is presented that takes into account the traffic induced flow rate and turbulence. The method is applied to pollutants dispersion in a street canyon. The approach is based on CFD calculations using Eulerian approach to the continuous phase and Lagrangian approach to the „discrete phase" of moving objects - vehicles. A commercial CFD code StarCD was used into which the Lagrangian model was integrated. As an example a street canyon is taken into consideration. It has the length of 50 m and the aspect ratio of 1.27. The speed of wind was assigned values of 4, 7 and 12 m/s at the altitude of 300 m. The total height of the domain is 115 m. In the study different traffic situations are considered, namely one-way and two-way traffic with different traffic rates per lane. The predictions show that different traffic situations affect pollutants dispersion in the street canyon and that there are also differences in the pollutants dispersion in case of one- and two-way traffic.

Keywords: pollutants dispersion, CFD modeling, Eulerian-Lagrangian approach, street canyon, traffic induced flow and turbulence

1. Introduction

The system of air quality control in big cities must be capable to provide information about short-term peak levels of pollution as well as about city background level. Namely peak levels are directly connected with pollution originating from traffic. Despite their short-distance reach, knowledge of the peak values and their dispersion is important as they most frequently overcome the threshold values.

To correctly predict situation around traffic constructions like road tunnels, large city crossroads and/or street canyons, relations between sources and receptors must be established. In this procedure, traffic plays a significant role. Without taking account of traffic will lead to neglecting one of the most important phenomenon that influences mixing processes in the proximity of traffic paths within canopy layer, namely in the situations of very low wind speed. Moving vehicles intensify both micro- and large-scale-mixing processes in the environment by inducing turbulence and enhancing advection by entraining masses of air in the direction of vehicle motion.

Dispersion of pollutants originating from traffic is rather short-distance and it is obvious that the actual geometry of the adjacent area plays an important

Environmental Monitoring and Assessment **65**: 343–351, 2000.

role. The canopy layer is strongly disturbed by buildings and other obstacles, which may influence the local concentrations, by more than one order. In such situations it is essential to establish the influence of principal phenomena on the pollutants distribution and to assess their relative contributions. Mostly it is the street canyon that serves as a basis for such studies. The principal phenomena can be classified as geometrical, traffic related and meteorological related (including climate and season conditions).

Among the geometrical phenomena we can consider the width and height of the street canyon, shape of buildings roofs and buildings fronts and different design of sidewalks (for instance trees along the roadway, etc.).

As for traffic related parameters we can mention the number of traffic lanes, one- or two-way traffic, traffic rate, and speed and type of cars.

Meteorological conditions play very significant role. It is mainly the speed and direction of the wind, solar irradiation of side walls and bottom of the canyon and thermal stratification of the atmosphere. The latter is probably less important in street canyons as the temperature distribution in the street canyon more reflects energy sources and sinks having origin in buildings, cars and solar radiation in an individual canyon. Thermal stratification rather plays a role of an „agent“ that either enhances or suppresses ventilation of the canyon in the upward direction in the situation of very low wind speed.

In the past some papers, e.g. Hider et al., 1997, deal with pollutant dispersion in the wake behind the vehicle but don't take account of global dispersion in the traffic lane where the wake structure is certainly different from that for an individual car. Some other papers, e.g. Kastner-Klein et al., 1999, make comparison between wind tunnel measurements and CFD calculations but this latter taking account only of traffic induced turbulence.

In this study the authors focus on traffic and its impact on the pollutants distribution in a street canyon. A model based on Eulerian-Lagrangian approach to moving objects has been developed and integrated into a commercial CFD code StarCD. The method is based on the solution of two-phase flow (continuous ambient air and discrete moving objects) to model the flow rate entrained by moving vehicles. The method has been recently applied to a road tunnel with a very good agreement (Jicha and Katolicky, 1999) compared to experiments.

2. Mathematical formulation and solution procedure

2.1. GAS PHASE EQUATIONS

The set of equations for the conservation of mass, momentum and energy is solved for the steady, incompressible turbulent flow. The equation for a general variable ϕ has the form:

$$\frac{\partial}{\partial x_i}(\rho u_i \phi) = \frac{\partial}{\partial x_i}\left(\Gamma \frac{\partial \phi}{\partial x_i}\right) + S_\phi + S_\phi^P \tag{1}$$

where ρ is the fluid density, Γ_{eff} is effective diffusion coefficient (sum of viscous and turbulent coefficients). S_ϕ represents additional terms like forces in the momentum equation including pressure gradients or a source of pollutants in the equation for concentration field. Additional source term S_ϕ^P results from the interaction between moving vehicles and ambient air. The interaction is treated using a modified Particle-Source-In-Cell (PSIC) technique by Crowe et al. (1977). The additional source term represents the momentum transported from the discrete moving objects into the ambient air. The set of equations (1) is solved using control volume method and SIMPLE algorithm.

2.2. Model of moving cars and their interaction with the ambient air

The quasi-steady approach to moving objects adopts a modified method used for liquid sprays. An individual moving vehicle is replaced with a continuous stream of vehicles, suppressing the role of the wake. Turbulent mixing in the wake and around the vehicle is compensated for by an extra source of kinetic energy of turbulence. Each vehicle is formed from several blocks of control volumes with a specified velocity and mass. The blocks pass through the mesh of the solution domain - see Fig.1. The velocities of all blocks entering a particular control volume are identical and equal the speed of the vehicle. Each vehicle is treated as a series of separate beams with identical velocities. Each beam is assigned a "concentration" factor *f(i)* related to the mass of the vehicle contained within one beam with a condition $\sum f(i) = 1$. Mass flow of the *i*-part of a vehicle contained in one *i* beam is defined as follows:

$$\dot{m}_{p,i} = \dot{Q}_{car} m_p f(i)$$

where $\dot{Q}_{car}$ is the traffic rate and m_p is the mass of a vehicle.

Assuming a high vehicle Reynolds number and a negligible ratio of density of air and the vehicle, Lagrangian equation for the change of velocity of the vehicle can be written as follows:

$$m_p \frac{d\vec{U}_p}{dt} = \frac{1}{2} \rho_\infty C_D A_p \left|\vec{U}_\infty - \vec{U}_p\right| \left(\vec{U}_\infty - \vec{U}_p\right) \tag{2}$$

where m_p, $\vec{U}_p$ and A_p are mass, velocity and cross section of moving objects, ρ_∞ is density of air, C_D is drag coefficient, $\vec{U}_\infty$ is velocity of the ambient air and

t is time. Introducing the so-called relaxation time τ_M of the block allows us to re-write Eq.(2) in the form:

$$\frac{d\vec{U}_p}{dt} = \frac{1}{\tau_M}\left(\vec{U}_\infty - \vec{U}_p\right) \tag{3}$$

Integrating the equation under the assumption that the relaxation time is constant over a particular control volume yields:

$$\vec{U}_p = \vec{U}_{p0}\exp\left(\frac{-\Delta t}{\tau_M}\right) + \vec{U}_\infty\left(1 - \exp\left(\frac{-\Delta t}{\tau_M}\right)\right) \tag{4}$$

where Δt is the time the block needs to cross a particular control volume and $\vec{U}_{p0}$ is an initial velocity of the block.

The change in the velocity of the block must be regarded as an apparent change since the vehicle moves with a constant speed. It means that we treat the vehicle as "a flying object" that reduces its velocity due to drag force. And this force is added as an additional momentum source into the ambient air. In practice when the block leaves a particular control volume that has just crossed, velocity is reset to its original value that the block had when entered into the control volume. From the difference of momentum, the additional source of momentum is then calculated according to the following formula:

$$S^p_{\phi,i} = \frac{\Delta H_i}{\Delta V} = \frac{1}{\Delta V}\dot{m}_{p,i}\left(\vec{U}_{p,i,out} - \vec{U}_{p,i,in}\right) \tag{5}$$

where the subscripts *out* and *in* correspond to the car leaving and entering the particular control volume, respectively. The total additional source term for the whole vehicle is then obtained by summing over all beams: $S^p_\phi = \sum_i S^p_{\phi,i}$.

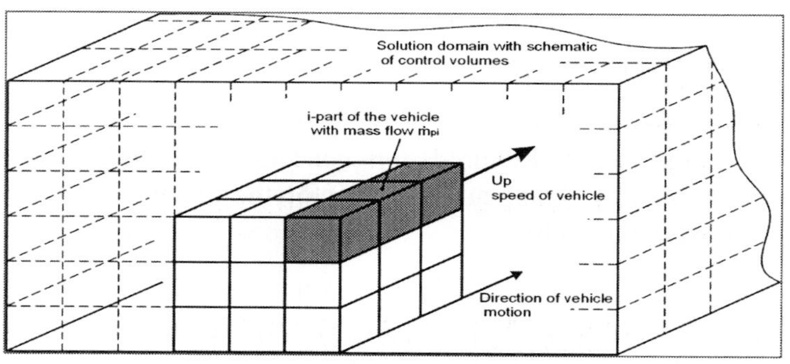

Figure 1 Schematic view of beams of moving objects in the solution domain

2.3. TURBULENCE MODELING AND TRAFFIC INDUCED TURBULENCE

A non-linear low-Reynolds k-ε model of turbulence after Shih et al.(1993) was used. The equation for the kinetic energy for steady incompressible non-buoyant flow reads:

$$(6)\qquad \frac{\partial}{\partial x_i}(\rho u_i k)=\frac{\partial}{\partial x_i}\left(\frac{\mu_{eff}}{\sigma_k}\frac{\partial k}{\partial x_i}\right)+\mu_t P-\rho\varepsilon+P_{NL}+S_k$$

where k is kinetic energy of turbulence, ε is its rate of dissipation and μ_{eff}. is effective viscosity. P is the production of kinetic energy of turbulence by shear and normal stresses, P_{NL} accounts for non-linear contribution. S_k is additional source term accounting for traffic induced turbulence. The equation for the dissipation rate ε has the form:

$$(7)\qquad \frac{\partial}{\partial x_i}(\rho u_i \varepsilon)=\frac{\partial}{\partial x_i}\left(\frac{\mu_{eff}}{\sigma_\varepsilon}\frac{\partial \varepsilon}{\partial x_i}\right)+C_1\frac{\varepsilon}{k}P-C_2\frac{\rho\varepsilon^2}{k}+C_1\frac{\varepsilon}{k}P_{NL}$$

Turbulent viscosity is calculated from the formula:

$$(8)\qquad \mu_t=C_\mu f_\mu \rho\frac{k^2}{\varepsilon}$$

Model constants are: C_1=1.44, C_2=1.92, C_μ=0.09, σ_k=1.0, σ_ε=1.3, f_μ is a damping function applied in the wall region.

Moving objects induce additional kinetic energy of turbulence that is added as an additional source S_k to the k-equation. From different studies e.g. Eskridge and Hunt (1979), Sedefian et al.,(1981) and Sini and Mestayer (1997) it follows that turbulence is induced mainly in the wake behind the vehicle. Therefore the additional sources were added only along the trajectories of the beams of blocks. After having tested different formulas for extra sources, the following one was chosen as the most appropriate:

$$(9)\qquad S_k=C_{car}\rho_\infty\left(U_{car}-U_\infty\right)^2\dot{Q}_{car}$$

where C_{car} is a model constant.

3. Results and their discussion.

The street canyon is solved in a 3D configuration as an infinite canyon with cyclic boundary conditions. The aspect ratio (width to the height) is 1.27. The upstream wind speed profile was calculated from a formula:

$$(10)\qquad U=U_{300}\left(\frac{z-15}{300}\right)^{0.3}$$

where U_{300} is the velocity at 300m height, z is the vertical distance from the roof level. Three values of the wind speed were investigated, namely 4, 7 and 12 m/s at the altitude of 300 m. Two different traffic situations were solved: one-way traffic in four traffic lanes and two-way traffic in total four lanes. The speed of vehicles was set to 50 km/hour for traffic rates of 360, 720 and 1440 cars/hour/lane. In the figure, blocks at the roadway level show control volumes where extra sources of momentum and turbulence were added and small black squares indicate control volumes where emission sources were prescribed.

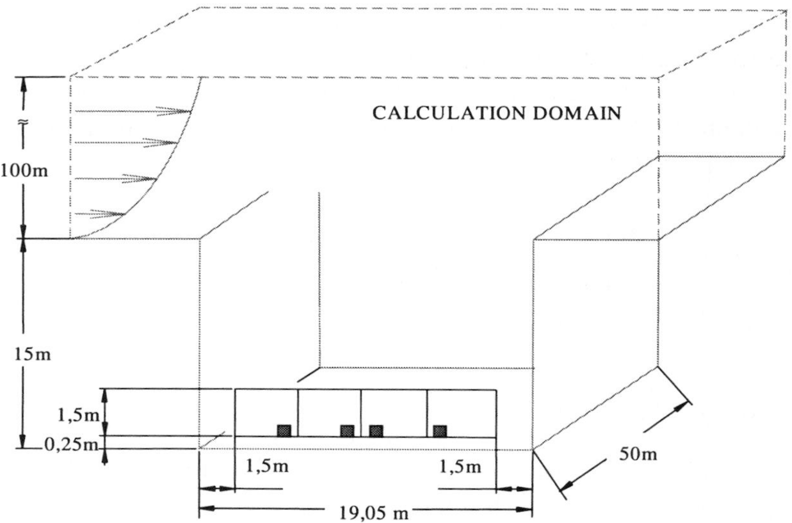

Figure 2. Schematic view of the street canyon

Results of the calculations are in Figures 3*a* to 3*d*. They show a vertical distribution of concentration in the mid-length of the canyon. Concentrations are calculated as a passive scalar representing smoke and are normalised with the smoke yield rate generated by one car. In figures *a.* and *b.* we can see concentrations in the middle of the width of the canyon (between two inner traffic lanes). In figures *c.* and *d.* there are concentrations in the position of a sidewalk 1m away from the leeward side of the canyon. We can see that concentrations are most influenced by the traffic rate (three distinctive groups of curves) and their increase with the traffic is almost linear. The character of the curves in the middle of the canyon is somewhat different from those close to the leeward wall. Just above the bottom surface of the canyon in the middle of it (Figs. *a.* and *b.*), concentrations are more sensitive to wind speed than close to the leeward wall (Figs. *c.* and *d.*). At the roof level the character is reversed, i.e. close to the leeward wall concentrations are more dependent on wind speed. A closer inspection of velocity field (not shown here) reveals that for low wind speeds the incoming fresh air flows around the leeward upper wall corner, penetrates a short depth into the canyon and dilutes the pollutants.

a) middle of the canyon, two-way traffic

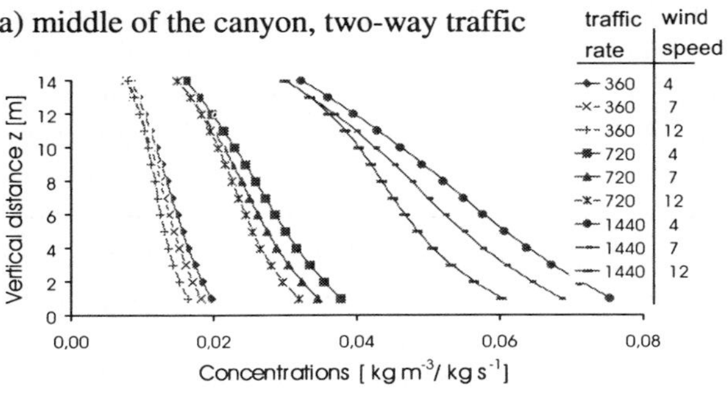

b) middle of the canyon, one-way traffic

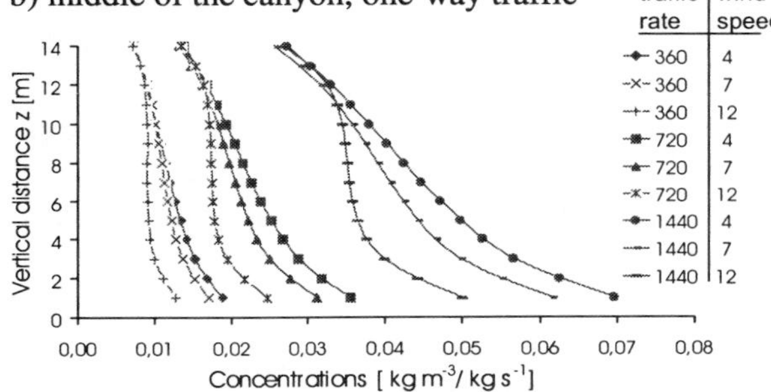

c) 1 m away from leeward wall, two-way traffic

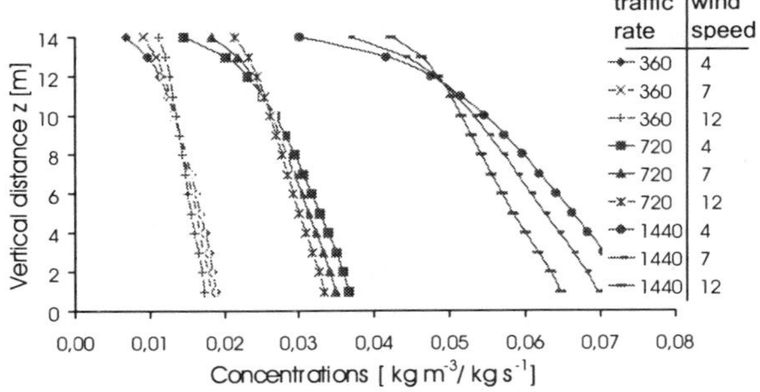

d) 1 m away from leeward wall, one-way traffic

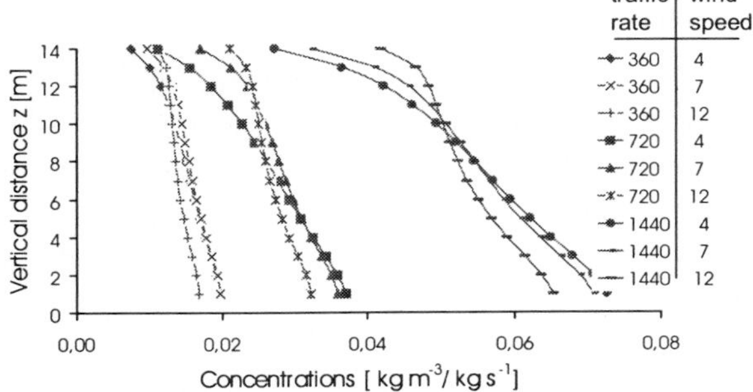

Fig. 3. Vertical concentration distribution in the middle of the canyon and close to leeward wall

At the roof level, concentrations in the middle of the canyon are dependent only on the wind speed what one would expect. However, close to the leeward wall concentrations reflects additional traffic dependence. This is certainly due to the plume flowing upward along the leeward wall and carrying traffic dependent pollutants from the road level. Let's now compare one-way and two-way traffic. In the middle of the canyon, (Figs. *a.* and *b.*) we can see a substantial difference in concentrations. One-way traffic exhibits lower concentrations than two-way traffic. And higher the wind speed and traffic rate, lower are concentrations compared to two-way traffic. For instance at z=1m, for wind speed of 12 m /s and traffic rate of 1440 car/hour the concentrations are 82% and for wind speed of 4 m/s and traffic of 360 car/hour 92% of values for two-way traffic. The possible explanation lies in the cross stream velocity of the vortex created by incoming wind. In the Fig. 4 we can see transverse velocity component in the canyon. At the roof level the velocity approaches the incoming wind speed, i.e. 4, 7 or 12 m/s. But close to the road level the speed of the vortex is distinctly higher in the case of one-way traffic than the two-way traffic and the peak value is even closer to the road level. That means in the one-way traffic the cross road ventilation is somewhat better than in the case of two-way traffic. In this sense we can conclude that one-way motion of vehicles enhances the circulation in the canyon and help ventilating the canyon.

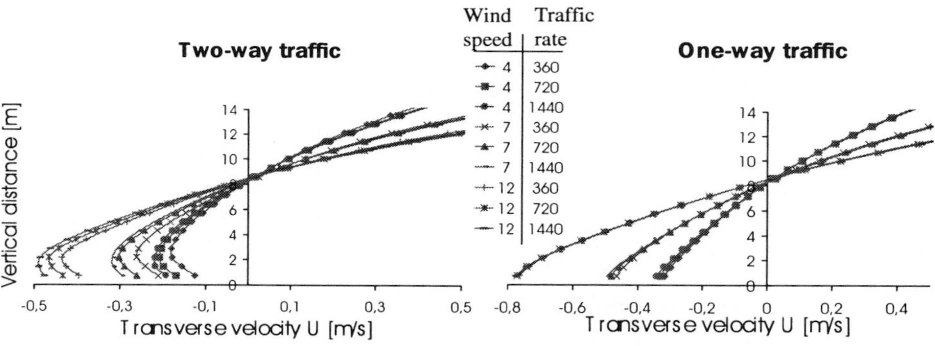

Fig. 4. Profiles of the vortex transverse velocity component

On the other hand, in the case of two-way traffic the total vertical gradient of concentrations is lower than in the case of one-way traffic, which probably results from a more intense turbulent mixing due to higher stresses in the middle of the roadway. Concentrations close to leeward wall (Figs. c. and d.) are both in one- and two-way traffic very similar and at z=1 the values are approximately the same. But mainly for highest traffic rate of 1440 car/hour/lane the concentrations in one-way traffic decrease more rapidly and in the range from z=3 to 10m are lower by approximately 10% than in two-way traffic. In the upper portion of the canyon the mixing process with the incoming air makes the differences in the concentration field negligible.

4. Conclusions

An Eulerian-Lagrangian method has been developed for traffic induced flow and was applied to concentration prediction in a street canyon. Results show that the traffic induced flow and turbulence in one-way traffic have a favourable impact on the concentration field. This latter show lower values than in two-way traffic. This can be seen mainly in the middle of the street close to the road level due to reinforced vortex motion and thus better cross road ventilation. Predictions show on one side results that one would expect (namely at the roof level), on the other side results that are somewhat surprising and need a thorough validation. Even the lower concentrations in one-way traffic compared to two-way traffic were confirmed by Kastner-Klein, (1999) from the wind tunnel experiments, a validation based on field measurements would be desirable.

Acknowledgment

This work is a part of Eurotrac-2 subproject SATURN and was supported by the Ministry of Education of the Czech republic under the grant OE32 /EU1489 and Brno University of Technology Development Fund 262100001.

References

Crowe, G.T., Sharma, M.P., Stock, D.E., (1977), The Particle-Source-In-Cell Model for Gas-Droplet Flows, *J. Fluid Eng.*, **99**, (325-332)

Eskridge R. E., Hunt J. C. R., (1979), Highway modelling. Part I: Prediction of velocity and turbulence fields in the wake of vehicles, *J. Applied Meteorology 18:387*

Hider, Z.; Hibberd, S.; Baker, C.J.,(1997), Modelling particulate dispersion in the wake of a road vehicle, *J. Wind Eng. Ind. Aerodyn.*, Vol 67/68, p733-744.

Jícha M., Katolicky J., (1999) Eulerian-Lagrangian computational model for traffic induced flow field and turbulence inside a vehicle tunnel, *Int. Journal for Wind Engineering and Industrial Aerodynamics* (sent for publication)

Kastner-Klein P., (1999) privite communication

Kastner-Klein P., Sini, J. F., Fedorovich, E., Mestayer, P. G., (1999), Similarity concept for dispersion of car exhaust gases in street canyon tested against widn-tunnel and numerical model data, 2nd Int. Conf. Urban Air Quality, paper RAP3.3, Madrid, Spain

Sedefian L., Rao S. T., Czapski U., (1981), Effects of traffic-generated turbulence on near-field dispersion, *Atmospheric Environment 15:527*

Shih, T. H., Zhu, J., Lumley, J. L. (1993), A realizable Reynolds stress algebraic equation model, *NASA TM-105993*

Sini J. F., Mestayer P. G., (1997), Traffic-induced urban pollution: A numerical simulation of street dispersion and net production, *22nd NATO/CCMS International Technical Meeting on Air Pollution Modelling and its Application, Clermont-Ferrand, France*

EXPERIMENTAL AND NUMERICAL VERIFICATION OF SIMILARITY CONCEPT FOR DISPERSION OF CAR EXHAUST GASES IN URBAN STREET CANYONS

PETRA KASTNER-KLEIN and EVGENI FEDOROVICH
Institute of Hydromechanics, University of Karlsruhe, 76128 Germany
e-mail: fedorovich@ifh.bau-verm.uni-karlsruhe.de

JEAN-FRANÇOIS SINI and PATRICE G. MESTAYER
Laboratory of Fluid Mechanics, Ecole Centrale de Nantes, 44321 France

Abstract. In urban conditions, car exhaust gases are often emitted inside poorly ventilated street canyons. One may suppose however that moving cars can themselves produce a certain ventilation effect in addition to natural air motions. Such ventilation mechanism is not sufficiently studied so far. A similarity criterion relating the vehicle- and wind-induced components of turbulent motion in an urban street canyon was proposed in 1982 by E. J. Plate for wind tunnel modelling purposes. The present study aims at further evaluation of the criterion and its applicability for a variety of wind and traffic conditions. This is accomplished by joint analyses of data from numerical simulations and wind tunnel measurements.

Key words: traffic emissions, street canyon, wind tunnel, numerical, modelling, similarity

1. Introduction

Car exhaust gases considerably contribute to the atmospheric pollution in urban areas. In many cases, these gases are emitted under conditions of poor ventilation in narrow street canyons of big cities. One may assume still that moving vehicles can themselves produce a certain ventilation effect in a street canyon.

The generation of secondary air motions by traffic was investigated in a number of *in situ* experiments (Eskridge and Hunt 1979, Eskridge and Rao 1983, Delaunay and Houseaux 1997), and also in numerical (Hertel and Berkowicz 1989, Stern and Yamartino 1998) and wind tunnel (Eskridge *et al.* 1979, Eskridge and Thompson 1982, Brilon *et al.* 1987, and Kastner-Klein *et al.* 1998) model studies. In the literature, the air motions produced by traffic are usually associated with the so-called vehicle-induced turbulence. However, in many instances the organised transport of pollutant by vehicle-induced mean air motion should be also taken into account (Eskridge and Hunt 1979, Delaunay and Houseaux 1997).

Environmental Monitoring and Assessment **65**: 353–361, 2000.

Diffusion and dispersion of gaseous pollutants in street canyons take place under the joint influence of natural and vehicle-induced air motions. The relations and interactions between these transport mechanisms are not sufficiently studied so far. In particular, there are very few field measurements that allow a separate quantification of natural and vehicle-induced components of pollutant transport. In this situation, wind tunnel and numerical modelling may be useful approaches towards understanding basic features of the combined transport phenomena.

A similarity criterion relating the wind- and vehicle-induced components of turbulent motion in an urban street canyon was proposed by Plate (1982). This criterion was originally formulated as criterion for the wind tunnel modelling of turbulent diffusion in a street canyon with traffic and will be hereafter referred to as PMC.

In the present paper, we try to interpret the PMC in a more general sense and verify it as similarity parameter for gas dispersion in a street canyon under the joint effect of natural ventilation and vehicle-induced turbulent air motions.

2. Similarity Concept

Plate (1982) suggested that the *traffic-to-wind turbulence kinetic energy (TKE) production ratio is a similarity criterion for the wind tunnel modelling of diffusion of car-exhaust gases in a street canyon with traffic.*

Based on the above idea, Kastner-Klein *et al.* (1998) expressed the wind-related TKE production per unit canyon volume as $G_w \propto u^3 / H$, where H is the canyon depth scale and u is the reference velocity of the external wind flow. The TKE production due to vehicle motions was estimated as $G_v \propto C_{dv} \cdot A_v \cdot v^3 \cdot n_v /(B \cdot H)$, where C_{dv} is the aerodynamic drag coefficient and A_v is the frontal area of an individual vehicle, v is the traffic velocity, n_v is the amount of vehicles per unit length of the canyon (traffic density), and B is the canyon width scale. Thus, the traffic-to-wind TKE production ratio is expressed as $P_t = G_v / G_w = n_p(v^3 / u^3)$, where $n_p = C_{dv} \cdot A_v \cdot n_v / B$ is the dimensionless traffic density.

The concept underlying the PMC reads that dimensionless values of pollutant concentration $c_* = c \cdot U \cdot L / E$ in the street canyon and in its wind tunnel analogue should be identical when P_t is the same in the model and in the prototype. In the expression for c_*, c is the actual concentration in ppm, U and L are characteristic velocity and length scales of the flow, respectively, and E $[L^2 \cdot T^{-1}]$ is the source strength per unit length. The car emissions are considered as line sources.

respect to the traffic and wind-flow characteristics. Both wind tunnel and numerical results display apparent and similar clustering tendency according to the P_t values. This provides one more proof to the idea that parameter P_t is a similarity number for the diffusion pattern in a street canyon with traffic.

The wind tunnel and numerical data on the pollutant concentration in an individual location within the street canyon (at the roadside level near the leeward building wall) are presented in Figure 4. It is easy to see that data for different model situations practically collapse to one curve when plotted against P_t. The data convergence, and agreement between the wind tunnel and numerical data are nearly perfect with P_t<8 (or $P_t^{1/3}$<2). The increase of discrepancies with the higher P_t values may be explained by residual flow entrainment due to moving vehicles. Such entrainment, which becomes stronger with larger traffic velocities and densities, is neither taken into account in the numerical model nor by the PMC. However, it apparently affects the concentration pattern in the considered wind tunnel measurement location when P_t gets large.

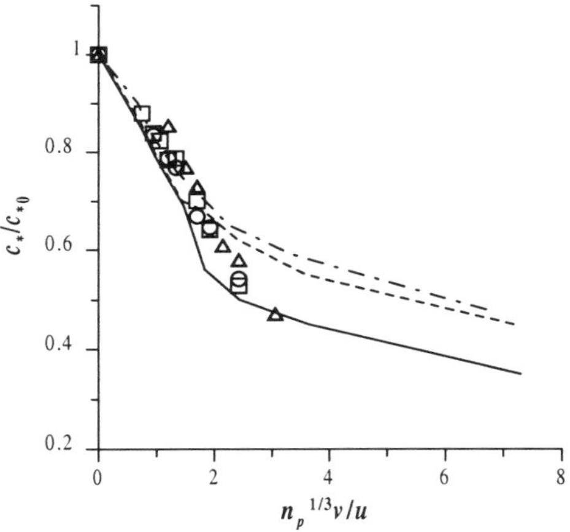

Figure 5. The wind tunnel data from Figure 4, points, compared with the numerical data of Stern and Yamartino (1998), lines. The numerical results refer to the situations with low traffic (solid line), medium traffic (dashed line), and high traffic (dashed and dotted line) according to classification by Stern and Yamartino (1998).

Additionally, we employ for comparison the data from a numerical model study by Stern and Yamartino (1998), who incorporated vehicle-induced turbulence effect in the CALGRID, a photochemical model for regulatory applications. In this model, the TKE production rate by traffic is estimated from the energy dissipated by a vehicle as it pushes its way through the ambient air.

In Figure 5, the wind tunnel data are plotted against P_t together with results of Stern and Yamartino (1998). In this case, the numerical data show less universal behaviour with respect to the similarity parameter than model data from the present study in Figure 4b. However, the independent numerical data in Figure 5 agree well with the wind tunnel measurements approximately in the same interval of the P_t number values as our numerical results, namely in the range $P_t < 8$.

5. Conclusions

Presented results from the combined wind tunnel and numerical model study generally confirm validity of the traffic-to-wind turbulence production ratio P_t as similarity number for the regime of turbulent diffusion in an urban street canyon with moving vehicles. In particular, clearly exposed local similarity have been observed in the concentration field at $P_t < 8$, which in the conducted experiments corresponded to moderate vehicle-to-wind velocity ratios ($v/u < 3$) and traffic densities less than 15 vehicles per 100 m.

Acknowledgements

The authors gratefully acknowledged support from the Air Pollution Prevention Measures Programme (PEF) funded by the Land of Baden-Württemberg (Germany) and the European Commission (EC), and from the TMR Programme of EC, Project TRAPOS. Computer support was provided by Institut de Développement et de Recherche pour l'Informatique Scientifique, CNRS.

References

Brilon, W., Niemann, H. J., and Romberg E.: 1987, Windkanaluntersuchungen zur Ausbreitung von Abgasen an Autobahnen, *Straßenverkehrstechnik*, **31**, 122-133.

Delaunay, D., and Houseaux, N.: 1997, Turbulence et quantité de mouvement induites par les véhicules a proximité d´une voie urbaine: mesures in situ, Centre Scientifique et Technique du Batiment (CSTB), Nantes, Report EN-AEC 97, 68pp.

Eskridge, R.E., and Hunt, J. C. R.: 1979, Highway modeling Part I: Prediction of velocity and turbulence fields in the wake of vehicles. *J. Appl. Meteorol.*, **18**, 387-400.

Eskridge, R.E., Binkowski, F. S., Hunt, J. C. R., Clark, T. L., and Demerjian, K. L.: 1979, Highway modeling Part II: Advection and diffusion of SF_6 tracer gas. *J. Appl. Meteorol.*, **18**, 401-412.

Eskridge, R.E., Thompson, R. S.: 1982, Experimental and Theoretical Study of the Wake of a Block-Shaped Vehicle in a Shear-Free Boundary Flow. *Atmos. Environ.,* **16**, 2821-2836.

Eskridge, R. E., and Rao, S. T.: 1983, Measurement and prediction of traffic-induced turbulence and velocity fields near roadways. *J. Climate Appl. Meteorol.*, **22**, 1431-1443.

Hertel, O., and Berkowicz R.: 1989, Modelling pollution from traffic in a street canyon. Evaluation of data and model development, *DMU Luft A-129*, 77pp.

Kastner-Klein, P., Berkowicz, R., Rastetter, A., and Plate E. J.: 1998, Modelling of vehicle induced turbulence in air pollution studies for streets, *Proc. 5th Workshop on Harmonisation within Atmospheric Dispersion Modelling,* Rhodes, Greece, 18-21 May 1998.

Plate, E. J.: 1982, Windkanalmodellierung von Ausbreitungsvorgängen in Stadtgebieten. *Kolloquiumsbericht Abgasbelastungen durch den Straßenverkehr*, Verlag TÜV Rheinland, 61-83.

Rao, S. T., Sistla, G., and Eskridge, R.E., and Petersen, W.B.: 1986, Turbulent diffusion behind vehicles: evaluation of roadway models. *Atmos. Environ.,* **20**, 1095-1103.

Sini, J.-F., Anquetin, S., and Mestayer P. G.: 1996, Pollutant dispersion and thermal effects in urban street canyons, *Atmos. Environ.*, **30**, 2659-2677.

Stern, R., and Yamartino R. J.: 1998, Development and initial application of the micro-calgrid photochemical model for high-resolution studies of urban environments, *Preprints 23rd NATO/CCMS ITM on Air Pollution Modelling and its Applications*, 28 September - 2 October 1998, Varna, Bulgaria.

Tennekes, H., and J. L. Lumley, 1972: *A First Course in Turbulence*, The MIT Press, 300pp.

COMPARISON OF NUMERICAL STREET DISPERSION MODELS WITH RESULTS FROM WIND TUNNEL AND FIELD MEASUREMENTS

Matthias Ketzel, Ruwim Berkowicz and Achim Lohmeyer[#]
National Environmental Research Institute, Fredriksborgvej 399, DK-4000 Roskilde,
[#]Lohmeyer Consulting Engineers, Mohrenstrasse 14, D-01445 Radebeul
E-mail: mke@dmu.dk

Abstract: Numerical dispersion models developed and validated in different European countries were applied to data sets from wind tunnel and field measurements. The comparison includes the Danish Operational Street Pollution Model (OSPM) and the microscale flow and dispersion model MISKAM. The latter is recommended for application in built-up areas in the draft of the new German guideline VDI 3782/8. In a first step the models were applied to simplified street configurations. Different parameters as length and height of adjacent buildings and the angle of the incoming flow were varied. The results were compared to recent wind tunnel measurements. In a second step the models were applied to two extensively investigated field data sets from Jagtvej, Copenhagen and Göttinger Straße, Hannover. Intensified and more transparent and accessible validation procedures would be helpful for the thorough user.
Keywords: air pollution modelling, comparison with measurements, MISKAM, model validation, OSPM, street pollution models, traffic pollution, wind tunnel studies

1 Introduction

Numerical microscale dispersion models are more and more frequently used for air quality and impact assessments, and the number of users increases continually. The final addressee of model results who hopes for support of economically and environmentally important decisions has often a great confidence in the model results. This confidence is not always justified due to incomplete validation of the models and missing availability of the validations. Model intercomparisons are mostly outdated when published, because of new (supposedly better) model versions published by the model developers. The situation for the user becomes even more confusing if he has to choose an appropriate model out of the dozens of available ones.

This paper presents first results of an intercomparison of results from two models and a comparison of these with wind tunnel and field measurements. One of these models is a semiempirical model with almost no free 'soft' parameters to be defined by the user. For the second more sophisticated microscale model the differences are shown that can appear in application even if it is done according to the recommendation in the manual or when applying different model versions. A suggestion for improving the above described situation is given.

Environmental Monitoring and Assessment **65**: 363–370, 2000.

2 The models

From the wide spectrum of available models two frequently used models were selected. One of these is the Operational Street Pollution Model [OSPM, 1, 2], a practical street pollution model, developed by the National Environmental Research Institute, Department of Atmospheric Environment. It is widely used in Denmark. The other is the microscale flow and dispersion model MISKAM [3, 4] in various versions. It is recommended for application in built-up areas in the draft of the new German guideline VDI 3782/8 [5]. Further models are expected to be included in the comparison.

In **OSPM** concentrations of exhaust gases are calculated using a combination of a plume model for the direct contribution and a box model for the recirculating part of the pollutants in the street. The turbulence within the canyon is calculated taking into account the traffic generated turbulence. The NO2 concentrations are calculated taking into account NO-NO2-O3 chemistry and the residence time of pollutants in the street. The model is designed to work with input and output in the form of one-hour averages.

MISKAM consists of a 3-dimensional non-hydrostatic flow model and a Eulerian dispersion model. The physical basis are the complete 3-dimensional equations of motion of the flow field and the advection-diffusion equation to determine the concentrations of substances with neutral density. The calculated result is the stationary flow and pressure field, diffusion coefficients and the concentration field in an area of typically 500 m x 500 m (100 x 100 cells or more, non equidistant grid).

3 The datasets

In the first step of validation, wind tunnel measurements for **simplified street configurations** from Kastner-Klein et al. [7] were used. The dataset contains measurements of a reference case of a two-dimensional street canyon with perpendicular approach flow as well as studies of the effects of building geometry, surrounding buildings and wind direction.

For comparison to field data the dataset from **Jagtvej in Copenhagen** was used, obtained from a permanent pollution monitoring station operated in the frame of the Danish National Urban Monitoring Programme [10]. Jagtvej is a busy street with about 22000 vehicles/day and is flanked on both sides with buildings about 25 m apart and about 18 m high.

The second used field data set was obtained from measurements in **Göttinger Straße in Hannover**. The State Environmental Agency of Lower Saxony operates a permanent monitoring station in a four-lane street canyon with a traffic load of ca. 30000 vehicles/day [11].

In addition to street and background concentrations and meteorological data from a 10 meter mast on top of a nearby building also continuous traffic counts are available for both field measurements.

The datasets were chosen due to their completeness and quality shown in previous studies [1,6]. Another advantage of the two field datasets is the additional availability of wind tunnel measurements both performed at the University of Hamburg [8,9]. Results were also used for comparison.

4 Results and discussion

4.1 NORMALISED CONCENTRATIONS

The non-dimensionalised concentration is defined as $c^* = C \bullet U_{Href} \bullet H / (Q/L)$, where C is the concentration, U_{Href} is a reference velocity taken at a height H_{ref} (e.g. 100 m), H is a characteristic length (e.g. the average height of the buildings 16 m) and (Q/L) is the source strength (traffic emissions).

Figure 1 shows the results obtained for the ideal square-shaped street canyon with a length of 90 m and a width and height of 16 m. Plotted are the normalised concentrations c* measured in the wind tunnel and calculated by the models. For reasons of symmetry only directions from 90° to 270° are given, 180° refers to flow parallel to the street. Best agreement with the wind tunnel results shows the old MISKAM version 1.1 used in Ref.[6]. The newer versions calculate either concentrations which are too high for the windward situation (90°, vers. 3.51) or values which are too low for the lee case (270°, vers. 3.6). OSPM gives too high values except for 270° (perpendicular approach flow) where the pronounced maximum is not reproduced.

This maximum is also absent in the field measurements in Jagtvej (see **Fig. 2**, perpendicular direction is 120°) but we find it in the data from Göttinger Straße (see **Fig. 3**, here perpendicular direction is 250°). In both field data sets we exclude night hours as well as low wind speed situations (lower 3 m/s) because of the uncertainties in emissions and in order to avoid the influence of traffic induced turbulence. We see in **Figure 2** a better agreement of MISKAM version 3.6 with the field data than for version 3.51.

Figure 3 shows that differences within a factor of 3 can appear while increasing the upper boundary of the grid model from 100 m to 500 m and enlarging the grid horizontally by 5 boxes without buildings. In this case we are as lucky as to get a much better agreement with field data after these changes. But this example shows how manipulable model results can be, just according to the choice of parameters which are free to choose for the user. Already the 100 m high grid model meets all recommendations given in the manual.

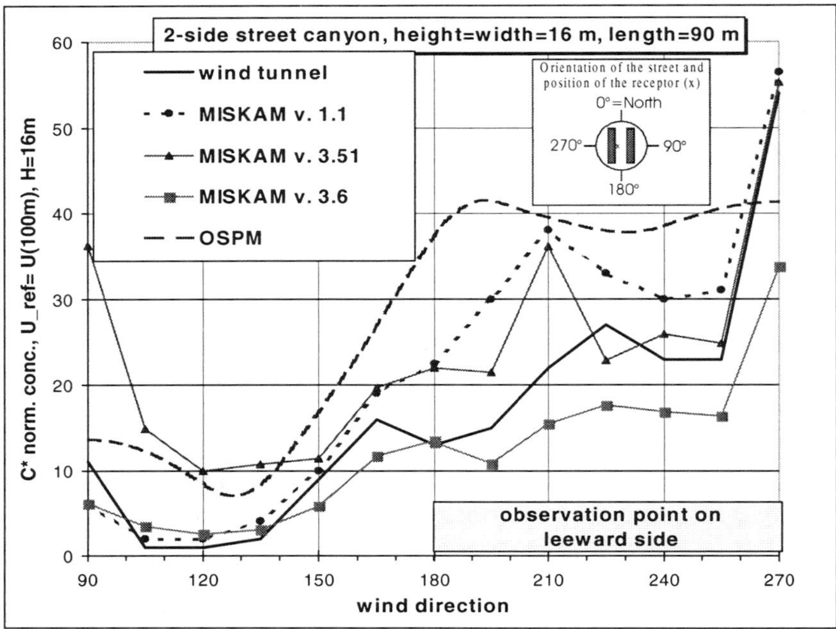

Fig.1: Comparison of normalised concentrations for the ideal street canyon

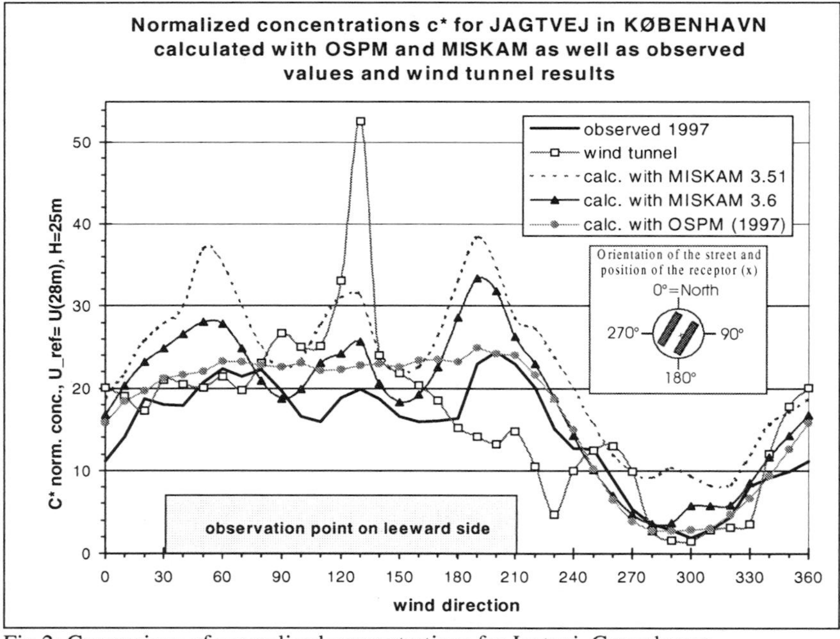

Fig.2: Comparison of normalised concentrations for Jagtvej, Copenhagen

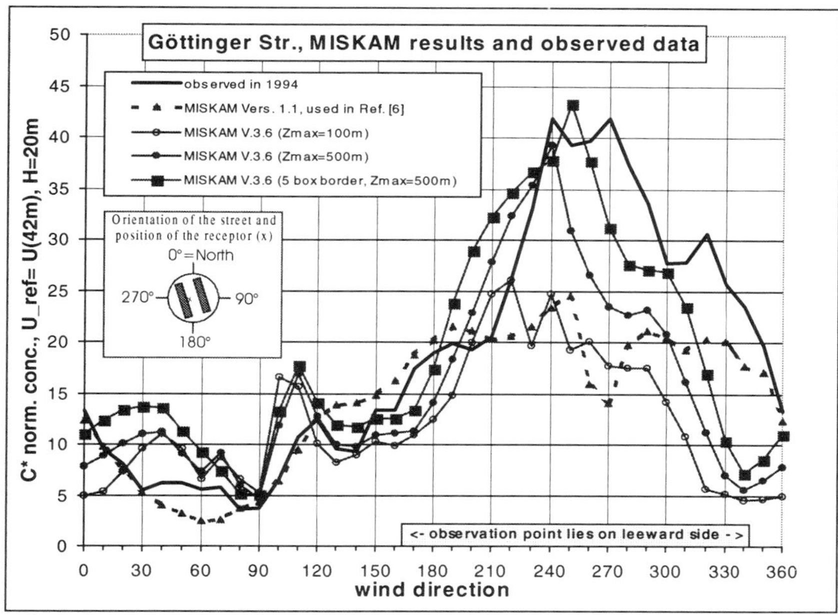

Fig.3: Comparison of MISKAM results using different configurations.

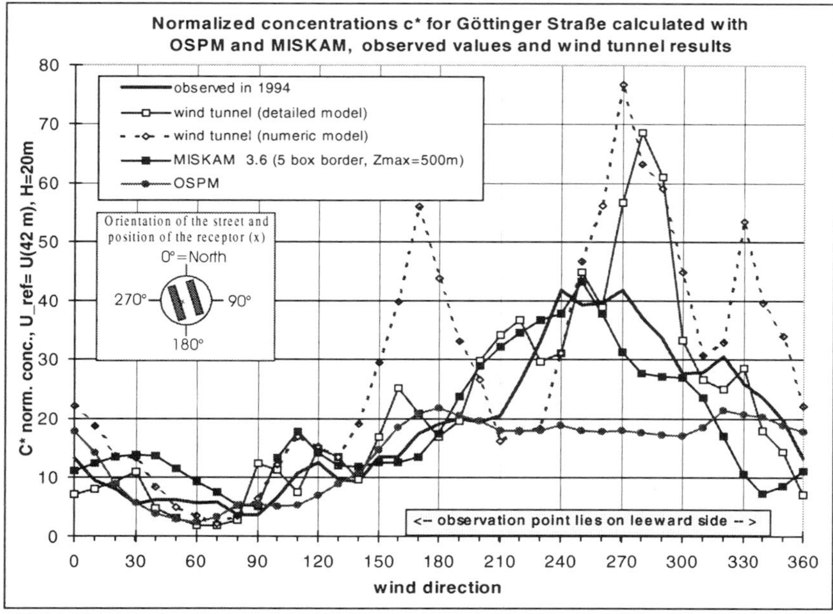

Fig.4: Comparison of normalised concentrations for Göttinger Straße.

The comparison of model results with field and wind tunnel data for Göttinger Straße is presented in **Figure 4**. MISKAM shows a reasonable agreement with the observed data after the mentioned grid changes and is now able to reproduce the maximum for a wind direction of around 250°. Shown are also results from two wind tunnel measurements. In the first one the model had exactly the shape of the real buildings (detailed model), while in the second model the shape of the buildings was adopted to the resolution of the MISKAM grid (called: 'numeric' model). Surprisingly the agreement with the MISKAM results is not improved at all in the second case. Both wind tunnel models significantly overestimate the observed maximum in the field data, while OSPM does not reproduce this maximum at all.

The essentially different behaviour around perpendicular direction in both field data sets can not yet fully be explained. Reasons might be found studying the details of the flow patterns for this direction and comparing the detailed building configurations of both sites.

4.2 STATISTICAL VALUES

For practical application, e.g. for a comparison with air pollution standards, the estimation of annual statistical values is required. As an example the calculations of annual average values is presented for Göttinger Straße.

The calculation of real concentrations from the normalised values (c*) follows the procedure used by Schädler et al. [6]: $C \cong c^{*} \bullet (Q/L) \bullet u^{-n}$; with $n = 1$ for $u > 3$m/s and $n = 0.35$ for $u \leq 3$m/s, using a measured half hourly time series of both wind speed / wind direction and traffic flow. This method is supposed to take into account the influence of the traffic induced turbulence for low wind conditions. Emission factors were from Schädler et al. [6], referred there as pattern '5 SK' or 'new'. The height of reference wind for all calculations was 42 m.

In **Table I** the calculated benzene annual mean concentrations are given for all c* curves plotted in Fig. 3 and 4. The absolute difference to the measured values and the relative difference of the street contribution (background subtracted) is shown as well. Best agreement is encountered for the c* curve obtained from the field data. This shows that the described method in principal works with acceptable error. Both wind tunnel results give an overestimation. The overestimation is larger for the 'numeric' model than for the 'detailed' model. MISKAM underpredicts the measured concentrations. The difference between the results obtained with the 'small' MISKAM grid (Zmax=100m) and the 'large' MISKAM grid (Zmax=500m, 5 box border) is quite substantial: 2.8 $\mu g/m^3$ or 30% of the street contribution. This emphasises again the need for more model validation and more stringent hints in the model handbook for grid definition etc.

Table I: Comparison of annual benzene concentrations calculated from normalised concentrations (c*) obtained by different methods with field data

	benzene annual mean [μg/m³]	absolute difference [μg/m³]	relative difference (backer. subtracted) [%]
measured values			
background	3.1		
street concentration	11.8		
calculated from c*			
averaged field data	11.3	-0.5	-6
wind tunnel - detail	13.1	1.3	15
wind tunnel - numeric	15.2	3.4	39
MIS. 1.1 used in Ref. [6]	9.6	-2.2	-25
MIS. 3.6 ; Zmax=100 m	8.2	-3.6	-41
MIS. 3.6 ; Zmax=500 m	9.3	-2.5	-29
MIS. 3.6; 5 box border Zmax=500 m	11.0	-0.8	-10
calculated directly with OSPM	8.6	-3.2	-37

5 Conclusion

The comparison of the model results in terms of normalised concentrations reveals discrepancies within a factor of 5 to 10 for single wind directions regarding all tested models as well numerical as physical. For annual mean concentrations the agreement is better because of the averaging procedure.

The microscale model MISKAM is expected to be more accurate than a simple model (like e.g. OSPM), but it contains also more free parameters to be set (partly a bit arbitrary) by the user that may lead to uncertainties in the model results. For example it could be shown, that the model results for a single wind direction can vary within a factor of 3 if the original grid model is surrounded by 5 additional empty grid boxes.

For the microscale models to be a useful practical tool, the procedure of validation needs to be improved and to be made more transparent. To achieve this goal, the model developers and the persons that make model comparisons should supply easy accesses (e.g. in the Internet) to their model input data and the results including the applied processing programs and plot files. It should be possible to ask the model developers to update previously made comparisons and give to the users easy access to these new comparisons before releasing a new version of a model to the user community. All users should be able to see how the agreement with the validation datasets develops from version to version. The users could also better predict changes in their previously made calculations and decide if the effort of a (sometimes very expensive) repetition of the calculations is required or not.

Acknowledgement

The authors wish to express their appreciation for support from the European Commission's Training and Mobility of Researchers Programme (TMR) within the frameworks of the European Research Network on 'Optimisation of Modelling Methods for Traffic Pollution in Streets' (TRAPOS).

Furthermore we gratefully acknowledge the provision of data and information by State Environmental Agency of Lower Saxony (NLÖ Hannover), Meteorological Institute - University of Hamburg and Institute for Hydromecanics - University of Karlsruhe.

References

[1] Berkowicz, R., Hertel, O.,Larsen, S.E., Sørensen, N.N. and Nielsen, M. (1997): Modelling traffic pollution in streets. National Environmental Research Institute, Roskilde, Denmark.

[2] Berkowicz, R. (1999): OSPM - A parameterised pollution model. This volume.

[3] Eichhorn, J. (1989): Entwicklung und Anwendung eines dreidimensionalen mikroskaligen Stadtklima-Modells. Dissertation, Universität Mainz, Germany.

[4] Eichhorn, J. (1998): MISKAM-Handbuch zur Version 3.xx. Giese-Eichhorn, Wackernheim, Germany. October 1998.

[5] German guideline VDI 3782/8 (draft from March 1998): Ausbreitungsrechnung für Kfz-Emissionen. VDI/DIN-Handbuch Reinhaltung der Luft, Beuth Verlag Berlin, Germany.

[6] Schädler, G., Bächlin, W., Lohmeyer, A. and van Wees, T. (1996): Vergleich und Bewertung derzeit ver-füg-barer mikro-skaliger Strömungs- und Ausbreitungsmodelle. In: Berichte Umweltforschung Baden-Württemberg, PEF 2 93 001 (FZKA-PEF 138).

[7] Kastner-Klein, P. and Plate, E.J. (1997): Windkanalversuche zur Verbesserung der Ermittlung von Kfz-bedingten Konzentrationsverteilungen in Stadtgebieten. Zwischenergebnisse zum Projekt PEF 2 95 001, In: Berichtsband 13. Statuskolloq. des PEF, (FZKA-PEF 153).

[8] Liedke, J., Leitl, B. and Schatzmann, M. (1998): Car exhaust dispersion in a street canyon - Wind tunnel data for validating numerical dispersion models. 2nd East European Conference on Wind Engineering (EECWE), Prague, 7-11 September 1998.

[9] Liedke, J. (1998a): Wind tunnel studies on a model of Jagtvej, København. personal information.

[10] NERI (1998): The Danish Air Quality Monitoring Programme. Annual Report for 1997. National Environmental Research Institute, Roskilde, Denmark. NERI Technical Report No. 245. ISSN 0905-815X

[11] NLÖ (1993): Lufthygienisches Überwachungssystem Niedersachsen - Luftschadstoff-belastungen in Straßenschluchten, edited by Niedersächsisches Landesamt für Ökologie, D-30449 Hannover, ISSN 0945-4187

MEASUREMENTS AND MODELLING OF AIR POLLUTION IN A STREET CANYON IN HELSINKI

JAAKKO KUKKONEN[1], ESKO VALKONEN[1], JARI WALDEN[1], TARJA KOSKENTALO[2], ARI KARPPINEN[1], RUWIM BERKOWICZ[3] and RAIMO KARTASTENPÄÄ[1]

1 Finnish Meteorological Institute, Air Quality Research, Sahaajankatu 20 E, 008100 Helsinki, Finland. E-mail: Jaakko.Kukkonen@fmi.fi

2 Helsinki Metropolitan Area Council, Opastinsilta 6 A, 00520 Helsinki, Finland

3 National Environmental Research Institute, Frederiksborgvej 399, DK-4000 Roskilde, Denmark

Abstract. A measuring campaign was conducted in the street canyon 'Runeberg street' in Helsinki in 1997. Hourly concentrations of carbon monoxide (CO), nitrogen oxides (NO_X), nitrogen dioxide (NO_2) and ozone (O_3) were measured at the street and roof levels, and the relevant hourly meteorological parameters were measured at the roof level. The hourly street level measurements and on-site electronic traffic counts were conducted during the whole year 1997, and roof level measurements were conducted during approximately two months, from 3 March to 30 April in 1997. The Operational Street Pollution Model (OSPM) was used to calculate the street concentrations and the results were compared with the measurements. The overall agreement between measured and predicted concentrations was good for CO and NO_x, but the model slightly overestimated the measured concentrations of NO_2. The database, which contains all measured and predicted data, is available for a further testing of other street canyon dispersion models.

Key words: air pollution, dispersion, street canyon, OSPM, nitrogen oxides, model validation

1. Introduction

There is an evident need for good-quality experimental data, measured in various urban environments, for model testing and validation purposes. For the purposes of model validation, air quality data has to be combined with simultaneous measurements of traffic flow and meteorological conditions, together with the information concerning buildings and local topography.

The main objective of this work was to produce good-quality data in a street canyon, which could be utilised for model validation. Concentrations of carbon monoxide (CO), nitrogen oxides (NO_x and NO_2) and ozone (O_3) were measured simultaneously at the street and roof levels, together with on-site meteorological parameters and traffic densities. The data were subsequently applied for test and validation of the OSPM street canyon dispersion model, developed by Hertel and Berkowicz (1989abc). The OSPM model has previously been validated in Copenhagen (Hertel and Berkowicz, 1989a; Berkowicz et al., 1996, Berkowicz et al., 1997b), in Utrecht (Hertel and Berkowicz, 1989b) and in Oslo (Hertel and Berkowicz, 1989c).

Environmental Monitoring and Assessment **65**: 371–379, 2000.

In Northern Europe, stable atmospheric stratification with a light wind speed may prevail for extensive periods. For instance, it is not uncommon for such conditions to last for several days in southern parts of Finland, particularly in winter and spring (Kukkonen et al., 1999, 2000). This study provides an opportunity to obtain experimental street canyon data in Northern climatic conditions, and validate the OSPM model against this dataset.

2. The measurement site and time period

The measurement campaign was conducted in Runeberg street (Runeberginkatu) in downtown Helsinki. Figure 1 shows a cross-section of the street canyon. The street level measurements were conducted from an inlet at the first floor. A meteorological mast and concentration monitors were located on the building roof. A service room near the meteorological mast could cause some disturbances on the meteorological measurements.

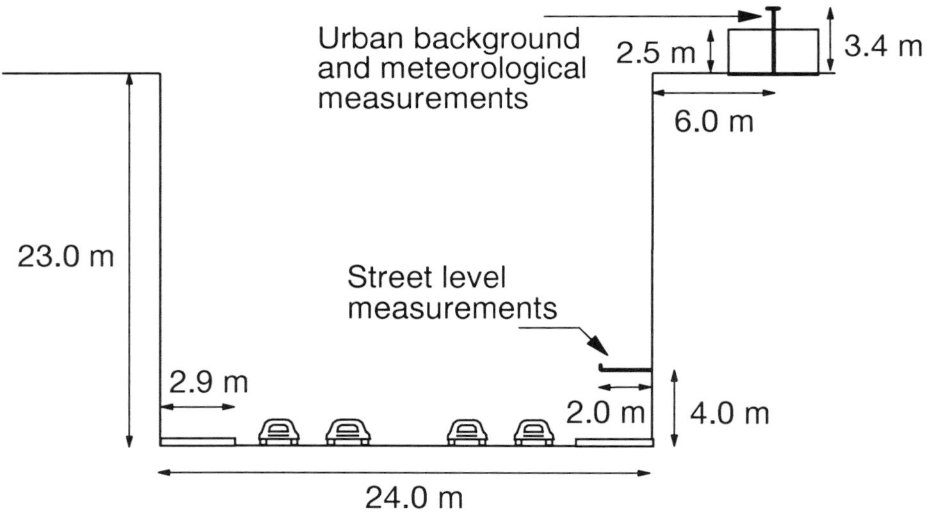

Figure 1. Vertical cross-section of the street canyon and the locations of the measuring points on the street and roof levels.

The hourly street level measurements and traffic counts were conducted during the whole year 1997. The time coverage of the urban background and meteorological measurements was nearly two months, from 3 March to 30 April, 1997. This time period was selected in order to include the stable, low wind speed conditions, which occur frequently in spring.

3. Methods

3.1 COMPUTATIONAL METHODS

3.1.1 Modelling of traffic volumes and emissions

The traffic counts were carried out during one year, 1997. The measurements were performed using electronic loops, which were installed under the street surface. The traffic data contains hourly traffic volumes, types of vehicles (light and heavy duty vehicles) and speeds of vehicles. The traffic counts show that traffic density in the street was substantial, on the average approximately 26 000 vehicles per day. There are four traffic lanes, two to both directions. The average driving speed within the street segment was 40 km/h.

We utilised the emission factors evaluated in a previous study by the Helsinki Metropolitan Area Council (YTV, 1997, Helsinki Metropolitan Area Board, 1999). The emission factors are dependent on vehicle classes and average driving speeds. We computed the hourly emissions of CO and NO_x originating from traffic in the street, using measured hourly traffic volumes and emission factors. The fraction of NO_x emitted as NO_2 was assumed to be 5,0 %.

3.1.2 Modelling of atmospheric dispersion

The OSPM model, developed by Hertel and Berkowicz (1989a,b,c) is a practical street pollution model based on simplified description of flow and dispersion conditions in street canyons. Concentrations of exhaust gases are computed using a combination of a plume model for the direct contribution from street traffic, and a box model for the recirculating part of pollutants in the street. The simplified parametrization of flow and dispersion conditions in a street canyon has been deduced from extensive analysis of experimental data and model tests (Berkowicz et al., 1996 and 1997b).

The emission field is treated as an area source at the street level. The wind direction at the street level is assumed to be mirror-reflected with respect to the roof level wind.

It is assumed in OSPM that the turbulence in the street is dominated by mechanical turbulence. This assumption is justified by the fact that the street air is usually well mixed and the thermal fluxes are small. However, at very low wind speeds, heating of building walls facing solar radiation can result in a thermally induced air circulation, which might modify the wind driven circulation (Nakamura and Oke, 1988; Sini et al., 1996). These effects, however, are not important in climate conditions of Northern Europe.

The mechanical turbulence is assumed to be generated by the wind and by the traffic in the street. The heat of traffic exhaust gases can also contribute to enhancement of the turbulence but this effect is minor compared with the mechanical component (Yamartino and Wiegand, 1986).

The contribution from the recirculation part is computed using a simple box model. It is assumed that the canyon vortex has the shape of a trapeze, with the maximum length of the upper edge being half of the vortex length. The ventilation of the recirculation zone takes place through the edges of the trapeze, but the ventilation can be limited by the presence of a downwind building, if the building intercepts one of the edges (Berkowicz, 1997b).

Considering the chemical transformation in a street canyon, only the fastest chemical reactions can have any significance. For nitrogen oxides, it is therefore sufficient to include only the three basic reactions involving NO, NO_2, O_2 and O_3.

3.2 EXPERIMENTAL METHODS

3.2.1 Concentration measurements

The concentrations of nitrogen oxides, O_3 and CO were analysed by chemiluminescence monitors, UV-absorption method, and IR-method, respectively. All the concentration monitors were calibrated regularly on site. The street-level and roof-level monitors were also intercalibrated against each other, in order to ensure the comparability of results.

3.2.2 Meteorological measurements

The meteorological parameters required as input by the OSPM model are: wind speed and direction, ambient air temperature and global solar radiation. The wind speed was measured using a Vaisala anemometer and the wind direction with a wind vane. The ambient temperature was measured by a Pt-100 sensor, and the global radiation using a pyranometer.

Detailed analysis of the measured meteorological data showed that the wind direction measurements at the Runeberg street were not sufficiently accurate. These were therefore replaced with corresponding measurements at the meteorological station in Kallio, located approximately 1.5 km north-east from the measurement site.

4. Results

4.1 MEASUREMENTS AT THE STREET LEVEL IN 1997

Figure 2 presents monthly averaged concentrations of NO_2, measured at three monitoring stations in downtown Helsinki in 1997. The values in Runeberg street are street-level concentrations. The other two measurement stations (Töölö and Vallila) represent busy traffic environments, but these are not located inside street canyons.

The seasonal variation of urban air pollution is to a large extent determined by the meteorological variables, which are important for air quality. The NO_2

concentrations are commonly highest in late autumn, winter and spring in Northern European conditions (Kukkonen et al., 1999a,b). However, the seasonal variation of NO_2 concentrations shown in Figure 2 is moderate for all the stations. The concentration levels and their variation are remarkably similar at the stations of Töölö and Runeberg street; both of these stations are located in the same district in downtown Helsinki.

Analysis of the meteorological conditions prevailing during the measurements shows that the dataset contains a clearly larger fraction of low wind speed cases, compared with datasets from previous street canyon measurement campaigns conducted, for instance, in Denmark, (Berkowicz et al., 1997b). These previous datasets have been utilised for development of OSPM. However, most of the low wind speed conditions prevailed at night, when the traffic flow and the emissions are low.

In the following, this paper addresses only results from the above mentioned intensive measurement period.

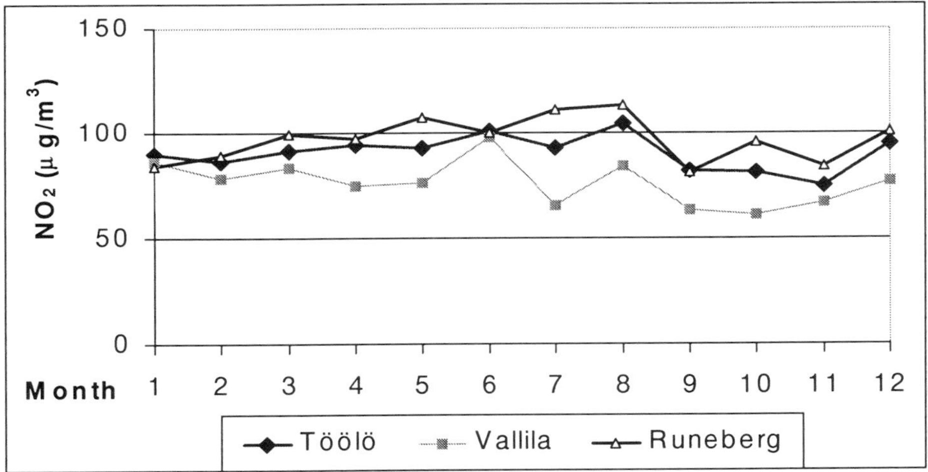

Figure 2. Measured monthly averaged concentrations of NO_2 at three monitoring stations in downtown Helsinki in 1997. The measurements at the Runeberg street are from the street level.

4.2 PREDICTED VERSUS MEASURED CONCENTRATIONS

The OSPM model utilises the hourly time series of street level emissions, urban background concentration data and meteorological data as input values, and predicts concentrations at the street level measurement location. The differences of measured and predicted values are caused by inaccuracies in (i) traffic emission estimates, (ii) the dispersion model, and (iii) the measured concentration and meteorological data. Part of differences are also due to the stochastic nature of atmospheric turbulence.

The measured and predicted NOx concentrations during the intensive measurement campaign have been presented in Figure 3. There are some periods with substantially high NOx concentrations, e.g., during 20-24 March (day numbers from 18 to 22). The agreement of predicted and measured daily average values tends to deteriorate during periods with the highest concentrations.

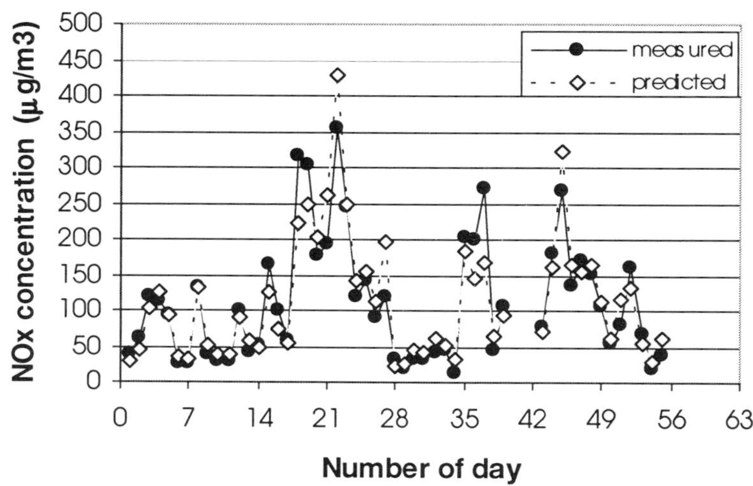

Figure 3. The measured and predicted daily average NOx concentrations during the intensive measurement campaign at the street level in Runeberg street, from 3 March to 30 April, 1997.

Figures 4a-c show a comparison of measured and predicted hourly averaged concentrations at the street level, during the intensive measurement period. The internal variation of data points in Figures 4a-c is substantial. Considering the CO and NO_x concentrations (Figures 4a-b), the scatter plots are symmetrical, i.e., the model predicts fairly well the average concentrations during the measurement campaign. Only a minor fraction of data points are located outside the "factor of two" lines.

However, the NO_2 results (Figure 4c), show a modest, but consistent bias. The model slightly overpredicts the measured NO_2 concentrations. In order to find out the reasons for this difference, we analysed the variation of the NO_x and NO_2 concentrations at the roof and street stations, against the wind direction. Both the NO_x and NO_2 roof station concentrations exhibit a similar (although less pronounced) variation against the wind direction, compared with the street station NO_x and NO_2 concentrations, respectively. We can therefore conclude that at least some of the concentrations measured at the roof station have been

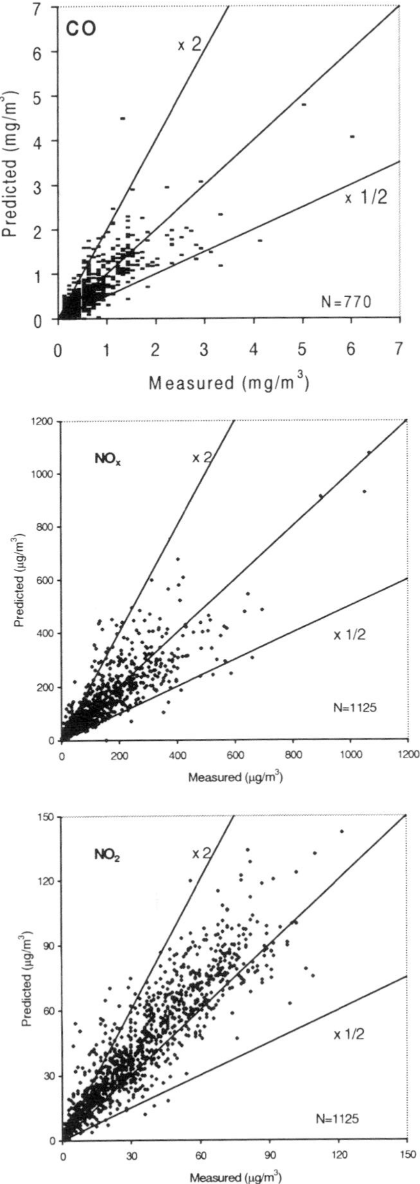

Figures 4a-c. Comparison between measured and predicted hourly concentrations of CO, NO_X and NO_2. The lines showing an agreement of predictions and data by a factor of two have also been indicated. N is the number of data points.

affected by street traffic emissions, and background contributions tend to be overestimated.

However, the NO_x concentration levels measured at the roof station are much lower than the street station levels, and the influence of urban background is therefore very small. On the other hand, the influence of urban background on the NO_2 concentrations on the street level is substantial. The model overprediction of NO_2 is therefore most likely caused by overestimation of the background contribution.

5. Conclusions

The dataset contains a clearly larger fraction of low wind speed cases, compared with datasets from previous street canyon measurement campaigns conducted, for instance, in Denmark. However, there were no 'episodic' conditions, with exceptionally high concentrations.

We have presented observed against predicted concentrations as scatter plots, and against the wind direction. The scatter plots show that the OSPM model predicts fairly well the average CO and NO_x concentrations during the measurement campaign; and there is a slight overprediction for the NO_2 concentrations. The reason for the moderate, but systematic overprediction of the NO_2 concentrations is most likely the influence of street emissions on the measured concentrations at the roof station.

The results also show that the OSPM model reproduces fairly well the mean concentration levels also for cases with very low wind speeds, although for those cases the scatter of data is substantial. In low wind speed conditions, modelling of the traffic-induced turbulence is crucial for the computed results.

6. Acknowledgements

We wish to express our thanks to Dr. Juhani Laurikko (Technical Research Center), and Ms. Marja Viiri and Mr. Matti Hämäläinen (Finnish Road Administration) for their valuable help on the analysis of the traffic densities and emissions. We also wish to thank Mr. Pasi Mäkinen (YTV), Mr. Timo Koskinen (FMI) and Mr. Kaj Lindgren (FMI) for their careful work in conducting the concentration and meteorological measurements. We thank Ms. Helena Saari (FMI) and Ms. Mia Pohjola for their valuable help in processing the data. Ms. Päivi Aarnio (YTV) is thanked for her comments on the manuscript. We also wish to thank the authorities and kind personnel of the private hospital Mehiläinen for the use of their facilities.

This study has been part of the national MOBILE and European SATURN projects. The financial support from the Technology Development Centre, the Ministry of the Environment and the Academy of Finland is gratefully acknowledged.

7. References

Berkowicz, R., Palmgren, F., Hertel, O. and Vignati, E., 1996. Using measurements of air pollution in streets for evaluation of urban air quality – meteorological analysis and model calculations. *The Science of the Total Environment 189/190*, pp. 259-265.

Berkowicz, R., 1997. Modelling street canyon pollution: model requirements and expectations. *Int. J. Environment and Pollution, Vol. 8, Nos. 3-6*, pp. 609-619.

Berkowicz, R., Hertel, O., Sørensen, N. and Michelsen, J., 1997a. Modelling air pollution from traffic in urban areas. In: Perkins, R. and Belcher, S. (eds.), *Flow and Dispersion through Groups of Obstacles*. Clarendon Press, Oxford, pp. 121-141.

Berkowicz et al., 1997b, Hertel, O., Larsen, S.E., Sorensen, N.N. and Nielsen, M., 1997b. Modelling traffic pollution in streets. *Ministry of Environment and Energy, National Environmental Research Institute*, Roskilde, 51 p.

Helsinki Metropolitan Area Board, 1999. Helsinki Metropolitan Area Transportation System Plan, PLJ 1998. *Helsinki Metropolitan Area Series A* 1999:4. YTV Helsinki Metropolitan Area Council, Helsinki, 25 p.

Hertel, O. and Berkowicz, R., 1989a. Modelling Pollution from Traffic in a Street Canyon. Evaluation of Data and Model Development. *DMU Luft-A129*. National Environmental Research Institute, Roskilde, 77 p.

Hertel, O. and Berkowicz, R., 1989b. Modelling NO_2 Concentrations in a Street Canyon. *DMU Luft A-131*. National Environmental Research Institute, Roskilde, 31 p.

Hertel, O. and Berkowicz, R., 1989c. Operational Street Pollution Model (OSPM). Evaluation of the Model on Data from St. Olavs Street in Oslo. *DMU Luft-A135*. National Environmental Research Institute, Roskilde, 34 p.

Kukkonen, J., Salmi, T., Saari, H., Konttinen, M. and Kartastenpää, R., 1999. Review of urban air quality in Finland. Boreal Environment Research, Vol. 4, No. 1, pp. 55-65.

Kukkonen, J., Konttinen, M., Bremer, P., Salmi, T. and Saari, H., 2000. The seasonal variation of urban air quality in northern European conditions. *International Journal of Environment and Pollution* (in print).

Nakamura, Y. and Oke, T.R., 1988. Wind, temperature and stability conditions in an E-W oriented canyon, *Atmospheric Environment* 22, 2691-2700.

Sini, J.F., Anquetin, S. and Mestayer, P.G., 1996. Pollutant dispersion and thermal effects in urban street canyons, Atmospheric Environment 30, 2659-2677.

Yamartino, R.J. and Wiegand, G., 1986. Development and evaluation of simple models for flow, turbulence and pollutant concentration fields within an urban street canyon, *Atmospheric Environment* 20, 2137-2156.

YTV, 1997. The effects of the transportation system upon air quality (in Finnish). *Publication Series of the Helsinki Metropolitan Area B 1997:11*. The Helsinki Metropolitan Area Council, Helsinki, 25 p. + appendices.

THE DRAFT OF THE NEW GERMAN GUIDELINE VDI 3782/8 TO MODEL AUTOMOBILE EXHAUST DISPERSION

ACHIM LOHMEYER, WOLFGANG BAECHLIN* and MATTHIAS KETZEL+

Lohmeyer Consulting Engineers, Mohrenstrasse 14, D-01445 Radebeul, Germany,
**Lohmeyer Consulting Engineers, An der Rossweid 3, D-76229 Karlsruhe, Germany,*
+National Environmental Research Institute, Frederiksborgvej 399, 4000 Roskilde, Denmark
E-mail: Achim.Lohmeyer@Lohmeyer.de

Abstract. The draft of the German guideline to calculate automobile exhaust dispersion is explained. It contains a two-stage-system: For first quick estimates the guideline contains the simple models MLuS and STREET. In case these models are not applicable or their results shows concentration levels close to the air quality standards, the more complex models PROKAS_V and MISKAM are recommended. PROKAS_V is a Gaussian plume model, MISKAM is a 3-dimensional microscale non hydrostatic flow model for built-up areas with an Eulerian dispersion model. The guideline comprises cases in rural areas without or with few adjacent buildings as well as urban areas with buildings near the roads. The contribution gives information about the models, typical results and some of the problems showing up presently.

Key words: air pollution, model, guideline, automobile, concentrations, MLuS, STREET, PROKAS, MISKAM, AIR-EIA.

1. Introduction

There exists a large variety of models to calculate atmospheric dispersion. A selection is possible for example in the Internet, using the „Model Documentation System" (Model Documentation System) of the European Environmental Agency. Restricting the choice to the calculation of concentrations in the vicinity of vehicle roads, about 25 serious models are displayed.

Not all models are equally suitable for practical purposes, but some of them seem quite similarly applicable. For that reason the users in the consulting businesses often have a problem in choosing the appropriate model. Also the environmental or licensing authorities have to invest a significant effort if different models are used, as they have to understand, interpret and sometimes reproduce expert studies prepared with these models. Additionally it is a legal problem if the determination of the air pollution is done with different models, as different models will yield different results.

The German Society of Engineers (VDI) has therefore picked up that problem and has presented in march 1998 as the result of the work of about 20 experts the draft of a guideline (VDI 3782/8) titled: Ausbreitungsrechnung fuer KFZ-Emissionen (Calculation of dispersion of automobile exhausts). The draft was

Environmental Monitoring and Assessment **65**: 381–387, 2000.

announced in the official federal German journals, objections had to be handed in till August 1998, the objections are not yet discussed finally.

The following chapters inform about the guideline, the models and typical results.

2. The choice of the models

Table 1 shows the 4 typical applications and the models which are recommended by the guideline. For each application, exactly one model is recommended. The selection of these models was based on

- their physical contend,
- the availability of a professional description and manual
- a successful comparison of calculated concentrations to concentrations in the wind tunnel and the field and
- the possibility to buy each of these models incl. a professional hotline service for a price of less than about 500.- Euro.

The table indicates the two-stage-solution: For estimations the models MLuS (MLuS-92, 1996) and STREET (Pfeifer et al., 1996) are provided, MLuS in areas without or with only a few buildings, STREET in built-up areas. If these simple models are not applicable or if the results show concentration levels close to the air quality standards and thus the need of more detailed studies, the use of PROKAS (Boesinger, 1996) and MISKAM (Eichhorn, 1996) is recommended. For further information about the models see AIR-EIA in the internet.

Table 1
Models to be used according to draft of the guideline (VDI 3782, Blatt 8, 1998)

Field of application (stage)	**Model to be used (depending on the presence of buildings)**	
	without or with few adjacent buildings	**densely spaced buildings near road, intersections**
1. Estimation of concentrations for planning purposes or to find the points with the major concentrations in a city. Method is simple, accuracy of result is reduced. According to speed v of vehicles:		
a) v > 50 km/h	MLuS-92	STREET
b) v ≤ 50 km/h	STREET	STREET
2. Detailed consideration of concentration distributions	PROKAS	MISKAM, statistic with PROKAS_S

3. Further contents of the guideline

In the guideline further information is given concerning the required input parameters and their need of accuracy, i.e. which input parameters are more important and which are less important and where to get these parameters. Additional advices are given dealing with the handling of the background concentration, the NO-NO_2-conversion, the transfer of wind information from further distant locations to the areas under consideration etc. and comparisons between the statistical parameters of calculated and measured concentration distributions. The emission modelling is not contained in the guideline, there exists a separate model, developed under supervision of the German Federal Environmental Agency (UBA), called MOBILEV (MOBILEV, 1998).

4. Description of the models

MLuS-92 is a regression model based on long-time measurements near streets, among them seven years time series from monitoring points at several distances at both sides from a motorway. MluS-92 has a restricted set of input parameters: Numbers of vehicles per day, truck content, driving speed, annual mean of the wind speed, the year under consideration and the distance of the point under consideration from the roadside. The calculated result are the vehicle induced concentrations (annual mean and 98-percentile). Currently MLuS-92 is in an upgrading process to additionally allow estimates of concentrations near the entrances of road tunnels and near intersections and for the influence of noise protection devices.

STREET is based on dimensionles results precalculated by MISKAM for 22 building configurations. It is a windows program, Fig. 1 displays the screen for input and results. Required are information about the classes of buildings adjacent to the street (Strassenkategorie), orientation of the street relative to north (Straßenrichtung), the annual distribution of the wind direction etc. The calculated result is the annual mean of benzene and soot and the 98 percentile of NO_2 at the point in 1.5 m distance from the buildings where the highest concentrations occur.

PROKAS is a windows program, containing a Gaussian plume model and a handling adapted to the needs of dealing with traffic induced concentrations. Modelling of up to 5000 line sources (reproduced by sets of point sources) is possible. Fig. 2 shows a typical result for the vicinity of a village. The squares indicate the positions of the points under consideration and the colour of the squares the calculated concentration.

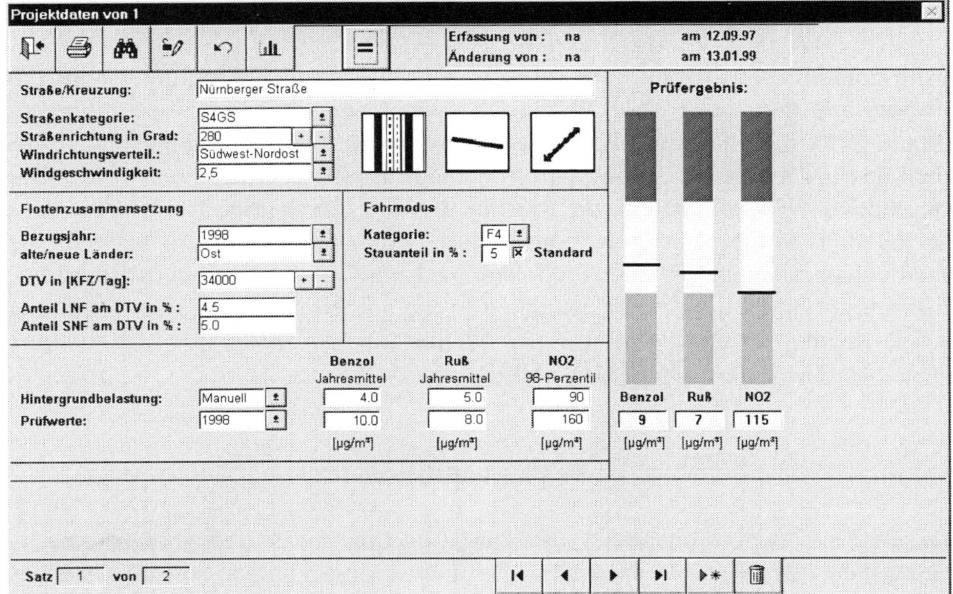

Figure 1. Screen for input and results of STREET.

MISKAM is a microscale 3-dimensional flow model for built-up areas with an Eulerian dispersion model. For MISKAM a comfortable windows surface is available allowing an improved handling of the model for practical cases and the determination of the statistics of the concentration distribution. Fig. 3 shows as a typical result an intersection with adjacent (white) buildings. The numbered points might be interesting positions for measurements of anticipated highest concentrations.

The guideline includes for each model a chapter with comparisons between calculated and measured concentrations (annual mean and 98 percentiles). Comparisons with several field measurements indicate deviations between calculated and measured values in a range of –40 % to +50 % depending on the specific pollutant and the model.

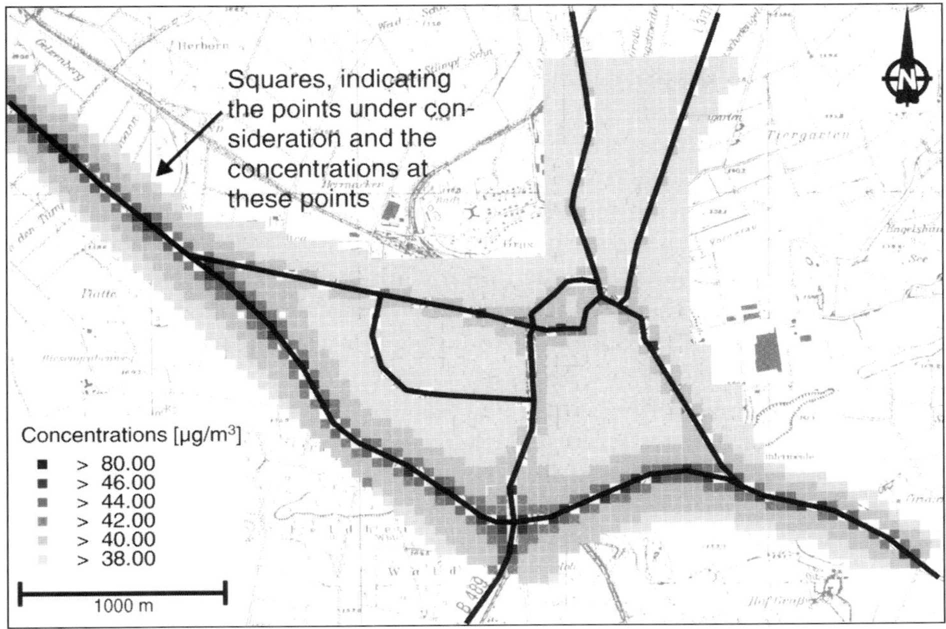

Figure 2. Typical result of annual mean of NO_2-concentration distribution calculated with PROKAS (Boesinger, 1996) for the vicinity of a village.

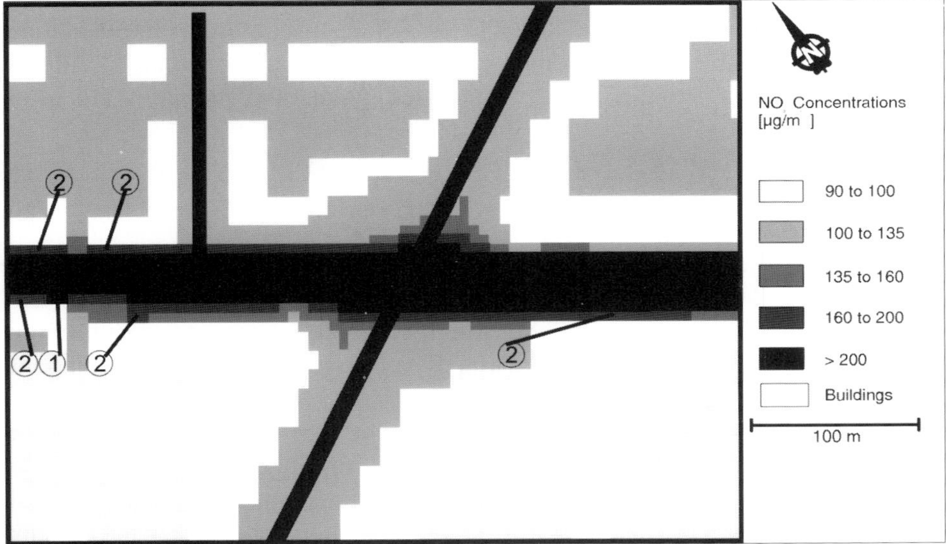

Figure 3. Typical result of concentrations in a city calculated with MISKAM (Eichhorn, 1996) at an intersection with adjacent (white) buildings. The numbered points might be interesting positions for measurements of anticipated highest concentrations.

5. Problems

The decision to set up the guideline and the choice of the models was made about 2 to 3 years ago, at the time of MISKAM version 1.1. In the meantime, the development of MISKAM went on. See for example Ketzel et al. (1999). It turns out, that MISKAM still changes permanently. On the other hand it is known, that the choice of the dispersion model is only one source for uncertainty for a calculated concentration. As shown by Schaedler et al. (1999), an equally important source is the emission modelling, the influence of the traffic induced turbulence, the wind statistics etc. Therefore it might at that stage not be advisable and necessary to fix a version in a guideline. Additionally we have learned in the meantime, that for a fixed model, it is necessary to have:

- A sufficient number of high quality „complete" wind tunnel and field measurements to compare the calculated results to these measurements.
- A complete documentation of the tests of the model, showing the validation for the successive versions. This documentation should be available to the user inclusive the data sets and input files, in order to enable the user to repeat the tests.
- A solution how to efficiently and quickly handle problems that arise during the use of a highly complex model fixed in the guideline. If a model is developed by a financially and personally weak enterprise there is the danger, that the guideline depends on just one person. Such a dependency supports the desire of the modeller community to have the well commented source code of the model with the guideline. The risk of a financial failure of the developer might be avoidable if the development is done by a major research center in cooperation with potential users, coordinated by the involved major federal or state agencies.
- A plenty of model runs for different test cases, sufficient time to test the model, and tests by many independent users.

In consequence of the last point, a project was launched in Germany, where all users are invited to calculate the concentrations in a street canyon in the city of Hannover. The measurements just started, so nobody knows the result before he has to hand in the result of his calculations. The user might apply the models and the procedure provided in the guideline (Richtlinie VDI 3782, Blatt 8 Entwurf, 1998) or he might use any other model or procedure. Thus it is expected to see how the deviations in the calculated results of different users arise and how the guideline has to be extended to get results which are as far as possible independent from the institution or person doing the job. Further information on this project can be found in BW-PLUS, see references.

Acknowledgment

The authors wish to express their appreciation for support from the European Commission's Training and Mobility of Researchers Programme (TMR) within the frameworks of the European Research Network on 'Optimisation of Modelling Methods for Traffic Pollution in Streets' (TRAPOS).

References

AIR-EIA: See http://aix.meng.auth.gr/AIR-EIA.

Boesinger, R.: 1996, Qualitaetssicherungspapier PROKAS, Hrsg.: Ingenieurbuero Dr.-Ing. Achim Lohmeyer, Karlsruhe.

BW-PLUS: See http://bwplus.fzk.de/aktuell/aufruf.htm.

Eichhorn, J.: 1996, Validation of a Microscale Pollution Dispersal Model, In Air Pollution Modelling and Its Application IX, Plenum Press, New York & London, 539-548.

Ketzel, M., Berkowicz, R. and Lohmeyer, A.: 1999, Dispersion of traffic emissions in street canyons - Comparison of European numerical models with each other as well as with results from wind tunnel and field measurements, Proceedings of the 2nd Intl. Conf. on Urban Air Quality - Measurement, Modelling and Managment, 3-5 March, 1999, Madrid, Institute of Physics, 76 Portland Place, London W1N 3DH.

MLuS-92: 1996, Merkblatt ueber Luftverunreinigungen an Strassen, Version 1992 mit Aktualisierung 1996, Teil: Straßen ohne oder mit lockerer Randbebauung, Neudruck 1996, Hrsg.:Forschungsgesellschaft fuer Strassen- und Verkehrswesen, Koeln.

MOBLEV: 1998, Massnahmenorientiertes Berechnungsinstrumentarium fuer die lokalenSchadstoffemissionen des Kfz-Verkehrs, Hrsg.: FIGE GmbH, Herzogenrath.

Model Documentation System: See http://aix.meng.auth.gr/lhtee/database.html.

Pfeifer, T., Frank, W., Kost, W. J. und Droescher, F.: 1996, Das Screening-Modell STREET, Wissenschaftliche Dokumentation, Hrsg.:TUEV Energie und Umwelt GmbH, Filderstadt.

Schaedler, G., Baechlin, W. and Lohmeyer, A.: 1999, Immissionsprognosen mit mikroscaligen-Modellen - Vergleich von berechneten mit gemessenen Groessen. Hrsg.: Forschungszentrum Karlsruhe GmbH, BW-PLUS.

VDI 3782, Blatt 8: 1998, Entwurf, Ausbreitungsrechnung fuer Kfz-Emissionen, Beuth Verlag, Berlin.

Turbulent ventilation of a street canyon

Morten Nielsen (n.m.nielsen@risoe.dk)
Risø National Laboratory, Wind Energy and Atmospheric Physics Department, P.O. Box 49, Roskilde 4000

Abstract. A selection of turbulence data corresponding to 185 days of field measurements has been analysed. The non-ideal building geometry influenced the circulation patterns in the street canyon and the largest average vertical velocities were observed in the wake of an unbroken line of buildings. The standard deviation of vertical velocity fluctuations normalised by the ambient wind speed was relatively insensitive to ambient wind direction and sensor position, and it was usually larger than the corresponding 1-hour average velocity. Cross-correlations of spatially separated velocity measurements were small, and this suggests that most of the velocity fluctuations were fairly local and not caused by unsteady street vortices. The observed velocities scaled with the ambient wind speed except under low-wind conditions.

Keywords: Canopy flow, street vortex, field experiment, turbulence, statistical analysis, spatial correlation

1. Introduction

The ventilation of an urban street canyon is influenced by circulation patterns induced by the topography. Fresh air penetrates the street canyon along the downwind wall and the highest concentrations of traffic pollutants are typically found in the wake of the upwind wall. The circulation is driven by the wind above the urban canopy, which is of fluctuating speed and direction. In this way the turbulent wind may cause pulsating circulation patterns and thereby enhance the efficiency of the ventilation, e.g. when the wind direction oscillates around the along-street direction. The organised circulation patterns, permanent or intermittent, co-exist with small-scale random eddies which assist in dispersing the pollutants. Often residual turbulence will be generated by moving vehicles or convection from heated surfaces even when the ambient wind ceases to blow. This additional turbulence is quite important since it moderates the pollution under unfavourable atmospheric conditions.

These conditions, the large variety of building shapes and the perhaps uncertain emission rates, make numerical or wind-tunnel modelling of dispersion in an urban environment rather difficult. Numerical models are based either on k-ϵ turbulence closure or simpler paramet-

Environmental Monitoring and Assessment **65**: 389–396, 2000.

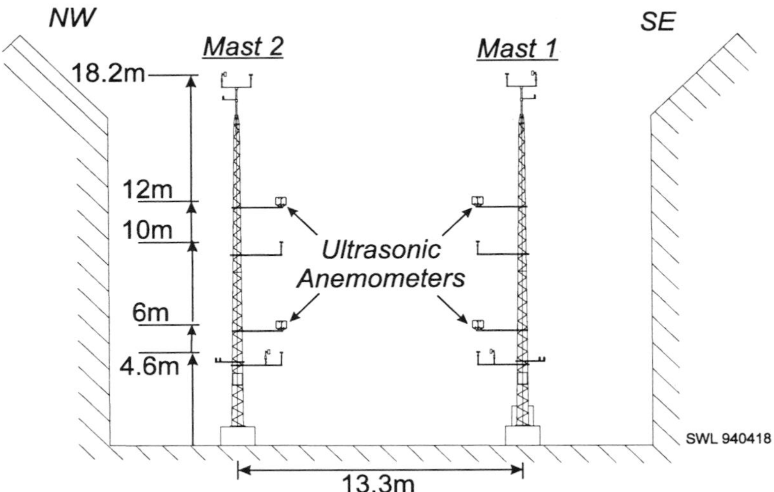

Figure 1. The two meteorological masts and position of the ultrasonic anemometers.

erisations of the plume from the vehichles, e.g. (Sørensen et al., 1994) and (Hertel et al., 1989) respectively.

This paper presents field measurements from the street *Jagtvej* in Copenhagen. It focuses on vertical velocities in the street canyon and shows that the typical standard deviations are much higher than their averages. In principle this variability might have been introduced by intermittent circulation patterns rather than small-scale turbulence. The latter question is addressed by presentation of cross-correlation statistics between spatially separated velocity measurements.

2. Experimental layout and available data

The measurements were Risø's contribution to the project *Air pollution from traffic in urban areas* led by the National Environmental Research Institute (NERI), see Berkowicz et al. (1997) for a general presentation. Two identical masts were erected on each side of the street near the site where NERI operates a long-term air quality measuring station. The masts carried a set of meteorological instruments, which measured flow, turbulence and temperatures for 12 months during 1994-95, see Figure 1. Ultrasonic anemometers were mounted at the 6- and 12-m levels on each mast and measured time series of the three-dimensional turbulent velocity. The wind above the canopy was measured on a reference mast situated on a flat roof 500 m away. The building heights in the neighbourhood are relatively uniform. The traffic on Jagtvej runs in two directions.

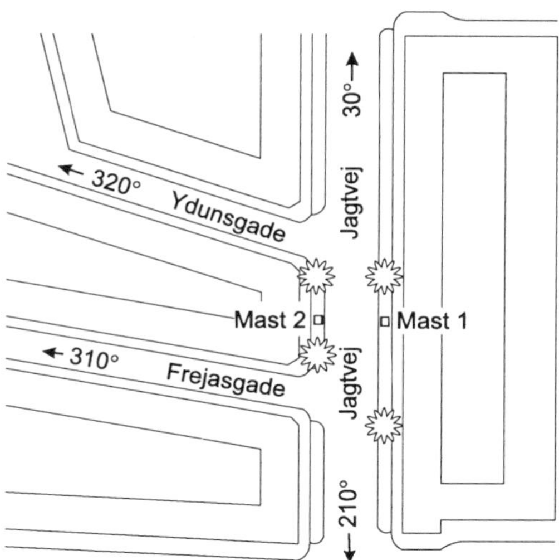

Figure 2. Map of the measuring site showing measuring masts, nearby streets and the nearest roadside trees. The indicated directions are measured clock-wise from the north.

A similar experiment including ultrasonic anemometers were previously done in a street in Zürich (Rotach, 1995) . The Zürich experiment experiment were designed to investigate turbulence in the roughness sublayer and deployed more instruments above the buildings and at roof-top level, i.e. at higher levels than in the Jagtvej experiment.

Figure 2 shows a map of the experimental site. The NERI air quality station is situated on the eastern side of the street just north of mast 1. With a street of infinite length and a uniform symmetric cross section the directional distribution of the shelter factor would have been symmetric around the cross-street axis (120-300°). In this ideal case the shelter at mast 2 would furthermore be a mirror of that at mast 1 relative to the along-street axis (30-210°). Perfect symmetry in the observations was however not to be expected, since the side streets break the row of buildings on the NW side of the road. To some extent the roadside trees probably also sheltered the sonic anemometers from along-street flows.

Block statistics based on 1-hour observation periods were calculated on-line by the data acquisition system and collected into a database, from which data for the present analysis have been extracted. The anemometers had certain periods with missing or erroneous signals. Data from these periods are avoided by requiring all average vertical velocities to be within the range ± 1.5 m/s and the corresponding variances normalised by the square of the reference wind speed σ_w/u_0 to be within the range 0.01 to 0.3. These selection criteria leave 4428 records

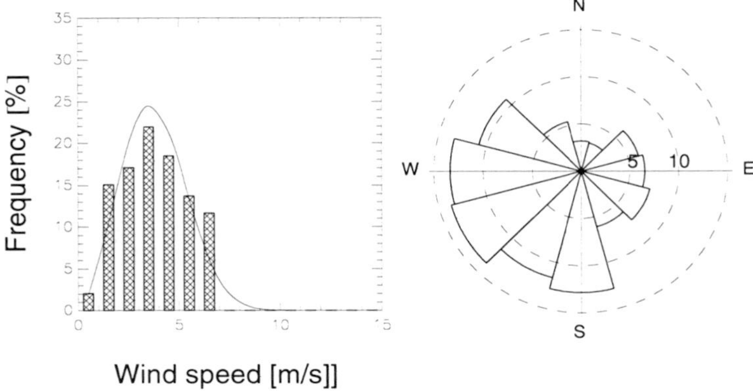

Figure 3. Wind speed histogram and directional distribution based on measurements at the reference station.

corresponding to 185 days of measurements for analysis or 83% of the available data. The main cause for data rejection was signal noise from the low-level anemometer on mast 1. The second screening criterion is a check of the measured turbulence intensity although it involves normalisation by the velocity above the buildings rather than street-canyon velocities with variable shelter. Figure 3 shows distributions of ambient wind speed and direction based on simultaneous data from the reference station. The average ambient wind speed is 4.4 m/s and the average wind direction is 230° or close to the along-street axis.

3. Results and Discussion

Figure 4 shows a composite plot of vertical velocities normalised by wind speeds measured at the reference station σ_w/u_0 as a function of the ambient wind direction. This ratio was preferred to local The scatter points and the solid lines indicate one-hourly average values, whereas the dashed curve above shows the standard deviation during one hour of observation. The curves plotted by solid lines were calculated as running mean values through data points sorted by direction. The scatter points give an impression of the statistical variability of hourly observations. The dashed vertical lines indicate the along-street axis (30–210°) and it is seen that along-street ambient winds generally produced weak vertical average velocities as expected. The largest vertical velocities were observed when the wind came from easterly directions, i.e. with the unbroken line of buildings on the upwind side of the street. A distinct wake on the upwind side of the street was observed for wind directions near 120°, whilst the side streets on the other side of the road probably prevented development of a corresponding downwash. At each

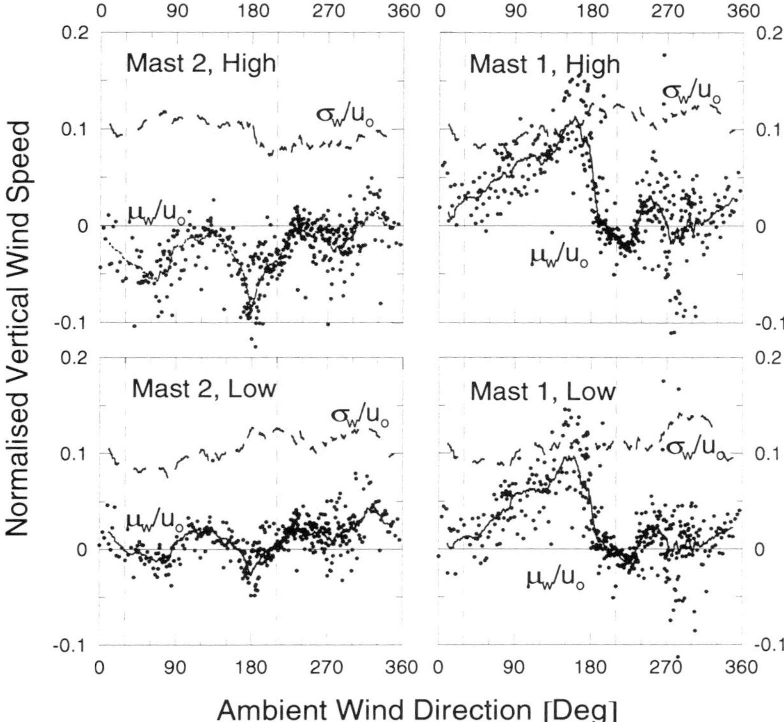

Figure 4. Normalised vertical wind speed statistics for each of the four ultrasonic anemometers. The scatter points show hourly average values as a function of ambient wind direction and the solid lines shows the running averages. The dashed lines above show similar running average values of the normalised standard deviations.

mast the directional variations of velocities at the two observation levels seem proportional. The normalised standard deviations shown by the dashed lines in Figure 4 are not very sensitive to the ambient wind direction or measuring position. They may all be approximated by a value of 0.1, which in most cases is large compared to the normalised average vertical velocity.

Are the results of Figure 4 sensitive to the ambient wind speed? Well, Figure 5 shows similar directional running average values calculated from observations by the lowest anemometer on mast 1 for four ranges of the ambient wind speed. The running averages for moderate wind speeds $2 < u_0 \leq 4$ m/s and high wind speeds $u_0 > 4$ m/s are not very different, and reverting to Figure 3 we see that these two ranges include a large fraction of the data set. The hourly normalised standard deviation increased under weak ambient wind conditions in agreement with the assumption of increasing significance of turbulence generation by traffic and heat convection. It is perhaps more surprising that nor-

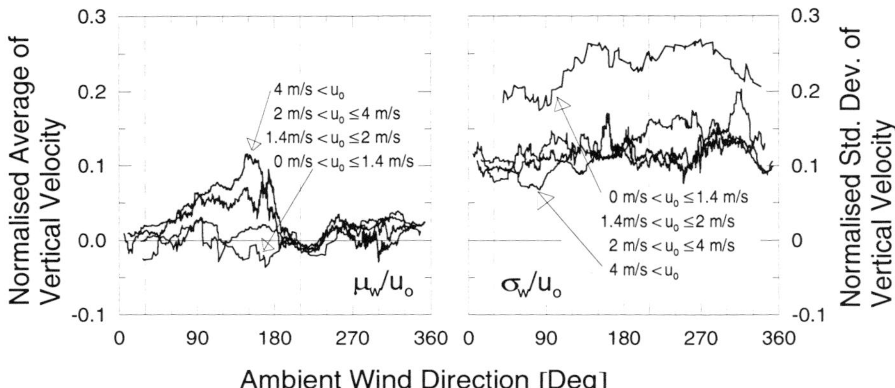

Figure 5. Wind-speed dependence of normalised average and standard deviation at the lowest anemometer of mast 1 as a function of ambient wind direction. Each plot shows four curves for different reference wind speeds.

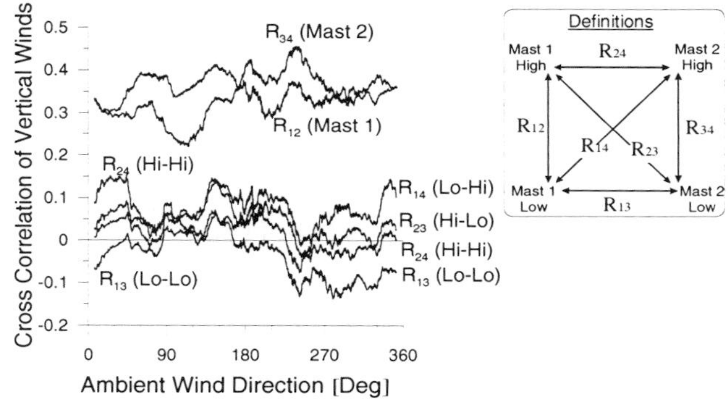

Figure 6. Correlation coefficients of the vertical motions observed by various instrument pairs as defined by the sketch inserted.

malised average vertical velocities decreased during weak ambient wind conditions, as if the ambient wind was too weak to drive the circulation patterns. Similar plots for the other ultrasonic anemomemters are in agreement with these observations.

As mentioned in the introduction, there may be two reasons why the standard deviation of the vertical motions is larger than the average values: 1) the vertical motions could have been associated with small-scale eddies or 2) the circulation patterns of the street canyon might have been unsteady. An ambient wind direction fluctuating around the along-street axis might even produce canopy circulation of alternating spin during an averaging period. Figure 6 shows correlation coefficients of vertical winds measured by pairs of anemometers. These correlations are calculated as the covariance of two signals normalised by their standard deviations. The most striking conclusions from Figure 6 are

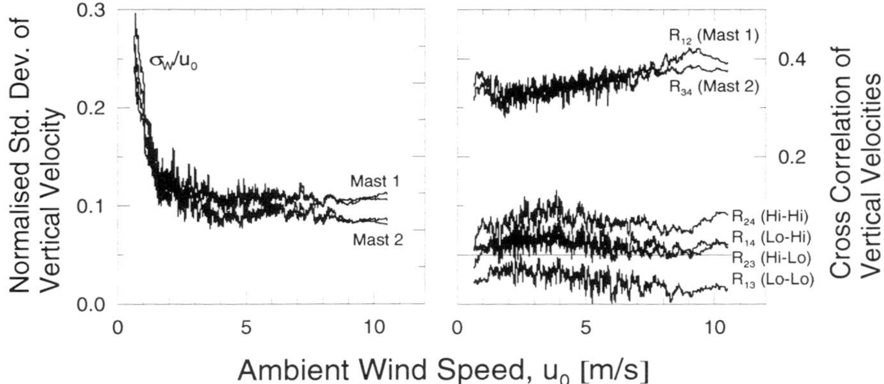

Figure 7. Normalised standard deviations and cross-correlations of spatially separated measurements plotted as a function of ambient wind speed.

that the correlations are insensitive to the ambient wind direction and the largest correlation $R_{12}, R_{34} \approx 0.35$ are found for instrument pairs on the same mast. This factor may be interpreted as the ratio between the energy of unsteady circulation patterns large enough to activate both instruments and the energy of all vertical velocity fluctuations. According to Figure 4 these fluctuations were larger than the 1-hour average velocities. Thus the organised circulation patterns were probably underestimated by mere average velocities, although most of the variability must have been caused by smaller eddies. The correlations for other instrument pairs on two masts are much smaller, and this suggests that intermittent circulation patterns rarely occupied the entire street canyon. It is noted that velocities at the 6-m level of the two masts generally were weakly counter-correlated $R_{13} < 0$ whilst the correlation at the 12-m level was positive $R_{24} > 0$.

Figure 7 examines how second-order statistics depend on the ambient wind speed. The plot on the left-hand side shows that the normalised standard deviation is rather insensitive to the ambient wind speed when this is larger than say $u_0 > 3$ m/s whilst it rapidly increases under low wind conditions. The four curves obtained by measurements with different instruments are remarkably similar. The correlation coefficients shown on the right-hand side of Figure 7 are relatively insensitive to the ambient wind speed with a slight tendency of increasing single-mast correlation with increasing ambient wind speed.

4. Conclusions

The non-ideal building geometry influenced the circulation patterns in the street canyon, and the largest average vertical velocities were observed in the wake of the unbroken line of buildings. Such complexities are probably common in most cities. The local average vertical velocities seemed to scale with the ambient wind speed unless this was very weak, say $u_0 < 1$ m/s.

The standard deviations of the vertical velocities in the street canyon σ_w were larger than the corresponding average values $\overline{w}$, and they were relatively insensitive to sensor position and the ambient wind direction. When the ambient wind was not too weak $u_0 > 3$ m/s, the variance of the vertical velocity perturbations was well approximated by $\sigma_w \simeq u_0$. This relation did not hold under weak wind conditions, probably disturbed by additional turbulence generated by moving vehicles or heated surfaces.

The spatial correlation of pairs of vertical velocity signals was $R \approx 0.35$ for measurements obtained at the 6- and 12-m levels of individual masts and within the range ± 0.1 for pairs of instruments distributed on two masts. These modest values suggest that the main contribution to turbulent kinetic energy was related to small-scale eddies rather than intermittent street vortices. The correlation factors were relatively insensitive to the ambient wind conditions.

Acknowledgements

The *Air pollution from traffic in urban areas* project was part of the *Strategic Environmental Programme* of the Danish government and led by the National Environmental Research Institute (NERI). Søren W. Lund installed and maintained the meteorological masts at Jagtvej, organised the database, and drew Figures 1 and 2. The nearby reference station is part of NERI's long-term monitoring network and data from this were provided by Kåre Kemp.

References

R. Berkowicz, O. Hertel, S. E. Larsen, N. N. Sørensen, and M. Nielsen: 1997, 'Modelling traffic polution in streets', *Technical report*, National Environmental Research Institute, Roskilde, Denmark.

O. Hertel and R. Berkowicz: 1989, 'Modelling pollution from traffic in a street canyon, evaluation of data and model development', *DMU Luft A-129*, National Environmental Research Institute, Roskilde, Denmark.

M. W. Rotach: 1995, 'Profiles of turbulence statistics in and above an urban street canyon', *Atmos. Environ.* **29**, 1473–1486.

N. N. Sørensen, Larsen S. E., and J. A. Michelsen: 1994, 'Computation of flow in a street canyon', *Risø-I-824(EN)*, Risø National Laboratory, Roskilde, Denmark.

THE EFFECT OF LARGE-SCALE TURBULENT STRUCTURES ON A SIMPLE 2-D CANYON-TYPE FLOW

SAVORY E.[1] and ABDELQARI A.[2]

[1]*Fluid Mechanics Research Group, School of Engineering in the Environment, University of Surrey, Guildford GU2 5XH, United Kingdom. E-mail: e.savory@surrey.ac.uk*

[2] *Equipe Dynamique de l'Atmosphere Habitee, Laboratoire de Mecanique des Fluides, Ecole Centrale de Nantes (ECN), 44072 Nantes Cedex 03, France*

Abstract. This paper is concerned with a preliminary experimental investigation of the interaction between large turbulent structures, generated in the wake of a circular cylinder, and the rough-wall turbulent boundary layer separated flow immediately downstream of a simple street canyon type geometry represented by backward-facing step. The motivation for the work was to provide some initial data for the validation of a 3-D k-ε turbulence model used for the prediction of flows and pollutant dispersion within the urban canopy. The aim has been to assess the extent of the perturbation of a simulated street canyon caused by regular large-scale eddies generated upstream. The research has involved the use of thermal anemometry to determine mean velocity and turbulence characteristics both upstream and downstream of the step, together with the mean reattachment length for the recirculating flow. The results indicate that the presence of the cylinder in the flow reduces the reattachment length. In addition, the periodic structures generated in the cylinder wake are rapidly mixed with the turbulence in the step shear layer such that no periodicity is detected at the reattachment zone.

Key words: large-scale turbulence, backward-facing step, shear layer, canyon flow

1. Introduction

A better understanding of the flow and pollutant dispersion characteristics within urban street canyons is important if air quality is to be improved and town centres designed to minimise adverse environmental conditions. The complexities of the urban climate have been extensively reviewed (Mestayer and Anquetin, 1995) and investigations of the problem have involved model scale experiments and numerical turbulence modelling of relatively simplified urban geometries and flows, coupled with limited field experiments.

In experimental and numerical modelling the street canyons are normally treated either as essentially 2-D rectangular cavities or backward-facing step flows, with the approaching fully-developed turbulent boundary layer separating at the leading edge. Such investigations have yielded useful results, particularly concerning the effects of canyon geometry (Oke, 1988, Mestayer *et al*, 1995, Sini *et al*, 1996) and solar-induced canyon wall temperatures on the flow velocities, pollutant levels and dilution times within the cavity (Mestayer *et al*, 1995, Sini *et al*, 1996). The general

Environmental Monitoring and Assessment **65**: 397–405, 2000.

perception of the canyon flow is that it is dominated by a single, stable, large, rotating flow vortex whose axis is parallel to the street direction. For narrower streets this becomes a series of two or three contrarotating vortices, one above another. However, the reality is much different. From investigations of 3-D rectangular cavities, associated with aircraft aerodynamics (Savory *et al*, 1993, 1997b) one of the present authors has noted how the flow is always highly 3-D in nature, particularly when the incident flow is not normal to the main canyon axis. Indeed, the cavity flow is strongly dependent upon geometrical aspect ratio, with particular cases giving rise to increased flow unsteadiness, enhanced momentum exchange with the external flow and strong circulations in the horizontal plane. It is possible that some of these features may be extrapolated to the larger-scale street canyon regime. In addition, in practical urban applications such flows are likely to be strongly perturbed by larger scale eddies either impinging on the boundary layer from higher levels in the urban atmosphere or generated by the flows over upstream canyons. This mesoscale flow field above the city may be modelled using a large eddy simulation, whilst conventional turbulence models may be used for the flow within the street canyons. It is anticipated that some form of coupling of these two numerical modelling approaches will be necessary for the development of a comprehensive urban canyon simulation.

Hence, the aim of the present work has been examine the flow over a rough-wall backward-facing step with a thick separating turbulent boundary layer both with and without the perturbations caused by the influence of large external flow structures. These structures were generated by a circular cylinder positioned in the freestream upstream of the step at a location that produces large eddies of a diameter equivalent to the boundary layer thickness at separation. This simple test case has been chosen in order to validate the urban canyon flow turbulence model CHENSI (Mestayer *et al*, 1993). Although it is recognised that the large-scale turbulence above the urban canopy is likely to be dominated by streamwise, rather than lateral, vorticity, the present test arrangement was used since it presented a convenient and valid method for introducing large-scale, periodic perturbations for model validation purposes.

The backward-facing step flow regime has been the subject of numerous investigations and is a standard test case for numerical modelling. In the case of smooth-wall flows, with a turbulent boundary layer at separation, a number of authors have studied the effects of Reynolds number on the reattachment length (Eaton *et al*, 1979, Durst and Tropea, 1982, Adams *et al*, 1984), whilst others have ascertained the influence of boundary layer thickness (Bradshaw and Wong, 1972, Adams and Johnston, 1988) and turbulence intensity at separation (Isomoto and Homani, 1989) on the subsequent reattachment. The flow downstream of a step with a rough-wall boundary layer has received relatively little attention, notably the research reported in (Badri Kusuma, 1993) and the present research is a continuation of that study. The next section of this paper briefly outlines the facilities and configurations used in the present investigation. The ensuing section discusses some of the results obtained from the study and, finally, some conclusions are drawn.

2. Experimental Details

2.1. WIND TUNNEL AND MODEL ARRANGEMENTS

The experiments were carried out in the low-speed boundary layer tunnel at ECN which has working section dimensions of 4m length x 2m width x 0.5m height. In the present experiments a nominal upstream freestream reference velocity (Uref) of 6.4m/s was used. The basic arrangement for the test geometry is shown in figure 1

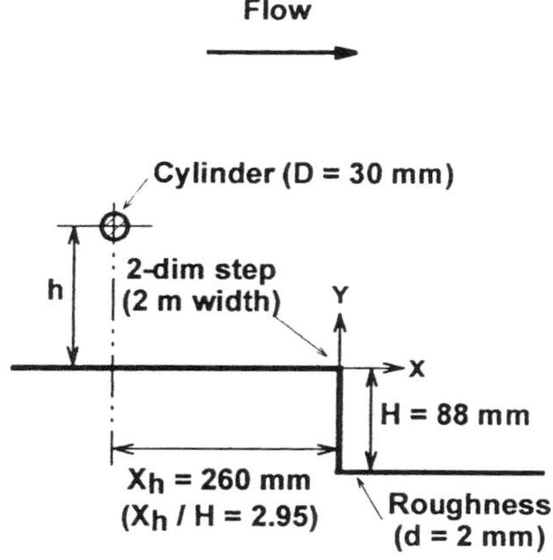

Figure 1. Diagrammatic layout of step and cylinder model (not to scale)

which also gives the principal dimensions. A step height (H) of 88mm was adopted since this was the value used in an earlier study (Badri Kasuma, 1993). The gravel roughness of 2mm diameter was utilised for the same reason. The following parameters for the boundary layer at the step agreed well with an analysis of the data from the earlier study; boundary layer thickness, δ=79.9mm, roughness length, y_o=0.65mm, shear velocity, U_*/Uref=0.0800. The cylinder model was a smooth-walled aluminium tube with an outside diameter (D) of 30mm, giving D/H=0.34. The cylinder spanned the width of the tunnel and graded, narrow supports were provided

provided to allow the model to be set at the desired height above the ground. The cylinder position for the initial measurements was chosen so that at the step the eddies shed by the cylinder had a diameter approximately the same as the thickness of the separating boundary layer and occupied the upper quarter of the boundary layer.

2.2. HOT-WIRE ANEMOMETRY

The measurements were carried out using a single hot-wire anemometer system, with the probe mounted in an aerofoil section, together with a pitot-static tube, which was itself attached to the end of a cylindrical rod vertical traversing mechanism. The hot-wire was operated with an overheat ratio of 0.5 and a freestream calibration against the pitot-static tube was carried out prior to each vertical traverse in order to minimise drift due to ambient temperature changes. Although data obtained from such an instrument in regions of high turbulence intensity or flow recirculation are grossly in error the results do permit an assessment of the effects of cylinder model location on the backward-facing step shear layer development.

2.3 FLOW REATTACHMENT PROBE

It is not possible to use a single hot-wire anemometer probe to determine the location of reattachment downstream of a step since such a device is not capable of resolving flow direction. A number of flush surface-mounted devices have been developed for determining flow direction and reattachment, such as a thermal tuft (Eaton *et al*, 1979). However, many of these instruments are not readily suited for use in rough-wall boundary layers and so a new probe was constructed for such a purpose (Savory *et al*, 1997a). Briefly, the reattachment probe consists of a conventional single hot-wire anemometer operated in constant temperature mode. At right angles to and approximately 1mm equidistant from the wire are the two junctions of a copper/constantan thermocouple. The principle of operation is simple and works on the basis that in a flow field there will be a positive or negative potential difference between the junctions if there is a net flow velocity in either of the two normal flow directions, which results in a net temperature difference at the junctions. In a reattachment zone the mean velocity near the surface tends towards zero and so the thermocouple e.m.f. also tends toward zero. However, in practice, it was necessary to develop an appropriate methodology in order to ensure that the probe gave reliable and repeatable results (Savory *et al*, 1997a).

2.4 SCOPE OF THE MEASUREMENTS

At the nominal freestream reference velocity of 6.4m/s the Reynolds number for the step was $Re_H=3.88\times10^4$ and the Reynolds number for the cylinder was $Re_D=1.32\times10^4$. The frequency of the periodic vortex shedding from the cylinder was

Re_D=1.32x10^4. The frequency of the periodic vortex shedding from the cylinder was 43 Hz. The range of measurements were; (1) Velocity time-histories in vertical profiles at X/H=-0.4, 0, 1, 3 and 5 without the cylinder and with the cylinder set at a heights h=90, 110 and 130mm (giving h/H=1.02, 1.25 and 1.48), (2) Reattachment probe data between X/H=1.5 and 7 without the cylinder and with the cylinder at heights of h=90, 100, 110, 120, 130 and 140mm (h/H =1.02 to 1.59).

3. Results and Discussion

From consideration of the step and cylinder geometry, together with an assumed half-angle for the growth of the cylinder wake of 17 degrees, it is estimated that an undisturbed cylinder wake would reach the ground plane downstream of the step at X/H=4.6 for the case where h/H=1.48 and at X/H=3.1 for the case where h/H=1.02.

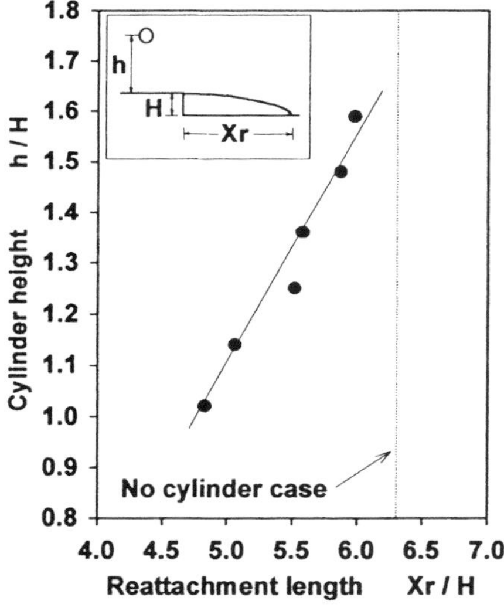

Figure 2. Variation of reattachment length with cylinder height

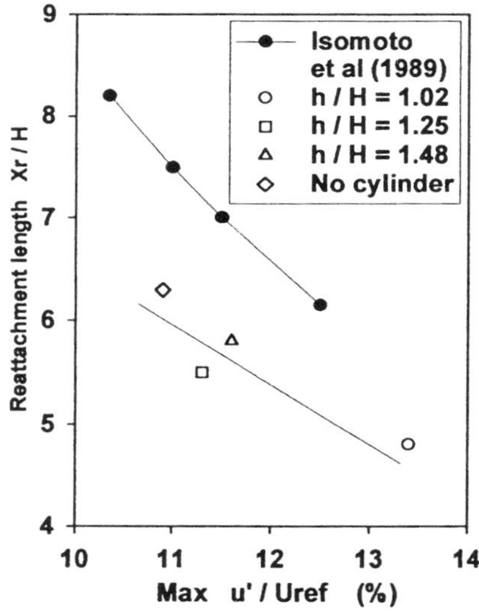

Figure 3. Variation of reattachment length with max. turbulence intensity at separation

The lowest cylinder height was chosen in order to produce a configuration which would, it was estimated, produce vortices that first impinge upon the ground plane at the upper edge of the step. The results from the reattachment probe for all of the cylinder locations are given in figure 2, together with the case of no-cylinder in the flow. The data suggest that the reattachment length of 6.3H in the absence of the cylinder is reduced slightly to 6.0H by the presence of the cylinder at h/H=1.59 and then reduced massively to 4.8H by the cylinder located at h/H=1.02. Previous work

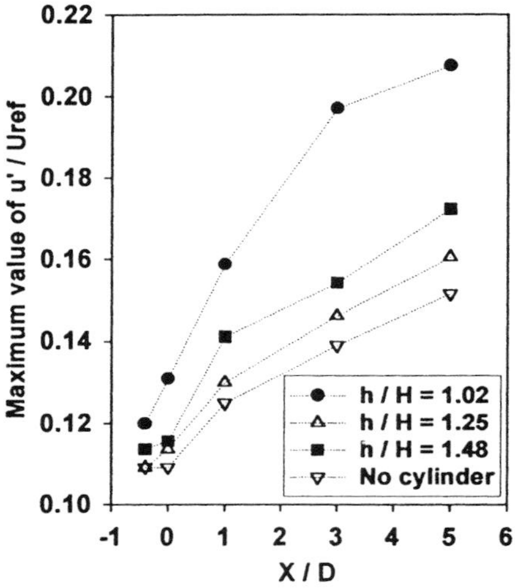

Figure 4. Variation of max. turbulence intensity with downstream distance for different cylinder locations

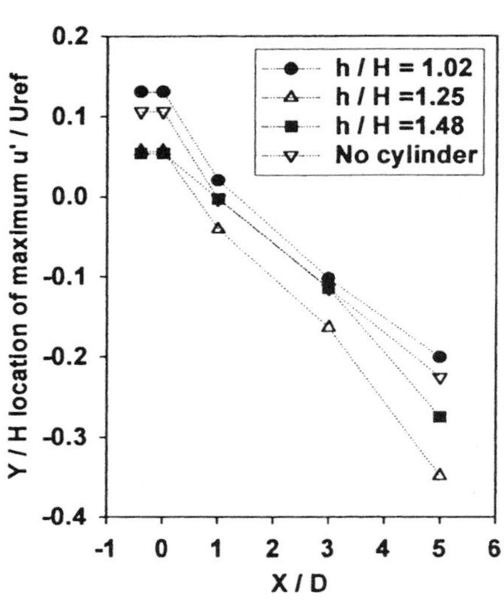

Figure 5. Variation of position of max. shear layer turb. intensity with downstream distance for different cylinder locations

(Isomoto and Honami, 1989) has shown that an increase of 2% in the boundary layer turbulence intensity at step height can reduce the reattachment length by as much as 2H. Figure 3 shows the results from the present work, together with the earlier data (Isomoto and Honami, 1989). It may be seen that the trends in the two data sets are broadly similar, with the difference in turbulence intensity between the no-cylinder case and the cylinder at h/H=1.02 being 2.5%, resulting in a reduction in reattachment length of 1.5H. The differences may be attributed to the fact that the cylinders used by the earlier authors were relatively small ($D/H \leq 0.12$) and that, in the present work, the large cylinder, coupled with the wall roughness, introduced much larger scale turbulence into the flow.

The development of the turbulence intensity profiles downstream of the step for the different cylinder locations are shown in figures 4 and 5. The maximum value of the streamwise turbulence intensity in each vertical profile is given in figure 4 and these clearly show that the shear layer turbulence is greatly increased throughout its length when the cylinder is at the lowest position (h/H = 1.02) when compared to the other cases. The other cylinder positions show small increases in the maximum turbulence levels, when compared to the no-cylinder case, although the rates of increase in the downstream direction are similar in all three cases. The location above the ground plane of the point of maximum turbulence intensity is shown in figure 5. It is surprising to note that the variations in vertical location of the turbulence maxima with downstream distance are very similar in all four cases, with no distinct

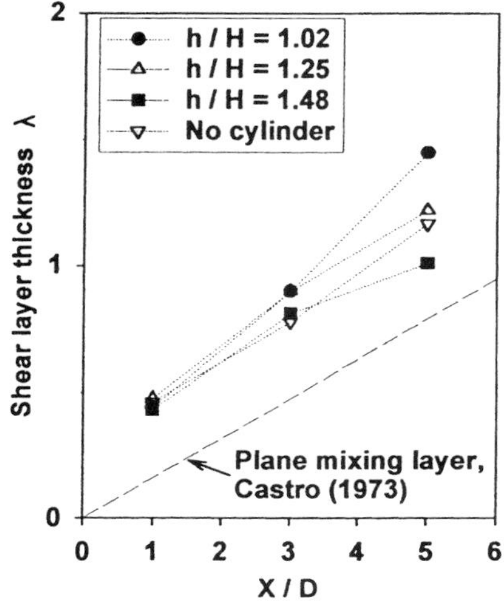

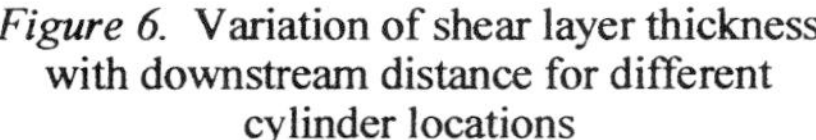

Figure 6. Variation of shear layer thickness with downstream distance for different cylinder locations

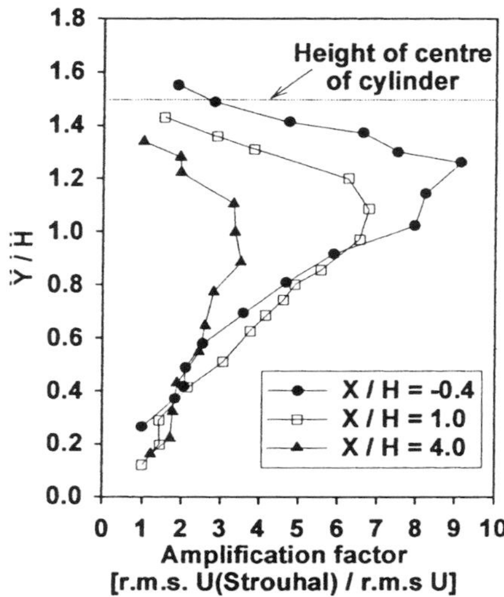

Figure 7. Profiles of Amplification Factor at vortex shedding frequency (43Hz) due to cylinder at h/H=1.48

trends with cylinder location. Hence, it would seem that the earlier reattachment which occurs is due to a small degree of shear layer thickening rather than any dramatic change in the curvature of the shear layer centre-line.

This appears to be borne out in figure 6 by the variation of shear layer thickness (λ) with downstream distance, estimated from the mean velocity profiles and defined as; $\lambda = (Uref / H).(\partial Y / \partial U)_{max}$, where $(\partial Y / \partial U)_{max}$ is the maximum slope in each profile. The lower two locations of the cylinder appear to produce a thicker shear layer as reattachment is approached, with growth rates that are slightly larger than a conventional plane mixing layer (Castro, 1973).

Figure 7 shows how the energy associated with the vortices shed from the cylinder attenuates rapidly through the recirculation region, such that it is not detectable at reattachment (X/H>4). Here, the energy is expressed as an amplification ratio, defined as the ratio of the r.m.s. streamwise velocity fluctuation at the shedding frequency to the r.m.s. velocity in the energy spectrum at that frequency but excluding the vortex shedding peak.

Although only the case of the cylinder at the highest location is shown here, the same result pertains for all the cylinder settings, demonstrating that these well-defined turbulent structures appear to be rapidly mixed into the shear layer since they are not detectable within the body of the shear layer or in the recirculation region.

4. Concluding Remarks

Clearly, the presence of the cylinder greatly alters the mean flow since it generates a large wake which mixes with the separated boundary layer. The effect of this is to increase the mean velocity beneath the wake and through the step shear layer. The turbulence intensities in the outer shear layer do not appear to be effected by the cylinder wake but the peak intensities in the centre of the shear layer are greater with the cylinder present. The presence of the cylinder, located at a height of Y/H=1.59 reduces the reattachment length from 6.3H to 6.0H. With the cylinder set at increasingly lower levels the reattachment length is reduced still further, to 4.8H when Y/H=1.02. It would appear that the increased turbulence intensities in the shear layer, coupled with the increased velocities, has the effect of causing earlier reattachment through a small shear layer thickening. The effect on reattachment of the step height turbulence intensity increase at separation is similar to that found in earlier work.

Acknowledgements

This work has been supported by EU funding, presently under the TMR programme TRAPOS. Thanks are due to Dr P G Mestayer and Dr J-F Sini at ECN for their valuable support and guidance. The authors are indebted to Mr Yves Lorin at ECN for technical support, particularly in constructing the models and reattachment probe.

References

Adams, E.W., Johnston, J.P., Eaton, J.K.: 1984, Experiments on the structure of turbulent reattaching flow, Rept. MD-43, Thermoscience Division., Dept of Mech. Eng., Stanford Univ., USA.

Adams, E.W., Johnston, J.P.: 1988, Effects of the separating shear layer on the reattachment flow structure. 1: Pressure & turbulent quantities, *Expts. in Fluids*, **6**, 400-408.

Badri Kusuma, S.M.: 1993, Etude experimentale d'un ecoulement turbulent en aval d'une marche decendante: Cas d'un jet parietal et de la couche limite, PhD thesis, Ecole Centrale Nantes, France.

Bradshaw, P., Wong, F.Y.F.: 1972, The reattachment and relaxation of a turbulent shear layer. *J. Fluid Mechanics*, **52**, 113-135.

Castro, I. P.: 1973, A highly distorted turbulent free shear layer, PhD Thesis, University of London.

Durst, F., Tropea, C.: 1982, Flow over two-dimensional backward facing steps, *Structure of complex turbulent shear flows*. IUTAM Symposium, 41-52, Marseille.

Eaton, J.K., Johnston, J.P., Jeans, A.H.: 1979, Measurements in a reattaching turbulent shear layer, *Proc. 2nd Symp. on Turbulent Shear Flows*, 16.7-16.12, Imp. Coll., London, July.

Isomoto, K., Honami, S.: 1989, The effect of inlet turbulence intensity on the reattachment process over a backward-facing step, *J. Fluids Engineering*, **111**, 87-92.

Mestayer, P.G., Sini J-F., Rey, C., Anquentin, S., Badri Kusuma, S.M., Lakehal, D., Moulinec, Ch.: 1993, Pollutant dispersion in the urban atmosphere: Simulation of turbulent flows using a k-ε model, *ERCOFTAC Bulletin*, **16**, 22-28.

Mestayer, P.G., Anquetin, S.: 1995, Climatology of cities, *Diffusion and Transport of Pollutants in Atmospheric Mesoscale Flow Fields* (A. Gyr, F-S. Rys, Eds), 165-189, Kluwer Academic Pubs.

Mestayer, P.G., Sini, J-F., Jobert, M.: 1995, Simulation of the wall temperature influence on flows and dispersion within street canyons, *Air Pollution 95, Vol. 1, Turbulence and Diffusion*, 109-116, Porto Carras, Greece, September.

Oke, T.R.: 1988, Street design and urban canopy layer climate, *Energy & Bdg.*, **11**, 103-113.

Savory, E., Toy, N., Disimile. P. J., DiMicco, R. G.: 1993, The drag of three-dimensional rectangular cavities, *J Applied Scientific Research*, **50**, 325-346.

Savory, E., Abdelqari, A., Mestayer, P. G., Sini, J.-F.: 1997a, The interaction between large-scale turbulent structures and a simulated street canyon flow, *Proc 2nd European and African Conf on Wind Engineering*, Genova, Italy, June, Vol. 1, 759-766.

Savory, E., Yamanishi, Y., Okamoto. S., Toy, N.: 1997b, Experimental investigation of the wakes of three-dimensional rectangular cavities, *Proc 3rd Int Conf on Experimental Fluid Mechanics*, Kaliningrad, Russia, June, 211-220.

Sini, J-F., Anquetin, S., Mestayer, P. G.: 1996, Pollutant dispersion and thermal effects in urban street canyons, *Atmospheric Environment*, **30**, 2659-2679.

INFLUENCE OF BUILDING DENSITY AND ROOF SHAPE ON THE WIND AND DISPERSION CHARACTERISTICS IN AN URBAN AREA: A NUMERICAL STUDY

GEORGIOS THEODORIDIS and NICOLAS MOUSSIOPOULOS
Laboratory of Heat Transfer and Environmental Engineering
Aristotle University Thessaloniki, Box 483, GR-540 06 Thessaloniki, Greece
E-mail:moussio@vergina.eng.auth.gr

(Received x May 1999; accepted xxx)

Abstract. The Computational Fluid Dynamics code CFX-TASCflow is used for simulating the wind flow and pollutant concentration patterns in two-dimensional wind-tunnel models of an urban area. Several two-dimensional multiple street canyon configurations are studied corresponding to different areal densities and roof shapes. A line source of a tracer gas is placed at the bottom of one street canyon for modelling street-level traffic emissions. The flow fields resulting from the simulations correspond to the patterns observed in street canyons. In particular and in good agreement with observations, a dual vortex system is predicted for a deep flat-roof street canyon configuration, while an even more complex vortex system is evidenced in the case of slanted-roof square street canyons. In agreement with measurement data, high pollutant concentration levels are predicted either on the leeward or the windward side of the street canyon, depending on the geometrical details of the surrounding buildings.

Key words: urban air-pollution, dispersion, computational fluid dynamics, turbulence modelling

1. Introduction

Atmospheric pollution in urban areas has, in the last decades, become a major hazard to public health. The building aggregates, placed within the atmospheric boundary layer, act as artificial obstacles to the wind flow and cause stagnant conditions in the city, even for relatively high ambient wind conditions.

Street canyons, formed along a street in densely built urban areas, have been recognised as very important configurations with regard to urban air pollution and have been the subject of extensive experimental investigations (Hussain and Lee, 1980; Hosker, 1985). The systematic study of available experimental data resulted in a classification of flow regimes in urban street canyons (Oke, 1988). More specifically, three flow types are observed, depending mainly on the ratio of the average building height H over the canyon width B. For widely spaced buildings, ($H/B<0.3$), the flow fields associated with the buildings do not interact, ("isolated roughness flow" regime). At closer spacing ($0.3<H/B<0.7$) the wake created by the upwind building is disturbed by the downwind building, ("wake interference flow"). Even closer spacing ($0.7<H/B$) results in the "skimming flow" regime: A stable vortex is established in the canyon and the ambient

flow is decoupled from the street flow. As the ratio *H/B* is further increased (*H/B*>1.5), an additional weak tertiary counter rotating vortex can develop at the bottom of the street canyon.

The development of microscale models, which can explicitly take into account the building structures, offers an appealing alternative for the study of the wind field and dispersion of pollutants in urban areas. Leitl and Meroney (1997) conducted a numerical investigation of the flow and dispersion characteristics in two- and three-dimensional street canyon configurations. Sini et al. (1996) used a microscale model for studying the flow and concentration patterns in two-dimensional street canyon configurations covering a wide range of *H/B* ratios. Differential heating of street surfaces was found to have a considerable influence on the flow structure.

In this study, the Computational Fluid Dynamics (CFD) code CFX-TASCflow is used for simulating the wind flow and pollutant concentration patterns in two-dimensional wind-tunnel models of an urban area.

2. The computational method

CFX-TASCflow is a general purpose CFD analysis system using a flexible multi-block grid system. Within CFX-TASCflow, the conservation equations for mass, momentum, and scalar quantities like temperature, turbulent kinetic energy and any number of species are solved in curvilinear co-ordinates. The numerical solution is based on first-order in time and second-order in space discretisation, applied on a co-located grid arrangement. The discrete momentum and continuity equations are solved with a coupled elliptic solver. An efficient algebraic multi-grid solution technique is adopted, giving a practically constant rate of convergence, regardless of the level of the grid refinement. A detailed description of the computational method can be found in Raw *et al.* (1989) and Raw (1994).

Turbulent diffusion can be described with the standard k-ε turbulence, or the two-layer model. The standard k-ε model uses the wall function approach to model near-wall viscous effects (cf. Launder and Spalding 1974). This approach assumes the universality of a logarithmic velocity profile in the near wall region and relies on the validity of near-wall turbulent equilibrium. An alternative to the use of wall functions is to employ a turbulence model that can more closely resolve the near-wall region, as the two-layer model (cf. Rodi, 1991). The two-layer model divides the computational domain into two regions: one away from walls and one near walls. The two-equation standard k-ε model is used away from the wall. In the near wall region, a one-equation model is employed to establish the turbulent kinetic energy while the length scale is specified by reasonably well-established algebraic equations. An interface is defined for matching the two- and one-equation regions. Generally it has been found that a relatively smooth eddy viscosity transition is obtained by placing this interface at a wall distance where the molecular viscosity becomes small relative to the turbulent viscosity.

3. Case specifications grids and boundary conditions

The cases considered in this contribution, have been studied experimentally by Rafailidis and Schatzmann (1995) and Rafailidis (1997) (see also the World Wide Web page: http://www.mi.uni-hamburg.de/technische_meteorologie/windtunnel/street02/overview.html). In these experiments, two-dimensional wind-tunnel models, corresponding to multiple street-canyon configurations with a variety of canyon aspect ratios (building height H to street width B) and roof shapes, were placed in a simulated deep urban boundary layer (see Figure 1, top). Velocity and turbulence intensity profiles were measured along vertical lines over the roofs at several locations. A tracer gas was emitted by a line source placed at the bottom of one street canyon. A flat plate was placed at a distance of 1mm over the line source in order to remove the upward momentum of the tracer gas. Concentrations were measured at various locations of the neighbouring walls as illustrated in Figure 1, bottom.

Non-dimensional concentrations $K=CUHL/Q$ were reported, were C [vol/vol] denotes the tracer concentration, U[m/s] is the free stream velocity, H[m] is the building height, L[m] is the length of the line source and Q[vol/s] the source strength.

From the available experimental cases, three two-dimensional multiple street canyon configurations were selected to be studied numerically:

- A square one ($H/B=1$) with all roofs flat,
- a deep one ($H/B=2$) with all roofs flat and
- a square one ($H/B=+1$) with two slanted roofs on the buildings surrounding the source.

In all cases the computational domain consisted of five street canyons. A typical computational grid with 201×41 grid points in the horizontal and vertical directions, respectively, over the roofs, 21×31 points within the central street canyon containing the line source and 21×21 points for each of the four upstream and downstream street canyons is illustrated in Figure 2. This grid was used for the slanted-roof case and the standard k-ε model, while similar grids were prepared for the flat-roof cases. For the cases studied with the two-layer model, a much higher grid resolution was necessary in the street canyons, in order to adequately resolve the viscous sub-layers close to the floor and the vertical building walls.

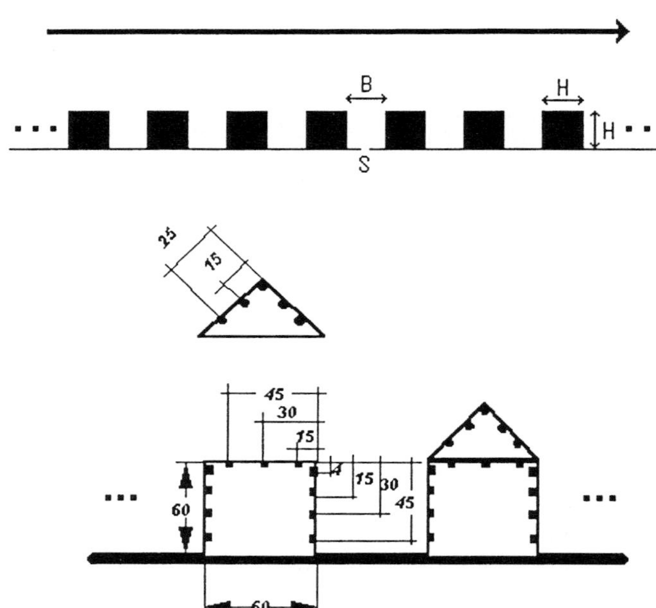

Figure 1. Experimental set-up of the two-dimensional urban configuration (top); locations of measuring tapping holes (bottom). From Rafailidis and Schatzmann (1995).

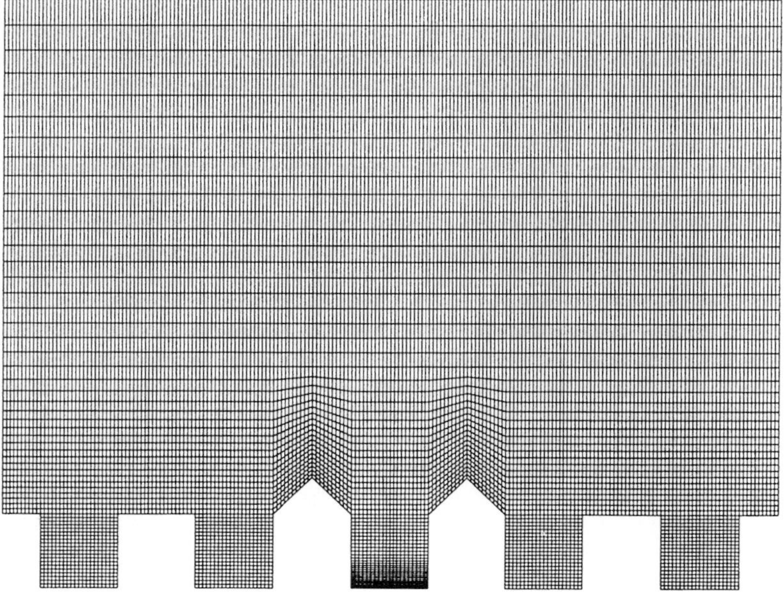

Figure 2. Typical grid used with the standard k-ε model for the slanted-roof configuration.

At the main inflow boundary (located at the leftmost vertical plane in Figure 2), the profiles of the horizontal velocity U, the turbulent kinetic energy k and the rate of dissipation ε are specified, such as to match the corresponding experimental conditions, while zero values are assigned to the vertical wind velocity V and the concentration C. The emission of the tracer gas is simulated by an inflow boundary of very low V velocity component (V=0.005m/s) at the bottom of the central street canyon. At the outflow boundary (located at the rightmost vertical plane in Figure 2), the gradients with respect to the streamwise direction are set to zero. At solid walls, the no-slip condition is applied. At the upper, free-stream, horizontal boundary, placed at a distance of $7H$ over the flat building roofs, symmetry boundary conditions are applied.

4. Results

As already mentioned in the previous section, the profiles for velocity, turbulent kinetic energy and rate of dissipation ε at the inflow boundary are specified with the aid of available experimental data from Rafailidis (1997). In these experiments, twenty street canyons were placed upstream and seven downstream of the canyon with the line source, to ensure that fully grown neutrally stratified boundary layers were established in the region of interest. Systematic measurements of the boundary layer profiles symmetrically upstream and downstream of the test section have shown that all parameters measured were repeatable between the various upstream and downstream positions. A comparison between the experimental inflow U velocity and turbulent kinetic energy k profiles with the corresponding computed profiles at the outflow boundary for the flat-roof case with the standard k-ε model is shown in Figure 3. As the agreement between the measured and computed data is excellent, especially for the U velocity profiles, it can be argued that the computations have managed to simulate the exact experimental conditions prevailing above the urban fetch.

Figure 4 illustrates the velocity fields, as computed with the standard k-ε model for the square (left) and deep (right) flat-roof cases. In the square-canyon case, a primary vortex is established covering most of the canyon region, accompanied by a small vortex at the leeward corner of the street canyon. In all the simulations presented in this study, the thin flat plate over the source, present in the experiments, was included in the computational domain. Preliminary computations for the flat-roof case without resolving this obstacle did not show any evidence of this secondary vortex, which is eventually caused by the presence of the flat plate. In the deep-canyon configuration, the corresponding primary vortex is weaker covering the top 2/3 of the canyon region, while almost stagnant conditions are established near street level, due to the existence of a weak counter-rotating vortex. This finding is in agreement with the observation of Rafailidis and Schatzmann (1995) who, besides concentration measurements, conducted systematic flow visualisation in the central street canyon region.

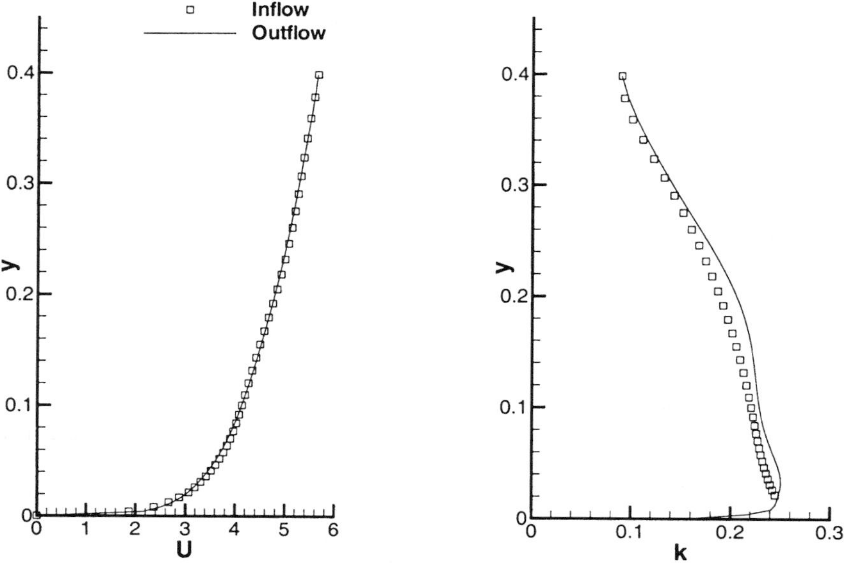

Figure 3. Comparison of inflow (measured) and outflow (computed) U (left) and k (right) profiles.

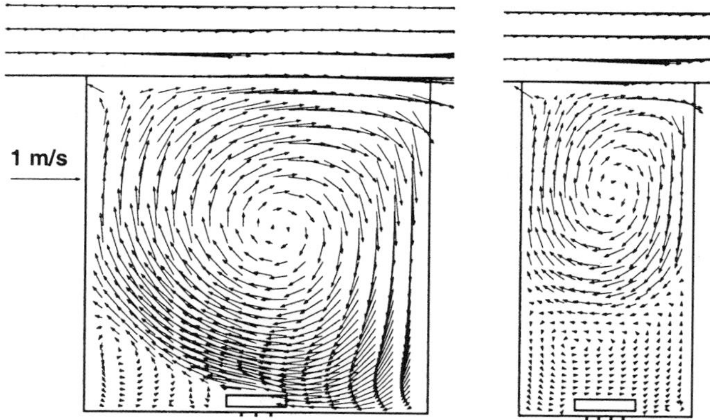

Figure 4. Velocity fields, as computed with the standard k-ε model for the square (left) and deep (right) flat-roof cases.

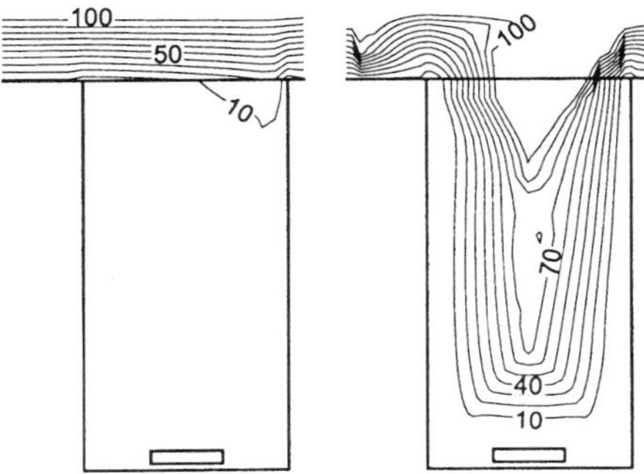

Figure 5. Normalised eddy viscosity $\mu_n=\mu_t/\mu$ as computed by the standard k-ε (left) and the two-layer (right) models.

Results for the normalised eddy viscosity $\mu_n=\mu_t/\mu$, where μ_t stands for the eddy viscosity and μ denotes the molecular viscosity, as computed by the standard k-ε (left) and the two-layer (right) models are presented in Figure 5 for the deep street-canyon case. The two-layer model is employed near the building walls, the flat plate walls over the source and the floor of the street canyons, i.e. regions of low local Reynolds number. The levels of μ_n as predicted by the two-layer model are, in average, one order of magnitude larger then the levels of μ_n predicted by the standard k-ε. This finding explains the well known trend of the standard k-ε model to underestimate the dispersion of pollutants, as eddy viscosity is the agent controlling the mechanism of turbulent diffusion in most numerical methods.

Figure 6 shows the comparison between measured and computed non-dimensional concentrations for the square (left) and the deep (right) flat-roof cases. In the case of the square canyon the maximum concentrations appear on the leeward wall (marked as side A in Figure 6 left). This is due to the combined action of the primary and corner vortices (see Figure 4 left), resulting in a wind direction from the windward to the leeward wall near the source. This picture is reversed in the deep street-canyon case (Figure 6, right), where the maximum concentrations appear on the windward wall (marked as side C in Figure 6, right). With the exception of the overestimation of the peak concentration levels by the standard k-ε model, the agreement between the computed and measured concentrations is good. In general, the concentration levels are much higher in the deep than in the square canyon configuration, the maximum levels in the former being twice as high as the maximum levels in the latter.

The corresponding results obtained with the standard k-ε model for the slanted-roof square-canyon case are presented in Figure 7. A more complex vortex system comprising a strong vortex at roof level, a moderate vortex covering most of the canyon region and a weak corner vortex is evidenced in this case (Figure 7 left), as compared with the flat-roof cases (Figure 4). In both the experiment and the computation the maximum concentrations appear on the windward wall (marked as side D in Figure 7, right). On the other hand, field measurements in street canyons with slanted roofs indicate that the maximum concentrations appear on the leeward side. This discrepancy between the real world and the simulations (both physical and numerical) is attributed to the fact that in the latter, only two buildings with slanted roofs were placed between several buildings with flat roofs, which is not a typical urban configuration. Nevertheless, with the exception of the overestimation of the peak concentration levels by the standard k-ε model, the agreement between the computed and measured concentrations is excellent.

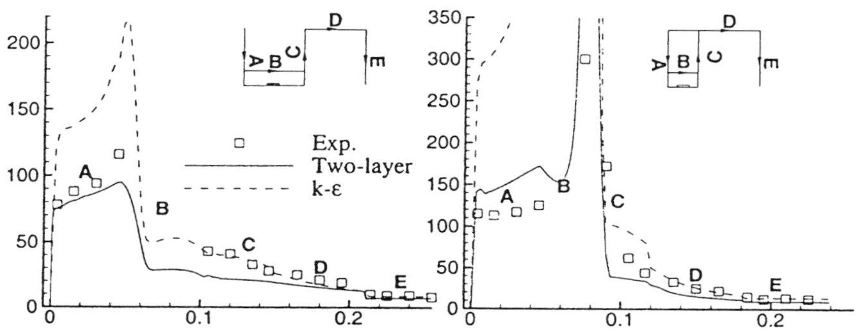

Figure 6. Comparison between measured and computed non-dimensional concentrations for the square (left) and deep (right) flat-roof cases.

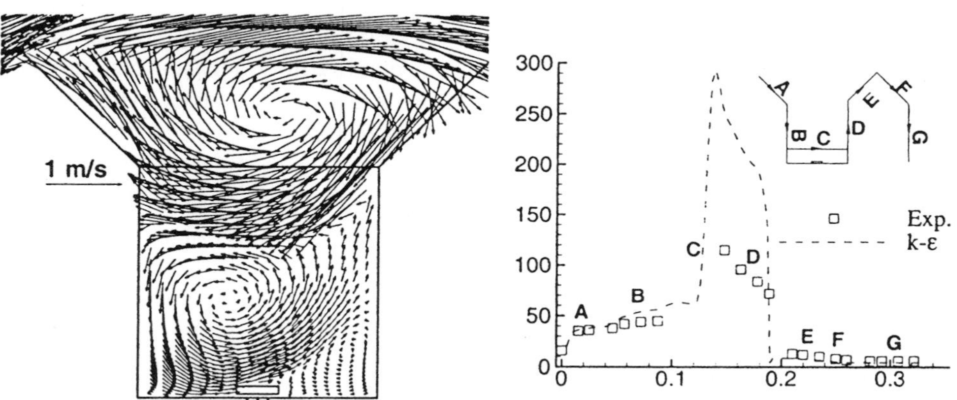

Figure 7. Results obtained with the standard k-ε model for the slanted-roof square-canyon case; velocity field (left); non-dimensional concentrations (right).

5. Conclusions

The CFD code CFX-TASCflow, was applied for the simulation of the wind fields and the dispersion of pollutants in street canyons. The flow field and concentration patterns predicted by CFX-TASCflow correspond to the patterns observed in street canyons and are in good agreement with measurement data. The overestimation of peak concentration levels by the standard k-ε is attributed to the rather low eddy viscosity levels predicted by the model in the canyon region.

Acknowledgements

This work is related to the DGXII-TMR project TRAPOS. The authors would like to thank Dr. Stilianos Rafailidis for providing all necessary information with regard to the experimental conditions and results.

References

Hosker R.P.: 1985, Flow around isolated structures and building clusters: A review, *ASHRAE Trans.* **91**, 1671-1692.

Hussain M. and Lee B.E.: 1980, An investigation of wind forces on three-dimensional roughness elements in simulated atmospheric boundary layer flow: Part II Flow over large arrays of identical roughness elements and the effect and side aspect ratio variations, Report BS 56, Dept. Building Science, University of Sheffield.

Launder, B. E., and Spalding, D. B.: 1974, The numerical computation of turbulent flows, *Comput. Meths. Appl. Mech. Eng.* **3**, 269-289.

Leitl, B.M. and Meroney, R.N.: 1997, Car exhaust dispersion in a street canyon. Numerical critique of a wind tunnel experiment, *J. of Wind Eng. and Ind. Aerodynamics* **67** & **68**, 293-304.

Oke T.R.: 1988, Street design and urban canopy layer climate, *Energy Building*, **11**, 103-113.

Rafailidis, S. and Schatzmann, M.: 1995, Concentration measurements with different roof patterns in street canyons with aspect ratios B/H=1/2 and B/H=1, Report, Meteorology Institute, University of Hamburg.

Rafailidis, S.: 1997, Influence of building areal density and roof shape on the wind characteristics above a town, *Boundary-Layer Meteorology* **85**, 255-271.

Raw, M.J., Galpin, P.F. and Hutchinson, B.R.: 1989, A collocated finite-volume method for solving the Navier-Stokes equations for incompressible and compressible flows in Turbomachinery: Results and applications, *Canadian Aeronautics and Space Journal* **35**, 189-196.

Raw, M.J.: 1994, A coupled algebraic multigrid method for the 3D Navier-Stockes equations, Proceedings of the 10th GAMM-Seminar Kiel, January 14-16, 1994.

Rodi, W.: 1991, Experience with two-layer models combining the k-ε model with a one-equation model near the wall, AIAA 91-0216, presented at the 29th Aerospace Sciences Meeting, Reno, Nevada, January 7-10, 1991.

Sini J.F., Anquetin S., and Mestayer P.G.: 1996, Pollutant dispersion and thermal effects in urban street canyons, *Atmospheric Environment*, **30** 2659-2677.

CARBON MONOXIDE CONCENTRATION IN A STREET CANYON OF BUENOS AIRES CITY (ARGENTINA)

LAURA E. VENEGAS and NICOLAS A. MAZZEO
National Scientific and Technological Research Council (CONICET)
Dept. of Atmospheric Sciences. Faculty of Sciences. University of Buenos Aires
Ciudad Universitaria. Pab. 2. 1428-Buenos Aires. Argentina

Abstract. The analysis of three years of 8-h CO concentration values registered in a deep street canyon downtown shows high frequency of values that exceed WHO health protection guidelines. An inverse relationship between opposing percentiles of the distributions of CO concentrations and mean wind speed could be found. Data also showed a variation of mean CO values with prevailing wind direction. The averaged concentration value obtained when the sampler probe is on the leeward side is lower than the obtained when it is on the windward wall. A preliminary explanation of this feature may be related to the advection of polluted air from a high traffic density area nearby.

Keywords. Street canyon; Air quality monitoring; Urban air pollution; Carbon monoxide

1. Introduction

Buenos Aires City is situated on a flat terrain, it has an extension of 200 km² and a population of three millions inhabitants. In Buenos Aires there are usually three millions vehicles, one of them comes to the city from the surroundings during working hours. Carbon monoxide (CO) is a product of incomplete combustion of carbon and its compounds. Gasoline powered motor vehicles are the major sources of carbon monoxide (CO) in urban atmosphere (Onursal and Gautam., 1997). This pollutant is a colourless, odourless and tasteless gas slightly lighter than air. It is considered a dangerous asphyxiant because it combines strongly with haemoglobin and reduces blood's ability to carry oxygen to cell tissues. Carbon monoxide air concentrations are measured in a highly populated area of Buenos Aires City with intense activity during working hours. Measurements are done by an automated continuous nondisperse infrared monitoring system. The available CO data include three years (1994-1996) of 8-h average concentration from 8a.m. to 4p.m. Hourly wind speed and wind direction are measured at a meteorological station of the National Weather Service located in the city. In this paper, we study the variation of 8-h CO concentrations with mean wind speed and prevailing wind direction considering the characteristics of the street canyon and of the monitoring site neighbourhood.

Environmental Monitoring and Assessment **65**: 417–424, 2000.

2. Sampling site

The sampling point of CO concentrations is located on the West side of a North-South street (see Figure 1). The street canyon is straight, 9 m width and 100 m length approximately. The sampler is situated outside a building near midblock and at 6m height above the ground. The building where the sampler is located is 13m high. The buildings on the West side of the street are of about 18-20 m high. In front of the sampler location there is a building of 6m high but the height of other buildings on the East side is 16-18 m. The buildings in the area situated West and Northwest from the sampler are taller than the buildings of the East and Southeast area. Crossing the Northern intersection there is a wide square of approximately 30000m² (Lavalle Square).

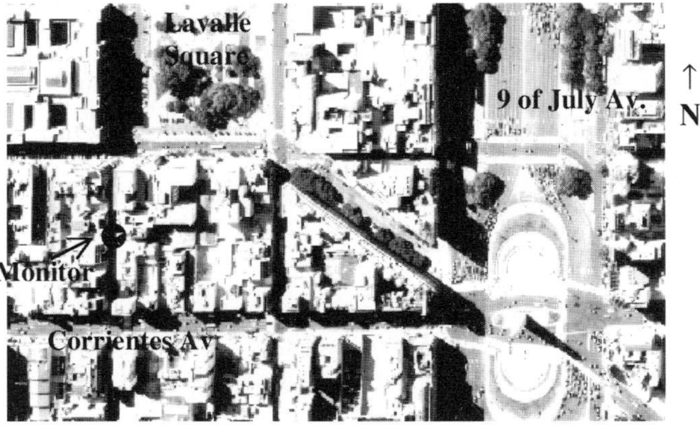

Figure 1. Sampling site and its surroundings

3. Data analysis

3.1. INTRODUCTION

In a recent study, Mazzeo and Venegas (1998) obtained that 8-h average concentration of CO observed at downtown during 1994-1996 could reach 19.2 ppm and the mean value of these data was 9.3 ppm. The cumulative percent curve on Figure 2 shows that 57% of the 8-h CO values were greater than 9 ppm (Air Quality Standard for CO, for an 8-h averaging period, according to the World Health Organisation and the United States Environmental Protection Agency, Murley, 1991).

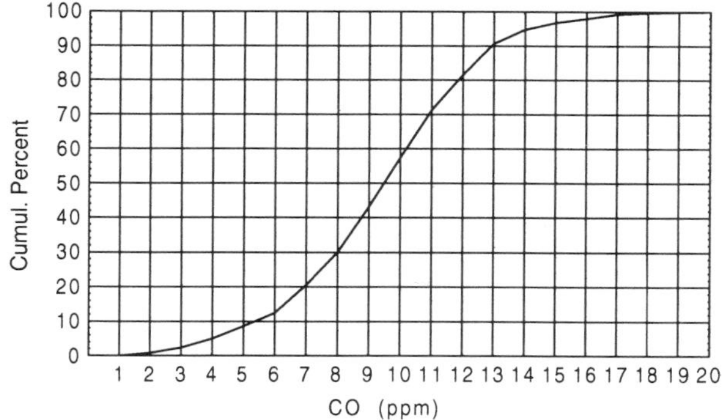

Figure 2. Cumulative percent frequency of 8-h CO

Figure 3 illustrates the relative frequency of 8-h average concentrations that exceeded 9 ppm during each month. The highest frequency (92.3%) of concentrations greater than 9 ppm was observed during October. In January (during summer vacations) 7.6% of the cases exceeded the limit of 9 ppm.

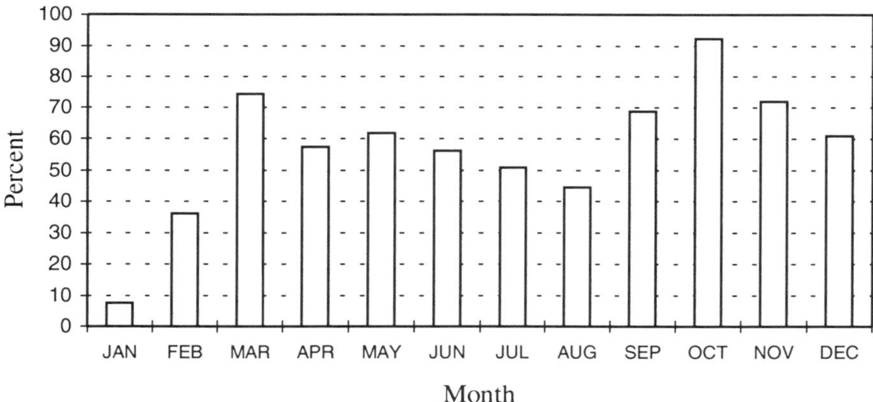

Figure 3. Relative frequency of 8-h CO greater than 9 ppm.

In order to consider no variation in emission rate, in this paper we analysed the 8-h average CO concentration values observed during working days only. Little variation of the mean 8-h CO value from Monday to Friday was found. Total traffic flow from 8a.m. to 4p.m. can be considered the same during working days. January and February data were not included in the analysis because some people were on summer holidays during these months. The data were grouped in the following way: autumn (March, April, May); winter (June, July , August) and spring (September, October, November). December was also excluded. Finally, 555 values were included in the analysis. The mean value of this data set was 10.2 ppm.

3.2. CO CONCENTRATION AND MEAN WIND SPEED

The dependence of observed 8-h concentrations on corresponding mean wind speed was not clear, as can be seen in Figure 4.

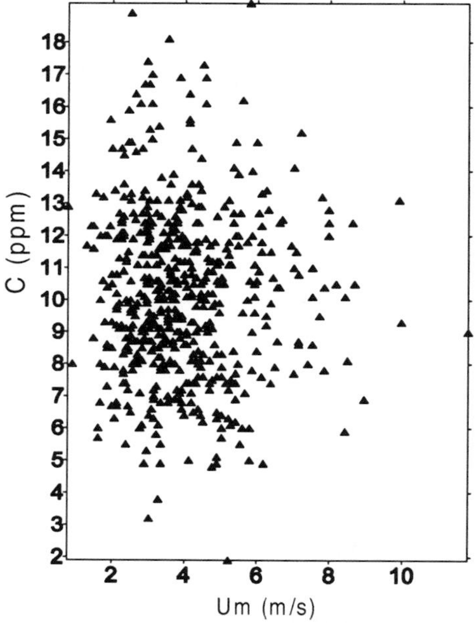

Figure 4. Variation of 8-h CO concentration with mean wind speed (Um)

Following Simpson et al. (1985) we studied the relationship between opposing percentiles of the distributions of 8-h CO concentration and mean wind speed. Previous studies (Daly and Steele, 1975, 1976; Simpson at al., 1983) have found that a simple relationship of the form $C = K\ (Um)^{-1}$ would appear to exist because of the high correlation between percentiles of mean wind speed (Um) and air pollution concentrations (C) at several sites. Simpson et al. (1983) have shown that C value corresponding to p-percentile (C_p) and mean wind speed value corresponding to (100 - p)-percentile ($Um_{(100-p)}$) are related by: $K=[C_p.Um_{(100-p)}]$. The values of K calculated from the cumulative distributions for 8-h CO data and mean wind speed data are shown in Figure 5. The mean value of K is Km= 6.76ppm.m.s^{-1}and its standard deviation is σ =3.73ppm.m.s^{-1}.

Figure 5 shows that for $15\% \leq p \leq 80\%$ the values of K are within the interval ($Km \pm \sigma$). These results show that the expression $C_p = K (Um_{(100-p)})^{-1}$ is a good representation of the relationship between opposing percentile-values in the statistical distributions for 8-h CO monitored in the canyon and mean wind speed, for the 15- to 80- percentile range. The linear regression fitting in Figure 5 is given by the expression K = 43.56 - 0.13 p (K is expressed in [ppm m s^{-1}] and p is expressed in [%]), with a coefficient of determination (r^2) of 0.964.

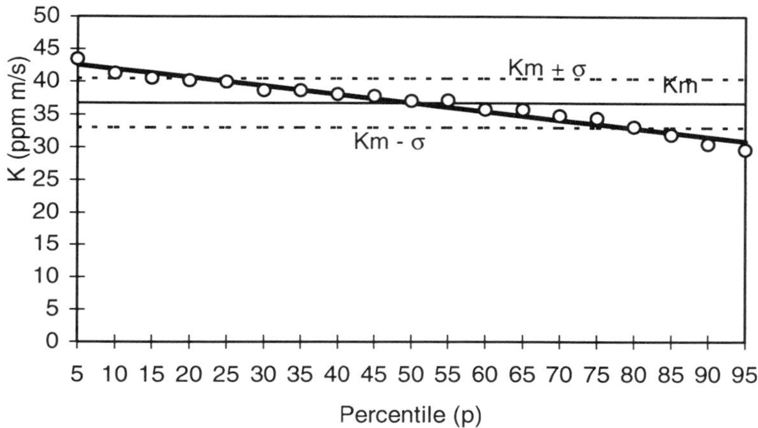

Figure 5. Values of $K = C_p.Um_{(100-p)}$. Thin solid line indicates mean value (Km); dashed lines ($Km \pm \sigma$); thick solid line is the calculated regression line.

3.3. CO CONCENTRATION AND PREVAILING WIND DIRECTION

The frequency of 8-h CO concentrations that exceeded 9 ppm was: 65.1% in autumn, 54.3% in winter and 77% in spring. The mean value for each season was: 10.0ppm (autumn), 9.8 ppm (winter) and 10.7 ppm (spring). Wind roses of the same period showed that during autumn and spring winds from NE, E and SE were very frequent. During winter the frequency of SW, W and NW winds increased. To study the variation of mean 8-h CO concentration with wind direction, we considered the cases with more than four hours of wind blowing from a given sector. Only these cases had a corresponding prevailing wind direction. Figure 6 shows the mean CO concentration calculated for prevailing wind directions grouped every 30° degrees. It can be seen a variation of mean 8-h CO concentrations with wind direction. Lower mean concentration values occurred when prevailing wind direction was from West sector. In these cases, the building where the sampler is located may affect the observations. Higher mean concentration values were observed when prevailing wind direction was from sector between E and S. The frequency of 8-h CO concentrations greater than 9 ppm was more than 80% with prevailing wind direction from this sector.

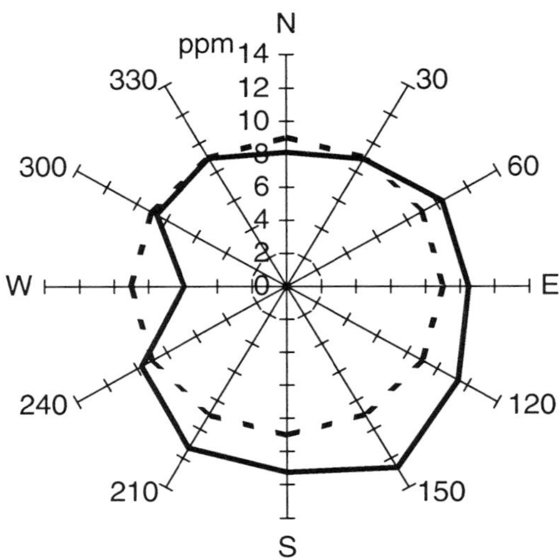

Figure 6. Variation of mean 8-h CO concentrations (ppm) with prevailing wind direction for working days (solid line). Dashed line indicates 9 ppm.

This deep and narrow street canyon has two lanes of one-way traffic. Assuming a classic vortex recirculation inside the canyon, under westerly winds it can be expected that the vehicle emissions are transported to the leeward canyon wall, towards the sampler. Only the effect of a classic recirculation pattern would not be able to explain the distribution of averaged concentration with wind direction showed in Figure 6. The averaged value obtained when the sampler probe is on the leeward side is lower than the obtained when it is on the windward wall. Simultaneous measurements of CO concentrations on both sides of the street canyon are needed to determine whether this pattern occur and how often.

Two major avenues are located S and E from the sampler. The first one is Corrientes Avenue with 6 lanes (24 m wide) and the second is 9 of July Avenue (the widest avenue of the City) with 20 lanes (133 m wide) at two blocks E from the sampling point (see Figure 1). There is a high traffic density in both avenues during working hours. The high mean values of 8-h CO obtained for prevailing wind direction within E→ S sector, may show that these cases can be affected by a greater advection of polluted air than the cases with prevailing wind from the opposite sector. There is a low building opposite the sampler site. Through this narrow passage the polluted air from SE, E can easily penetrate and be caught by the sampler probe and thus influence substantially measurement. The advection of polluted air from the West towards the sampling point could be

hindered because of the presence of tall buildings and narrow streets at W and NW. On the other hand, Northern winds advect less polluted air from Lavalle Square to the sampler.
In addition, sometimes small-scale circulations inside the street-canyon might also be responsible of the high concentration values observed on the windward side. According to some studies (Theodoridis and Moussiopoulos, 1999; Scaperdas and Colvile, 1999) the flow patterns inside a deep street canyon (street width/ building height = 0.5) may lead to high concentration values on the windward side building. In a deep canyon two counterrotating vortices can be formed. A primary vortex covering the top 2/3 of the canyon region and a counter rotating vortex in the bottom. The opposite rotation of the two vortices leads to a wind direction at street level being the same as the wind at roof top level.
In the canyon considered in this paper, dense traffic may induce strong momentum and turbulence in the carriageway so small-scale vortex may have very little chance to form. Additional field measurements of CO concentrations and air flow parameters are needed to study the presence of a small-scale vortex at the bottom of this canyon and its effect on concentration distributions at windward and leeward canyon walls.

4. Summary

In this paper we analysed three years of 8-h (8 a.m. to 4 p.m.) CO values registered at a sampler point located at 6m height in a deep street canyon. Only working days observations were considered. These cases had a mean value of 10.2 ppm. and showed high frequency of 8-h CO values greater than 9ppm. Dependence of 8-h CO on mean wind speed was not clear but data showed an inverse relationship between opposing percentiles of the distributions of concentration and mean wind speed. Also a variation with prevailing wind direction could be observed. Considering the two main avenues located upwind in these cases, the advection of polluted air from an area with high traffic density seems to be important. Further studies on local recirculations inside this canyon are needed to evaluate the relative contribution of different scale motions to high concentration values.

Acknowledgement

The authors would like to thank Dr. N. Moussiopoulos and Dr. A. Scaperdas for their comments and suggestions. The authors are also grateful to the National Weather Service of Argentina for the meteorological data used in this work. This work was supported by the Project: Study of the atmosphere of Buenos Aires. (University of Buenos Aires TX03 - CONICET PIP 0424/98)

References

Daly, N.J. and Steele, L.P.: 1975, Air quality in Canberra, Report to the Department of the Capital Territory, Camberra, Australia.

Daly, N.J. and Steele, L.P.: 1976, A predictive model for CO in Canberra, Symposium on air pollution diffusion modeling, Australian Environment Council, Canberra, Australia, p.264.

Mazzeo, N.A. and Venegas, L.E.: 1998, An Analysis of Carbon monoxide Concentration in the Atmosphere of Buenos Aires City (Argentine Republic). Proceedings of the 11th World Clean Air and Environment Congress and Exhibition. IUAPPA. Durban. South Africa.

Murley, L. (ed.): 1995, *Clean Air Around the World*, Third Edition, IUAPPA.

Onursal, B. and Gautam, S.: 1997, Vehicular air pollution. Series: World Bank Technical Paper, N° 3738, (Spanish), pp.306.

Scaperdas, A. and Colvile, R.N.: 1999, Assessing the representativeness of monitoring data from an urban intersection site in central London, UK. *Atmospheric Environment*, **33**, 661-674

Simpson, R.W.; Daly, N.J. and Jakeman, A.J.: 1983, The prediction of maximum air pollution concentrations for TSP and CO using Larsen's Model and the ATDL Model, *Atmospheric Environment,* **17**(12), 2497-2503.

Simpson, R.W.; Jakeman, A.J. and Daly, N.J. : 1985, The relationship between the ATDL Model and the statistical distributions of wind speed and pollution data, *Atmospheric Environment,* **19**(1), 75-82.

Theodoridis, G. and Moussiopoulos, N.: 1999, Influence of building density and roof shape on the wind and dispersion characteristics in an urban area: a numerical study. 2nd International Conference on Urban Air Quality. 79-80.

AIR QUALITY IN NORWICH, UK
MULTI-SCALE MODELLING TO ASSESS THE SIGNIFICANCE OF CITY, COUNTY AND REGIONAL POLLUTION SOURCES

T.CHATTERTON, S.DORLING, A.LOVETT AND M.STEPHENSON*

*School of Environmental Sciences, University of East Anglia, Norwich, UK, NR4 7TJ. *Environmental Health Department, Norwich City Council, Norwich, UK, NR1 3AG.*
E-mail: T.Chatterton@uea.ac.uk

Abstract. Norwich is the eastern most city in the United Kingdom. Despite a population of only 100,000 and very little local industry, studies have shown that the city experiences levels of nitrogen dioxide, ozone, particulates and sulphur dioxide exceeding the UK Air Quality Standards. Because of Norwich's situation within a large, predominantly rural area a large non-resident workforce is one factor that contributes to large, often very congested traffic flows. The city's location close to the European mainland also exposes it to polluted airmasses transported from the continent, especially in the case of particulates and ozone. In order to assess the relative contributions of local and regional sources, data from rural and urban monitoring sites are to be used in conjunction with ADMS-Urban and the UK Meteorological Office's NAME model.

Keywords. PM10, urban air quality, background, NAME, ADMS, long range transport, air pollution, particulates, nitrogen dioxide, sulphur dioxide

1 Introduction

In 1997 Norwich City Council established a partnership with the University of East Anglia (UEA) to investigate local patterns of air pollution. Since then the project has expanded and involves the Norfolk Environmental Protection Group (a consortium of local authorities within the county of Norfolk). The theme of the work is an examination of the relative contributions to local air pollution that are made by local sources of pollution within Norwich compared to those outside the city; in the rest of the county, the UK and beyond.

The work will focus upon the use of the UK Meteorological Office's Nuclear Accident Model (NAME) (Maryon *et.al.*, 1991) to retrospectively analyse air quality in the region. Although originally designed as a single tool for lagrangian based modelling of radioactive particles released from nuclear accidents, it has subsequently been adapted by the Meteorological Office to perform as a regional-scale dispersion model for a wide range of pollutants (APEG, 1999). Although this model is currently being used by a number of local authorities within the UK to provide forecasts of air quality, this is the first time local authorities have used it as a source apportionment tool to help interpret past pollution scenarios.

The purpose of this paper is to describe the rationale for using the NAME model in this way. The paper therefore comprises of a description of the general

Environmental Monitoring and Assessment **65**: 425–433, 2000.

character of the City of Norwich and the surrounding region, and summarises the current state of knowledge regarding observed patterns of pollution within Norwich. Finally the position and use of the NAME model within a wider range of modelling techniques is outlined.

2 The Local Area

Norwich is a relatively small city compared to many in the UK, having a population of around 100,000. However, it is situated as the 'hub' of a large, surrounding rural hinterland comprising mainly of the county of Norfolk. The area is very remote compared to the rest of the country (for example, Norwich is the only city in England without a Motorway within 50 miles). This accentuates the focus placed on Norwich as the centre for both economic and social activity within the region; UK 1991 Census data indicates that the daytime population of Norwich is increased by around 50% due to an influx of non-resident workers alone. Norwich is one of the few urban areas in England that still relies on a medieval street layout, with a great number of narrow canyon streets not suited to modern traffic levels. Consequently, many areas of Norwich become overloaded with cars, particularly during peak hours causing pollution hotspots.

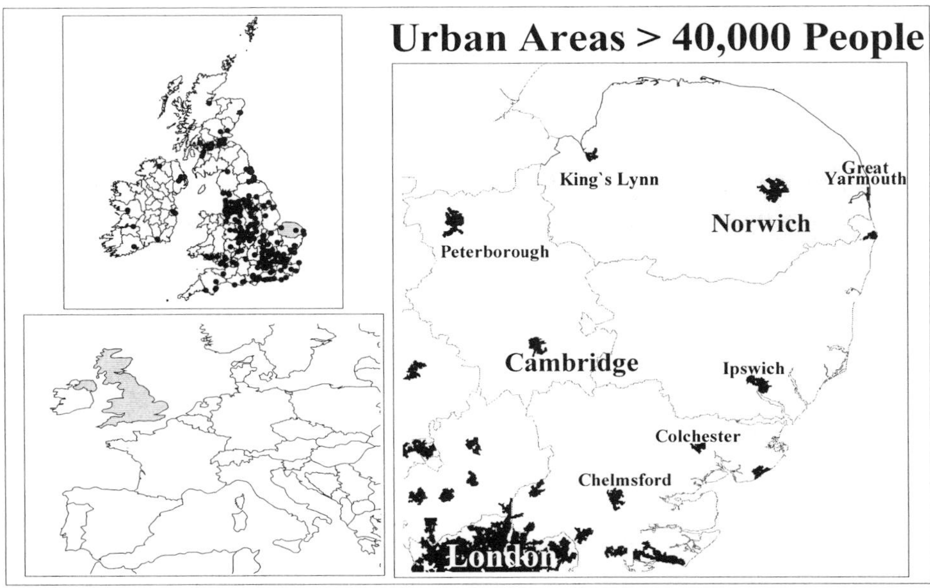

Figure1: Norwich, UK

Table 1: Population and emissions densities for Norfolk and South East England

Region	Population Density per km^2	Area (km^2)	Average emissions per km^2 (t/y)[4]		
			NO_x	PM_{10}	SO_2
Norwich	3100[1]	40	51.3	11.5	18.3
Norfolk (excl. Norwich)	121[1]	5389	2.8	0.4	1.0
London	4480[2]	1573	92.6	7.8	10.8
South East (excl. London)	411[3]	18352	11.3	0.9	6.0
[1] Shaw, J.M. 1994		[2] Office for National Statistics 1997a			
[3] Office for National Statistics 1997b		[4] NETCEN 1999			

As Norwich has changed from a centre of industrial production to an economy based mainly on service industries transport has become the predominant source of air pollution. The city only has two Part A registered processes; a gas-turbine power station and a factory manufacturing agricultural chemicals, neither of which are significant producers of the principle urban pollutants dealt with in the UK National Air Quality Strategy.

Despite the local population falling well below the 250,000 level that has formed a loose borderline (at least in terms of UK government funding) above which an urban area can be expected to suffer from high levels of air pollution, in the recent Stage One Review and Assessment of Air Quality carried out by Norwich City Council (Norwich City Council, 1999) it was found that, as things stand, the City's air quality is unlikely to comply with UK standards by 2005. This is based on the generic guidance issued to local authorities by the UK government (DETR 1998a) which takes into account the predicted changes in air quality that will come about due to national measures and sets these against current statistics for an area such as traffic flows and current pollution levels. The work outlined in this paper will be used to help assess the probability that air quality will meet both EU and UK standards by 2005.

The principle pollutants that afflict the city are Sulphur Dioxide (SO_2), Nitrogen Dioxide (NO_2), Particulates (PM_{10}) and Ozone (O_3). The UK government has placed O_3 control outside the remit of Local Authorities "*until the outcome of forthcoming European negotiations*" (DETR, 1998b) the study of O_3 is, therefore, beyond the central focus of this study.

As will be shown below, the problem caused by SO_2 in Norwich is extremely localised and, as yet, unattributed to a source. The other two pollutants, NO_2 and PM_{10} both have a common vehicular source. However, levels of NO_2 are almost entirely related to local sources whereas it has been suggested that particulates also have a large component that comes from long-range transport (King and Dorling,1997; Stedman 1997). Norwich is the Eastern most city in the UK (Figure 1) and is likely to be most impacted by continental pollution sources under certain meteorological conditions. Therefore, despite limited local sources of particulate emissions there is a strong probability of particulates continuing to rise to levels defined as being damaging to human health.

3 Monitoring

In order to assess the contribution of local sources to air pollution in Norwich, pollution is being monitored at three scales; Regional, Urban and Local.

3.1 REGIONAL MONITIORING

In order to establish a general background level for pollution in Norfolk, data is being collected from DETR Rural sites, additional particulate monitors (PM_{10} & $PM_{2.5}$) run by UEA and other collaborators for a six month period in 1998 in West Norfolk and from monitoring at the Weybourne Atmospheric Observatory on the North Coast where long-range transport of various pollutants originating from both London and mainland Europe have already been traced (Cardenas *et al, 1998*).

3.2 URBAN MONITORING

Norwich has two automatic, continuous monitoring sites affiliated to the Department of the Environment, Transport and the Region's (DETR's) Automated Urban Network (AUN). The main site is currently designated an Urban Centre site (although there are repeated attempts made to have it reclassified as an Urban background site). This site provides data showing the air quality in the city away from the presence of any major roads or other sources of pollution and can, to a certain extent, be used to estimate background levels throughout the city. This site monitors carbon monoxide (CO), nitrogen oxides (NO_x), O_3, particulates (PM_{10}) and SO_2.

3.3 LOCAL MONITORING

Norwich's second AUN site is a simple NO_x monitor sited at a roadside location near the city centre. The City Council also operate a mobile continuous monitor currently sited at a roadside location, measuring $PM_{2.5}$ in addition to the five pollutants also measured by the Urban Centre site (see 3.2). This monitoring unit is to be used both to examine in detail pollution levels at suspected hotspots as well as to assess the effect of major traffic management measures on air quality. All three continuous monitors are quality controlled under the NETCEN AUN\UK Calibration Club regime (NETCEN 1996).

In addition to the continuous monitors, the city council also operate a number of non-continuous, automatic monitors for SO_2, black smoke and lead, and a large number of diffusion tubes measuring levels of benzene, NO_2, O_3 and SO_2.

These tubes are used to provide indications of pollution levels at localised hotspots across the city.

Table 2: Data from Norwich Centre AUN monitoring site 7/97 - 12/98

		CO ppm	NO_2 ppb	O3 ppb	PM_{10} µg/m3	SO_2 ppb	CO 8hr ppm	PM_{10} 24hr µg/m3
24/7-31/12 1997	Mean	0.4	15	14	22	6	0.4	22
	Maximum	8.9	85	96	161	65	6.1	76
	Exceedences	-	0	141h/15d	-	0	0	64h/5d
1/1-31/12 1998	Mean	0.4	13	17	20	4	0	20
	Maximum	10.9	67	72	105	37	5	47
	Exceedences	-		77h/13d	-	0	0	0

4 Pollutants

4.1 SULPHUR DIOXIDE

Table 3 shows monthly averages derived from daily average measurements of SO_2 at five locations in Norwich. Although four of the five sites show very low levels of SO_2 corresponding to the average figures recorded by Norwich Centre AUN site in Table 2, Site 5 at the Guildhall (in the very centre of Norwich) shows significantly higher readings suggesting that, especially in summer, the UK Air Quality Standard for SO_2 (100ppb 15-minute average) is being regularly exceeded. Daily maxima show levels recorded as high as 169ppb. So far no localised combustion sources have been identified as the potential emitters. The site is however near a taxi rank, although it is unclear as to whether this is likely to produce the levels of SO_2 being recorded. A network of SO_2 diffusion tubes will be employed in the area to identify the location of the source more clearly.

Table 3: SO_2 (Bubbler) Data, Norwich 1997 (Monthly Averages ppb)

	Jan	Feb	Mar	Apr	May	Jun	Jul	Aug	Sep	Oct	Nov	Dec
1 Churchill Rd	4	2	3	4	0	2	3	5	0	3	3	3
2 Hardy Rd	6	4	3	6	7	6	8	9	6	6	5	6
3 Tuckswood	7	6	8	5	8	5	6	8	8	7	6	6
4 Rouen Rd	5	5	6	4	6	5	5	7	7	5	5	6
5 Guildhall	12	15	22	23	46	63	90	**101**	63	44	32	24

4.2 NITROGEN DIOXIDE

Table 2 shows that no exceedences of either the hourly or annual UK Air Quality Standard for NO_2 (150ppb hourly average, 21ppb annual average) occurred at the Norwich Centre monitoring site during its first seventeen months. However, Table 4 shows that eight out of fifteen diffusion tube sites have shown exceedences of the 21ppb annual average level. This strongly reinforces the findings of Kirby et al (1998) in Cambridge that NO_2 'hotspots' are generally very localised phenomena. Although there is evidence of occasional long-range transport of NO_x to rural Norfolk locations (Cardenas *et al, 1998*) this phenomenon would not appear to account for the high long-term average NO_2 measurements recorded in Norwich. It is anticipated, therefore, that further modelling and monitoring currently being undertaken will conclude that high levels of NO_2 are caused almost entirely by local emissions (predominantly road transport).

Table 4: NO_2 Diffusion Tube Data for Norwich (Annual Averages ppb)

	1995	1996	1997	1998
1 Guildhall	20.4	19.9	20.5	16.5
2 Churchhill Road	19.1	17.8	14.6	12.8
3 St Augustines	27.3	31.1	36.0	33.0
4 Vulcan Road	21.0	26.3	26.7	24.0
5 Tombland	25.2	29.8	29.2	26.1
6 Colman Road	20.2	23.7	22.3	21.1
7 Gentlemans Walk	15.1	17.6	18.0	15.3
8 Exchange St	20.7	23.7	24.3	21.3
9 St Stephens	23.2	27.8	26.4	20.7
10 Surrey St	15.7	16.8	22.8	22.5
11 Ipswich Road			24.7	22.7
12 Castle Meadow			31.4	24.8
13 Riverside			27.9	21.6
14 Heartsease Lane			21.0	19.2
15 Elliott House			20.5	16.0
Average	20.78	23.44	24.43	21.18

4.3 PARTICULATES

Until January 1999 the only Particulate monitor in Norwich has been a TEOM measuring PM_{10} at the Norwich Centre AUN site. This site has been 'on-line' since July 1997 and has recorded very few exceedences of the UK Air Quality Standard (50µg/m^3-running 24-hour mean). However, during this period there have been no large-scale (spatial or temporal) particulate episodes comparable to those experienced across the UK in early 1996 (King and Dorling,1997; Stedman 1997). Analyses have been carried out comparing particulate levels during 1998 in Norwich with those recorded at similar (e.g. Urban Centre or Urban Background) sites across the country. Although no data for Norwich is available for 1996, similarities between Norwich Centre and other sites suggest that it is very likely that prolonged periods of particulate pollution were experienced in the City at this time. It is also expected that roadside levels of PM_{10} will frequently exceed UK Standards. This latter hypothesis forms the central thread to this work.

PM_{10} are comprised of three fractions: Primary - those particles formed by combustion; Secondary - those formed by chemical reactions in the atmosphere (e.g. sulphates and nitrates); Coarse - bits of matter, such as windblown grit, sea-salt or entrained road-dust. The relative proportions of these fractions will vary according to a number of factors, particularly location and meteorological conditions. Sample filters from the Norwich Centre site are currently being analysed for the precise composition of the sampled material. This information will be contrasted with additional chemical analyses from the rural, West Norfolk monitoring site. Preliminary comparisons between these two sites have shown an average urban excess in PM_{10} levels of only 3 $\mu g/m^3$ during this time. This suggests that the particulates on both sets of filters will contain a significant proportion from regional sources of PM_{10}.

In early 1999 a Mobile Monitoring Unit containing two TEOMs measuring PM_{10} and $PM_{2.5}$ was set up at a site adjacent to a major roundabout not far from the Norwich Centre AUN site. This site will provide an indication of the degree to which primary particles are being generated by road transport at this particular location and adding to the general background levels.

The early data from this site over a three-week period in April 1999 shows an average level of PM_{10} at the Roadside location 2.3$\mu g/m^3$ above that at the Norwich Centre AUN site (sited roughly 300m away in an urban background setting). During this period the roadside site registered between 66 $\mu g/m^3$ more and 16 $\mu g/m^3$ than the Urban Centre site, suggesting that traffic does have a considerable influence on PM_{10} levels depending on location.

5 Modelling

In order to further investigate the patterns of locally generated and transported pollution experienced in Norwich a multi-layer-modelling programme is being undertaken.

5.1 REGIONAL SCALE MODELLING

In order to assess the regional component of pollution, modelling is being carried out in partnership with the UK Meteorological Office using their NAME, a lagrangian multiple particle model incorporating sulphate chemistry. The model will be set up to produce estimates of a range of pollutants at a number of point receptor sites within the county of Norfolk and for a grid system across the East Anglian region over a one-year period. Using a variety of emission inventories of different resolutions covering the whole of Europe the transboundary component of three different pollutants will be estimated (NO_x, Particulates (primary and secondary sulphur aerosols) and SO_2).

5.2 URBAN SCALE MODELLING

Pollution generated within Norwich and the immediate surrounding area is being modelled using ADMS-Urban (McHugh et.al., 1997) This will help identify pollution 'hotspots' within Norwich for the targeting of further monitoring but will also allow comparisons to be made between observed pollution levels and modelled estimates based solely on local emissions. It is anticipated that, by using results from the regional scale work with the NAME model to produce estimates of regional background levels, improvements can be made to the accuracy of ADMS-Urban results.

5.3 LOCAL SCALE MODELLING

Analyses of particular hotspots or significant emissions sources will be carried out using both 'pen and paper' techniques from UK government documents D1 and the Design Manual for Roads and Bridges and also street-canyon and point source modelling using ADMS-Urban. This will allow detailed analysis of problem areas and also provide a check with which to compare results from the urban-scale modelling.

The integration of modelling and monitoring data is being carried out using an ArcView based Geographical Information System. Work is also underway to communicate air quality information for the region to the general public. This establishment of the Norwich City Air Quality Website on the Internet (http://www.uea.ac.uk/env/aq/welcome.html).

Acknowledgements:

This work is being carried out as part of a PhD Studentship funded by the Natural Environment Research Council and Norwich City Council. Additional funding for modelling work has been provided by members of Norwich Environmental Protection Group.

References

APEG: 1999, Source Apportionment of Airborne Particulate Matter in the United Kingdom, *Report of the Airborne Particles Expert Group*, January 1999, ISBN 0-7058-1771-7

Cardenas, L. M., Austin, J. F., Burgess, R. A., Clemitshaw, K. C., Dorling, S., Penkett, S. A., Harrison, A. R. M.: 1998, Correlations between CO, NOy, O-3, and non-methane hydrocarbons and their relationships with meteorology during winter 1993 on the North Norfolk Coast, UK, *Atmospheric Environment*, **32**, No.19, pp3339-3351.

Department of the Environment Transport and the Regions: 1998a, Review and Assessment: Pollutant-Specific Guidance, HMSO, London.

Department of the Environment Transport and the Regions: 1998b, Review of the United Kingdom National Air Quality Strategy (a consultation document), HMSO, London.

King, A., Dorling, S.: 1997, PM10 particulate matter - The significance of ambient levels, *Atmospheric Environment*, **31**, No 15, pp2379-2381.

Kirby, C., Greig, A., Drye, T.: 1998, Temporal and spatial variations in nitrogen dioxide concentrations across an urban landscape: Cambridge, UK, *Environmental Monitoring and Assessment*, **52**, pp65-82.

Maryon, R.H., Smith, F.B., Conway, B.J., Goddard, D.M.: 1991, The UK Nuclear Accident Model, *Progress in Nuclear Energy*, **26**, No.2, pp85-104.

McHugh, C.A., Carruthers, D.J. and Edmunds, H.A.: 1996, ADMS-Urban: an air quality management system for traffic, domestic and industrial pollution, *International Journal of Environment and Pollution*, **8**(3-6): p. 666-674.

National Environmental Technology Centre: 1996, Local Site Operators Manual, http://www.aeat.co.uk/netcen/airqual/reports/lsoman/LSOmanual.html

National Environmental Technology Centre: 1999, UK National Atmospheric Emissions Inventory, http://www.aeat.co.uk/netcen/airqual/naei/naeimain.htm

Norwich City Council: 1999, Air Quality Review and Assessment - Stage 1, *Norwich City Council Report*, January 1999.

Office for National Statistics: 1997a, Focus on the South East, The Stationary Office, UK.

Office for National Statistics: 1997b, Focus on London, The Stationary Office, UK.

Shaw, J.M: 1994, A Norfolk Census Atlas, Norfolk County Council, UK.

Stedman, J.R: 1997, A UK-wide episode of elevated particle (PM10) concentration in March 1996, *Atmospheric Environment*, **31**, No 15, pp2381-2383.

ASSESSMENT OF AIR QUALITY, EMISSIONS AND MANAGEMENT IN A LOCAL URBAN ENVIRONMENT.

[a] CRABBE, H., [a] BEAUMONT, R. AND [b] NORTON, D.
[a] *Urban Pollution Research Centre, Middlesex University, Bounds Green Road, London. N11 2NQ. U.K.*
Tel: 0181 362 6361. Fax: 0181 362 6580. E mail: h.crabbe@mdx.ac.uk.
[b] *Environmental Health Department, London Borough of Barnet.*

Abstract:
This paper describes one example of how the UK National Air Quality Strategy (NAQS) is implemented in a local urban environment. The paper reviews the beginning of this process, by examining the review and assessment procedures of the NAQS in the London Borough of Barnet. By the application of available UK tools of local air quality management (LAQM), the process began through analysis of the levels of local emissions and progressed onto modelling of current and future air quality. A map showing combined emission hotspot areas for the Borough indicated that higher emission rates occur in the south of the Borough and along the major transport corridors, as road sources dominate emissions. Dispersion modelling studies were also conducted for this purpose, using the screening models GRAM, PGRAM and ADMS Urban for an in-depth assessment. These analyses found that some local point sources and the majority of Borough roads with over 20,000 vehicles per day produced exceedances of the future objectives for air quality for some pollutants.

Recommendations for the progression of LAQM in the Borough are made and include the update and expansion of the emissions information held for use in future modelling studies. The paper demonstrates the experience of implementing the Strategy, using the tools and procedures available for this purpose, in a local urban environment that is similar to many in the UK.

Key words: Local Air Quality Management, National Air Quality Strategy, dispersion modelling, emission inventories, review and assessment procedures, London Borough of Barnet.

1. Introduction

A system of Local Air Quality Management (LAQM) introduced recently in the UK requires local authorities as a statutory duty to review and assess air quality under the National Air Quality Strategy (NAQS) (DoE, 1997). This involves a review process with the aim to identify areas of concern where health-based national air quality targets and objectives are predicted to be breached in the future. The Review and Assessment (R&A) procedure is a three stage approach of increasing complexity of analysis, outlined in the Government's guidance note LAQM G1(97) (DETR, 1997a), supported with published technical guidance notes. The aim is to identify all significant sources of air pollution within unitary

Environmental Monitoring and Assessment **65**: 435–442, 2000.

boundaries and establish the current and future levels of air quality resulting from them by dates given for selected air pollutants.

The new UK Government has recently reviewed the Strategy with changes to the attainment of future objective dates and levels for some pollutants (DETR, 1998, 2000). The first round of the R&A process undertaken by local authorities is well underway, due for completion by the summer of 2000. Current activities however, should aim for the original 2005 target objective date, being recognised by law in the Air Quality Regulations (HM Government, 1997).

The London Borough of Barnet (LBB) in the north of the UK capital city is one of thirty-three London borough's, each with a responsibility for developing its own air quality management plan, reporting to the Secretary of State. With a population of some 312,000 people, the Borough is characterised by being a large urban residential area with local shopping centres and a few pockets of light industry. A large proportion of the population commute into central London along the many heavily trafficked roads. Middlesex University in partnership with the LBB, initiated the R&A process for the Borough by reviewing air quality and emissions within its boundaries. This paper focuses on the experience of implementing these ideas in a local urban environment, by reviewing the introduction of LAQM assessment and emission audit procedures in the LBB. The advances in the science of LAQM and assessment tools will be discussed.

2. LAQM and emissions assessment in the LBB.

As a first step towards identifying air pollution problems within the Borough, selected LAQM practices and tools were implemented to assess air quality and emissions. Nationally available tools include in-depth urban emission inventories, monitoring networks and dispersion modelling techniques. In any case, information on the significant sources of air pollutants is needed in order to manage emission levels and improve air quality. Emission inventories can be used as an integrated tool by identifying the intensity and spatial distribution of emissions. By mapping emission rates spatially, the areas with the highest levels of emissions are determined and these could hence be possible areas of concern for air quality and given priority consideration. The emissions inventory will also provide data for in-depth dispersion modelling studies, to determine the dispersion of emissions to predict levels of air pollution at specific locations.

In this study, a detailed local emissions audit of the area was first applied by the use of a recently updated emission inventory of pollutants and their sources within London, compiled by the London Research Centre (LRC, 1997). For each pollutant, a 'bottom-up' approach to compiling emissions was taken. This involved compiling activity data for each polluting source, e.g. vehicle flows, composition and speeds, industrial activity, population density, intensity of domestic sources and diesel train activity. Emission factors were taken from the UK Emission Factor Database (LRC, 1998). The process identified several specific emission hotspot areas within the Borough from local point, area and line sources. This emissions audit contributed to the information gathering requirements of pollution sources for Stage 1 of the R & A process. Source apportionment for the pollutants emitted within the LBB identified that emissions came from road, rail, area and point sources (table I). It was concluded that, for the Borough, road source emissions are by far the largest contributor to the key pollutants, giving 84% of sulphur dioxide (SO_2), 84% of nitrogen oxides (NO_x), 99% of carbon monoxide (CO), 87% of benzene, 96% of 1,3 - butadiene, and 91% of particulate matter (PM_{10}) of the Borough's annual average emissions. As a result, emission maps for all sources show that higher emission rates occur following a pattern along the major road network. Due to the high density of major commuter roads passing through the Borough and the lack of other significant sources of pollutants (e.g. major industrial plants and airports), the LBB has a higher proportion of emissions coming from road sources than most other urban areas and London Boroughs (table II). A map locating these 'emission hotspot' areas allowed for the concentration of air quality management resources and tools such as dispersion modelling and monitoring (Fig.1).

3. Modelling of air quality in the Borough for Review and Assessment.

For the Second R & A stage, a sample of these emission hotspot areas were chosen to have screening dispersion modelling assessments to predict levels of ambient and future air quality at these sites. This included several major road sources and point sources identified from the local inventory as having the potential to emit high levels of pollution. The screening models GRAM (University of Greenwich Review of Air quality Model, for roads sources) and PGRAM (for point sources) were utilised for this purpose (Fisher, 1997). These models were chosen for this purpose as they are designed to screen out situations that will clearly achieve the NAQS objectives, and retain those that do not. At this screening level, some point sources were predicted to have exceedances of the NAQS objectives by 2005 for selective industrial pollutants under 'worst case' scenarios. The majority of major roads with current daily flows over 20,000 vehicles were also predicted to produce pollution levels that breech the objectives, for the pollutants nitrogen dioxide (NO_2), PM_{10} and CO (Crabbe and Beaumont, 1998). As a result these sources were referred to be investigated further through in-depth modelling studies and validatory monitoring at Stage 3 R&A.

Table I Source apportionment for pollutants emitted in the London Borough of Barnet (derived from the LRC LAEI database, 1997).

	Pollutants emitted (percentages)										
Source types:	SO_2	NO_x	CO	CO_2	NMVOC	benzene	butadiene	smoke	TSP	PM_{10}	methane
Road emissions	84	84	99	34	71	87	96	96	90	91	1
Rail emissions	6	3	0	1	0	2	4	0	4	4	0
Area emissions:	8	13	1	65	28	11	0	4	3	3	99
Point source emissions:	2	0	0	0	0	0	0	0	3	2	0

Table II Relative proportion of road emissions in the London Borough of Barnet in comparison to those for London as a whole (derived from LRC, (1997)).

	SO_2	NO_x	CO	CO_2	NMVOC	Benzene	1,3-Butadiene	PM_{10}
Road emissions as a % of total emissions in Barnet	84	84	99	34	71	87	96	91
Road emissions as a % of total emissions for the whole of London	23	76	97	30	62	86	97	77

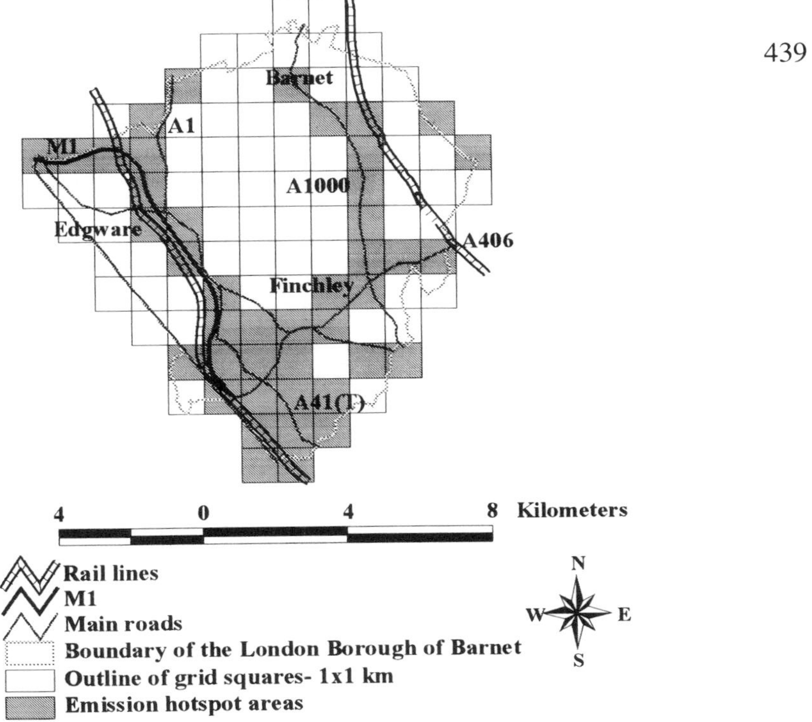

Fig. 1. The London Borough of Barnet and major transport routes. The highlighted squares show where the highest emissions occur for a combination of the key pollutants. These 'emission hotspot' areas are also indicative of where the AQMA's are likely to be designated, within which breaches of the NAQS objectives for air quality are predicted to occur.

A preliminary Third-stage in-depth modelling analysis of one of these areas used the 'new generation' ADMS Urban air quality modelling system. This was suitable to model the air quality impacts of selected point, area and mobile activities. The model is integrated with an emissions inventory relational database within a GIS system (Carruthers, et.al., 1997). This allows for the interrogation of sources within a particular area and the resulting air quality impacts can be modelled within a spatial framework. The model was selected to find out where possible breaches of the NAQS objectives were likely to occur within the Borough. For this assessment, an update of the local emissions database was necessary for the more advanced information required by the modelling system. This demanded more accurate temporal and detailed traffic data on a local scale (approximately within a specified 3x3km square area). To calculate exposure levels that the residential population is likely to be subject to, receptors were selected at the nearest residential housing to the main roads in the area. A meteorological data file consisting of a year's hourly values of wind speed and direction, temperature and cloud cover (to calculate boundary layer heights) was needed to calculate annual average values. A representative dataset was used from the UK's Meteorological Office site at the London Weather Centre, some 10km distance away, as this represents an urban built-up microenvironment. Scenarios also accounted for traffic growth in future years, taking growth figures from the DETR's National Road Traffic Forecast (1997b) for urban areas. Ambient levels of pollution were predicted for the year 2005, so predictions could be compared to the relevant standards over their averaging time. This latter study confirmed the screening model predictions, with breeches of the NO_2, PM_{10} and CO objectives along the major road routes within this area (Crabbe and Beaumont, 1998). This study also tested the methodology of assessing air quality at Stage 3 R&A and determined what data and level of detail was necessary for future studies. Only after validatory monitoring to confirm this prediction, will it be possible to consider if this part of the Borough is to be declared as an Air Quality Management Area (AQMA). Figure 1 suggests the parts of the Borough that are most likely to be designated as AQMA's, probably being areas *within* the defined 'emission hotspot' areas. Air quality action plans and emission reduction methods such as traffic management measures will then be required on a local scale in 'air quality hotspots' to manage the local air quality, in addition to existing national policy measures and improvements in technology designed to reduce air pollution. The LAQM process within the LBB is still in progress and any decisions and outcomes on action plans will be scrutinised through forthcoming public consultation.

4. Conclusions: recommendations and limitations.

This paper highlights the development in urban LAQM in the UK, by focusing on its implementation so far in one urban environment. The research has reviewed the application of an emissions inventory database, its use, update and interpretation through the R & A procedures of the NAQS in the LBB. Implementation of this information through screening and in-depth modelling studies have been outlined here. The results from this research lead to an enhanced knowledge of the application of the R&A process in a typical London Borough and recommendations for the management of air quality within the

Borough were given. This included the need for further investigation of specific polluters impacting on the levels of local air quality (e.g. arising from diesel trains passing through the Borough) and the necessity to quantify the impact of imported pollution. Recommendations to increase the use of simple monitoring techniques in emission hotspot areas (e.g. by passive diffusion tube samplers) and advice on the requirement and positioning of advanced monitoring to validate the ADMS model in predicting air quality hotspot areas, were also given.

It is necessary to note here the limitations of using emissions inventory annual average data in dispersion modelling as the latter requires data with a temporal variation and finer spatial resolution to be used successfully to predict ambient air quality levels. There are potential deficiencies in both line and point source data compiled for this purpose (Vawda, 1998). The data obtained are average **estimates** of the activity's emissions, and it is important to recognise the limitations in their preparation to their accuracy and sensitivity (Mann and Sokhi, 1997). In this study, more detailed emissions information was collated for the Stage 3 assessment study, requiring considerable time and resources to compile. This could be a limiting factor in the detail and accuracy of assessments that local authorities can undertake given the required time frame of completion of the R & A process by the Regulations, by the middle of 2000.

This research ultimately helps to demonstrate the transferability of the application of LAQM tools to other local authorities wanting to assess their air quality, emissions and management in similar urban environments. The paper also describes the experience of using the latest scientific tools available to LAQM on a local urban scale.

Acknowledgements

The authors wish to thank the London Research Centre for the use of data from the London Atmospheric Emissions Inventory database and responding to queries and requests for information. We would also like to acknowledge the help from Officers of the LBB, Directorate of Environmental Services, in providing information for the development of research for this project. The UK Meteorological Office is acknowledged for the provision of suitable meteorological data under license used in the modelling studies.

References

Carruthers, D.J., Edmunds, H.A. McHugh, C.A. Riches, P.J. and Singles, R.J. (1997) ADMS Urban- an integrated air quality modelling system for local government. In Power, H. Tirabassi, T. Brebbia, C.A. (Eds) Air Pollution V, Modelling, Monitoring and Management, Southampton, CMP. pp45-58.

Crabbe, H. and Beaumont, R. (1998) Air quality assessment in the London Borough of Barnet. Unpublished research report, Middlesex University.

DETR (Department of the Environment, Transport and the Regions)(1997a) Framework for review and assessment of air quality. LAQM.G1(97). Stationary Office, London.

DETR (1997b) National Road Traffic Forecast, web address: http://www.roads.detr.gov.uk/roadnetwork/

DETR (1998) Review of the National Air Quality Strategy, a Consultation Document. Stationary Office, London.

DETR (2000) The Air Quality Strategy for England, Scotland and Wales. Stationary Office, London.

DoE (Department of the Environment) (1997). The United Kingdom National Air Quality Strategy, Stationary Office, London.

Fisher, B. (1997) Carrying out air quality reviews and assessments under the 1995 Environment Act. Training Booklet. University of Greenwich.

HM Government (1997) The Air Quality Regulations 1997. Statutory Instruments, 1997 No. 3043, Environmental Protection. The Stationary Office, London.

LRC (London Research Centre), (1997). The London Atmospheric Emission Inventory. LRC, London.

LRC (1998) The UK Emission Factor Database web address: http://www.london-research.gov.uk/emission/main.htm

Mann, R.C. and Sokhi, R.S. (1997) A high resolution atmospheric emissions inventory for road vehicle sources and the factors influencing its accuracy and sensitivity. In Power, H. Tirabassi, T. Brebbia, C.A. (Eds) Air Pollution V, Modelling, Monitoring and Management, Southampton, Computational Mechanics Publications. pp983-992.

Vawda. Y. (1998). Urban Emission inventories- their uses and limitations in dispersion modelling. *Clean Air,* Vol **28**, No.5, pp149-153.

AIR QUALITY MONITORING AND MANAGEMENT IN LISBON

Francisco Ferreira, Hugo Tente, Pedro Torres, Sérgio Cardoso and José M. Palma-Oliveira*
Faculdade de Ciências e Tecnologia, Quinta da Torre, 2825-114 MONTE CAPARICA, Portugal
**Faculdade de Psicologia e Ciências da Educação, Alameda da Universidade, 1600 LISBOA, Portugal*
E-mail: ff@mail.fct.unl.pt

Abstract. The environmental decision-making process is related with the interpretation of data both in spatial and temporal dimensions. This paper presents a methodology that integrates the time-space framework of air quality data to infer the temporal pattern and spatial variability that could be interpreted for environmental decision purposes. Variograms that accommodate time and space lags were used for the analysis and proved to be effective. Its environmental meaning, in particular its relationship with traffic patterns is discussed. Data from air quality monitoring stations located in the central part of Lisbon were used in this study. It describes a strategy to identify the type of vehicles responsible for certain pollutant levels, particularly for nitrogen oxides, and discusses the application of new air quality European legislation to the city of Lisbon, Portugal.

Key words: air quality management, air quality monitoring, traffic pollution, kriging, demand side management.

1. Introduction

Environmental management must include three sequential tasks: data capture, data analysis and decision making. Since any environmental issue is dynamic by nature and manifests itself on a certain spatial scale, the process of environmental data capture and analysis has to accommodate its twofold nature, the spatial and the temporal dimension. Ideally, environmental monitoring should be proceeded on a continuous-space and continuous-time frame, considering the appropriate spatial and temporal scale for the phenomena under study. Point source observations provide only small, sometimes inadequate clues to a much larger problem. Appropriate exploratory data analysis must answer questions about patches, defined as some spatial extension of significant data, besides point data, as well as questions regarding meaningful periods of time. The goal of this paper is to present a methodology to assess simultaneously the spatio-temporal patterns of air quality at the city of Lisbon. This patterns are clearly related with traffic, and the identification of the type of vehicles and circulation patterns responsible for the high level concentrations detected for certain pollutants are evaluated. The response of different monitoring stations is also analysed, and strategies to reduce the urban pollution in Lisbon are discussed based in the data presented.

Environmental Monitoring and Assessment **65**: 443–450, 2000.

2. Application

2.1. AIR QUALITY KRIGING

The sampling scheme of air quality data usually includes a set of monitoring stations at specific points, measuring pollutant concentration on a quasi-continuous fashion. Therefore, air quality monitoring is extremely exhaustive in the time domain but sparse in the space one. Some experiments on air quality analysis have been unsuccessful if only the spatial dimension of the data is considered. Spatial patterns can be described quantitatively in terms of the semivariance function, which is based on the idea that statistical variation of data is a function of distance (Cressie, 1991). The consideration of time within the spatial data analysis framework as an improvement has been proposed by several authors (Miller, 1996). Variograms that accommodate time and space lags were developed and used for the analysis and proved to be effective.

The sampling scheme of air quality data usually includes a set of monitoring stations at specific points, measuring pollutant concentration on a quasi-continuous fashion. Therefore, air quality monitoring is extremely exhaustive in the time domain but sparse in the space one. Some experiments on air quality analysis have been unsuccessful if only the spatial dimension of the data is considered.

Z(x,t) is the air quality concentration at a given time t at the monitoring station located at x, usually defined by a two-coordinate pair on a plane reference system. Adopting the rational beyond spatial analysis, observations close in time tend to be more similar than those further apart. Thus, for a specific location (x_0) in the space domain, the semivariance of z (x_0,t) can be computed, considering pairs of data at successive increasing temporal lags N(d), and using the following expression:

$$\gamma(d) = \frac{1}{2N(d)} \sum_{t=1}^{N(d)} \left[z(x,t) - z(x,t+d)\right]^2 \tag{1}$$

If there is a semivariance function, the variogram parameters can reveal some meaningful temporal patterns of air quality, mainly the time range. This analysis has to be conducted for all the spatial locations available for the study. The temporal range (d_t) concluded from the above model, and for the condition of its similarity for all the spatial locations, can now be incorporated for the spatial variability assessment, by the expression:

$$\gamma(h) = \frac{1}{2d_t N(h)} \sum_{x=1}^{N(h)} \sum_{t=1}^{d_t} \left[z(x,t) - z(x+h,t)\right]^2 \tag{2}$$

The proposed methodology was applied for Lisbon, and for NO_2, which derives mainly from the traffic activity. Figure 1 shows the spatial distribution of the air quality monitoring stations referenced on a 10 m panchromatic SPOT image along the major roads.

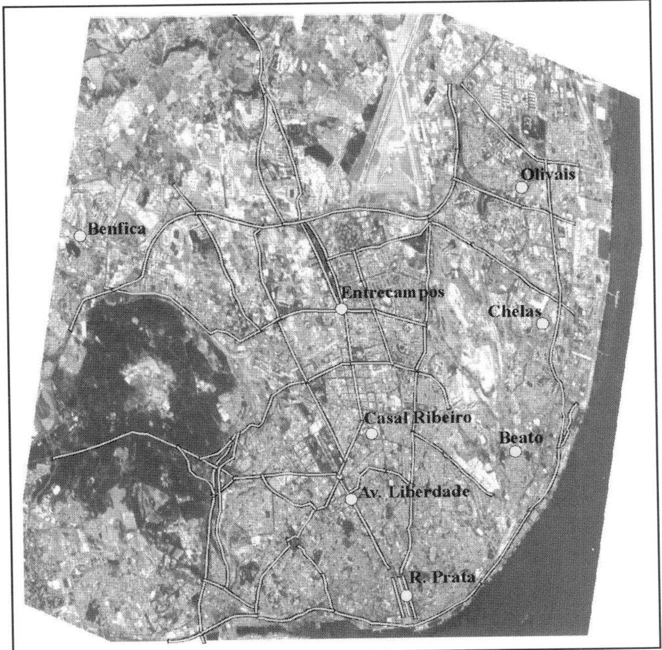

Figure 1. Air quality monitoring stations and main traffic roads

One NO_2 data set with the hourly average from April 18, 1995 was used to assess the air quality behaviour during a working day. The time range found in the experimental temporal variogram was used to compute the spatial variogram. Figure 2 a) presents the temporal variogram for the NO_2 hourly data. A final experimental variogram was developed using a spatial range around 3.5 Km, where spatially correlated data follow a spherical model, as presented in Figure 2 b). This might be related with the similarities between traffic patterns among the different air quality monitoring stations in Lisbon during the day (Ferreira *et al.*, 1998).

For kriging interpolation the experimental variogram was fitted to the spherical model, according to equation model presented.

$$\gamma(h) = 4500 + 4241.2\left(1.5\frac{h}{4000} - 0.5\frac{h^3}{4000^3}\right), h < 4000 \tag{3}$$

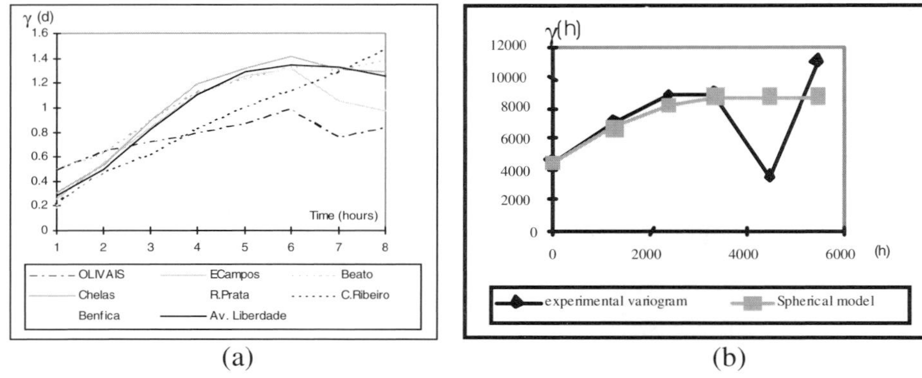

(a) (b)

Figure 2. NO_2 hourly data from Lisbon: (a) temporal, and (b) spatial variograms.

The kriged NO_2 concentrations as well as its standard deviation were estimated for a 10 km grid. The spatial resolution was set at 50m in both the northings and the eastings from the sampling grid. Figure 3 displays an example of air quality isosurface for NO_2 for the city of Lisbon , during the morning rush hour - 8:00am. It is quite visible (left) the movement of higher concentrations towards South along the major avenues, which is related with higher traffic from Lisbon entrances to downtown. Figure 3 b) shows the kriged standard deviations for NO_2. The lower values are related with the existence of a higher number of monitoring stations, therefore decreasing the error.

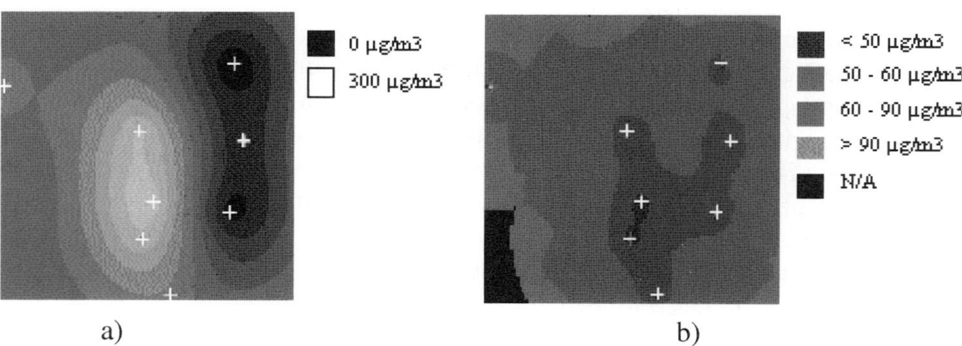

a) b)

Figure 3. Map of kriged NO_2 values (a) and correspondent standard deviations (b).

2.2. POLLUTANT'S MONITORING RESPONSE TO TRAFFIC STRUCTURE

Air quality data collected between January 1, 1996 and March 31, 1997 from all monitoring stations in Lisbon were used to find which ones are more clearly responding to traffic pollution. The aim was to understand spatial and temporal

trends and similarities between pollutant's levels measured at monitoring stations, either at street canyons or in other locations in the city.
Figure 4 shows indeed that monitoring stations closer to roads present higher average NO values then NO_2, as expected. due to traffic pollution. Beato, Chelas and Olivais are located in the Eastern part of Lisbon, where traffic is lower. These stations where located a few years ago in this area to monitor industrial pollution. The major industries were closed in the last years and since they are not located in street canyons their response is clearly different.

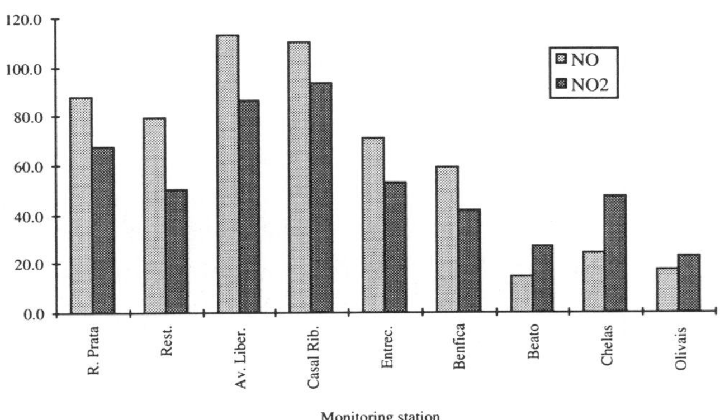

Figure 4. NO and NO_2 average hourly levels

Two air quality monitoring stations in Lisbon (Av. Liberdade and Entrecampos) were selected for a detailed analysis during different times of the day. Two one-hour periods in working days in June, 1998 were chosen considering two different traffic patterns (off-peak and peak hour). The experiments took place between 1:00 pm and 2:00 pm, and between 6:00 pm and 7:00 pm. A video camera was used to record the traffic nearby the stations. The type of vehicles and fuel were identified for all cases. Three pollutants (CO, NO, and NO_2) were measured on a minute-by-minute basis.
Correlations between the traffic patterns and the observed pollutant levels were calculated to evaluate the role of vehicles and fuel with the collected air quality values. For Av. Liberdade, during the lunch-time/off-peak period, the roads are clear and vehicles circulate with speeds up to 50-60 km/h. In this period, the main relationships found are between cars and gasoline vehicles and the pollutants measured (correlation coefficients higher than 0.75, particularly with NO_2 and CO).
For the peak-hour period, when the circulation speed is particularly low except for the bus lane, the higher correlations obtained are for buses and NO (r=0.914). This is probable due to the fact that most of the other vehicles are almost stopped and only buses circulate. The same effect identified for the

lunch-time period is detected but with lower significance. For Entrecampos, the data obtained do not show any particular pattern for both periods. This area is close to a roundabout which may cause the analysis to be more difficult to perform and establish correlations.

2.3. EVALUATION OF NEW EUROPEAN AIR QUALITY MONITORING LEGISLATION

An evaluation of the European Council Directive 1999/30/CE air quality limit values (European Union, 1999) was performed for Lisbon monitoring stations using 1996 data. Using the population criteria of the Directive, the number of required stations is three. By analysing the spatial correlations between the existing stations and identifying the areas that are currently not covered, a minimum of four stations are required.

For three of the four pollutants regulated by the proposal (sulphur dioxide, particulate matter, and nitrogen dioxide), the measured values were compared with the proposed ones. Particulate matter is measured only in one location. All the proposed limits would have been exceeded if they were effective.

For nitrogen dioxide, there are three stations where both hourly and annual limits would have been exceeded. Also upper and lower assessment threshold limit values would have been exceeded. For sulphur dioxide, even only emitted by diesel vehicles, upper and lower assessment threshold limit values would have been exceeded in one monitoring station. Also, two of the monitoring stations had an efficiency lower than 90%. The correct location of each monitoring station according to the Directive guidelines was evaluated. Some of the requirements are not currently achieved, in particular microscale sitting criteria.

2.4. STRATEGIES FOR A MORE SUSTAINABLE TRANSPORTATION

From the previous analysis it is clear that traffic is the major source of pollution in Lisbon, reaching concentrations above the future European legislation limits. In part, this is due to a limited use of demand side measures in transportation in Lisbon. However, these policies are used to control the traffic in several cities across Europe (Harman, 1996).

Until now there is no clear integrated methods of impact evaluation of the demand measures. The work by the European Commission is more directed to the abatement of air pollution by technical ways as seen in the Auto-Oil program I. However the new air quality framework directive and the work under the program Auto-Oil II could be instrumental in providing the necessity of accessing the impact of the different structural measures. This last program is trying to evaluate, in an quantitative way the direct impact of demand side measures in the air quality and other externalities.

There are different measures that one could refer as demand side measures. Some of those measures are outside the scope of this paper because are long term political measures (i.e., incentives for fleet modernization , land use). The ones that we are about to describe are the ones that can have a direct and immediate impact in the driver behavior and in the air quality, particularly for Lisbon: improving traffic flows through signaling/guidance, removal of infrastructure bottlenecks, speed limit (non-urban), and speed limit (urban); reduce traffic in specified areas by parking restrictions, zoning/restricted access; improve attractiveness of other modes such as public transport prioritization, promote cycling; public transport improvement by extension of services, pricing measures (subsidized fares), and car pooling initiatives; road pricing: parking charges, cordon pricing - undifferentiated and time differentiated, and vehicle differentiated.

All these measures are aimed to accomplish a switching of modes and, at the same time, a complete behavior change for the users and, in our case, the car commuters. However our behavior as a consumer and a citizen is determined for a diversified array of factors that, in this case, almost all contribute for the individual transport choice. Thus, although we have, in our modern societies a high positive attitude towards the environment the degree that we behave accordingly with that attitude is low. Several reasons are usual introduced in order to explain this gap:

- Extreme masking and delay of cause-effect gradients. Experimental research shows that humans are very immediate in the relations of learning from the environment (i.e., they tend to learn only when the cause is very near to the effect). Thus when we are talking about air quality of the air or health effects is difficult to connect our behavior with those effects.
- Low subjectivity cost-effectiveness of environment-protecting behavior. There are multiple instances of a behavior detrimental to the environmental (i.e., driving a car) may prove more cost effective from the actor point of view than a less detrimental one. This configures a social dilemma situation where the individual rationality is at odds with the collective rationality (Pawlik, 1991; Palma-Oliveira, 1992).

4. Conclusion

Spatial and temporal kriging can be an excellent tool to understand air quality over a broader area such as the city of Lisbon. Traffic was identified as the major cause of air pollution and a specific study was developed to find the circulation pattern and the type of vehicles most responsible for the high concentrations detected. During off-peak cars are the vehicles responsible for the high levels of carbon monoxide and nitrogen dioxide detected, while for peak hours buses should be considered the nitrogen monoxide major source. The requirements for the application of the new European Framework Directive were stated both as a need to reduce pollution levels and by optimising the location of the air quality monitoring stations. A group of measures to reduce the impact of air pollution from traffic were proposed for the specific case of Lisbon. Since some of them are being implemented further evaluation is necessary to track changes.

Acknowledgements

The authors would like to thank the Comissão de Gestão do Ar de Lisboa who provided the air pollution data, and the Centro Nacional de Informação Geográfica for the SPOT image of Lisbon.

References

Cressie, N.: 1991, *Statistics for Spatial Data*, John Wiley & Sons, Inc., New York.

European Union, 1999. Council Directive 1999/30/EC of 22 April 1999 relating to limit values for sulphur dioxide, nitrogen dioxide and oxides of nitrogen, particulate matter and lead in ambient air, OJ L 163 29.06.99 p.41

Ferreira, F., Seixas J., Nunes, C. and Silva, J. P.: 1998, Air Pollution Space-Time Analysis, in I Linkov and R Wilson (eds.), Air Pollution in the Ural Mountains - Environmental, Health and Policy Aspects, Kluwer Academic Publishers, Dordrecht, 357-360.

Harman, R.: 1996, New Directions: a manual of European best practice in transport planning. London: Transport 2000.

Miller, E. 1996, Toward a Four-Dimensional GIS: Four-Dimensional Interpolation Utilizing Kriging, GIS Research UK, University of Canterbury.

Palma-Oliveira, J.M.: 1992, Stress Ambiental: Um Selectivo Ponto da Situação. Modelo Explicativo, Revista da Sociedade Portuguesa da Psicologia, **28**, 13-77.

Pawlik, K.; 1991, The Psychology of Global Environmental Change: Some Basic Data and Agenda for Cooperative International Research. International Journal of Psychology, **26** (5), 547-563.

DEVELOPMENT AND USE OF INTEGRATED AIR QUALITY MANAGEMENT TOOLS IN URBAN AREAS WITH THE AID OF ENVIRONMENTAL TELEMATICS

KOSTAS KARATZAS and NICOLAS MOUSSIOPOULOS

Laboratory of Heat Transfer and Environmental Engineering
Aristotle University Thessaloniki, Box 483, GR-540 06 Thessaloniki, Greece
E-mail:kostas@aix.meng.auth.gr

Abstract. Contemporary urban air quality management requires the use of appropriate systems which include air quality models, a Geographical Information System (GIS) and a combination of expert systems and decision support tools, while at the same time possessing the capability to receive information from in situ measurements. Until recently, the relation between Information Technology capabilities and the system's design and architecture were poorly addressed, mainly due to technological limitations posed. Moreover, air quality management scenario design issues were partially considered, because of the difficulty in aggregating complex, air quality related issues, in a comprehensive and effective manner, from the end users point of view. In the present paper the use of Environmental Telematics is discussed as a framework for the development of urban air quality management systems, while a comprehensive approach for the application and evaluation of relevant scenarios is presented.

Key words: urban air quality management system, telematics, scenarios, World Wide Web

1. Introduction

Air quality management (AQM) is one of the main environmental problems currently confronting urban agglomerations. Its complicated nature and the numerous interrelations between parameters and actors involved in AQM, makes the use of a system approach a necessity. A contemporary Urban Air Quality Management System (UAQMS) requires the integration of a heterogeneous set of resources, which include (Fedra and Haurie, 1999):

- Multiple sources of information, including on-line monitoring systems.
- A dynamic and spatially distributed structure with multiple temporal and spatial scales for the complex dispersion and transformation processes that translates emissions into ambient air quality, which is the domain of air quality modelling proper.
- Distributed (and mobile) emission sources with pronounced temporal patterns that include industry, households, and traffic sector that can be modelled as a network (dynamic) equilibrium process.
- Direct regulatory and indirect economic control on emission sources.

Environmental Monitoring and Assessment **65**: 451–458, 2000.

- Multiple objectives and criteria at different spatial and temporal scales for the different actors and the regulatory framework.

To provide relevant information to the management and decision making process, information technology and analytical tools such as simulation models can be combined into powerful information and decision support systems.

2. Main characteristics of an UAQMS

A contemporary UAQMS should address all information relevant with the problem at hand, provide access to appropriate tools and support effective decision making. More specifically, such a system should include the following activities (Larssen, 1996):

- Selection of information on air pollution activities and emissions.
- Monitoring of air pollution and dispersion parameters.
- Engagement of appropriate air pollution models for the calculation of air pollution concentrations and exposure parameters.
- Inventory of population, materials and urban development and
- Establishment and improvement of air pollution regulations.

The former activities can be integrated to an UAQMS on the basis of an appropriate system module and applications as follows (Fedra *et al.*, 1999; Karatzas, *et al.*, 1998):

- A monitoring network facility, preferably operating in line with tele-transmission techniques (Kouroumlis *et al.*, 1999).
- A database for the collection and management of data.
- A Geographical Information System (GIS) for handling all spatially distributed data.
- A set of models for the appropriate and validated representation of air quality and dispersion parameters.
- An expert system for supporting the formation and interpretation of control strategies and the design for alternative scenarios to be analysed.
- A network communication infrastructure and a potential for dissemination of air quality related information to the public. The World Wide Web communication technologies and capabilities for publishing information is widely used.
- An integration information platform allowing effective collaboration of all system components with advance information flow. It is very common that a client/server type of integration platform is being used (Swobota, 1998).Based on the above, a typical UAQMS, developed in the frame of the ECOSIM Environmental Telematics Applications Programme (URL 1) is shown in Figure 1, as applied for the city of Athens, Greece.

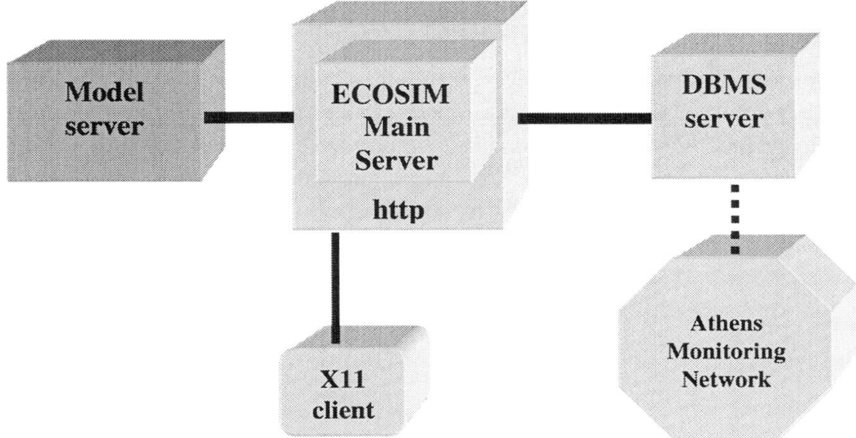

Figure 1. The UAQMS ECOSIM as applied for the city of Athens.

The ECOSIM system is characterised by its technological framework and architecture. The main elements of this architecture include:

- A flexible client-server implementation for distributed and decentralised use of information resources.
- A communication architecture based on the http protocol which is used to integrate real-time data acquisition from monitoring sites, as well as optional high-performance computing resources such as supercomputers or workstation clusters; primary consideration here is the scalability of applications over a wide range of performance requirements.
- Multimedia user interface design to support an intuitive understanding of results.
- Integration of GIS with data bases, monitoring results, and spatially explicit simulation modelling.
- Embedded rule-based expert systems for logical modelling and user support.

All above features are designed to address difficult analytical problems, and at the same time provide a convenient and easy to use intuitive user interface (Fedra *et al.,* 1999; Karatzas *et al.,* 1998).

2.1. REMOTE USE RESOURCES AND INFORMATION IN UAQMS

An additional key parameter for designing and developing UAQMS is related to the fact that some of the system components (like air pollution models) are usually not available locally, while a large percentage of the information required for the system operation is by nature heterogeneous and distributed geographically (e.g. emission data and air quality monitoring data). This distributed nature of data and heterogeneity of information requires an intermediate layer interfering the client and server relationship mentioned above. Middleware systems are designed to serve such needs and their integration into Web applications has been explored in various directions recently (Beitz and Woo, 1995; Beitz *et al.,* 1997). Moreover, an object oriented approach in UAQMS can resolve data handling problems, which involve client and server data change synchronisation, while the use of an object request brokers (a mediator between applications, including distributed ones) like CORBA (Common Object Request Broker Architecture, URL 2) is foreseen. CORBA is a standard for open distributed systems defined by the Object Management Group (OMG, URL 3). It defines ways for objects and clients to interact within a distributed environment, based on the idea of an Object Request Broker, which allows objects to communicate with one another.

In addition to the above, the development of programming languages like JAVA, can serve as an optimum solution for the development of interactive air quality management applications on the Web. This approach is currently being tested and evaluated via the environmental telematics project group (URL 4).

3. The scenario approach

In order for the decision-maker to formulate and submit to the UAQMS queries concerning alternatives in urban AQM, a scenario generation tool is required. This tool will be the provider of all available 'tuneable', variables according to a pre-selected and predefined, hierarchically structured, set of realistic alternatives. These alternatives reflect dependencies on two major air quality related issues: meteorology and emissions. Predefined meteorological scenarios are proposed representing classified meteorological conditions and relevant probabilities for the occurrence of high-level air pollution concentration values. Moreover, for the emission abatement scenarios, the use of emission multipliers for each emission source category is proposed (multidimensional scenario approach). This approach has already proved to be efficient serving user needs and requirements (Kuruvilla *et al.,* 1994), while being also simple, thus overcoming the known problem of complexity in scenario formulation.

4. Emission reduction scenario formation and evaluation in UAQMS

The use of the multidimensional scenario array in the UAQMS can provide the decision-maker with qualitative and quantitative information (e.g. which emissions should be reduced in which percentage in a specific territory-site of the city, in order to achieve a certain improvement in air quality). Such an investigation requires all the actors (emitters) to be taken into account in emission abatement strategies. Moreover the use of such a system gives the decision maker new capabilities in studying these scenarios, the benefit of which is demonstrated with the aid of game theory, according to a methodology applied in environmental decision making by Adams, 1996.

Let us consider the case where a decrease in emissions of a specific polluter like NO_x, equal to R (kg/day) is being studied with the help of an UAQMS, for a specific site. Emitters located within the latter can be identified in

- *M*=low emission proportion agents (like central heating, concerning NO_x)
- *L*=high emission proportion agents (like traffic, concerning NO_x)
- *O*=unknown emission proportion agents (like unregistered activities), this being the 'orphan share' of all emissions

where *M, L,* and *O* are the contribution of each emitter category to the total emission reduction *R*. The first agents (*M*) are characterised as *de minimis* parties, as they are contributing only a relatively small amount to the total air pollution. The total emission reduction will then be *M+O+L=R* or, if *R* is normalised to 1, *M+O+L=1*. Each emitter contributes to the total emission cut discussed according to Equation 1:

$$\sum_{i\in M} w_i = M \qquad \sum_{j\in L} w_j = L \qquad \sum_{k\in O} w_k = O \tag{1}$$

where $w_{i\in M, j\in L, k\in O}$ = emission reduction demand per agent category.

It is logical to assume that firstly an agreement between the authorities and the de minimis emitters will be undertaken, as the latter are considered more acceptable in such scenario applications. For this reason a take it or leave it offer for a reduction of emissions equal to $w_i \cdot t$ will be presented to all $i \in M$, where t represents the portion of reduction compared with the agents 'fair share' in emission ($t=1$). All $m \in M$ simultaneously accept or reject this offer (let *M'* denote these agents). For the next negotiation stage, let us assume that the authorities choose a parameter *a* which represents the percentage of the 'orphan share' which will be charged to the de minimis agents, the rest being charged to the large percentage emitters. Let D_m denote the difference between the emission settling acceptance (first negotiation round) and the expected emission reduction requirements at the second negotiation phase. It has been proved (Adams, 1996; Karatzas, 1998), that when the emission reduction demand is positive, it increases as the settlement rate increases, thus the agents are asked to 'pay' an increasing part of the 'orphan share'. In this case, the benefit of a settlement decreases as the rate of settlement increases. These techniques are easily applicable to an UAQMS, as the former methodology can be implemented to it on the basis of an expert system. Thus, the existence of such a system can be very beneficial for the decision-maker, giving him/her the ability to proceed into a thorough analysis of emission related scenarios and their application consequences in urban air quality management issues.

5. A generalised AQMS

Taking into account all previously mentioned modules of an UAQMS the proposed architecture is given in Figure 2.

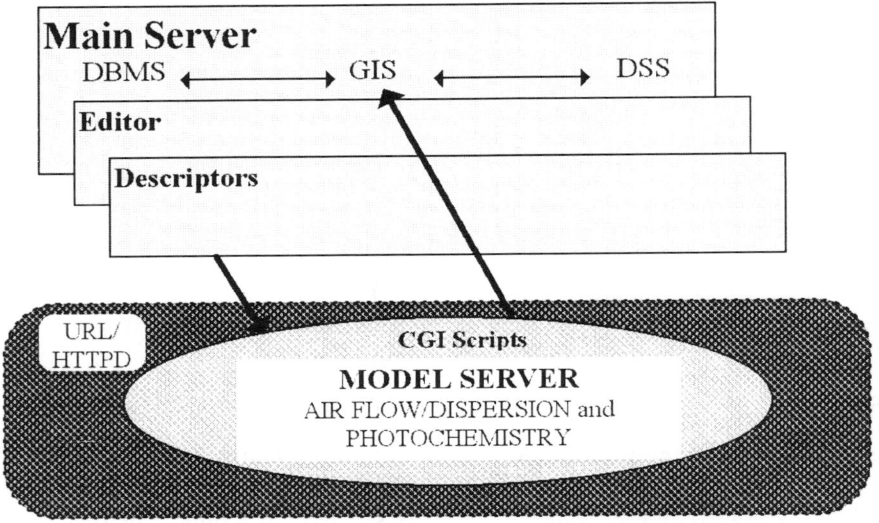

Figure 2. The Architecture of the proposed Web-based UAQMS.

The scenario editor will receive input from the Decision Support System modules to formulate descriptors (meta-information in object oriented description format) which will be passed to the model server, where CGI scripts (Deep and Holfelder, 1996) will interpret them and activate models appropriately. Various hardware platforms (including supercomputers) can be used as servers, distributed throughout the Internet, to allow appropriate model and computer applications environments to be integrated into one system. Alternatively CORBA applications can link scenarios with models and all other services. The speed of the process is directly proportional to the throughput and latency of the network connection between the servers.

5. Conclusions

Contemporary UAQMS are making use of advanced information technologies combining state of the art air quality modelling, GIS, databases and expert systems. The latter can profit from the use of advanced decision support tools and methods focusing on emission reduction scenario formation and evaluation. The use of Web infrastructures and communication can strengthen such systems, while future developments in Web based, distributed system building, are expected to result in really interactive and independent UAQMS.

Acknowledgements

This paper is related to the European Union - DGXIII projects ECOSIM, IRENIE and AIR-EIA and the project DESPOTIS funded by the General Secretariat of Research and Technology, Greece.

References

Adams, G.D.: 1994, *Three's a crowd: Multilateral game theoretical analysis of environmental policy*, PhD thesis, UMI Dissertation Services, No 6375, 111 pp.

Beiz, A. and Woo, T.: 1995, *'Integrating WWW and middleware'* http://www.scu.edu.au/ sponsored/ausweb/ausweb95/papers/management/beitz.

Beitz, A., Iannela, R., Vogel, A., Yang Zh. and Woo, T.: 1997, *'Integrating WWW and Middleware'*, http://www.scu.edu.au/sponsored/ausweb/ausweb95/papers/ management/beitz.

Deep, J. and Holfelder, P.: 1996, *'Developing CGI applications with PERL'*, Wiley Computer Publishing, 584 pp.

Fedra, K., and Haurie., A.: 1999, 'A Decision Support System for Air Quality Management Combining GIS and Optimization Techniques' *Environmental Modeling and Assessment* (in print).

Fedra, K., Karatzas, K. and Moussiopoulos, N.: 1999, *Integrated urban environmental management: monitoring, simulation, decision support*, 3rd International Exhibition and Conference HELECO99, 3-6 June, Thessaloniki.

Karatzas, K.: 1998, *Use of Integrated Air Quality Management Tools in Urban Areas as an Improvement in Environmental Decision Making Process from the Game Theory Point of View*, poster presentation at the 91st annual meeting and exhibition of AWMA, San Diego.

Karatzas, K., Moussiopoulos, N., Fatta, D. Loizidou, M., Perivoliotis, L. and Lascaratos, A.: 1998, *Final validation report for the ECOSIM Athens demonstrator*, ECOSIM Project Deliverable, DGXIII, Environmental Telematics Sector, http://www.ess.co.at/ECOSIM/Deliverables/ D0701.doc.gz.

Kouroumlis, Ch., Karatzas, K., Moussiopoulos, N., Kalognomou, E. and Naneris, Ch.: 1999, *Development of a hierarchical system for the tele-transmission of environmental data,* 3rd International Exhibition and Conference HELECO99, 3-6 June, Thessaloniki.

Kuruvilla J., Rao, S.T., Sistla, G., Zhou, N., Hao, W., Schere, K., Roselle, S., Possiel, N. and Scheffe, R.: 1994, 'Examination of the Efficacy of VOC and NO_x Emissions Reductions on Ozone Improvement in the New York Metropolitan Area', in S.E. Gryning and M. Millan (eds), *Air Pollution Modelling and Its Applications X*, Plenum Publishing Corporation, pp. 559-568, New York.

Larssen, S.: 1996, '*Air Quality Management Strategy Planning Tool*', Norwegian Institute of Air Research TR 4/96, ISBN 82-425-0753-8.

Swobota, W.: 1998, *Generic tools and Information Basis*, Area Report No. 4, Telematics Application Programme, Environmental Telematics, DG XIII.

URL 1: http://www.ess.co.at/ECOSIM/

URL 2: http://www.acl.lanl.gov/CORBA/

URL 3: http://www.omg.org

URL 4: http://concord.escde.be

APPLICATION OF THE UAM-V AND USE OF INDICATOR SPECIES TO ASSESS CONTROL STRATEGIES FOR OZONE REDUCTION IN THE LOWER FRASER VALLEY OF BRITISH COLUMBIA

JOANNE L. POTTIER[1] , HANS P. DEUEL[2] , SARA C. PRYOR[3]

1 Aquatic and Atmospheric Sciences Division, Environment Canada, 7th floor, 1200 West 73rd Ave., Vancouver, B.C. Canada V6P 6H9.

2 Systems Applications Intl., 101 Lucas Valley Rd., San Rafael, CA 94903.

3 Atmospheric Science Program, Indiana University, Bloomington, IN 47405.

Abstract – Modelled and measured indicator species ratios of O_3/NOz, O_3/NOy, H_2O_2/HNO_3, HCHO/NOy for the Lower Fraser Valley were compared with VOC-NO_x-O_3 sensitivity threshold values reported in previous studies. Modelled region - specific indicator ratio thresholds derived from 50% NOx and 50% VOC reduction scenarios are provided. They show strongest agreement with the H_2O_2/HNO_3 ratio values found elsewhere. A clear transition region for the LFV from VOC to NOx sensitivity could not be identified, but there is evidence that O_3 concentrations in the western valley, dominated by metropolitan Vancouver, are VOC sensitive, and the more rural eastern valley O_3 concentrations exhibit greater NO_x sensitivity. The UAM-V Process Analysis utility was used to identify physical and chemical mechanisms which contributed to O_3 formation and destruction and indicate the key importance of entrainment from elevated layers generated by the highly complex meteorological conditions in determining near surface O_3 concentrations.

Keywords: indicator species, VOC-NOx-O_3 sensitivity, UAM-V, control strategies

1. Introduction

The Lower Fraser Valley (LFV) surrounding the city of Vancouver, in southwestern British Columbia, occasionally experiences ground-level ozone (O_3) concentrations of health concern. The area has been the subject of modelling studies (SAI, 1997; Pottier *et al.,* 1997) following an intensive field measurement program in July and August 1993 (Pacific 93) (Steyn *et al.*, 1997). Application of UAM-V/SAIMM and the Calgrid - MC2 systems by Jiang *et al.* (1998) provided preliminary evidence of VOC sensitive conditions in the western urban area and more NOx sensitive conditions in downwind rural and agricultural areas.

The non-linearity of the response of O_3 concentrations to reductions of oxides of nitrogen (NOx) and volatile organic compounds (VOC) (Lin *et al.*, 1998), and limitations in the use of ambient VOC/NOx ratios for control strategy assessment (Singh, 1995) has prompted studies which attempt to link indicator species ratios to NOx and VOC sensitivity. Sillman *et al.* (1997) derived threshold values for several ratios, including O_3/NOy, O_3/NOz, H_2O_2/HNO_3, and HCHO/NOy. Ratios higher (lower) than criteria values were linked to NOx (VOC) sensitive environments, implying a greater response to NOx (VOC) controls. Middle range values suggested a transition between VOC and NOx

Environmental Monitoring and Assessment **65**: 459–467, 2000.

sensitivity, a significant region in control strategy design. Other studies (Lu and Chang, 1997; Andreami-Aksoyoglu and Keller, 1997) assessed the universal applicability of threshold ranges and found different ranges (Table I) and local influences. Their results re-emphasize the general applicability of the concepts developed by Sillman (1995) and Sillman *et al.* (1997) but also indicate regional specificity of the absolute threshold values, and the potential role of specific model formulation in determining transition threshold levels (Sillman, 1999). This study compares UAM-V modelled ratio values in the LFV with the above values, and measurements taken at the Harris chemistry site (Figure 1) operated during Pacific 93. Based on earlier model results, input emissions, statistical analyses (Pryor, 1998) and wind flow patterns (Figure 1), it was expected that O_3 concentrations in the western urban NOx rich areas would show VOC sensitivity (lower than threshold range ratio values). A transition towards NOx sensitivity would be expected eastward in downwind rural areas in the central LFV, which is subject to complex channelled and local meso-scale flows, leaving the exact boundaries of a 'transition' area in question. In an attempt to identify this region, locally applicable ratios were developed with similar techniques to those used in previous studies.

Table I. Transition range values from Sillman et al. (1997) and Lu and Chang (1997) compared to LFV threshold ranges (taken from Figure 3). The last two columns compare measured and modelled values at Harris Road. Ratio values are concurrent with peak O_3 on August 4th, 1993.

Ratio	Sillman et al (various locations)	Lu and Chang (California)	LFV Transition Range	Harris Rd Measured (1500 PST)	Harris Cell Modelled (1300 PST)
H_2O_2/HNO_3	0.3 - 0.6	0.8 - 1.0	0.2 - 0.7	0.05	0.06
HCHO/NOy	0.2 – 0.4	0.5 – 0.7	0.45 - 0.75	0.35	0.27
O_3/NOz	8 – 10	25 – 30	9 - 13	9.64	7.39
O_3/NOy	6 – 7.5	N/A	5 - 10	3.50	2.44

2. Methodology

The UAM-V employs updated isoprene chemistry (Whitten *et al.*, 1997), changes to PAN temperature effects, radical-radical chemistry, and a new dry deposition algorithm based on Wesley (1989). Varying mixing height calculations are achieved from 3 dimensional input of temperature, vertical diffusion, and pressure. The SAIMM prognostic meteorological preprocessor consisted of terrain following coordinates and 18 vertical levels. A four-dimensional data assimilation nudging technique (Fast, 1995) was used for wind and potential temperature. The model was implemented in a nested grid formulation with eight vertical layers. The outer grid has a resolution of 10 km^2 and an area of 200 x 230 km. The inner domain has a horizontal resolution of 5 km^2 (grid lines in Figure1).

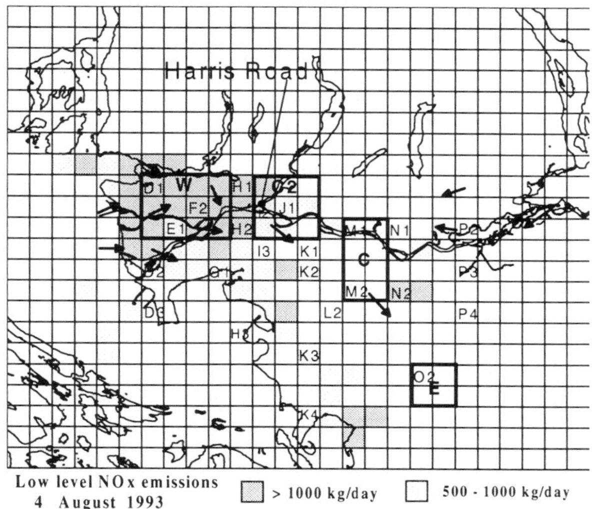

Figure 1. The LFV inner domain. Individual grid cells used in plotting indicator species ratios are indicated by the alpha-numerical values (e.g. K3). Aggregated cell areas W, C, C2, and E were used in the analysis. Dark arrows indicate the wind direction (not proportional to wind speed).

Measurements of O_3, speciated hydrocarbons and nitrogen compounds at Harris Road (Figure 1) were obtained as described in Li *et al.* (1997). Ozone was measured with a TECO model 49 UV photometric O_3 analyzer, nitric oxide (NO) and nitrogen dioxide (NO_2) with chemiluminescence analyzers, and the sum of nitrogen oxides (NOy) was measured with a TECO model 42S outfitted with a 'gold-converter' to obtain an independent measurement. NOz was calculated from NOy – NOx. Gas phase nitric acid (HNO_3) was measured using a denuder-filterpack sampling system, and formaldehyde (HCHO) and hydrogen peroxide (H_2O_2) were obtained with a Unisearch TDL-laser system. Modelled HNO_3 is assumed equal to the sum of measured HNO_3 and particle nitrate. Chemical measurements were concentrated at Harris Road because this location was expected to be; characterized by relatively active air mass chemistry and in the transition between VOC and NOx limited regimes. Model extraction areas W, C, and E (Figure 1) were selected along the sea-breeze path in order to follow a chemically aging air-mass downwind from the NOx-rich urban centre.

3. Results

3.1 MEASURED AND MODELLED RATIOS AT HARRIS ROAD

Modelled and measured data for the Harris Road site on August 4, 1993 are shown in Table I. Peak measured O_3 concentration (74 ppb) was underpredicted (53 ppb) and occurred 2 hours earlier in the model. However, all modelled ratios for the 5 km^2 cell containing this site were within 23% of measured values. Relating measured values to Sillman ratios in Table I, the HCHO/NOy, and O_3/NOz ratios for Harris Road indicate a transition site while lower-than-threshold values for H_2O_2/HNO_3 and O_3/NOy ratios indicate VOC limited conditions. Comparison with values from Lu and Chang (1997) indicates VOC sensitivity for all modelled and measured ratios. While there is an apparent best 'fit' with Sillman *et al.* (1997) transition values for two ratios, local applicability is not assured; arguments can be made for a VOC sensitive 'fit' with Lu and Chang (1997). Comparisons suggest either a VOC sensitive or a 'transition' site.

3.2 MODELLED GRID CELL RATIOS

To test local sensitivity of O_3 behaviour, modelled base case ratios for 25 grid cells (Figure 1) were plotted against west - east distance in the LFV (Figure 2).

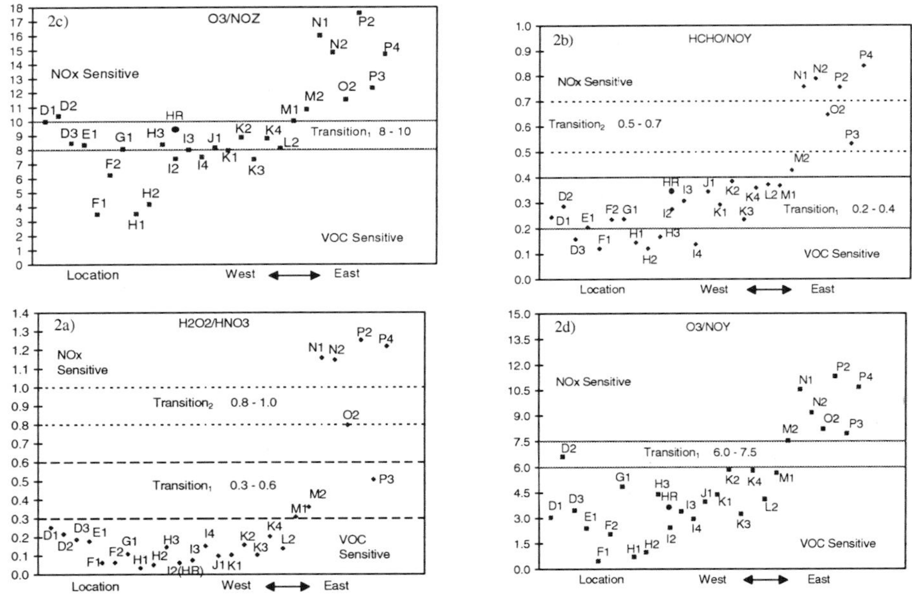

Figure 2. Cell ratios for 5km² grid cells, west - east across the LFV. 'HR' = Harris Road, actually located in Cell I2. Transition₁ = Sillman et al. (1997), Transition₂ = Lu and Chang (1997).

The H_2O_2/HNO_3 ratio (Figure 2a) indicates VOC (NOx) sensitivity in the western and central (eastern) valley. The HCHO/NOy (Figure 2b) and O_3/NOz (Figure 2c) plots indicate more complex behaviour. Transition₁ values appear low for the LFV, including sites which were expected to show more VOC sensitivity, while Transition$_2$ values appear high. The Transition$_1$ range in all plots suggests a crossover from VOC to NOx near cell M1, examined further in section 3.3.

3.3 DERIVATION OF LOCAL THRESHOLD VALUES USING EMISSION REDUCTION SCENARIOS

To derive local threshold values, plots of O_3 change versus ratio values for 50% NOx and 50% VOC reduction scenarios were generated (Figure 3). Large increases in O_3 are predicted from NOx reductions at lower ratios (VOC limited regime). A reduction of NOx in a VOC limited regime can drive the VOC/NOx ratio toward optimal value for O_3 production, providing a disbenefit for such an emissions control. Figure 3 plots suggest transition ranges slightly higher than those found by Sillman *et al.* (1997) (Table I). Threshold range values determined from Figure 3 are approximately (0.2-0.7), (0.45-0.75), (9-13), and (5-10) for the H_2O_2/HNO_3, HCHO/NOy, O_3/NOz, and O_3/NOy ratios, respectively. In comparison with Lu and Chang (1997) ranges in Table I, the local HCHO/NOy range falls in close agreement while local threshold values for H_2O_2/HNO_3 and O_3/NOz ratios are lower. Crossover from VOC to NOx sensitivity is not readily apparent at values characterized by those near the M1 model grid cell (Figure 1), somewhat inconsistent with Figure 2 indications. Figures 3a-d indicate that for each of the transition thresholds there is a range of possible values and while there is qualitative agreement between values derived here and those from other environments, these results re-emphasize the need for region-specific thresholds.

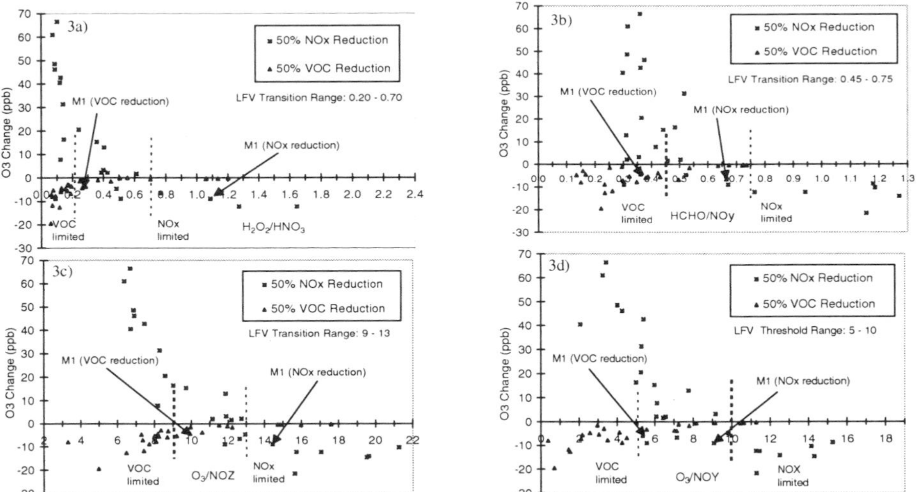

Figure 3. Change in local peak O_3 concentration resulting from 50% reduction in NOx and 50% reduction in VOC, plotted against concurrent indicator species ratios.

3.4 INVESTIGATING PHYSICAL AND CHEMICAL MECHANISMS WITH PROCESS ANALYSIS

The use of indicator species is a powerful tool for analysis of the chemical controls on O_3 production but they neglect the role of other processes such as vertical exchange of O_3 or precursor compounds. The process analysis (PA) tool was used to identify the dominant mechanisms contributing to O_3 concentration across the LFV (Table II). All terms in the model that contribute to changes in the O_3 concentration from selected cells in grid groups; W, C and E in Figure 1 were extracted and composited. Since the objective was to study a transition from VOC to NOx sensitivity in an advected and chemically aging air mass, cells on the western edge of the NOx emission urban plume (Figure 1) were excluded in order to avoid dilution, and central 'C' cells were selected west of fresh NOx emissions in order to avoid plume injection. The east 'E' cells were selected to examine an area suspected of being highly NOx sensitive. The results are shown in Table II.

Table II. Mechanisms providing positive and negative contributions to O_3 concentration in aggregated cell groups W, C, and E across the LFV, for layers 1 and 4, concurrent with peak O_3 for the afternoon of Aug. 4th, 1993. Percentage contribution is shown in brackets.

	W (West) Cells	C (Central) Cells	E (East) Cells
Layer 1 (0-50 m)	Positive: Vert. Diffusion (+73.3) Vert. Transport (+26.7)	Positive: Vert. Diffusion (+47.6) Vert. Transport (+25.7) Chemistry (+26.7)(+ 6 ppb)	Positive: Vert. Diffusion (+60.0) Horiz. Transport (+10.0) Chem. (+30.0) (+ 8 ppb)
	Negative: Chem. (-48.6) (-10 ppb) Dry dep. (-46.8) Horiz. Transport (-4.6)	Negative: Dry dep. (-77.4) Horiz. Transport (-22.6)	Negative: Dry dep. (-96.7) Vert. Transport (-3.3)
Layer 4 (350-600 m)	Positive: Chem. (+49.0) (+ 4 ppb) Vert. Transport (+51.0)	Positive: Chem. (+100) (+ 4 ppb)	Positive: Chem.(+89.3) (+ 8 ppb) Vert. Transport (+10.7)
	Negative: Horiz. Transport (-82.1) Vert. Diffusion (-17.9)	Negative: Vert. Diffusion (-54.2) Vert. Transport (-37.5) Chem. (- 8.3)	Negative: Vert. Diffusion (-53.4) Horiz. Transport (-45.7)

In the 'W' (VOC sensitive) area, chemistry is the most significant loss mechanism for O_3 at the surface, indicating a net O_3 loss through consumption of approximately 10 ppb (mostly from titration with NO), and a net production of 4 ppb aloft. The gain mechanisms of vertical transport and diffusion indicate entrainment of O_3 from aloft. O_3 concentrations of 100-200 ppb were measured at 100 m above the surface in the central LFV during Pacific 93 (McKendry *et al.*

(1997). The likely source is a 'return - flow' aloft which occurs when the surface sea breeze flow is forced up valley walls in the east, returning a westward flow aloft and transporting O_3 westward, confirmed by a loss of O_3 from horizontal transport out of area 'E' in layer 4. This agrees with advection terms aloft in the model output and with westerly flows aloft reported by Banta *et al.* (1997).

The 3-dimensional analysis highlights several features in the LFV. In all three areas, photochemical production of O_3 is more significant aloft than at the surface, and vertical and horizontal transport play a large role in determining the O_3 profile. Vertical mixing is suggested in the C (central) area, with vertical transport responsible for moving O_3 from layer 4 into the surface layer. The advection of surface O_3 eastward is shown by negative contributions of horizontal transport in both the W and C areas, becoming a gain mechanism in the east E area. Loss through dry deposition is significant in all areas. These processes indicate that residual O_3 aloft can contribute to O_3 build-up during extensive episodes, suggesting early control of precursor emissions can be crucial.

4.0 Summary and conclusions

An assessment of the applicability of indicator species ratios to the LFV was performed. A qualitative assessment of modelled ratios across the air shed showed strongest agreement with transition values for the H_2O_2/HNO_3 and O_3/NOy ratios developed by Sillman *et al.* (1997). Locally derived threshold ranges based on techniques used in other studies indicated that higher transition values are likely more appropriate for the HCHO/NOy and O_3/NOz ratios in this region than those derived by Sillman *et al.* (1997) for several United States locations (Table I). With the exception of the HCHO/NOy ratio, locally derived ranges appear lower than those found by Lu and Chang (1997) in their California study; results emphasize the need for region - specific derivation of indicator thresholds.

This region - specific study indicates that much of the western valley shows VOC sensitivity while the eastern valley is characterized by NOx sensitivity. Clear delineation of a central 'transition' area, however, is not provided by indicator species threshold ratio values. Specific grid cells selected for this study (Figure 1) do not show ratio values which consistently fall within VOC or NOx sensitivity suggested by four locally derived 'transition' ranges.

A 3-dimensional representation of mechanisms which contribute to O_3 formation (or consumption) was explored, showing significant photochemical O_3 production aloft. In this environment the geography and complexity of meteorological flows lead to re-circulation of pollutants and establishment of elevated layers aloft from the surface. Vertical exchange from these elevated layers is shown to be a key factor in determining surface O_3 concentrations. The analysis therefore emphasizes a need for early control of precursor emissions at the onset of an episode since feedback of O_3 returning into west and central areas can contribute to build-up of O_3 concentrations. Both radical propagation and vertical transport and diffusion would therefore exacerbate the O_3 concentration levels over extended episodes.

Acknowledgement

The authors are grateful to Wayne Belzer and Jan Bottenheim for their comments and advice. SP gratefully acknowledges support from the National Science Foundation, Atmospheric Chemistry Division (grant # ATM 9711755).

References

Andreami-Aksoyoglu, S. and Keller, J.: 1997, Indicator species for O_3 sensitivity relative to NOx and VOC in Switzerland and their dependence on meteorology, *Air Pollution V*, Modelling, Monitoring and Management, Computational Mechanical Publications, Southampton, 883-891.

Banta, R., Shepson, P., Bottenheim, H., Anlauf, K., Wiebe, H., Gallant, A., Biesenthal, T., Olivier, L., Zhu, C., McKendry, I., and Steyn, D.: 1997, Nocturnal cleansing flows in a tributary valley, *Atmos. Environ.* **31**, 2147- 2162.

Fast, J. D.: 1995, Mesoscale modeling and four-dimensional data assimilation in areas of highly complex terrain, *J. Appl. Meteor.* **34,** 2762-2782.

Jiang, W., Hedley, M. and Singleton, D. L.: 1998, Comparison of the MC2/Calgrid and SAIMM/UAM-V photochemical modelling systems in the Lower Fraser Valley, British Columbia, *Atmos. Environ.* **32**, 2969-2980.

Li, S., Anlauf, K., Wiebe, H., Bottenheim, J., Shepson, P. and Bisenthal, T.: 1997, Emission ratios and photochemical production efficiencies of nitrogen oxides, ketones, and aldehydes in the LFV Pacific 93 oxidant study, *Atmos. Environ.* **31**, 2037-2048.

Lin, X., Trainer, M. and Lui, S.: 1998, On the non-linearity of the tropospheric ozone production, *J. Geophys. Res.* **93**, 15879-15888.

Lu, C. H. and Chang, J. S.: 1997, On the indicator-based approach to assess ozone sensitivities to reductions in VOC and NOx emissions, Air Pollution V, Modelling, Monitoring and Management, Computational Mechanical

Publications, Southampton, 913-922.

McKendry I., Steyn, D., Ludgren, J., Hoff, R., Strapp, W., Anlauf, K., Froude, F., Martin, J., Banta, R. and Olivier, L.: 1997, Elevated ozone layers and vertical down-mixing over the Lower Fraser Valley, BC, *Atmos. Environ.* **31,** 2135-2146.

Pottier, J. L. , Haney, .J, and Deuel, H. P.: 1997, Modelling the future - an application of the Variable Grid Urban Airshed Model (UAM-V) to the Fraser Valley of British Columbia, Canada, *Air Pollution V*, Modelling, Monitoring and Management, Computational Mechanical Publications, Southampton, 465-474.

Pryor, S. C.: 1998, A case study of emission changes and ozone responses, *Atmos. Environ.* **32,** 123-131.

Sillman, S.: 1995, The use of NOy, H2O2 and HNO3 as indicators for O3-NOx-VOC sensitivity in urban locations, *J. Geophys. Res.* **100**, 14175-14188.

Sillman, S., He, D., Cardelino, C., Imhoff, R.E.: 1997, The use of photochemical indicators to evaluate ozone-NOx-hydrocarbon sensitivity: case studies from Atlanta, New York, and Los Angeles, *J. Air & Waste Manage. Assoc.* **47**, 1030-1040.

Sillman, S.: 1999, The relation between ozone, NOx and hydrocarbons in urban and polluted rural environments, *Atmos. Environ.* **33**, 1821-1845.

Singh, H.B. (Ed.): 1995, Composition, chemistry, and climate of the atmosphere, van Nostrand Reinhold, New York, USA pp. 527.

Steyn, D., Bottenheim, J. and Thomson, R.: 1997, Overview of tropospheric O_3 in the Lower Fraser Valley, and the Pacific '93 field study, *Atmos. Environ.* **31**, 2025-2035.

Systems Applications International (SAI): 1997, Application of the UAM-V modeling system with process analysis to the Lower Fraser Valley for the 1-6 August 1993 (Pacific 93) episode, San Rafael, CA, SYSAPP-97/27d.

Wesely, M. L.: 1989, Parameterization of surface resistances to gaseous dry deposition in regional-scale numerical models, *Atmos. Environ.* **23**, 1293-1304.

Whitten, G. Z., Deuel, H. P., Burton, C. S. and Haney, J.L.: 1997, Overview of the implementation of an updated isoprene chemistry mechanism in CB4/UAM-V, Systems Applications International, San Rafael, CA, SYSAPP-97/33.

THE IMPACT OF ROAD TRAFFIC IN MADRID COMMUNITY BY USING A NESTED GAUSSIAN-MESOSCALE AIR QUALITY MODEL (GAMA)

R. San José, L.D. Pedraza, I. Salas and R.M. González[(1)]
Environmental Software and Modelling Group, Computer Science School, Technical University of Madrid, Boadilla del Monte 28660 Madrid (Spain)
[(1)]Department of Geophysics and Meteorology, Faculty of Physics, Complutense University of Madrid (Spain)

Abstract

In this contribution we show the integration of a mesoscale air quality model OPANA with the ISCST3 Gaussian model (EPA) in order to analyze the impact of different emission sources and particularly the traffic emission into the different gridboxes which define the OPANA Eulerian structure. The application is done over the Madrid (Spain) regional area with 80 x 100 km and gridboxes of about 5 km. Thousands of Gaussian runs over interested gridboxes are executed in order to simulate the traffic emissions from each gridbox. Each mobile unit is represented by a Gaussian point emitter. Input meteorological variables for the ISCST3 are taken from the OPANA mesoscale air quality model. Results shows that it is possible to model the impact of traffic emissions over each gridbox. A short comparison with air quality monitoring in each gridbox is also shown.

Key words: Eulerian, mesoscale air quality, ISCST3, Gaussian, traffic emissions.

Introduction

During the last years the Environmental Offices in many European cities are becoming more interested on the application of operational air quality models and particularly on the local impact of the different emission sources such as mobile sources (traffic). The air quality scientific community has increased the research on this area accordingly. In this contribution we show the coupling of a complex operational mesoscale air quality model OPANA (San José et al. 1994, 1996, 1997a,b) and the US. EPA Gaussian model ISCST3 (ISCST3, 1997) in order to study in detail local effects of local emissions.

OPANA is an operational version of ANA model (Atmospheric mesoscale Numerical pollution model for regional and urban Areas). ANA model is composed by a non-hydrostatic mesoscale meteorological module REMEST (based on the MEMO – Flassak and Moussiopoulos (1987) and MM5 (Grell et al. (1994) well known models), a chemical module CHEMA (based on the SMVGEAR (Jacobson and Turco, 1994) numerical solver) with the CBM-IV chemical mechanism (Gery et. Al. (1989)), an emission model EMIMA – which accounts for the antropogenic and biogenic emissions in the model domain;

Environmental Monitoring and Assessment **65**: 469–476, 2000.

biogenic emissions are based on the landuse classification from LANDSAT-5 satellite data for isoprene, monoterpene and natural NOx emissions – and a deposition module DEPO which is based on the resistance approach (Wesely 1989) and the experience of our group of deposition flux field experiments funded by DGXII (European Commission) (1993-1998). OPANA model was properly applied into the EMMA project (DGXIII – EC, 1996-1998) and it is running operationally at the Madrid Community Environmental Office. OPANA is based on solving the Navier-Stokes equation system for the atmospheric flow and as a consequence it requires a 3D grid domain approach. The numerical accuracy of such a type of models is quite high, however, the results are limited by the grid cell sizes since the meteorological and air concentrations are given as averages over the grid cell. Grid cell size is limited by the computer power since the Courant law limits the time step for the meteorological section of the air quality model OPANA. Figure 1 shows a scheme of the OPANA model.

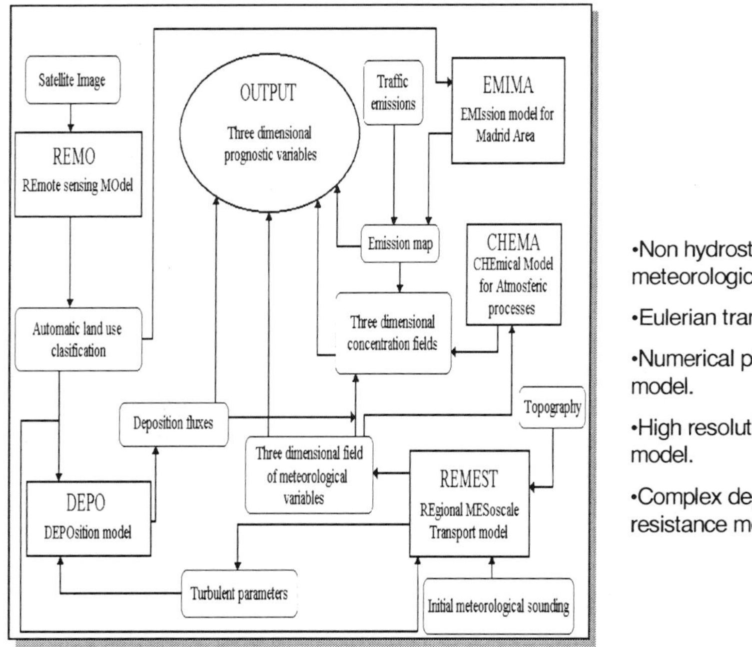

•Non hydrostatic mesoscale meteorological model.

•Eulerian transport model.

•Numerical photochemical model.

•High resolution emission model.

•Complex deposition resistance model.

Figure 1.- OPANA model structure

ISCST3 model is the Industrial Source Complex Short Term modelling system from EPA (Environmental Protection Agency, USA) – last version released by US Environmental Protection Agency on December, 23, 1998 – which is a

Gaussian approach model for treating point, line and area sources. Gaussian models are based on analytical solution of the transport equation of the atmospheric flow. In order to have an analytical solution important assumptions should be done. Input data for ISCST3 is the wind speed at one level, mixing heights, terrain and emission sources. ISCST3 model has been quite successful on modelling point sources over short distances (a few kilometers) since the Gaussian assumption of the plume is kept over short periods of time. ISCST3 model operates on one hour time steps. If meteorology is properly provided then results of ISCST3 model can compare satisfactory with observations. If forecasting meteorology is given the ISCST3 model can operate as a forecasting model provided that enough accurate meteorology is provided as input data. We have developed the IMW (ISCST3 Manager for windows) software tool with Visual Basic and a UNIX version (IMXW) with Tcl/Tk 8.0. The model is already running operationally in two industrial complex in the North of Spain. Figure 2 shows a scheme of the IMW with all the components.

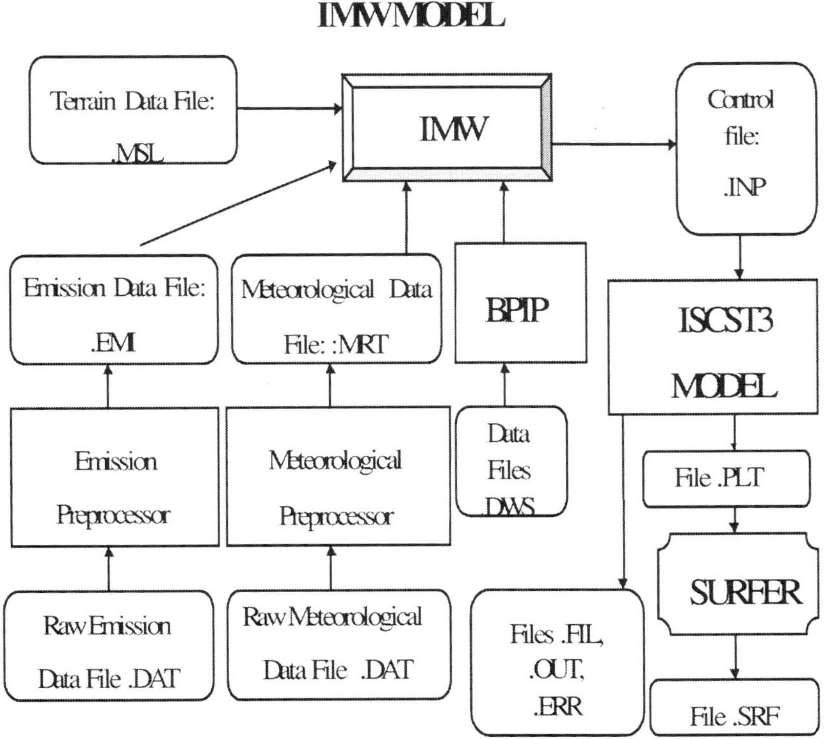

Figure 2.- IMW model structure.

The GAMA model (GAussian Mesoscale Air quality model) is a software platform which integrates the OPANA model and the ISCST3 model. The integrated platform is designed to quantify the impact of local emission sources (such as mobile sources) into the gridbox domain (or several gridboxes). Figure 3 shows the application over four of these grid boxes over the Madrid Community and on the surrounding area of Móstoles (Madrid). The total model domain is 80 x 100 km. In this case the user can select this domain and run the ISCST3 model for the different mobile sources which are expected to be in the grid cell during 24 hours or more. The mobile sources can be estimated by using COPERT II (Road traffic emission model from Environmental Topic Center) which uses a statistical approach like EMIMA model or by using a Traffic Demand Model. In this application we used the results of the EMME/2 model which is a traffic demand model (stocastical approach) which takes as input the accounting values at different points in Madrid Community and the matrix of origin/destiny for different trips. When having this information, the emission factors (CORINAIR) are properly applied to estimate the pollution emissions for each mobile source. Figure 3 shows the gridbox over the full model domain.

The traffic model is integrated into the platform by using the GRASS GIS system (Geographic Resource Analysis Support System). This tool provides a friendly user interface to manage the different road segments over the Madrid domain. Each mobile unit –which is prognostic by the traffic model – is located in a exact location (UTM coordinates) and this unit is finally associated to an emission point source into the IMW package. Up to several thousands of point sources should be treated simultaneously (one processor). The input meteorology is taken from the OPANA model. In our application the OPANA model can provide three dimensional input wind fields every a few seconds since the Courant law control the time step (by taking into account the maximum wind speed and the grid cell size). Because of the accuracy of the meteorological information, the use of ISCST3 becomes competitive. The Madrid Community Environmental Office supports seven air quality monitoring stations located outside of the Madrid City Area.

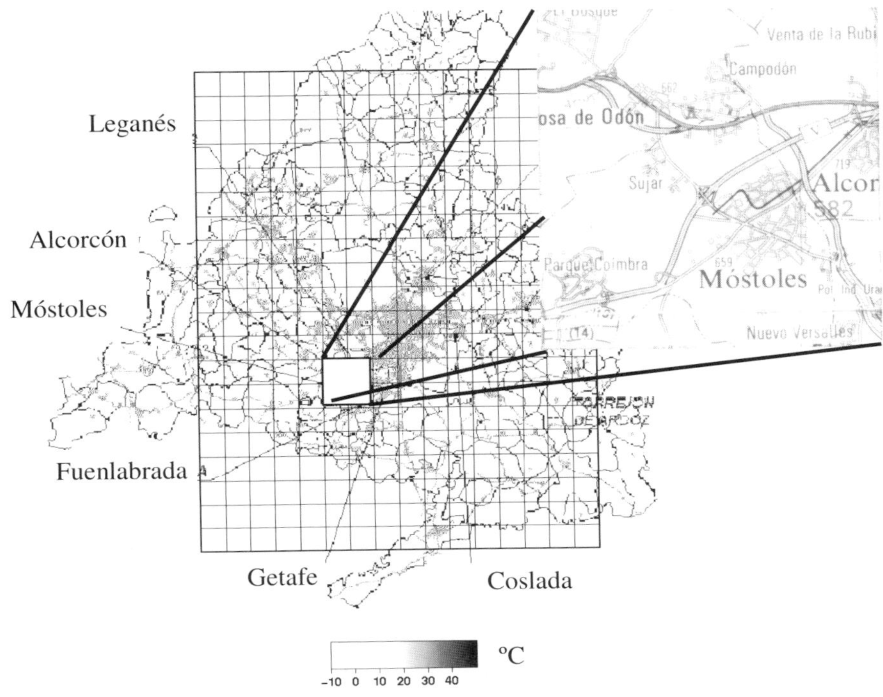

Figure 3.- GAMA model. Application over four grid-boxes. GRIDCELL (3,4)MÓSTOLES AREA 200000 People.

The nesting domain is the Móstoles area where one of the monitoring stations is located (Móstoles station, 426829,4464396 UTM). After running all the Gaussian models - by assuming each mobile unit is a point source – a Kriging interpolation technique is used to visualize the isoconcentrations over the gridbox domain (in our case four gridboxes are used). The user selects the exact location of the monitoring station in order to have the accumulated one hour air concentration on the exact point. Figure 4 shows the traffic NOx pattern at 8h00 in the Móstoles gridbox. Figure 5 shows the comparison between the NOx concentrations at Móstoles monitoring station – exact location – measured by the monitoring station and produced by the mobile units along 120 hour of simulation starting on 31 August, 1998. During the morning hours the traffic impact on air concentrations is reduced because of the atmospheric transport and on the contrary during the evening hours the atmospheric transport enhances the air concentrations due to reduced vertical mixing height – stable boundary layer is less thicker than the unstable boundary layer during the day time hours. These results show the importance of meteorology on air quality studies and also show that the integration of simple Gaussian models such as ISCST3 can provide valuable information to the user provided that the proper physical interpretation is carried out. We have made different tests on different locations over different periods of the year and results show that it is difficult to provide with a "typical" meteorological scenario for such a experiments since the local meteorological conditions are one of the most important control components on the local air quality.

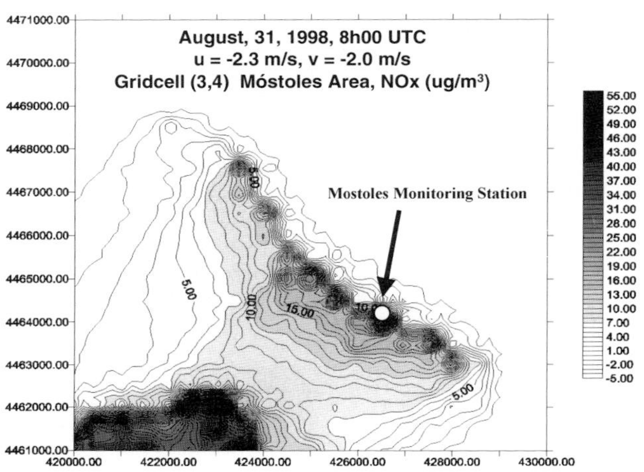

Figure 4.- NOx isoconcentration patterns produced by traffic mobile sources at 10 x 10 km Móstoles gridbox.

Conclusions

A preliminary version Gaussian Mesoscale Air quality model (GAMA) has been presented. GAMA is an integrated mesoscale Eulerian air quality model (OPANA) and a Gaussian model (ISCST3). The Gaussian approach has been applied over each mobile source at every gridbox – in our case the gridbox has been selected to be 10 x 10 km although the OPANA model run over 5 km griboxes in this specific simulation because the average wind speed corresponds with a domain of about 10 km better than over 5 km domain since the average Gaussian time step is one hour – every hour during a 120 hour simulation starting on August, 31, 1998 over the Madrid domain (80 x 100 km). In Figure 6 we show the comparison between NOx concentrations produced with total mobile emission sources in the "Móstoles" (3,4) gridcell and the observed concentrations in the "Móstoles" monitoring station. Results show quantitatively the impact of road traffic on air concentrations and the importance of local meteorology (turbulent vertical and horizontal transport) on air quality studies. Future further improvements of the GAMA model are related to consider chemical transformations on the Gaussian simulations and also using Lagrangian model (LAGMO) for source assessment impact studies.

Comparison between NOx concentrations produced with total mobile emission sources in the "Móstoles" (3,4) gridcell and the observed concentrations in the "Móstoles" monitoring station

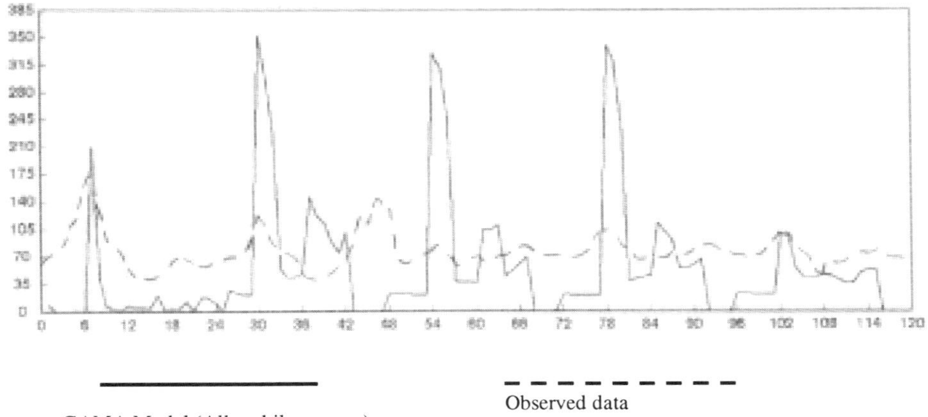

GAMA Model (All mobile sources)
Receptor: (426829,4464396)
NOX (ug/m3)
Model:Regional
Date: 31/08/1998

Observed data
Mostoles monitoring station
(426829,4464396)
NOX (ug/m3)
Model:Regional
Date: 31/08/1998

Figure 5.- Monitoring NOx air concentrations and simulated air concentrations produced by traffic mobile sources at Móstoles gridbox (10 x 10 km).

Acknowledgements

We would like to thank DGXIII-EU for funding partially the EMMA project. The Madrid Community Environmental Office for providing full support on monitoring data. SICE S.A. company for funding partially the project. The Road Department of the Technical University of Madrid for providing the output of the EMME/2 model for our study.

References

Flassak T. and Moussiopoulos N. (1987) An application of an efficient non-hydrostatic mesoscale model, Boundary Layer Meteorology, 41, pp. 135-147.

Gery M.W., Whitten G.Z., Killus J.P. and Dodge M.C (1989), A photochemical kinetics mechanism for urban and regional scale computer modelling, *Journal of Geophysical Research,* 94, D10, pp. 12925-12956.

Grell G.A., Dudhia J. And Stauffer D.R. (1994) A description of the Fifth-Generation Penn State/NCAR Mesoscale Model (MM5). NCAR/TN-398+STR. NCAR Technical Note.

Jacobson M.Z. and Turco R.P. (1994) SMVGEAR: A sparse-matrix, vectorized gear code for atmospheric models", *Atmospheric Environment,* 28, 2, pp.273-284.

San José R., Rodriguez L., Moreno J., Palacios M., Sanz M.A. and Delgado M. (1994) Eulerian and photochemical modelling over Madrid area in a mesoscale context, *Air Pollution II, Vol. 1Computer Simulation, Computational Mechanics Publications, Ed. Baldasano, Brebbia, Power and Zannetti.*, pp. 209-217.

San José R., Cortés J., Moreno J., Prieto J.F. and González R.M. (1996) Ozone modelling over a large city by using a mesoscale Eulerian model: Madrid case study, *Development and Application of Computer Techniques to Environmental Studies, Computational Mechanics Publications, Ed. Zannetti and Brebbia,* pp. 309-319.

San José R., Prieto J.F., Martín J., Delgado L., Jimenez E. And González R.M. (1997)a A Integrated Environmental Monitoring, Forecasting and Warning Systems in Metropolitan Areas (EMMA): Madrid application, *Measurements and Modelling in Environmental Pollution, Ed. San José and Brebbia,* pp. 313-323.

San José R., Prieto J.F., Castellanos N. and Arranz J.M. (1997)b Sensitivity study of dry deposition fluxes in ANA air quality model over Madrid mesoscale area, *Measurements and Modelling in Environmental Pollution, Ed. San José and Brebbia,* pp. 119-130.

Wesely M.L. (1989) Parameterization of surface resistances to gaseous dry deposition in regional-scale numerical models. *Atmospheric Environment,* 23, pp. 1293-1304.

ON THE USE OF MRF/AVN GLOBAL INFORMATION TO IMPROVE THE OPERATIONAL AIR QUALITY MODEL OPANA

R. San José, M.A. Rodríguez, I. Salas and R.M. González[1]
Environmental Software and Modelling Group, Computer Science School, Technical University of Madrid, Boadilla del Monte 28660 Madrid (Spain)
[1]Department of Geophysics and Meteorology, Faculty of Physics, Complutense University of Madrid (Spain)

Abstract

Operational air quality models have become an important tool to assist the decision makers in European Environmental Offices at different levels: cities, regional and state. Because of the important advance on computing capabilities during the last few years the possibility of incorporating the complex research and academic mesoscale air quality models under routine operational basis has become a reality. OPANA model is the operational version of the research model ANA (Atmospheric mesoscale Numerical pollution model for regional and urban Areas). This model is a limited area model (mesoscale beta) and the capability to extend the prediction horizon is limited unless proper boundary conditions are provided during long simulations. In this contribution we show how AVN/MRF (NOAA) vertical numerical meteorological soundings are incorporated to the OPANA system by using JAVA technology. This new feature helps to keep the air quality model into medium power workstations and the performance is improved accordingly. This technology avoids running mesoscale models over larger areas (continental scale) to accordingly increase the forecasting temporal horizon.

Key words: Eulerian, mesoscale air quality, AVN/MRF, JAVA.

1. Introduction

Operational Air Quality Models are becoming an important tool to provide help to policy maker technicians in European Environmental Offices at different levels: cities, regional and states. The impressive advance on computer capabilities during the last decade has pushed the possibility to use former academic and research codes for air quality modelling under routine operational basis. These circumstances have increased the interest of the city or regional authorities on the quality assurance and quality control of such a models. In spite of these above considerations, most operational air quality models are simple tools with many limitations in comparison to research applications. On the contrary, the ANA model (Atmospheric mesoscale Numerical pollution model for regional and urban areas) has a sophisticated demand on computer memory and power. It was developed during the last few years (San José et al.1994, 1996, 1997a,b) and an operational version of this model, the so-called OPANA was installed at different Spanish Environmental Offices (Madrid Community,

Environmental Monitoring and Assessment **65**: 477–484, 2000.

Asturias and Bilbao) under the umbrella of different projects (EMMA, EQUAL, etc.). OPANA was one of the first models to be incorporated into the city and regional authority environmental office without reducing the scientific capabilities but just adjusting the model to the computer architecture which was to be available in the City Environmental Office.

ANA model is composed of a non-hydrostatic mesoscale meteorological module REMEST (based on the MEMO – Flassak and Moussiopoulos (1987) and MM5 (Grell et al. (1994) well known models), a chemical module CHEMA (based on the SMVGEAR (Jacobson and Turco, 1994) numerical solver) with the CBM-IV chemical mechanism (Gery et. Al. (1989)), an emission model EMIMA – which accounts for the antropogenic and biogenic emissions in the model domain; biogenic emissions are based on the landuse classification from LANDSAT-5 satellite data for isoprene, monoterpene and natural NOx emissions – and a deposition module DEPO which is based on the resistance approach (Wesely 1989) and the experience of our group of deposition flux field experiments funded by DGXII (European Commission) (1993-1998). OPANA model was properly applied into the EMMA project (DGXIII – EC, 1996-1998) and it is running operationally at the Madrid Community Environmental Office. OPANA is based on solving the Navier-Stokes equation system for the atmospheric flow and as a consequence it requires a 3D grid domain approach. The numerical accuracy of such a type of models is quite high, however, the results are limited by the grid cell sizes since the meteorological and air concentrations are given as averages over the grid cell. Grid cell size is limited by the computer power since the Courant law limits the time step for the meteorological section of the air quality model OPANA. A sophisticated graphical user interface developed in Tcl/Tk 8.0 which makes use of the VIS5D tool which was developed at the University of Wisconsin-Madison, Space Science and Engineering Center (SSEC) and also supported by NASA and EPA. VIS5D on-line with the OPANA-VIS package is capable to visualize the 3D field for all meteorological variables and air concentrations and fluxes. Figure 1 shows a scheme of the OPANA model.

2. AVN/MRF Global meteorological models

AVN/MRF model is a National Meteorological Center (NMC) global spectral Medium Range Forecast (MRF) model which was developed by Sela (1982). AVN is the Aviation version of MRF model. MRF model has 191 km output spatial resolution and covers both hemispheres. It runs every 24 hours (00h00) and has a forecast duration of 288 hours and 13 vertical layers. AVN model has two versions for 191 km and 111 km output spatial resolution.

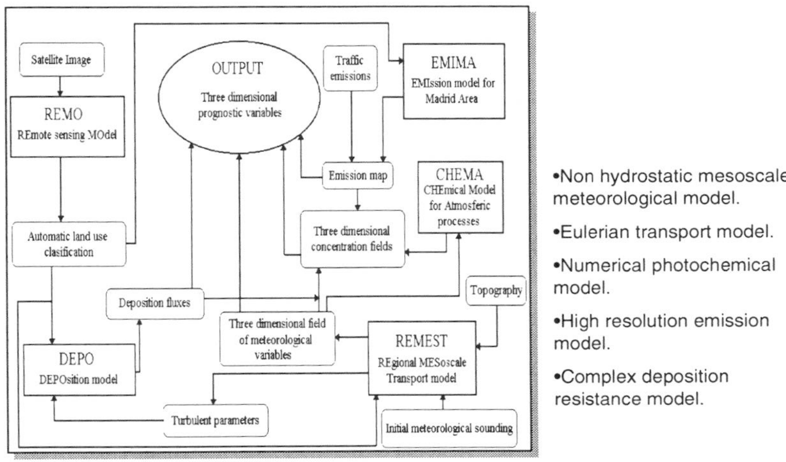

Figure 1. – OPANA model.

AVN191 model covers both hemispheres, it runs every six hours (00/06/12/18) with a forecast duration of 72 hours and 13 vertical layers. AVN111 covers the Northern hemisphere and runs every 12 hours (00/12). It has a forecast duration of 48 hours and 23 vertical layers. All this information is found in the ARL web site. Table 1 shows the ARL (Air Resources Laboratory) timetable of the different model runs.

MODEL	MODEL DOMAIN	TIME RUN (UTC)	Forecast Duration (H)	Temporal Resolution (H)	Spatial Resolution (km)	Output Resolution (km)	MODELS LEVELS
AVN	NH/SH	00/06/ 12/18	72	6	~ 106	191	13
AVN	NH	00/12	48	6	~ 106	111	23
MRF	NH/SH	00	288	12	~ 106	191	13

Table 1.- ARL Web communication scheme and time table model runs.

AVN/MRF model releases all the information on the Internet and the vertical meteorological soundings and meteograms can be extracted and interpolated by using the Web facility which has been developed by ARL (Air Resources Laboratory). The user can select the position and the period of simulation and the system activates the proper software to extract from the database the selected information. All these procedures can be carried out for different locations in our model domain and this information is provided to the OPANA numerical model to improve the simulation since the "future" values for specific times and locations are known from the AVN/MRF output. These procedures have been performed automatically by developing the proper JAVA tool which has been implemented into the OPANA tool to download operationally the required information to initialize the OPANA model under daily basis. The procedure takes just a few minutes (depending on the Internet connection speed) to download automatically this information and locate it in the proper depository to be accessed by the OPANA model when running. Figure 2 shows the vertical meteorological numerical soundings which can be downloaded for incorporation into the initial set of data of the numerical model (OPANA).

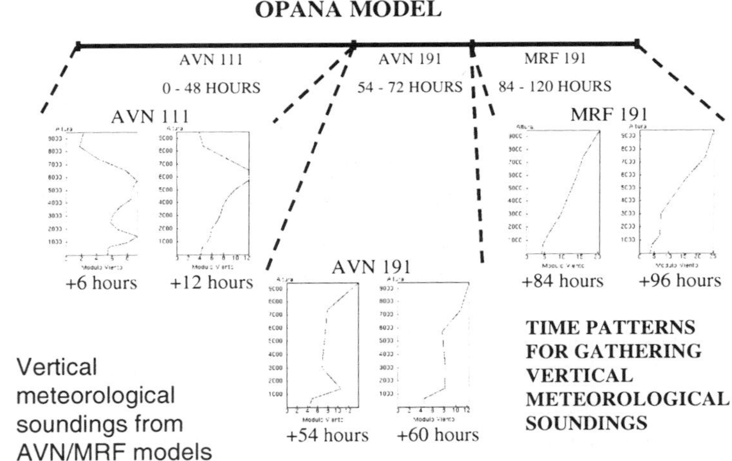

Figure 2.- Scheme of the vertical numerical meteorological soundings which are automatically download from the ARL Web server by using the JAVA/OPANA application.

3. Analysis of results

In this section we will analyze the impact of having the AVN/MRF vertical meteorological soundings to initialize the OPANA model. Figure 3 and 4 shows the comparison of ozone air concentrations at Alcalá de Henares (Madrid) monitoring station with the forecasted ozone data when using only the Madrid International Airport meteorological sounding during the first 24 hours (two meteorological soundings everyday) and when using AVN/MRF data following the scheme shown in Figure 2. Correlation coefficient is improved from 0.79 to 0.84 and the y-intercept changes from –0.81 to +1.95 which suggests a change from underprediction to overprediction data. It is always important to keep in mind that the data corresponds to 5 km gridcell over a domain of 80 x 100 km and with 15 vertical layers non-equally spaced up to 6000 m in height. The simulation corresponds to the period of August, 31, 1998 and during 120 hours. Figure 5 shows the comparison between both simulations with a correlation coefficient of 0.89. Similar information can be obtained for other pollutants although the correlation coefficient will be lower for those pollutants with a strong local emission factor. This is closely related to the emission model itself but independent to our objectives in this contribution. In all cases the change in correlation coefficient is meant to improve it.

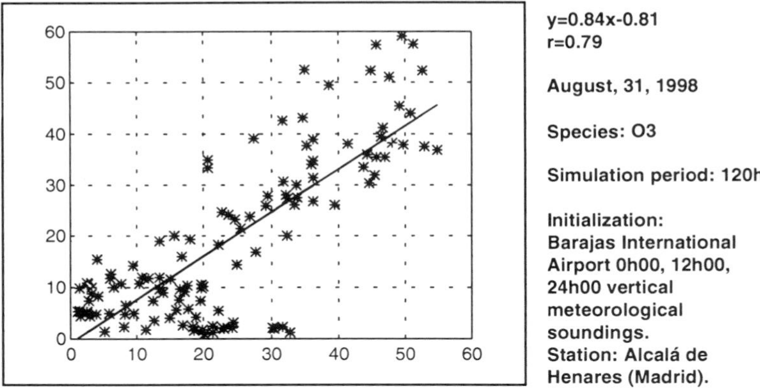

Figure 3.- Comparison between forecasted O3 concentrations by using Madrid International Airport vertical meteorological sounding and observations.

OPANA MODEL

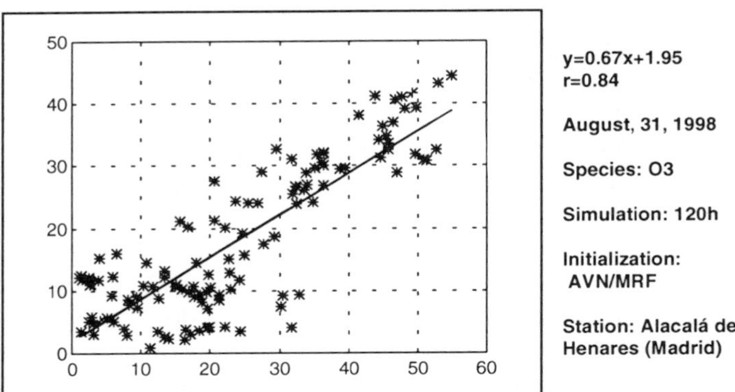

Figure 4.- Comparison between forecasted O3 concentrations by using AVN/MRF vertical numerical meteorological sounding and observations

OPANA MODEL

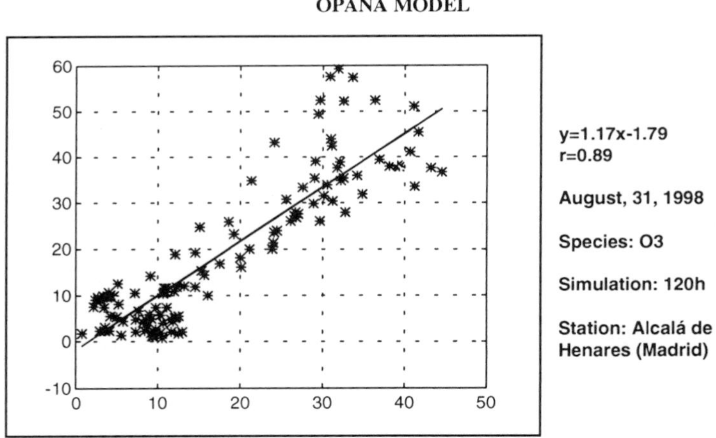

Figure 5- Comparison between forecasted O3 concentrations by using AVN/MRF vertical numerical meteorological sounding and those forecasted when using Madrid International Airport vertical meteorological soundings (only first 24 hours).

Results show that after implementing the AVN/MRF vertical meteorological soundings, the physics is improved since the model is provided with updated meteorological information, but this improvement is not as spectacular as we thought since the results just change a small amount in the correlation coefficient. This could be explained by the strong local meteorology influence in our application and in this particular period of the year. These results enforce the idea that local meteorology – which is driven by the local topography and land-use types, continue to be the most important factor for air quality simulations, in addition to the chemical activity.

4. Conclusions

In this contribution we have shown how to incorporate meteorological information from global operational meteorological model such as AVN and MRF NOAA/NCEP models for improving the air quality forecasts from OPANA model (an operational Air Pollution Numerical model for urban and regional Areas). Formally speaking, the boundary conditions 3D files provided every 6 or 12 hours - depending on the model domain - from larger domain models assure that the atmospheric flow is updated properly into the nested mesoscale model domain. This process, however, obligates the user to run larger and complex meteorological and air quality models - models with the prohibited computational costs for the urban or regional user (Environmental Management Offices). In this contribution we show that by using a simple JAVA application through Internet Services, the information corresponding to forecasting vertical meteorological soundings obtained by running the AVN/MRF model in the CRAY systems by NOAA allows one to "drive" the higher spatial and temporal resolution application (OPANA) to obtain improved results. Further work should focus on analysing the results over longer periods (preferentially over one year) to confirm the general improvement of mesocale air quality model applications. We believe that this technique can be widely used to update other subsystems of the model such as landuse data and emission data.

Acknowledgements

We would like to thank the Madrid Community Environmental Office for providing the air quality monitoring network data. We would also like to thank ARL (NCEP, USA) for the AVN/MRF meteorological data.

References

Flassak T. and Moussiopoulos N. (1987) An application of an efficient non-hydrostatic mesoscale model, Boundary Layer Meteorology, 41, pp. 135-147.

Gery M.W., Whitten G.Z., Killus J.P. and Dodge M.C (1989), A photochemical kinetics mechanism for urban and regional scale computer modelling, *Journal of Geophysical Research,* 94, D10, pp. 12925-12956.

Grell G.A., Dudhia J. And Stauffer D.R. (1994) A description of the Fifth-Generation Penn State/NCAR Mesoscale Model (MM5). NCAR/TN-398+STR. NCAR Technical Note.

Jacobson M.Z. and Turco R.P. (1994) SMVGEAR: A sparse-matrix, vectorized gear code for atmospheric models", *Atmospheric Environment,* 28, 2, pp.273-284.

San José R., Rodriguez L., Moreno J., Palacios M., Sanz M.A. and Delgado M. (1994) Eulerian and photochemical modelling over Madrid area in a mesoscale context, *Air Pollution II, Vol. 1Computer Simulation, Computational Mechanics Publications, Ed. Baldasano, Brebbia, Power and Zannetti.*, pp. 209-217.

San José R., Cortés J., Moreno J., Prieto J.F. and González R.M. (1996) Ozone modelling over a large city by using a mesoscale Eulerian model: Madrid case study, *Development and Application of Computer Techniques to Environmental Studies, Computational Mechanics Publications, Ed. Zannetti and Brebbia,* pp. 309-319.

San José R., Prieto J.F., Martín J., Delgado L., Jimenez E. And González R.M. (1997)a A Integrated Environmental Monitoring, Forecasting and Warning Systems in Metropolitan Areas (EMMA): Madrid application, *Measurements and Modelling in Environmental Pollution, Ed. San José and Brebbia,* pp. 313-323.

San José R., Prieto J.F., Castellanos N. and Arranz J.M. (1997)b Sensitivity study of dry deposition fluxes in ANA air quality model over Madrid mesoscale area, *Measurements and Modelling in Environmental Pollution, Ed. San José and Brebbia,* pp. 119-130.

Sela J. The NMC Spectral Model. NOAA Technical Report. NW5-30, 36 (1982).

Wesely M.L. (1989) Parameterization of surface resistances to gaseous dry deposition in regional-scale numerical models. *Atmospheric Environment,* 23, pp. 1293-1304.